D1711533

Optical Nonlinearities and Instabilities in Semiconductors

Optical Nonlinearities and Instabilities in Semiconductors

Edited by

Hartmut Haug

INSTITUT FÜR THEORETISCHE PHYSIK
UNIVERSITÄT FRANKFURT
FRANKFURT, FEDERAL REPUBLIC OF GERMANY

ACADEMIC PRESS, INC.
Harcourt Brace Jovanovich, Publishers
Boston San Diego New York
Berkeley London Sydney
Tokyo Toronto

ACADEMIC PRESS, INC.
1250 Sixth Avenue, San Diego, CA 92101

United Kingdom Edition published by
ACADEMIC PRESS, INC. (LONDON) LTD.
24-28 Oval Road, London NW1 7DX

Library of Congress Cataloging-in-Publication Data

Optical nonlinearities and instabilities in semiconductors/Hartmut Haug, editor.
p. cm.
Includes bibliographies and index.
ISBN 0-12-332915-9
1. Semiconductors—Optical properties. 2. Optical bistability.
3. Nonlinear optics. I. Haug, Hartmut.
QC611.6.06065 1988
537.6′23—dc19

87-17643
CIP

88 89 90 91 9 8 7 6 5 4 3 2 1
Printed in the United States of America

CONTENTS

PREFACE

This book contains, in addition to the introduction, 16 review papers, written by leading scientists in the field, on various aspects of nonlinear optical phenomena and related optical instabilities in semiconductors. The papers in the first part of the book deal mainly with the measurements and explanations of the optical nonlinearities of various semiconductor materials and structures. Nearly all investigations are concerned with the spectral region close to the band gap, in order to exploit the resonance enhancement of the nonlinear optical behaviour. The papers in the second part of the book deal with the resulting optical instabilities, particularly with optical bistability, and, to a minor degree, with diode laser thresholds, self-oscillations and chaos. Optical bistability has stimulated most of the research described in this book because of its potential technological importance for optical data processing. Nucleation centers for these developments have been: In Europe, the EJOB project (European Joint Optical Bistability), in which several European research institutes collaborated under the sponsorship of the Council of the European Communities; in the United States, the OCC (Optical Circuitry Cooperative)—a National Science Foundation, university, and industry cooperative research activity—at the Optical Sciences Center in Tucson, and the AT&T Bell Laboratories, particularly with their work on semiconductor quantum well structures and devices.

A characteristic feature of this book is that experimentalists and theoreticians have contributed equally. The book is arranged mostly in closely related pairs of experimental and theoretical contributions, either in successive review papers or in jointly written reports.

This book describes the state of the art in this still rapidly moving field. It is hoped that these reviews will help research students and active research workers who are interested in the basic physics or in the device applications of these effects. Finally, it is a pleasure for me to thank all the contributors and the publisher for the fruitful collaboration.

Hartmut Haug

CONTRIBUTORS

Numbers in parentheses indicate chapter numbers.

M. J. ADAMS (15), *British Telecom Research Laboratories, Martlesham Heath, Ipswich IP5 7RE, England*

L. L. CHASE (9), *Lawrence Livermore National Laboratory, University of California, Livermore, California 95550*

D. S. CHEMLA (4, 13), *AT&T Bell Laboratories, Holmdel, New Jersey 07733*

M. L. CLAUDE (9), *Groupe de Physique des Solides de l'E.N.S., Université Paris VII, Tour 23, place Jussieu, 75251 Paris Cedex 05, France*

A. FORCHEL (7), *4. Physikalisches Institut, Universität Stuttgart, Pfaffenwaldring 57, D-7000 Stuttgart-80, Federal Republic of Germany*

F. FORSMANN (14), *Institut für Angewandte Physik, Universität Münster, Corrensst. 2-4, D-4400 Münster, Federal Republic of Germany*

H. M. GIBBS (12), *Optical Sciences Center, University of Arizona, Tucson, Arizona 85721*

J. B. GRUN (8), *Laboratoire de Spectroscopie et d'Optique du Corps Solide Unité Associée au C.N.R.S. no. 232, Université Louis Pasteur, 5, rue de l'Université, 67084 Strasbourg Cedex, France*

CH. HARDER (16), *IBM Research Division, Zürich Research Laboratory, 8803 Rüschlikon, Switzerland*

HARTMUT HAUG (1, 3), *Institut für Theoretische Physik, Universität Frankfurt, Robert-Mayer-Str. 8, D-6000 Frankfurt, Federal Republic of Germany*

H. HILLMER (7), *4. Physikalisches Institut, Universität Stuttgart, Pfaffenwaldring 57, D-7000 Stuttgart-80, Federal Republic of Germany*

B. Hönerlage (8), *Laboratoire de Spectroscopie et d'Optique du Corps Solide Unité Associée au C.N.R.S. no. 232, Université Louis Pasteur, 5, rue de l'Université, 67084 Strasbourg Cedex, France*

D. Hulin (9), *Groupe de Physique des Solides de l'E.N.S., Université Paris VII, Tour 23, place Jussieu, 75251 Paris Cedex 05, France*

D. Jäger (14), *Institut für Angewandte Physik, Universität Münster, Correns-str. 2-4, D-4400 Münster, Federal Republic of Germany*

C. Klingshirn (2), *Physikalisches Institut der Universität, Robert-Mayer-Str. 2-4, D-6000 Frankfurt am Main, Federal Republic of Germany*

S. W. Koch (11), *Optical Sciences Center and Physics Department, University of Arizona, Tucson, Arizona 85721*

T. Kuhn (7), *Institut für Theoretische Physik, Universität Stuttgart, Pfaffenwaldring 57, D-7000 Stuttgart-80, Federal Republic of Germany*

R. Lévy (8), *Laboratoire de Spectroscopie et d'Optique du Corps Solide Unité Associée au C.N.R.S. no. 232, Université Louis Pasteur, 5, rue de l'Université, 67084 Strasbourg Cedex, France*

G. Mahler (7), *Institut für Theoretische Physik, Universität Stuttgart, Pfaffenwaldring 57, D-7000 Stuttgart-80, Federal Republic of Germany*

D. A. B. Miller (4, 13), *AT&T Bell Laboratories, Holmdel, New Jersey 07733*

A. Mysyrowicz (9), *Groupe de Physique des Solides de l'E.N.S., Université Paris VII, Tour 23, place Jussieu, 75251 Paris Cedex 05, France*

M. J. O'Mahony (15), *British Telecom Research Laboratories, Martlesham Heath, Ipswich IP5 7RE, England*

N. Peyghambarian (12), *Optical Sciences Center, University of Arizona, Tucson, Arizona 85721*

T. Rozzi (17), *School of Electrical Engineering, University of Bath, Cleverton Down, Bath BA2 7AY United Kingdom*

S. Schmitt-Rink (4, 13), *AT&T Bell Laboratories, Murray Hill, New Jersey 07974*

W. Schäfer (6), *Institut für Physik, Universität Dortmund, Postfach 50 05 00, D-4600 Dortmund, Federal Republic of Germany*

K. Shore (17), *School of Electrical Engineering, University of Bath, Cleverton Down, Bath BA2 7AY, United Kingdom*

F. A. P. TOOLEY (10), *Department of Physics, Heriot-Watt University, Edinburgh EH14 4AS, United Kingdom*

R. G. ULBRICH (5), *Institut für Physik, Universität Dortmund, Postfach 50 05 00, D-4600 Dortmund, Federal Republic of Germany*

A. C. WALKER (10), *Department of Physics, Heriot-Watt University, Edinburgh EH14 4AS, United Kingdom*

H. J. WESTLAKE (15), *British Telecom Research Laboratories, Martlesham Heath, Ipswich IP5 7RE, England*

B. S. WHERRETT (10), *Department of Physics, Heriot-Watt University, Edinburgh EH14 4AS, United Kingdom*

A. YARIV (16), *Department of Electrical Engineering and Applied Physics, T. J. Watson Laboratory, California Institute of Technology, Pasadena, California 91125*

1 INTRODUCTION

Hartmut Haug

INSTITUT FÜR THEORETISCHE PHYSIK
UNIVERSITÄT FRANKFURT, ROBERT-MAYER-STR. 8
D-6000 FRANKFURT, FEDERAL REPUBLIC OF GERMANY

I. HISTORICAL REMARKS

After quantum mechanics provided an understanding of the basic nature of semiconductors, many solid state physicists became engaged in spectroscopic studies of semiconductors. Most of these studies concentrated either on the rest-strahl bands of the lattice vibrations or on the electronic interband transitions between the filled valence bands and the first empty conduction band. For most semiconductors, the absorption edge of the electronic transitions lies in the convenient region of the near infra-red, visible or near ultra-violet. The basic electronic interband absorption process with which this book is mainly concerned is the excitation of an electron above the band gap. In this process, a hole is left behind in the valence bands and an electron is put into the lowest conduction band. Both the electron and the hole can be treated as charged quasi-particles that move through the periodic lattice with an effective mass which is typically

OPTICAL NONLINEARITIES
AND INSTABILITIES IN
SEMICONDUCTORS

ISBN 0-12-332915-9

only a small fraction of a free electron mass. The electron or hole can be bound to impurities, or the electron and hole can form a bound pair state, called an exciton. These bound pair states have energies typically several meV below the band gap and cause a series of resonances below the band edge. The smallness of the binding energy is mainly due to the weakening of the Coulomb forces by a strong electronic and ionic polarization. The interaction of the electron–hole pairs with phonons broadens these resonance lines and gives them a characteristic lineshape. For several decades, the conventional semiconductor spectroscopy was, and to some degree still is, concerned with the identification of resonance lines in pure and doped semiconductors and with the study of the interactions of the electrons with phonons and impurities through lineshape measurements and analysis (see, e.g., Pankove, 1971).

In the sixties and seventies, laser spectroscopy opened the possibility to generate and maintain at least for some short time interval such a large concentration of electron–hole pairs that their mutual interaction became more important in determining the shape of the optical spectra than the interaction of the electron–hole pairs with impurities or phonons. In highly excited semiconductors, one was able to identify (for review articles see, e.g., Seraphin, 1966; Nishina and Ueta, 1976; Hanamura and Haug, 1977; and Klingshirn and Haug, 1981) exciton–exciton scattering processes and the formation of an exciton molecule. Large concentrations of optically generated electron–hole pairs cause a further increase in the screening of the Coulomb forces. At a critical intensity the sharp resonances of the excitons cease to exist because the screened Coulomb potential is too weak to support a bound state. The electronic excitations then form a two-component plasma. In effect, the semiconductor has changed from an insulator to a medium with metallic properties solely by intense laser excitation. The theory of these strongly changing optical spectra has been reviewed by Haug and Schmitt-Rink, 1984 and 1985. In semiconductors with an indirect band gap and correspondingly large electron–hole lifetimes, the plasma at low temperatures forms a stable thermodynamic phase: the electron–hole plasma liquid (for review articles see, e.g., Jeffries, 1975; Rice, 1977; and Hensel et al., 1977). In the seventies, the quantum-statistical properties of dense electron–hole systems were investigated. At the same time lasers were used to study the nonlinear optical properties of semiconductors far away from resonances. E.g., two-photon spectroscopy has been used to study transitions that are forbidden for one-photon transitions (for review articles see Fröhlich, 1981; Chemla and Jerphagnon, 1980). In these experiments the population of the excited states was always small and did not determine the investigated nonlinear optical properties. Hanamura (1973) predicted that exciton molecules should have a "giant two-photon oscillator strength" mainly due to the fact that the exciton serves as a nearly resonant intermediate state. A further enhancement stems from the large spatial extent of the excitons, which increases the chance for the absorption of a second photon in the large volume of the first virtually excited exciton. Gale and Mysyrowics (1974) found indeed a large two-photon absorption of the exciton molecule in CuCl[1]. On the other hand, Woerdmann (1970) demonstrated quite early that the changes of the optical constants by the generated free-carriers in Si are sufficiently large to form laser-induced optical gratings which can be used for real-time holography. However, by and large, semiconductors were not yet considered to be media with pronounced nonlinear optical properties.

The situation changed at the end of the seventies when Gibbs et al. (1979) and Miller et al. (1979) showed that the intensity-dependent changes of the absorption and refraction coefficients were large enough to observe optical bistability in GaAs and InSb, respectively. The main idea was to exploit the resonance enhancement close to the fundamental absorption edge. In optical bistability (see, e.g., Gibbs, 1985) the intensity of the light field in the nonlinear medium can change discontinuously between two values. This instability, most often observed in a resonator geometry, requires large optical nonlinearities. With the interest in optical bistability (both because of its fundamental aspects as a discontinuous nonequilibrium phase transition and because of its possible applications in optical data processing), the well-known changing spectra of highly excited semiconductors were now reconsidered as manifestations of large optical nonlinearities. While the work in the seventies concentrated on the low-temperature properties of highly excited semiconductors to observe most clearly the quantum statistical properties of high-density electron–hole systems, the work in the past few years has been increasingly concerned with the room-temperature optical nonlinearities of semiconductors because of their potential device applications. Optical bistability[2], four-wave mixing, laser-induced lattices[3], phase conjugation[4], self-focusing and -defocusing, oscillatory and chaotic instabilities[5], more-photon transitions and instabilities have since been observed in many semiconductors. The present book attempts to bring together all these investigations of the resonance-enhanced nonlinear optical properties of semiconductors and their manifestations in optical instabilities.

In recent years, new crystal growth techniques, such as molecular beam epitaxy, made it possible to form semiconductor microstructures, such as multi-quantum-wells. The smallness of the structures removes one degree of freedom from the electrons which results in a reduced dimensionality of the electron motion. The optical properties of these low-dimensional semiconductors are remarkably different from those of bulk semiconductors. With this new technique of material engineering, new devices with freely designed linear and nonlinear optical properties can be produced. Again, these new developments are included in this book[6].

Another line of research was related to the highly successful development of semiconductor lasers. These devices again depend on the nonlinear optical properties of semiconductors. It is well known that the onset of laser action can be described by a nonlinear equation for the field in the resonator, which is in its simplest form the so-called Van der Pol equation (see, e.g., Haken, 1984). The absorptive and dispersive nonlinearities of the medium determine the saturation behaviour, the mode properties and the stability of the devices. More specifically, higher instabilities such as bistability, self-pulsing, oscillatory and chaotic output of semiconductor lasers are again direct consequences of the nonlinear response of the semiconducting medium. The book attempts to draw attention to this fact[7] and to connect the research in the instabilities of laser diodes with that of optical nonlinearities and instabilities in passive semiconductor systems.

II. THE NONLINEAR OPTICAL RESPONSE OF SEMICONDUCTORS

The wave equation for the electric field of light in an optical medium, which follows from the macroscopic Maxwell equations is given by

$$\left(\frac{\partial^2}{\partial t^2} - \frac{1}{c^2}\frac{\partial^2}{\partial z^2}\right)\mathbf{E} = \frac{4\pi}{c^2}\frac{\partial^2}{\partial t^2}\mathbf{P}, \tag{1}$$

where $\mathbf{P}(z, t)$ is the polarization of the medium. In nonlinear optics (see, e.g., Bloembergen, 1965; and Shen, 1984) far away from any resonances, this optically induced polarization is usually evaluated by expanding it in powers of the electric field. The interaction Hamilton operator for the field $\mathbf{E}(t)$ and the system is given in the dipole approximation and in the interaction representation by

$$\mathrm{H}_{\mathrm{int}}(t) = -\mathbf{P}(t)\mathbf{E}(t),$$

where

$$\mathbf{P}(t) = e^{i/\hbar \mathrm{H}_0 t}\mathbf{P}e^{-i/\hbar \mathrm{H}_0 t} \tag{2}$$

is the time-dependent polarization operator. H_0 is the Hamilton operator of the unperturbed system.

In the interaction representation, the Liouville equation for the density matrix ρ of the system is given by

$$\frac{\partial \rho}{\partial t} = -i/\hbar[H_{\text{int}}(t), \rho] \tag{3}$$

with the formal solution

$$\rho(t) = T\exp\left(\frac{i}{\hbar}\int_0^t H_{\text{int}}(t')\,dt'\right)\rho_0, \tag{4}$$

where T is the time ordering operator.

Calculating the expectation value of the polarization, one finds

$$\begin{aligned}
\langle \mathbf{P}(t)\rangle &= \operatorname{tr}\rho(t)\mathbf{P}(t)\\
&= -(i/\hbar)\int_0^t dt_1 \langle [\mathbf{P}(t_1), H_{\text{int}}(t_1)]\rangle_0\\
&\quad + (-i/\hbar)^2 \int_0^t dt_1 \int_0^{t_1} dt_2 \langle [\mathbf{P}(t_1), [H_{\text{int}}(t_1), H_{\text{int}}(t_2)]]\rangle_0\\
&\quad + (-i/\hbar)^3 \int_0^t dt_1 \int_0^{t_1} dt_2 \int_0^{t_2} dt_3 \langle [\mathbf{P}(t_1), [H_{\text{int}}(t_1),\\
&\qquad [H_{\text{int}}(t_2), H_{\text{int}}(t_3)]]]\rangle_0 + \cdots .
\end{aligned} \tag{5}$$

For a field that consists of a sum of sharp frequency components $E_k(\omega_i)$, a Fourier transform of the expansion (5) yields

$$\begin{aligned}
P_m(\omega_i) &= \sum \chi^{(1)}_{mn}(\omega_j) E_n(\omega_j)\, \delta_{\omega_i, \omega_j}\\
&\quad + \sum \chi^{(2)}_{mnp}(\omega_j, \omega_k) E_n(\omega_j) E_p(\omega_k)\, \delta_{\omega_i, \omega_j + \omega_k}\\
&\quad + \sum \chi^{(3)}_{mnpq}(\omega_j, \omega_k, \omega_l) E_n(\omega_j) E_p(\omega_k) E_q(\omega_l)\, \delta_{\omega_i, \omega_j + \omega_k + \omega_l} + \cdots
\end{aligned} \tag{6}$$

The tensor $\chi^{(n)}$ is called the nth order nonlinear susceptibility. The Fourier transforms of the field obey the relation $E_k(\omega) = E_k^*(-\omega)$. Thus, $\mathbf{P}_l^{(2)}(2\omega_i) = \Sigma\chi^{(2)}_{lmn}(\omega_i, \omega_i) E_m(\omega_i) E_n(\omega_i)$ describes e.g. the generation of a polarization at frequency $2\omega_i$, which gives rise to second harmonic generation. The term $P_l^{(2)}(0) = \Sigma\chi^{(2)}_{lmn}(\omega_i, -\omega_i)\mathbf{E}_n(\omega_i) E_l^*(\omega_i)$ gives rise to a static polarization. In systems with inversion symmetry, $\chi^{(2)}$ vanishes, so that $\chi^{(3)}$ is the first nonlinear response term. This third-order term describes in general the generation of a field at frequency $\omega_i + \omega_j + \omega_k$ under the influence of three incident fields with the frequencies ω_i, ω_j and ω_k. In the degenerate case with $\omega_i = -\omega_j = \omega_k = \omega$, one gets for linearly polarized light $\mathbf{E} = E\mathbf{e}$ an intensity-dependent polarization at the frequency ω

$$P(\omega) = \chi^{(3)}(\omega, -\omega, \omega)|E(\omega)|^2\,\mathbf{E}(\omega), \tag{7}$$

where the tensor indices are omitted. This result describes an intensity-dependent correction to the linear susceptibility

$$P(\omega) = \{\chi^{(1)} + \chi^{(3)}|E(\omega)|^2\}E(\omega) = \chi(\omega, I)E(\omega), \tag{8}$$

which in turn gives rise to an intensity-dependent change of the optical dielectric function $\epsilon(\omega, I) = 1 + 4\pi\chi(\omega, I)$ and the related coefficients of absorption α and refraction n

$$\alpha(\omega, I) = \frac{\omega\epsilon''(\omega, I)}{n(\omega, I)c} \tag{9}$$

$$n(\omega, I) = \sqrt{\tfrac{1}{2}\left(\epsilon'(\omega, I) + \sqrt{\epsilon'^2(\omega, I) + \epsilon''^2(\omega, I)}\right)}\,. \tag{10}$$

This special type of optical nonlinearity is particularly interesting in the context of this book, because if the nonlinear optical polarization is inserted into the macroscopic Maxwell-equations, the resulting nonlinear partial differential equation can exhibit instabilities (see, e.g., Boyd et al., 1985) of various kinds. New stationary solutions can appear as is the case when laser action sets in or when optical bistability occurs. These new stationary solutions which correspond to fixed points in a phase-space diagram $E'(\omega)$ versus $E''(\omega)$ can themselves become unstable, and oscillatory solutions which correspond to limit cycles or chaotic solutions which correspond to strange attractors can appear.

Nonlinear processes that can be analyzed in the above described conventional frame of nonlinear optics indeed exist in semiconductors, e.g., the two-photon transitions in which the excitonic molecule is virtually excited via an intermediate exciton level. The two-photon transition to an exciton level which is forbidden for one-photon transitions is another example.

However, most of the recently observed large optical nonlinearities in semiconductors are of a different type. They arise because the applied light field does not only cause virtual transitions but generates a finite concentration of electron–hole pairs, which alter via several physical processes the probability for successive optical transitions. These nonlinearities cannot be calculated within the framework of intensity-dependent nonlinear susceptibilities.

In contrast, one often can calculate these nonlinear optical properties of a semiconductor close to the band edge by using linear response theory for the interaction of the electronic system and the applied light field, (i.e. the

$\chi^{(1)}$ term in eq. 6) but the quantum statistical average has to be taken for a system with a finite number N of interacting electron–hole pair excitations (Haug and Schmitt-Rink, 1984 and 1985) so that $\chi^{(1)}(\omega)$ becomes $\chi^{(1)}(\omega, N)$. The concentration of these electronic excitations has to be determined self-consistently via a rate equation for the number of electron–hole pairs, in which the concentration-dependent optical constants appear again[8]. Clearly, the wave equation for the field plus the rate equation for the electronic excitations form again a nonlinear system of equations, which, e.g., is able to describe the various instabilities in semiconductors. This approach is similar to the description of a two-level system in terms of the Maxwell–Bloch equations for the electric field, the polarization, and the population difference between the upper and lower levels. If the polarization has a fast relaxation process, one can eliminate this variable to reduce the description to one for the field and the population difference. In semiconductors, the polarization does indeed decay very rapidly due to interparticle collisions, so that the adiabatic elimination is often justified. For the same reason, the distribution of the excitation is, under most experimental conditions, a quasi-thermal one, which can be characterized by a temperature and a quasi-chemical potential. The temperature of the excitation is, however, normally different from the lattice temperature. Due to the fact that the electronic excitations are mobile and interact strongly with each other, the semiconductor has nonlinear optical properties that are not present in a two-level system. The natural formalism with which the laser-excited semiconductor can be described is the nonequilibrium Green's function method of Keldysh (1965). In this formalism, both the electron and hole propagators as well as that of the interband polarization are treated together in a way reminiscent of the treatment of inversion and polarization in the Maxwell–Bloch equations. The nonequilibrium Green's function formalism allows one to calculate systematically the renormalized energies and the population of these states. Two contributions in this book[9] make use of this technique, which also offers the framework to unify the treatments of coherent nonresonant nonlinearities and the resonant population-induced ones.

The most obvious many-body effect in optically excited semiconductors is the band-filling effect[10]. According to Fermi's golden rule, the transition probability per unit time for the absorption of a photon in a direct-gap semiconductor is given by

$$P_a = 2\pi\hbar \sum_k m_{cv}^2\, \delta(\epsilon_{ck} - \epsilon_{vk} - \hbar\omega)\big[(1 - f_{ck}) f_{vk} - f_{ck}(1 - f_{vk})\big] \quad (11)$$

where $\hbar m_{cv} = (\hbar p_{cv}/m_0)\sqrt{2\pi e^2/\hbar\omega V}$ is the photon–electron matrix element, p_{cv} is the momentum matrix element for the interband transition, and m_0 and e are the bare mass and charge of the electron, respectively. $\epsilon_{ck} = E_g + \hbar^2k^2/2m_e$ and $\epsilon_{vk} = -\hbar^2k^2/2m_{mv}$ are the energies of the electrons around the band extrema in both bands. The first term proportional to $(1 - f_{ck})f_{vk}$ gives the probability of absorption, the second term that of induced emission. Obviously, the difference between both terms is just the negative inversion $f_{kv} - f_{kc}$. In a quasi-thermal equilibrium situation, f_{kc} and f_{kv}ar Fermi functions, e.g. $f_{kc} = 1/\{\exp[(\epsilon_{ck} - \mu_c)\beta] + 1\}$, with $\beta = 1/kT$ and the quasi-chemical potential μ_c which is given by $N = \Sigma f_{ck}$. Often it is more convenient to introduce the hole picture for the missing electrons in the valence bands: $m_h = -m_v$, $\epsilon_{hk} = -\epsilon_{vk}$, $\mu_h = -\mu_v$ and $1 - f_{hk} = f_{kv}$. Furthermore, the conduction band electrons are just called electrons $c \to e$. In this picture the absorption coefficient, which is the absorption probability of the photon per unit length, is given by

$$\alpha(\omega, N) = \frac{(2\pi e)^2}{\omega cn(\omega, N)V}\sum_k \frac{p^2cv}{m_0}\delta(\epsilon_{ek} + \epsilon_{hk} - \hbar\omega)(1 - f_{ek} - f_{hk}), \quad (12)$$

where $n(\omega, N)$ is the index of refraction, which has to be determined self-consistently via the Kramers–Kronig transformation

$$n(\omega, N) = \frac{c}{\pi}P\int_0^\infty d\omega' \frac{\alpha(\omega', N)}{\omega'^2 - \omega^2}. \quad (13)$$

Already, in this free-particle transition picture both the absorption and dispersion are functions of density N of electronic excitations and thus give rise to optical nonlinearities. The total number of electrons is in turn given by a rate equation

$$\frac{dN}{dt} = \frac{\alpha(L, N)I}{\hbar\omega} - R(N), \quad (14)$$

where $I = v|E|^2/8\pi$. Here, v is the energy velocity which in a weakly absorbing medium can be approximated by the phase velocity $c/n(\omega, N)$. $R(N)$ is the recombination rate.

In a dense plasma, the single-particle picture has to be modified by replacing the free electron and hole energies by those of the interacting particles. The energies in a plasma change because screening weakens the Coulomb interactions, which leads to a gap shrinkage with increasing plasma density. A simple heuristic argument to understand this gap shrinkage $\Delta E_g(N)$ qualitatively is to consider the self-energy of a point charge in a plasma

$$\Delta E_g(N) = \lim_{r\to 0}\left[V_s(r) - V(r)\right], \quad (15)$$

with $V(r) = \dfrac{e^2}{\epsilon_0 r}$, where ϵ_0 is the background dielectric constant in the unexcited semiconductor. The screened Coulomb potential is $V_s = V(r)\exp(-k_s r)$, where $k_s = k_s(N)$ is the density-dependent screening wave number in the plasma. Thus, the band-gap reduction is

$$\Delta E_g(N) = -\frac{e^2 k_s}{\epsilon_0}. \tag{16}$$

As the screening wave number increases with density, the band gap shrinks. Actually, at higher densities, there is an additional term due to the band-gap filling, which will be discussed in Chapter 3. A microscopic justification of the simple arguments just presented comes from the calculation of the real part of the single-particle self-energy (Haug and Schmitt-Rink, 1984). The imaginary part describes a density-dependent collision broadening, which also is important for the calculation of high-density spectra (Klingshirn and Haug, 1981). In addition to the interband contributions to the nonlinear optical response, there is also an additional intraband contribution of the free carriers which can be written as a Drude term

$$\epsilon(\omega, N) = \epsilon_0\left[1 - \frac{\omega_{pl}^2}{(\omega + i\delta)^2}\right]. \tag{17}$$

Here, ω_{pl} is the plasma frequency

$$\omega_{pl}^2 = \frac{4\pi e^2 N}{\epsilon_0 m_r}, \tag{18}$$

where m_r is the reduced electron–hole mass. Particularly in narrow-gap semiconductors, its contribution to density-dependent dispersive changes can become important.

In the low-density regime, where the attractive Coulomb forces are still strong enough to support bound states, i.e. the exciton, the spectra deviate strongly from that of the free-particle transition picture. It is obvious that the density-dependent ionization of the exciton due to screening is a source of large optical nonlinearity, because in that region the sharp resonance lines of the bound states vanish and a broad-band plasma spectrum appears. The description of these changing spectra requires the treatment of the electron–hole pair state in the presence of other excitations[11].

The fact that an exciton exhibits much larger optical nonlinearities than a hydrogen atom can be understood easily. Nonlinearities are expected if the optical field E becomes comparable to the internal electric field $E_i = e/(\epsilon_0 a_0^2)$, where a_0 is the Bohr radius $a_0 = \hbar^2\epsilon_0/(m_r e^2)$. Thus, the internal field in an exciton is

$$E_i^{(x)} = m_r \frac{e^5}{\epsilon_0^3 \hbar^4} = E_i^{(\mathrm{H})} \left(\frac{M_r}{m_0} \right) \left(\frac{1}{\epsilon_0^3} \right). \tag{19}$$

Taking as typical parameters the reduced exciton mass $m = m_0/10$ and $\epsilon_0 = 5$, one sees that the critical fields in an exciton are more than three orders lower than in a hydrogen atom. Actually, the exciton is not ionized by the radiation field itself but by the presence of other charged excitations which favours the exciton system even more in comparison with atomic systems as far as their nonlinearities are concerned.

Let us briefly consider some trends in the optical properties of semiconductors[12]. The atoms of the semiconductors of group IV are arranged in a diamond lattice with a tetrahedral symmetry. All four outer valence electrons take part in the binding, but can be easily excited. The gaps of the two most frequently investigated and used semiconductors, namely Si and Ge, are indirect, i.e. the extrema of the valence and conduction bands are at different points in momentum space. The gap energies are of the order of 1 eV. Si in particular has been found to have a large nonlinear response, leading to laser-induced lattices and optical bistability (Jäger, 1984)[13], which has been interpreted in terms of the free-carrier intraband absorption. If the group IV atoms are replaced by pairs of atoms of the group III and V, one gets the binary semiconductors like GaAs, InSb or GaP, where the first two are examples of direct-gap semiconductors while the last one is an indirect-gap semiconductor. The gaps still are of the order of 1 or 1.5 eV. These relative small gaps result in correspondingly large background dielectric constants so that the Coulomb forces are weak. In InSb, e.g., the exciton resonance can only be observed in very pure samples at very low temperatures. The large optical nonlinearities of this material are mainly due to band-filling effects, while in GaAs, excitonic effects are more pronounced[14]. If one goes on to the more polar II–VI and I–VII compound semiconductors, the band-gap increases further up to about 3.5 eV. Correspondingly, the background dielectric function decreases, so that any effect due to Coulomb interaction becomes more important. The magnitudes of the optical nonlinearities decrease with increasing band-gap; however the stronger Coulomb forces cause excitonic resonances around which the optical nonlinearities are enhanced. The linear and nonlinear optical properties of CdS, e.g., are strongly influenced by the exciton resonance, while in CuCl, e.g., the nonlinear response is mainly determined by the exciton molecule[15].

Recently, new crystal growth techniques allowed to grow quasi–two-dimensional layers of semiconductors so-called multiple quantum wells.

The best-known example of this is GaAs sandwiched by GaAlAs, which has a wider band gap. For layers that are of the order of or thinner than an exciton Bohr radius, the electron motion perpendicular to the layer is quantized and often the optical properties are mainly determined by the highest-valence- and the lowest-conduction-subband alone. Under these conditions, one has a quasi-two dimensional semiconductor. Due to the local confinement of the optically generated electron–hole pair, all excitonic effects are greatly enhanced (see, e.g., Chemla, 1983). Thus, e.g., the exciton in GaAs–GaAlAs structures can still be observed at room temperatures. Hence, the ionization of the exciton in these structures is the most important source of optical nonlinearity. The flexibility in the design of these microstructures introduced a new degree of freedom of material engineering also of nonlinear optical properties[16].

Finally, large optical nonlinearities can also arise due to thermal effects, particularly at room temperature and for long excitation pulses. Under these conditions the electron–hole excitations decay partly nonradiatively and transfer parts of their energy to the lattice vibrations. An increase in the lattice temperature changes the band gap due to the self-energy of an electron–hole pair in the field of the excited phonons and therefore changes the optical constants. The absorption coefficient becomes a function of the temperature $\alpha(\omega, T)$ and the temperature is determined by a rate equation similar to that of the electron–hole density N. The generation term is again proportional to $\alpha(\omega, T)I$, causing a thermally induced optical nonlinearity in the same way as the electronically induced one for short-pulse excitation and at lower temperatures[17]. These photothermal nonlinearities have also been used to obtain bistable semiconductor elements as described e.g. in the contributions of Klingshirn[18] and of Peyghambarian and Gibbs[19].

REFERENCES

Bloembergen, N. (1965). *Nonlinear Optics*, W. A. Benjamin, New York.

Boyd, R. W., Raymer, M. G., and Narducci, L., eds. (1985). *Optical Instabilities*. Cambridge University Press, Cambridge, England.

Chemla, D. S. (1983). *Helv. Phys. Acta* **56,** 607.

Chemla, D. S., and Jerphagnon, J. (1980). In "Optical Properties of Solids" (M. Balkanski, ed.), *Handbook on Semiconductors*, Vol. 2, p. 545, ed. T. S. Moss. North-Holland, Amsterdam.

Fröhlich, D. (1981). *Festkörperprobleme* **XXI** (Advances in Solid State Physics **21**), 363.

Gale, G. M. and Mysyrowicz, A. (1974). *Phys. Rev. Lett.* **32,** 727.

Gibbs, H. M. (1985). *Optical Bistability: Controlling Light with Light*. Academic Press, New York.

Gibbs, H. M., McCall, S. L., Venkatesan, T. N. C., Gossard, A. C., Passner, A., and Wiegmann, W. (1979). *Appl. Phys. Lett.* **35,** 451.
Haken, H. (1984). *Laser Theory*. Springer-Verlag, Heidelberg.
Hanamura, E. (1973). *Solid State Commun.* **12,** 951.
Hanamura, E., and Haug, H. (1977). *Phys. Rep.* **C33,** 209.
Haug, H., and Schmitt-Rink, S. (1984). *Progr. Quantum Electron.* **9,** 3.
Haug, H., and Schmitt-Rink, S. (1985). *J. Opt. Soc. Am. B* **2,** 1135.
Hensel, J. C., Philips, T. G., and Thomas, G. A. (1977). In *Solid State Phys.* **32,** 88.
Jäger, D., Forsmann, M., and Wedding, B. (1985). *IEEE J. Quantum Electron.* **QE-21,** 1453.
Jeffries, C. D. (1975). *Science* **185,** 955.
Keldysh, L. V. (1965). *Sov. Phys. JETP* **20,** 1018.
Klingshirn, C., and Haug, H. (1981). *Phys. Rep.* **70,** 315 (1981).
Landau, L. D., and Lifshitz, E. M. (1985). *Theoretical Physics*, vol. 10, *Kinetics*. Pergamon Press, New York.
Miller, D. A. B., Smith, S. D., and Johnston, A. (1979). *Appl. Phys. Lett.* **35,** 658.
Nishina, Y., and Ueta, M., eds. (1976). *Physics of highly excited States in Solids.* Japan, Lecture Notes in Physics, vol. 57. Springer-Verlag, Berlin.
Pankove, J. I. (1971). *Optical Processes in Semiconductors*. Dover, New York.
Rice, T. M. (1977). In *Solid State Phys.* **32,** 1.
Seraphin, B. O., ed. (1976). *Optical Properties of Solids, New Developments*. North-Holland, Amsterdam.
Shen, Y. R. (1984). *The Principles of Nonlinear Optics*. J. Wiley, New York.
Woerdmann, J. P. (1970). *Opt. Commun.* **2,** 212.

NOTES

1. See Chapter 8.
2. See Chapters 10 through 14.
3. See Chapter 2.
4. See Chapter 9.
5. See Chapters 11 and 14.
6. See Chapters 4, 12, and 13.
7. See Chapters 15 and 16.
8. For a treatment of the kinetics and transport properties see Chapter 7.
9. See Chapters 3 and 6.
10. See Chapter 10.
11. See Chapters 2 through 6.
12. For a more detailed discussion see Chapters 2 and 10.
13. See Chapter 14.
14. See Chapters 3 through 6.
15. See Chapters 2 and 8.
16. See Chapters 4 and 13.
17. See Chapters 2 and 10.
18. See Chapter 2.
19. See Chapter 12.

2 SURVEY OF EXPERIMENTALLY OBSERVED OPTICAL NONLINEARITIES IN HOMOGENEOUS SEMICONDUCTORS

C. Klingshirn

FACHBEREICH PHYSIK DER UNIVERSITÄT
ERWIN SCHRÖDINGER STRAßE
D-6750 KAISERSLAUTERN, FEDERAL REPUBLIC OF GERMANY

I. INTRODUCTION

In this contribution, the experimentally observed optical nonlinearities in homogeneous semiconductors are reviewed. In the introduction, the linear optical properties of semiconductors are briefly outlined, proceeding from the more ionic bound, large and direct gap I–VII and II–VI materials with pronounced excitonic features to the direct and indirect gap III–V and elementary semiconductors and finally ending with the narrow-gap materials, where excitonic effects are almost negligible. In section II, the optical nonlinearities occurring in those various types of materials are reviewed,

OPTICAL NONLINEARITIES
AND INSTABILITIES IN
SEMICONDUCTORS

ISBN 0-12-332915-9

adapting the experimentalist's points of view, thus completing the theoretical contributions. In section III, selected experimental techniques to investigate the optical nonlinearities are described. Typical results concerning the various contributions to the nonlinearities as well as the various classes of materials are presented in section IV. The concluding section V is used to mention shortly possible applications of the optical nonlinearities, e.g., in optical bistability.

Due to the finite length of this contribution, it is obviously not possible to include or even to cite all work that has been done in this field and we apologize for this. The selection we have to make is of course somewhat arbitrary, but the main guidelines for our choice are the following: we shall concentrate on homogeneous semiconductors. Optical nonlinearities in p–n junction lasers will be reviewed in various contributions to this book. The linear and nonlinear optical properties of quantum well structures have recently been reviewed, e.g., by Hegarty and Sturge (1985) and by Chemla and Miller (1985) and will be described in two of the next chapters.

Concerning the physical origin of the nonlinearities, we shall stress those effects that are connected with excitations in the electronic subsystem of the material. This choice is simply due to the fact that photons of energies larger than the one of the phonon Reststrahlbande interact primarily with the electronic system by real or virtual excitations. However, since the luminescence yield is usually considerably smaller than one, some fraction of the energy deposited in the electronic system by the incident light field will be transferred into heat of the lattice. Subsequently, we shall also shortly review photothermal optical nonlinearities. Concerning our selection of the discussed materials, we have preferentially chosen those of the more common semiconductors to which no separate chapters are devoted in this book.

Many of the well-known semiconducting materials are tetrahedrally coordinated. This leads for the elementary semiconductors Si and Ge to the diamond lattice (pointgroup O_h) and for the binary compounds either to the blende (T_d)- or wurtzite (C_{6v}) structures. Fig. 1 shows schematically the bandstructures for O_h, T_d and C_{6v}. The first two of them are rather similar. The difference between T_d and C_{6v} structures appears in real space only in the position of the next nearest neighbours. The unit cell of C_{6v} is twice as big as for T_d. Correspondingly, the first Brillouin zone for C_{6v} is only one half of that of T_d. The direction $\Gamma \rightarrow \Lambda \rightarrow L$ of T_d is folded back in C_{6v} to give $\Gamma \rightarrow \Delta \rightarrow A \rightarrow \Delta \rightarrow \Gamma$ as shown in Fig. 1c. In C_{6v} the valence band is split by the hexagonal crystal-field and spin-orbit coupling into three

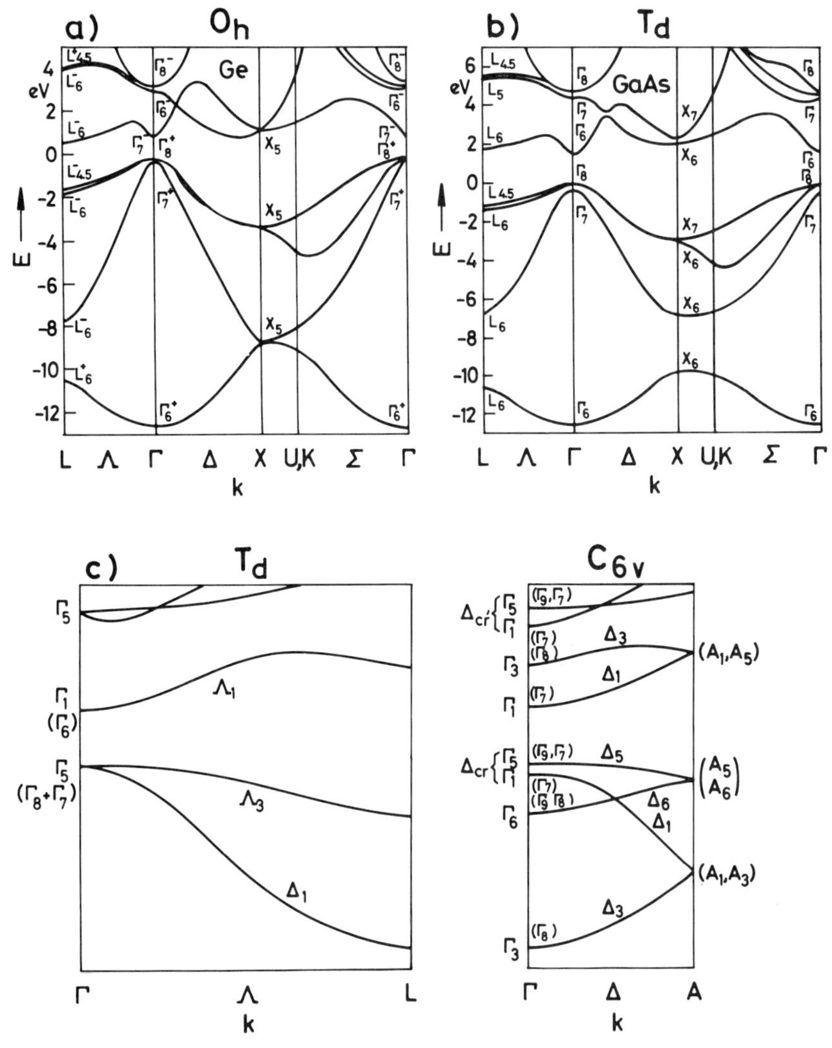

Figure 1. Schematic bandstructures for semiconductors with O_h (a), T_d (b) and C_{6v} symmetries. In (c) we show the origin of the bands of C_{6v} in the direction $\Gamma \rightarrow \Delta \rightarrow A \rightarrow \Delta \rightarrow \Gamma$ from folding the corresponding bands in T_d symmetry ($\Gamma \rightarrow \Lambda \rightarrow L$). Only bands close to the forbidden gap are shown in (c), neglecting for simplicity spin orbit coupling. The irreducible representations resulting from spin are given in brackets. (a) and (b) from Chelikowski, J. R. and Cohen, M. L. (1976), (c) according to Segall, B. (1967).

subbands Γ_9, Γ_7, Γ_7, while one has only a twofold splitting at the Γ point in T_d ($\Gamma_7 + \Gamma_8$).

The band structure describes one-particle states of electrons and/or holes. In optical transitions, two-particle states are always involved, namely electron and hole-pair states. The two particles can be bound together due to their attractive Coulomb interaction, leading thus to the concept of excitons. They determine the optical properties close to the absorption edge of all pure semiconductors with a gap $\gtrsim 0.5$ eV. In the semiconductors we are usually dealing with Wannier excitons, i.e., the binding energy of the exciton is small as compared to the width of the forbidden gap and the Bohr radius is much larger than the lattice constant. In the case of isotropic, nondegenerate and parabolic bands, the energy states E_x, the binding energy E_x^{b}, the exciton Bohr radius in the ground state a_x and the wave function ϕ_x can be described by Eq. 1.

$$
\begin{aligned}
&E_x(\vec{k}, n_B) = E_g - E_x^{\mathrm{b}} \frac{1}{n_{\mathrm{B}}^2} + \frac{\hbar^2 \vec{k}^2}{2M} \\
&E_x^{\mathrm{b}} = Ry \frac{\mu}{\epsilon^2}; \qquad a_x = a_{\mathrm{B}} \frac{\epsilon}{\mu} \\
&\phi_x = \frac{1}{\sqrt{\Omega}} \varphi_{\mathrm{e}}(\vec{r}_{\mathrm{e}}) \varphi_{\mathrm{h}}(\vec{r}_{\mathrm{h}}) \phi_{n_{\mathrm{B}},l,m}(\vec{r}_{\mathrm{e}} - \vec{r}_{\mathrm{h}}) e^{i\vec{k}\vec{R}} \\
&M = m_e + m_h; \qquad \mu = \frac{1}{m_0} \frac{m_e m_h}{m_e + m_h}
\end{aligned}
\tag{1}
$$

where n_B, l, m are the hydrogen quantum numbers, $\vec{k}$ is the wavevector of the center of mass $\vec{R}$, Ry is the Rydberg energy and a_{B} is the Bohr radius. $\varphi_{\mathrm{e}}(\vec{r}_{\mathrm{e}})$, $\varphi_{\mathrm{h}}(\vec{r}_{\mathrm{h}})$ are the Wannier functions of electron and hole, respectively. $\phi_{n_{\mathrm{B}},l,m}$ is the envelope function, m_0 is the free electron mass, μ the dimensionless reduced mass and M is the translational mass.

If there is a direct, dipole-allowed band-to-band transition, the excitons with s-envelope function ($l = 0$) are strongly coupling to the radiation field. The quanta of the resulting mixed state of electronic excitation (i.e. excitons) and the electromagnetic field (photons) are called excitonic polaritons. Their dispersion relation $E(\vec{k})$ results from the ones of photons and of excitons and the noncrossing rule. This dispersion relation is shown for negligible damping around $E_x(n_{\mathrm{B}} = 1, \vec{k})$ in Fig. 2a, together with the resulting reflection and absorption spectra (Figs. 2b and c). The lower polariton branch (LPB) starts with an almost vertical, photonlike dispersion and then bends over to an excitonlike behaviour. The transition region is

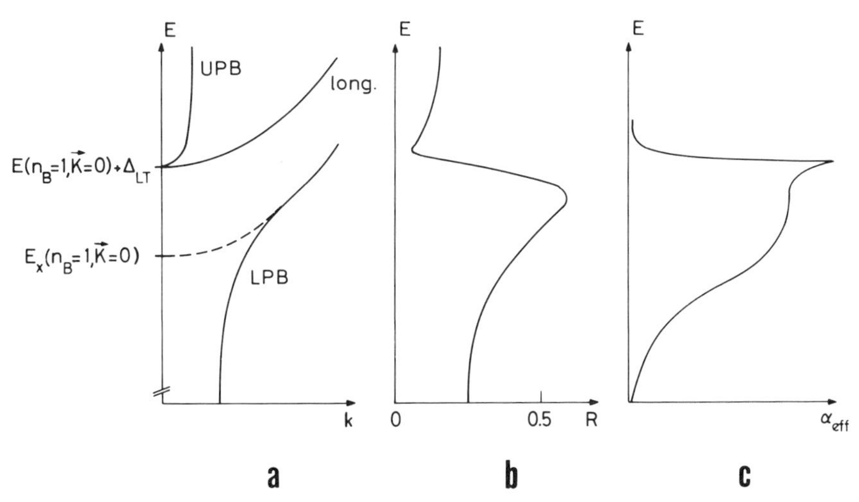

Figure 2. The dispersion of excitonic polaritons for negligible damping in the vicinity of the $n_B = 1$ resonance (a) and the resulting spectra of reflectivity R (b) and of the effective absorption coefficient α_{eff}. From Klingshirn, C. (1984).

called "bottleneck". There is a finite longitudinal transverse splitting Δ_{LT} proportional to the oscillator strength of the exciton resonance, a longitudinal exciton mode and an upper polariton branch (UPB) which bends over to a photonlike behaviour. It can be seen that for certain energies are more than one polariton branches due to the $\vec{k}$-dependence of the eigenenergy (spatial dispersion). This causes a lot of complications, e.g., the question about additional boundary conditions. This topic is, however, beyond the scope of this contribution. The reader is referred, e.g., to the review by Hönerlage et al. (1985) and the references given therein. Similar resonances appear for the higher states, however with an oscillator strength decreasing like n_B^{-3}. In the continuum, which corresponds to the band-to-band transitions, one has a rather smooth behaviour of the real and imaginary parts of the refractive index $\tilde{n} = n + i\varkappa$. The oscillator strength is however still enhanced by the Coulomb interaction, resulting in a rather constant absorption coefficient α instead of a square root dependence $\alpha \approx (\hbar\omega - E_g)^{1/2}$. More detailed information on excitons in real crystals is found, e.g., in the review articles by Klingshirn (1984) or Hönerlage et al. (1985). Apart from these free excitons described above, which are characterized by a plane wave factor $e^{i\vec{k}\vec{R}}$, there are excitons bound to impurities. A^0X and D^0X, e.g., mean excitons bound to neutral acceptors and donors, respectively. At low temperature, they show up as narrow absorption and emission bands situated energetically below the free exciton resonances. With increasing temperature, they tend to disappear between 20 K and 70 K. For bound excitons, details are found, e.g., in the review by Dean and Herbert (1979).

With increasing temperature, the damping of the free exciton resonances increases, too, depending on $k_B T$ and on the strength of the exciton–phonon coupling. Consequently, the distinct reflection and absorption structures are smeared out, and one often observes only a hump on the low-energy side of the absorption spectrum. Nevertheless, there is still the Coulomb interaction between electron and hole present and the resulting transfer of oscillator strength toward the band edge.

The low-energy wing of the absorption edge develops with temperature according to the Urbach–Martienssen rule

$$\alpha(\hbar\omega) = \alpha_0 \exp[\sigma(\hbar\omega - E_0)/k_B T] \tag{2}$$

The energy E_0 equals roughly the $n_B = 1$ exciton energy; α_0 and σ are material parameters, too. For a discussion of the microscopic models for the description of eq. 2, see, e.g., Liebler et al. (1985).

There are also direct gap materials with a dipole-forbidden band-to-band transition such as Cu_2O, SnO_2 or TiO_2 (rutil). In these substances, one has a weak coupling between radiation field and excitons (preferentially p-excitons, i.e. $l = 1$), and it is not necessary to introduce the polariton concept. Since these materials are much less studied in the direction of optical nonlinearities, they will not be treated further. The same holds, e.g., for layered compounds of the type HgI_2 or PbI_2 which have strongly dipole allowed exciton transitions.

While the Cu-halides and most of the II–VI compound semiconductors have a direct band-to-band transition with both band extrema situated at the Γ-point, one finds among the III–V materials both direct and indirect gap semiconductors. In the latter case, the conduction band minimum either in the Δ-direction between Γ and X-point or directly at the X-point falls below the minimum at Γ (Fig. 1b). The same is true for Si, while in Ge the global conduction band minima are situated at the L-points. In ternary solid solutions between two different III–V compounds, it is possible to go continuously from a direct-gap material to an indirect one by changing the composition (e.g., in $Ga_{1-x}Al_xAs$ or in $Ga_{1-x}In_xP$). Excitons are also formed by an electron in the indirect conduction band minimum and a hole around Γ. In order to couple these indirect gap excitons to the radiation field, a momentum-conserving phonon has to be involved. Consequently, the oscillator strength of these transitions is reduced and the shape of the absorption spectrum changes. At low temperatures, indirect excitons can be formed only under emission of phonons. This process leads to a steplike increase of the absorption coefficient below the direct exciton transition at energies given by

$$\hbar\omega_{\text{step}} = E_{x,\,\text{indirect}} + \sum n_i \hbar\omega_i \tag{3}$$

At higher temperatures, transitions are also possible under emission of phonons, resulting in terms with a negative sign in the sum of Eq. 3. While absorption coefficients are of the order of 10^4 to 10^6 cm^{-1} in the resonances of direct, dipole allowed excitons, the values are ranging from 1 cm^{-1} to 10 cm^{-1} for indirect excitons. A rather exhaustive collection of semiconductor parameters is given by Madelung et al. (1982). There is a general trend that the binding energy E_x^{b} of excitons and their oscillator strength Δ_{LT} are decreasing with decreasing width of the forbidden gap E_g. Examples are CuCl: $E_g \approx 3.5$ eV, E_x^{b}y $= 200$ meV, $\Delta_{\text{LT}} \approx 5$ meV, CdS: $E_g \approx 2.5$ eV, $E_x^{\text{b}} \approx 30$ meV, $\Delta_{\text{LT}} = 2$ meV, GaAs: $E_g \approx 1.5$ eV, $E_x^{\text{b}} \approx 4$ meV, $\Delta_{\text{LT}} \approx 0.1$ meV.

With decreasing E_x^{b} and Δ_{LT}, it becomes increasingly difficult to observe distinct exciton resonances even at low temperature and in pure materials. In the narrow-gap semiconductors ($E_g \leq 0.5$ eV), excitons are virtually absent as distinct resonance structures. The Coulomb interaction just leads to a slight modification of the absorption edge due to band-to-band transitions. Some of the most prominent representatives of these narrow-gap

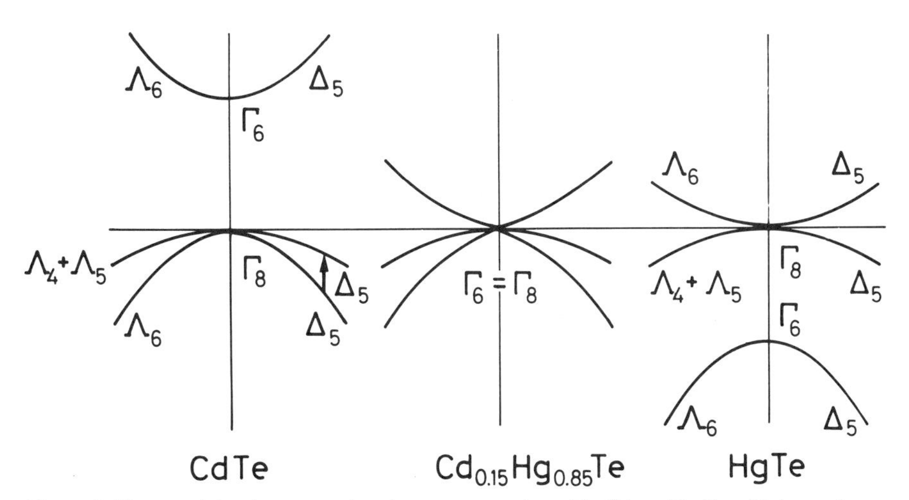

Figure 3. The transition from a semiconductor to a semimetal in $Cd_{1-x}Hg_xTe$ with increasing x. From Madelung, O. (1970).

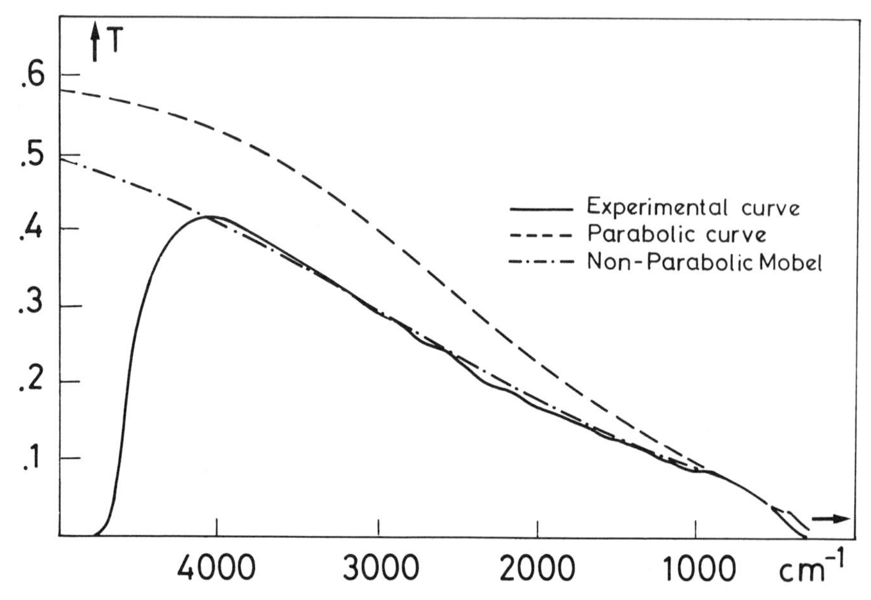

Figure 4. Absorption spectrum of the p-doped narrow-gap semiconductor $Cd_{1-x}Hg_xTe$. From Brossat, T. and Raymond, F. (1985).

materials are the III–V compound InSb with a direct gap at the Γ point, the lead salts PbS, PbSe and PbTe with a direct gap at the L-point and the mixed crystal $Cd_{1-x}Hg_xTe$ (CMT). The transition from the semiconductor CdTe ($E_g \approx 1.6$ eV) to the semimetal HgTe ($E_g \approx -0.3$ eV) can be continuously followed in this material by varying x (Fig. 3). With increasing x, the conduction band Γ approaches the valence band Γ_8; for $x = 0.85$, they are degenerate at the Γ point; and for $x < 0.85$, the Γ_6 band is situated below the Γ_8 band and changes the sign of its curvature. The Γ_8 band, which splits for $x < 0.85$ into the light and heavy hole bands for $k \neq 0$, forms the new conduction and the uppermost valence band which are degenerate at Γ. The transmission spectrum for $x < 0.85$ is determined by the band-to-band transitions above E_g and the Reststrahlenbande at the optical phonon frequencies. (See Fig. 4.) If the material contains free carriers, there is some additional contribution to the optical properties by free carrier absorption, and, in the case of p-type conductivity, by intersub-valence-band transitions indicated schematically by an arrow in Fig. 3. Fig. 4 shows a fit to the experimental data assuming parabolic and nonparabolic valence bands. From spectroscopy in the window between E_g and the optical phonons, it is possible to deduce the carrier concentration.

II. OPTICAL NONLINEARITIES

In linear optics, one expects that the optical properties of matter such as the transmission $T(\omega)$- and reflection spectra $R(\omega)$ or the complex dielectric function $\epsilon(\omega) = \tilde{n}^2(\omega)$ depend on the frequency of the incident radiation field, and, in the case of birefringent materials, also on the direction of polarization and propagation relative to the crystallographic axes, but definitely not on the intensity I or field amplitude $\vec{E}$ of the radiation field. Consequently, the polarization $\vec{P}$ of matter oscillates with the same frequency ω as the incident radiation, and two light beams, which cross each other in matter, will not interact.

All these assumptions are no longer valid in nonlinear optics. This means T, R or ϵ depend in a reversible way on the light intensity and lightbeams start to interact with each other in matter. There are two approaches to describe these types of phenomena. In one case, one assumes that the response of the medium to the incident light field, i.e. the polarization, depends only on the instantaneously present field amplitudes. This condition is fulfilled if (electronic) excitations are created only virtually. In this case $\vec{P} = \epsilon_0(\epsilon - 1)\vec{E} = \epsilon_0 \chi \vec{E}$ can be developed into a power series of the incident field amplitudes as shown schematically in Eq. 4:

$$\frac{1}{\epsilon_0} P_i = \sum_j \chi_{ij}^{(1)} E_j + \sum_{j,k} \chi_{ijk}^{(2)} E_j E_k + \sum_{j,k,l} \chi_{ijkl}^{(3)} E_j E_k E_l + \cdots \tag{4}$$

The first term on the right-hand side describes the linear optical properties; $\chi^{(2)}$ gives phenomena such as second harmonic-, sum-, and difference-frequency generation or the dc-effect. $\chi^{(3)}$ describes four-wave mixing, Hyper-Raman–scattering (HRS), coherent anti-Stokes–Raman scattering (CARS), etc. For a detailed review of these effects see, e.g., the publications by Fröhlich (1981) and Shen (1984).

In the second case, the optical properties are modified by real excitations of some quasiparticles with finite lifetime. In this case, one has

$$\vec{P} = \epsilon_0 \chi(N) \vec{E} \tag{5}$$

where N stands in the moment, e.g., for an increase of the electron–hole pair density, or of the phonon population number or of the lattice temperature. Due to the finite lifetime, N depends not only on the instantaneous intensity but also on the generation rate G in the past weighted with some decay function, e.g.

$$N(t) \simeq \int_{-\infty}^{t} G(t') e^{-(t-t')/\tau} \, dt' \tag{6}$$

G itself is connected with the intensity at time t', e.g., in the presence of one- and two-photon excitation by

$$G(t') = \alpha I(t') + \beta I^2(t'). \tag{7}$$

The coefficients τ, α and β may in turn depend on N, leading thus to a rather complex set of equations. In the following, we shall find examples for both types of nonlinearities.

The variation of the optical properties, e.g., of ϵ at a frequency ω by excitation at another frequency ω_{exc} with intensity I_{exc} are often called renormalization effects. With intensity we mean the energy flux density, i.e. the Poynting vector $\vec{S} = \vec{E} \times \vec{H}$ averaged over a period of light. If one considers the variation of ϵ or $\tilde{n}$ produced by I_{exc} at $\hbar\omega_{exc}$ one usually speaks about self-renormalization.

In the rest of this section, we shall review some selected renormalization processes. As before, we start with excitonic and biexcitonic effects, which are more characteristic for the large-gap materials. Then we discuss the electron–hole plasma which has been found in a large variety of both direct- and indirect-gap materials, including narrow-gap semiconductors. While such nonlinearities are caused by real or virtual excitations of the electronic system, we shall say some words about thermal nonlinearities in the last subsection.

If we increase the incident light intensity and thus the generation rate (e.g. via Eqs. 6 and 7), we may reach densities of excitons or more generally of electron–hole pairs n_p at which these quasiparticles start to interact with each other. The simplest interaction mechanisms (which have been discussed already over many years) are elastic and inelastic scattering processes (see, e.g., Benoit à la Guillaume et al. 1969, Elkomoss and Munschy, 1977, Klingshirn and Haug, 1981). Two excitonlike polaritons in the $n_B = 1$ state may collide. One of the inelastic processes involves a scattering of one of the excitons on the photonlike part of the LPB, the other one under energy and momentum conservation into states with $n_B = 2, 3, \ldots \infty$. This leads to the well-known luminescence bands P_2, P_3, P_∞. Depending on the losses and on the distribution function of the polaritons in the various excitonlike branches, this scattering may result in optical amplification (gain) and/or in excitation-induced absorption. The various P-bands have been found in several semiconductors such as CdSe, CdS and ZnO as shown, e.g., by Klingshirn and Haug (1981) or more recently in ZnS by Shevel and Kunz (1986) in the appropriate spectral positions. However, recently Razbirin (1978) and Broser and Gutowski (1985) proposed for CdS different interpretations for the P-bands, mainly in terms of biexciton decay. (Biexcitons are discussed later in this chapter.) Though the processes discussed especially by Broser and Gutowski (1985) are likely to contribute to the emission in the spectral region of the P_2, P_3 and P_∞ bands in CdS under appropriate excitations, they are probably not of sufficiently general nature to explain the corresponding results in the other compounds.

While elastic and inelastic exciton-exciton scattering is predominant at low temperatures, one may imagine similar processes between one exciton-like polariton and a free carrier at higher temperatures, where a certain fraction of the excitons are ionized thermally. This scattering gives rise to a luminescence band and to excitation-induced absorption and gain structures characterized by the fact that the maxima of gain and luminescence shift with increasing temperature faster to smaller photon energies than the band gap does. Models developed to describe this phenomenon are reviewed by Klingshirn and Haug (1981); a more recent calculation is given by Lindwurm and Haug (1983), which nicely coincides with experiment. Yoshikumi et al. (1979) claim that these emission bands may be due to recombination in an electron–hole plasma (to be discussed later), but this interpretation has the following shortcoming. At room temperature, the electron and hole populations in the plasma are no longer degenerate. This means that there is no gain due to direct band-to-band recombination in the plasma. On the other hand, stimulated emission has clearly been observed up to room temperature in the emission band presently under consideration, e.g., by Bille (1973). As indicated by Klingshirn (1985) a possible bridge between these two models is to assume inelastic scattering also in a plasma or in a state that is between an exciton gas and a plasma. This idea seems to be favoured by the fact that there is a rather continuous transition from the exciton to the plasma state in direct-gap materials even at low temperature and in indirect ones above the critical temperature T_c for electron–hole liquid formation. All the scattering processes just oulined, which involve excitons, will contribute to a collision broadening of the free exciton resonance apart from the individual structures they produce, e.g., the P-bands. This excitation-induced collision broadening of the free exciton resonance is shown schematically in Fig. 5.

Bound exciton complexes may also lead to excitation-induced variations of the optical properties, e.g., by a change of the charge state of the impurity.

A rather rich variety of optical nonlinearities in large-gap semiconductors is connected at low temperatures with transitions involving biexcitons. A biexciton (or excitonic molecule) is a bound state consisting of two electrons and two holes. While the exciton can be treated in analogy to the

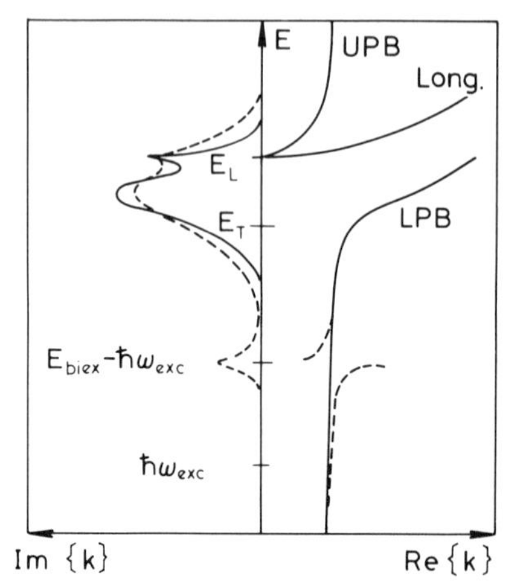

Figure 5. Schematic drawing of the real and imaginary parts of the wavevector (or dispersion and absorption) as a function of energy close to the $n_B = 1$ exciton resonance for low excitation (solid lines) and modifications induced by collision broadening and by two-polariton transitions to the biexciton (dashed lines). From Kalt, H. et al. (1985).

hydrogen (or the positronium) atom, a biexciton corresponds to the hydrogen (or positronium) molecule. It forms a bound state situated by a certain binding energy E^b_{biex} below twice the energy of the lowest free exciton and has a $E(\vec{k})$ relation characterized by an effective mass of roughly twice that of the free exciton

$$E_{\text{biex}}(\vec{k}) = 2E_x(n_B = 1, k = 0) + \hbar^2 k^2 / 2M_{xx} \tag{8}$$

While one-photon transitions from the crystal ground state to the biexciton level are not possible in first-order perturbation theory and are partly also forbidden by the dipole selection rules (e.g., in T_d symmetry), one finds a giant oscillator strength for two-photon (more precisely, two-polariton) absorption (TPA). The reason for the large TPA coefficient β in the case of biexcitons as compared to excitons can mainly be seen in the resonance denominator appearing in second-order perturbation and partly in the larger spatial extension of the biexciton wavefunction as compared to the exciton one as described, e.g., by Klingshirn and Haug (1981).

This TPA to the biexciton state can be described in slightly different words. Assume we shine a laser beam with $\hbar\omega_{\text{exc}}$, I_{exc} on a sample, i.e., we populate in the sample the polariton state at $\hbar\omega_{\text{exc}}$. For simplicity, we assume it to be on the LPB of Fig. 5. If the state at $\hbar\omega_{\text{exc}}$ is populated, polaritons with an energy $\hbar\omega_{\text{abs}} = E_{\text{biex}} - \hbar\omega_{\text{exc}}$ can be absorbed, converting the polariton into a really excited biexciton. This means, by shining light at $\hbar\omega_{exc}$ on the sample, we create an absorption peak at $\hbar\omega_{abs}$ and via Kramers–Kronig relations a resonance structure in the real part of $\tilde{n}$ or $\vec{k}$. See Fig. 5. The oscillator strength of this resonance is directly proportional to the polariton density at $\hbar\omega_{exc}$. In simplest approximation, this resonance can be treated as an additive term in $\epsilon(\omega)$ as shown in the paper by Klingshirn (1985).

The situation becomes more complex if $\hbar\omega_{exc}$ is tuned to the region around half the biexciton energy. In this case, the laser beam at $\hbar\omega_{exc}$ falls on the resonance at $\hbar\omega_{abs}$, i.e., we get strong self-renormalization phenomena. Various theoretical approaches to this self-renormalization produced a rather surprising result, namely a quadratic pole in the dispersion relation of the excitonic polariton, together with a strong deformation at the exciton polariton dispersion at the resonance itself. The solution of this problem was found only recently by Haug (1985) and by Kranz and Haug (1986). It is necessary to include the spatial dispersion, i.e. the $\vec{k}$-dependence of the eigenenergies (here of excitons and biexcitons) and to treat the electronic excitations and the radiation field on an equal footing. In doing so, one gets a "normal" resonance with some additional polariton branches. In cubic materials (T_d symmetry), left and right circularly polarized polaritons are "good" eigenmodes of the system. The crystal groundstate and the biexciton groundstate have both symmetry Γ_1. These two facts allow one to produce induced circular dichroism by TPA. If one shines, e.g., σ^+ polarized light at $\hbar\omega_{exc}$ on the sample, then a TPA process is—according to the selection rules—possible only if the other quantum is polarized σ^-. This means that one gets the additional resonance at $\hbar\omega_{abs}$ (Fig. 5) only for this polarization, while the dispersion for σ^+ light is not influenced around $\hbar\omega_{abs}$. If one shines linearly polarized light around $\hbar\omega_{abs}$ on the sample, the beam is decomposed into two beams of initially equal intensities being polarized σ^+ and σ^-, respectively. The σ^+ polarized polaritons propagate in the sample according to the unrenormalized dispersion curve, while the σ^- ones are absorbed and propagate with a different velocity. Consequently, one finds that the probe beam around $\hbar\omega_{abs}$ is elliptically polarized after passing through the sample, and the long axis of the ellipse is turned with respect to the original linear polarization. Corresponding experiments and calculations have been summarized, e.g., by Haug (1982). A similar effect should be observable also in uniaxial materials (e.g., C_{6v} symmetry) when both beams are propagating parallel to the crystallographic axis.

Apart from the TPA processes just described, there is also induced absorption. By this term, one means transitions to the biexciton state from excitonlike polaritons distributed in the bottleneck or thermally on the excitonlike part of the LPB, the UPB or on the longitudinal exciton branch. Obviously, one needs some resonant or nonresonant pump source to produce the excitons. The spectral position of the induced absorption is rather independent of $\hbar\omega_{\text{exc}}$ and occurs around $\hbar\omega = E_{\text{biex}} - E_x$. The shape of the absorption peak is determined by the distribution of the excitons on the various branches.

There is also the inverse process, i.e. the decay of biexcitons into a photonlike polariton appearing as luminescence and an excitonlike polariton. The shape of this luminescence band depends on the distribution of the biexcitons. The spectra of gain and/or absorption are determined, in addition, by the population of the various exciton and biexciton branches.

In TPA and induced absorption, a real excitation of biexcitons takes place. It is also possible to create biexcitons with two incident quanta only virtually, meaning at an energy different from the eigenenergy. Such an excitation can exist only for times τ given by the uncertainty principle. Then it must disappear again. One radiative decay process is a decomposition in a photonlike polariton $\hbar\omega_{\text{R}}$ and a longitudinal exciton at energy $\hbar\omega_f$. Energy and momentum have to be conserved in the hole process. If the biexciton is virtually created from two identical quanta $\hbar\omega_{\text{exc}}$, one finds

$$\begin{aligned} \hbar\omega_{\text{R}} &= 2\hbar\omega_{\text{exc}} - \hbar\omega_f \\ \vec{k}_{\text{R}} &= 2\vec{k}_{\text{exc}} - \vec{k}_f \end{aligned} \tag{9}$$

Since the eigenenergy of the longitudinal branch is only weakly dependent on k, a plot of $\hbar\omega_{\text{R}}$ against $\hbar\omega_{\text{exc}}$ gives a straight line with slope of two. Therefore, this process is called two-photon or hyper-Raman scattering (HRS). If the final state $\hbar\omega_f$ is a polariton, the energy can depend sensitively on $\vec{k}$ (see Fig. 2). As a consequence, one finds all slopes between zero and two in a plot $\hbar\omega_{\text{R}}$ over $\hbar\omega_{\text{exc}}$ depending on the scattering geometry. The relation between $\hbar\omega_{\text{exc}}$ and $\hbar\omega_{\text{R}}$ is usually a rather smooth and montoneous one. If, however, either $\hbar\omega_{\text{exc}}$, $\hbar\omega_f$ or $\hbar\omega_{\text{R}}$ fall on a sharp resonance, then some discontinuities appear. Thus, HRS can be used as a means to detect and analyse excitation-induced resonances such as the one at $\hbar\omega_{\text{abs}}$ shown in Fig. 5. If the sample has resonatorlike shape (formed, e.g., by two plan-parallel surfaces), the HRS process can result in laser emission. The emission wavelength can be tuned by changing the scattering geometry as shown by Baumert and Broser (1981). On the other hand, it is also possible to stimulate the decay of the virtually excited biexcitons, by sending an additional, monochromatic beam with energy $\hbar\omega_f$ and wave

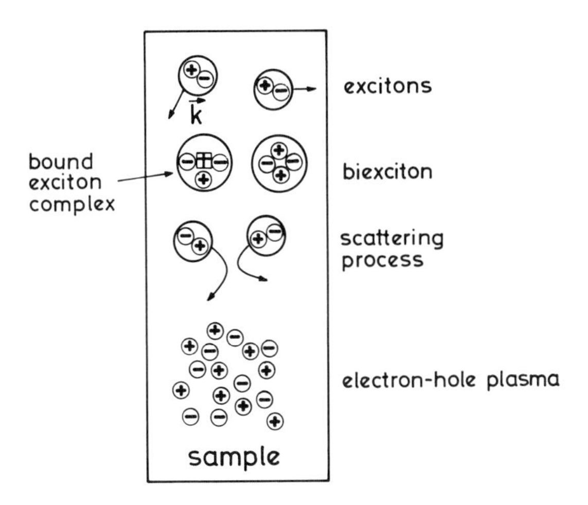

Figure 6. Schematic drawing of a gas of excitons, including a D^0X complex, a scattering process and a biexciton (upper part) and an electron–hole plasma (lower part).

vector $\vec{k}_f$ into the sample. The virtually excited biexciton then decays under energy and momentum conservation into a quantum $\hbar\omega_f$, $\vec{k}_f$ and another one $\hbar\omega_R$, $\vec{k}_R$. This process is identical with degenerate or nondegenerate four-wave mixing depending on whether $\omega_{exc} = \omega_R = \omega_f$ or not. Alternatively, this $\chi^{(3)}$-process can be also considered as diffraction from a dynamic, laser-induced grating. These analogies have been stressed, e.g., by Mita and Nagasawa (1978), by Masumoto and Shionoya (1980), by Maruani and Chemla (1981), by Hönerlage et al. (1982), and by Kalt et al. (1985 and 1986).

Until now, we considered optical nonlinearities in a range of excitation densities where excitons and biexcitons are still good quasiparticles. With increasing electron–hole pair density, this picture breaks down as shown schematically in Fig. 6. In the upper half of the "sample," we have visualized free and bound excitons, a scattering process and a biexciton. In the lower half, the excitation is so strong that the mean distance between excitons is comparable to their Bohr radius. Under these conditions, it is no longer possible to say that one electron is bound to one hole. Instead, one gets a new collective phase of electrons and holes, the so-called electron–hole plasma EHP. Another point of view is to state that the Coulomb interaction between an electron and a hole is screened by the other carriers to an extent that there are no more bound states. The formation of an EHP has drastic consequences on the optical spectra of semiconductors. This is visualized for low and high temperatures in Fig. 7.

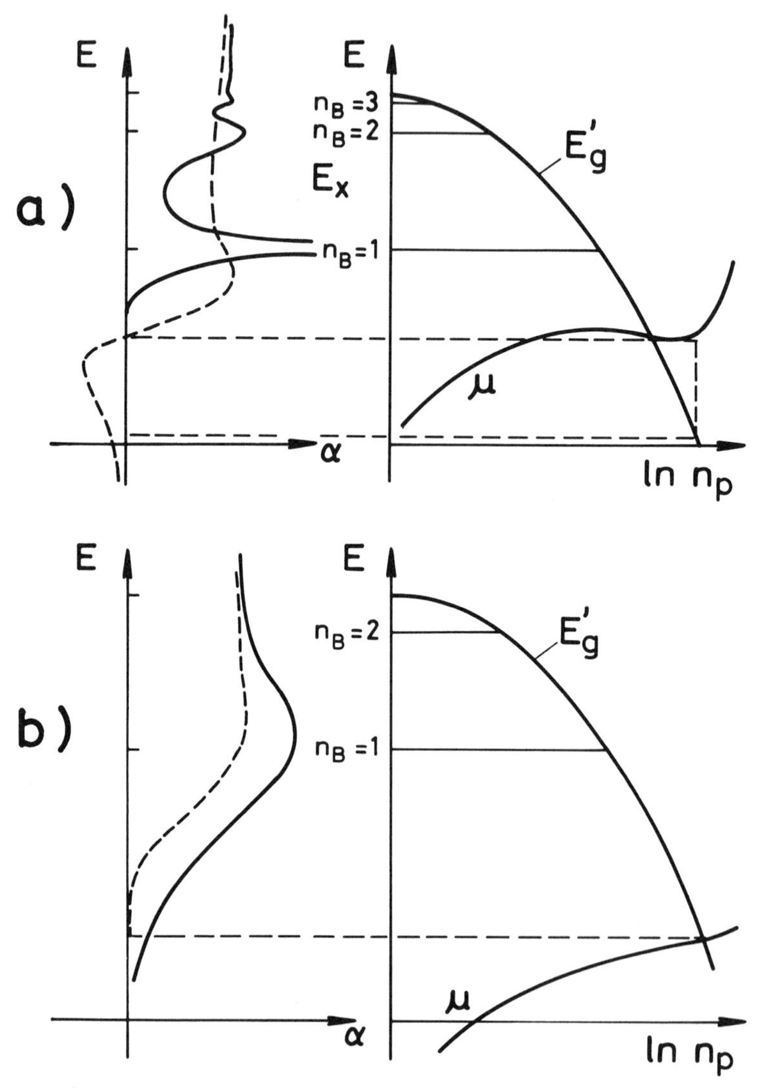

Figure 7. Schematic drawing of the optical absorption α (left) of a low-density exciton gas (solid lines) and of an electron–hole plasma (dashed lines) for low temperature (a) and high temperature (b). The right part gives the dependence of various energies as a function of the electron–hole pair density n_p. E'_g : renormalized gap; μ: chemical potential of the electron–hole pairs; and E_x : exciton energies.

The electron–hole pair density n_p which is the crucial parameter (besides the temperature) in Fig. 7 can often be assumed to increase linearly with I_{exc}. Sometimes one finds a saturation of the curve $n_p = n_p(I_{exc})$, e.g., connected with the onset of stimulated emission as found, e.g., by Swoboda, E. from time-resolved gain spectroscopy in CdS. In some materials, the dominant recombination process is of the Auger type. In this case, the recombination rate increases roughly with the third power of the carrier density. By simple reaction kinetics, it can be shown that this results in a dependence $n_p \sim I_{exc}^{1/3}$.

We show in Fig. 7a the low-temperature case. On the left-hand side, one sees the absorption spectrum at low excitation, revealing the series of exciton states, and at higher energies the transitions into the continuum. The right-hand part gives various energies as a function of the electron–hole pair density n_p. The width of the forbidden gap E_g is a monotonously decreasing function of n_p mainly due to exchange and correlation energies. The exciton energies are rather independent of n_p since the decrease of the renormalized gap $E_g'(n_p)$ and the decrease of the binding energy due to screening of the Coulomb potential are compensating each other. The density where E_x^b tends to zero is the Mott density. The oscillator strength of the exciton resonance is also decreasing with increasing n_p, but even for vanishing binding energy, there is still some electron–hole pair correlation, leading to the "excitonic enhancement" in the plasma recombination probability around the chemical potential μ. This chemical potential μ is the energetic distance between the quasi-Fermi-levels describing the population of electrons and holes. It is also shown in Fig. 7. With increasing n_p, μ can be situated above $E_g'(n_p)$. This situation describes population inversion between conduction and valence bands, i.e., the occurrence of optical amplification (gain) by direct band-to-band recombination. The gain and absorption spectrum for such a situation is plotted again on the left-hand side of Fig. 7. The sample is transparent roughly for photon energies below $E_g'(n_p)$. Between E_g' and μ, one has gain and, above μ, absorption due to band-to-band transitions. The fact that the gain spectrum extends somewhat below E_g' is due to a broadening connected with final state damping.

The structure of $\mu(n_p)$ shows with increasing n_p a maximum and then a minimum and is thus similar to the van der Waals equation for real gases. One can apply Maxwell's construction, resulting below a critical temperature T_c in a first-order phase transition from a low-density gas consisting mainly of excitons to liquidlike plasma (EHL). This phase transition has been found experimentally in indirect-gap semiconductors such as Ge, Si or

GaP. For a review see, e.g., Rice (1977) and Hensel et al. (1977). In direct-gap materials, the lifetime of carriers τ_p is below 1 ns and thus so short that a clear phase-separation in EHL-droplets surrounded by a gas cannot develop, as demonstrated by Bohnert et al. (1981) and confirmed later on by Yoshida and Shionoya (1983), by Unuma et al. (1984) and by Saito (1985). The situation at high temperatures is depicted in Fig. 7b. At low values of n_p, the exciton resonances are thermally broadened, depending on $k_B T / E_x^b$ and on the exciton–phonon coupling. Only the $n_b = 1$ exciton is seen as individual structure, if there are structures at all. The low-energy tail can be described by the Urbach–Martienssen rule in Eq. 2. E_g' again decreases monotonously with n_p. Since we are now above T_c, one finds that $\mu(n_p)$ is just a monotonously increasing function of n_p. The effective density of states that corresponds to the onset of population degeneracy is increasing as $T^{3/2}$ so μ stays below E_g' in many cases. Nevertheless, the Coulomb interaction is screened by the carriers, i.e., one gets on EHP. The absorption spectrum $\alpha(\omega, n_p)$ contains apart from the combined density of states $D(\omega, n_p)$, excitonic enhancement $\rho(\omega, n_p, T_p)$ a population term:

$$\alpha(\omega, n_p, T_p) \sim D(\omega, n_p)\rho(\omega, n_p, T_p)\Gamma(\omega, n_p, T_p)(1 - f_e - f_h) \quad (10)$$

where f_e and f_h are the quasi-Fermi functions of electrons and holes, respectively. Since we have $\mu \lesssim E_g'$ in the EHP we find for $\hbar\omega \approx E_g$: $1 - f_e - f_h \approx 0$. Since the Fermi-functions are at higher temperatures rather smoothly varying functions of energy, we get weak absorption over a rather large range of energies around E_g' as shown on the left-hand side of Fig. 7b as a dashed line. This means that one gets a blue shift of the absorption edge as compared to the low-density case. At low temperatures, we have a red shift as long as μ stays below the exciton energy. For $\mu > E_x$ one also finds a blue shift. The blue shift of the absorption edge, which occurs at high temperatures and which may appear also at low temperatures depending on the generation rate and the material parameters, has sometimes been interpreted in terms of a simple dynamic Burstein–Mass shift, i.e. as a band-filling effect. Actually, one has to include for a quantitative description always band-filling and many-body effects such as band-gap renormalization and screening of the Coulomb interaction.

Until now, we discussed the EHP-formation mainly with respect to direct-gap materials. In indirect-gap materials, there is no measurable gain even in case of population inversion. The participation of momentum-conserving phonons reduces the transition probabilities and, thus, the gain to values around 1 cm^{-1}. This is below the detection threshold. Information about the plasma is obtained mainly from luminescence spectroscopy. The indirect absorption edge is modified by the creation of a plasma and the direct exciton behaves in indirect-gap materials similarly as in the direct-gap ones.

Due to the high hole-concentration in an EHP, intervalence band transitions become possible. The corresponding induced-absorption bands are situated in the IR part of the spectrum.

In the last part of this section, we shall shortly discuss changes of the optical properties connected with sample heating due to the absorbed part of the incident light. In most semiconductors, one has a red shift of the gap with increasing temperature which starts quadratic with T and then develops to a linear T dependence as listed by Madelung et al. (1982). Free and bound exciton energies tend to shift parallel with the gap. If $\hbar\omega$ is situated below the free exciton resonance, one usually finds an increase of α for constant ω due to this phenomenon and the Urbach–Martienssen rule in Eq. 2.

Exceptions may occur at very low temperatures where narrow absorption structures connected with bound exciton complexes may shift over $\hbar\omega$ with increasing incident intensity, i.e. increasing sample temperature, resulting in $\partial\alpha/\partial I = (\partial\alpha/\partial T)(\partial T/\partial I) \gtrless 0$ depending on the position of $\hbar\omega$ relative to the eigenenergy of the BEC. Furthermore, one finds usually for fixed $\hbar\omega < E_x$ an increase of the refractive index with increasing I_{exc}, because the resonance comes with increasing I_{exc} closer to $\hbar\omega$.

There are some semiconductors that show an increase of the gap with temperature; these include CuCl, CuBr and some of the narrow-gap materials (the lead salts and CMT). In these cases, an increase of the incident intensity can result in a decrease of the absorption and of the refractive index.

In samples with rather long carrier lifetimes (indirect or narrow-gap materials) or under illumination with rather long pulses, optical nonlinearities due to electronic and thermal excitation can be simultaneously present. Depending on the material and the excitation conditions, both effects may have the same or opposite signs.

III. EXPERIMENTAL TECHNIQUES

In this section, we shall outline the experimental techniques. In the next one, we shall describe experimental results, again proceeding from large-gap to narrow-gap materials.

The first experiments, performed in the investigation of highly excited semiconductors, concerned luminescence spectroscopy. In indirect-gap materials, they reveal a lot of details about the EHP and the phase diagram of the EHL as summarized by Rice (1977), by Hensel (1977) or by Wolfe (1985). Some information was found concerning biexcitons, e.g., by Kulakovskii et al. (1981). In the direct-gap materials, high excitation effects are usually accompanied by stimulated emission. The observation of stimulated emission is on one hand a proof for gain, i.e., for an excitation-induced nonlinearity; on the other hand, stimulated emission is strongly distorting the spectral shape of luminescence spectra, thus making quantitative evaluations rather difficult. Since the gain may reach maximum values around 10^4 cm^{-1}, i.e., 1 μm^{-1}, it is often difficult to avoid stimulated emission, e.g., by reducing the spot size. Therefore, we shall not go into details of luminescence spectroscopy, but shall concentrate on two main spectroscopic techniques, namely the excite-and-probe beam spectroscopy (EPB) and the spectroscopy with dynamic, laser-induced gratings (LIG). Some other methods that are related to EPB or LIG will also be mentioned. The main idea in EPB is to excite the sample with an intense, spectrally narrow pump-beam (I_{exc}, $\hbar\omega_{exc}$) and to measure the transmission and/or reflection of a weak probe beam, once with pump on and once with pump off. The probe beam intensity I_{Pr} must be so low that I_{Pr} itself does not introduce any nonlinearities. Eventually, it is necessary to subtract from the transmitted or reflected part of I_{Pr} the luminescence. The setup is shown schematically in Fig. 8a. The probe beam can be spectrally narrow (e.g., a tunable dye-laser) and is scanned over the region of interest. This method is useful if no spectrometer is used and if the intensities of both luminescence and stray light of I_{exc} are negligible as compared to the transmitted or reflected intensities of I_{Pr}. More often, one uses a spectrally broad probe beam, which is dispersed by a spectrometer and detected, e.g., by an optical multichannel analyszer. If the duration of the excitation pulse τ_{exc} is long as compared to the lifetime of the electron–hole pairs τ_p, it is possible to center the probe beam temporally and spatially with respect of the pump beam (i.e., $\tau_{Pr} < \tau_{exc}$; $\phi_{Pr} < \phi_{exc}$) and to measure under quasi-stationary conditions. For $\phi_{Pr} > \phi_{exc}$ one can get information about the lateral spatial expansion of the cloud of the excited electron–hole pairs by spatially resolved measurements as done by Majumder et al. (1985). For $\tau_{exc} \ll \tau_p$ one can study the decay kinetics of n_p if the probe beam is either delayed with respect to the pump or if $\tau_{Pr} > \tau_p$ is used together with a time-dispersive detector, e.g., a streak-camera. In all these variants of the EPB, one measures the transmitted and reflected intensities I_T, $I_T{}^*$ and $I_R{}^*$, I_R of

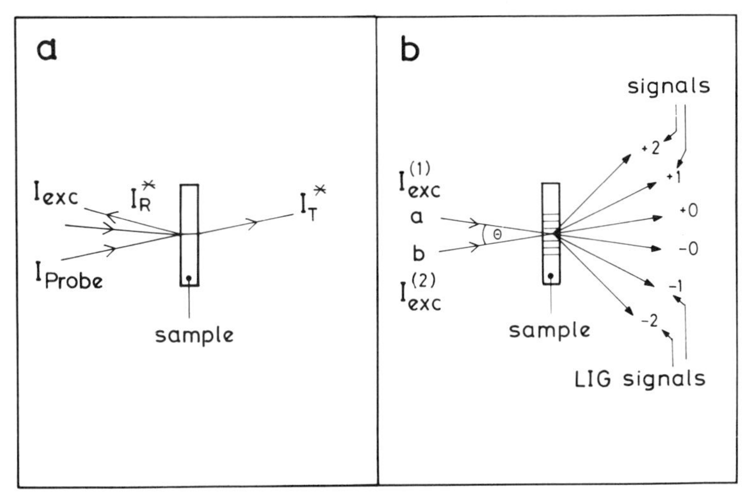

Figure 8. Schematic sketch of one of the possible arrangements for the excite-and-probe beam spectroscopy (a) and for the (self)-diffraction from laser-induced gratings (b).

the probe without and with excitation, respectively. The comparison of I_T and $I_T{}^*$ eventually corrected by I_R and $I_R{}^*$ gives information about the excitation-induced variation of the absorption spectrum

$$\alpha = \alpha_0(\omega_{\mathrm{Pr}}) + \Delta\alpha(\omega_{\mathrm{Pr}}, I_{\mathrm{exc}}, \omega_{\mathrm{exc}}) \tag{11}$$

The investigation of the reflection signals, e.g., of the excitonic resonances in EPB experiments gives information about the behaviour to these quasi-particles. If the sample has plan-parallel surfaces, one usually observes in the transparent spectral region Fabry–Perot modes. From the shift of these modes with I_{exc}, one may get information about the variation of the refractive index

$$n = n_0(\omega_{\mathrm{Pr}}) + \Delta n(\omega_{\mathrm{Pr}}, I_{\mathrm{exc}}, \omega_{\mathrm{exc}}) \tag{12}$$

This means that EPB gives via Eqs. 11 and 12 the renormalization processes. The dependences on ω_{exc}, I_{exc} stands schematic for both real and virtual excitations (see section 2). By extrapolating ω_{Pr} to ω_{exc}, Eqs. 11 and 12 give also the self-renormalization. Variation of $\hbar\omega_{\mathrm{exc}}$ with constant I_{exc} allows excitation spectroscopy of the optical nonlinearities.

In some experiments, the weak, spectrally broad probe beam has been replaced by a luminescence band of the sample itself, which is excited by the pump. This modification of the EPB, known as LATS (luminescence-assisted, two-photon spectroscopy) has been used, e.g., by Schrey et al.

(1979) and Hvam (1983). In other experiments, a striplike part of the sample is excited. The spontaneous emission of the sample is reabsorbed or amplified in the direction of the strip. By varying the strip length l and measuring the luminescence spectra as a function of l, it is possible to evaluate the absorption—or gain spectra. For more details of this variation of the strip length (VSL) method, see, e.g., the work by Hvam (1978) and the literature cited therein. We shall find in the following examples of most of the experiments possible with EPB.

The basic idea of LIG experiments is depicted in Fig. 8b. Two coherent, monochromatic laser beams $I_{exc}^{(1)}$ and $I_{exc}^{(2)}$ of equal polarization and photon energy intersect under a small angle θ. In the interference field, one gets a periodic modulation of the total intensity I_{tot} according to

$$\begin{aligned} I_{tot} &= I_{exc}^{(1)} + I_{exc}^{(2)} + 2\sqrt{I_{exc}^{(1)} I_{exc}^{(2)}} \cos \varphi \\ \varphi &= \vec{k}_{exc}^{(1)} \vec{r} - k_{exc}^{(2)} \vec{r} + \rho_a^0 + \rho_b^0 \end{aligned} \tag{13}$$

If a sample with optically nonlinear properties is brought into the interference field, a phase or an amplitude grating is formed, depending on whether the variation of the real or of the imaginary part of the complex refractive index $\tilde{n}$ is predominant. In the simplest experiments, one can investigate the diffraction of $I_{exc}^{(1)}$ and $I_{exc}^{(2)}$ by the grating they produce themselves. These experiments are called self-diffraction from LIG and yield information about self-renormalization processes.

Two limiting cases are possible in LIG spectroscopy. For thick gratings, where the grating thickness d, the wavelength λ and the grating spacing Λ fulfill the inequality 14a

$$d \gg 2\Lambda^2\lambda^{-1} \tag{14a}$$

$$d \ll 2\Lambda^2\lambda^{-1} \qquad \text{with } \Lambda = \frac{\lambda}{2 \sin \frac{\theta}{2}} \tag{14b}$$

the diffracted beam has to obey the Bragg conditions. This means in our case that there is only intensity diffracted from beam (1) into the direction of beam (2) and vice versa. In the case of a thin grating, defined by formula 14b, one can have several diffracted orders, depending on Λ and the structure factor of the unit cell of the grating. This case is shown in Fig. 8b. For thin gratings that have weak total absorption and a spatial modulation $\Delta\tilde{n}$ of $\tilde{n}$ that is close to sinusoidal, one finds for the efficiency η of the first order

$$\eta \approx \left| \frac{\pi \Delta \tilde{n} d}{\lambda} \right|^2 \tag{15}$$

If one uses as a probe or read-out beam an additional beam with frequency $\omega_{\mathrm{Pr}} \neq \omega_{\mathrm{exc}}$ one can study renormalization processes introduced by $I_{\mathrm{exc}}, \omega_{\mathrm{exc}}$ at ω_{Pr}. By varying the duration of the pump pulses with respect to the decay time of the grating and by shifting the read-out pulse temporally with respect to the pump, one has similar possibilities to get quasi-stationary or dynamic results as already discussed in connection with the EPB technique. In LIG experiments, one has two additional degrees of freedom. One is the spacing of the grating Λ. The decay process of the grating depends on Λ for the following reason. If Λ is large as compared to the diffusion or drift length of the excited species, the decay of Λ is determined only by radiative and radiationless recombination and eventually by surface effects. In the opposite limit, the grating is additionally smeared out by drift or diffusion processes. Measurements of the grating efficiency in which Λ is varied by changing θ with all other parameters kept constant yield information about both decay mechanisms. The other degree of freedom is to make $\hbar\omega_{\mathrm{exc}}^{(1)}$ and $\hbar\omega_{\mathrm{exc}}^{(2)}$ equal—as discussed so far—or different. If $\omega_{\mathrm{exc}}^{(1)} \neq \omega_{\mathrm{exc}}^{(2)}$ we get a moving grating which produces via the Doppler effect also a frequency shift of the diffracted beam with respect to ω_{Pr}. If the grating moves over a distance Λ in a time comparable to τ_p, the efficiency drops. Scattering geometries of LIG experiments that deviate from the one shown in Fig. 8b are found, e.g., in a review by Eichler (1985). The LIG-experiments are just another way to describe four-wave mixing: self-diffraction experiments with $\omega_{\mathrm{exc}}^{(1)} = \omega_{\mathrm{exc}}^{(2)}$ correspond to degenerate four-wave mixing (FWM), while the LIG experiments with $\omega_{\mathrm{exc}}^{(1)} \neq \omega_{\mathrm{exc}}^{(2)}$ are the equivalent to nondegenerate FWM. The appearance of the diffracted orders can be described in some cases also in the quasi-particle picture. Two quanta $\hbar\omega_{\mathrm{exc}}^{(1)}$ and/or $\hbar\omega_{\mathrm{exc}}^{(2)}$ virtually excite an intermediate state, which decays stimulated by the probe beam. This results in an additional quantum in the probe beam and a "signal" quantum which fulfills energy and momentum conservation. These laws result in the same frequency shifts (for $\omega_{\mathrm{exc}}^{(1)} \neq \omega_{\mathrm{exc}}^{(2)}$) and directional dependences of intensity as deduced from the laws of diffraction. The hyper-Raman scattering process described in section II is the spontaneous decay process as compared to the stimulated one described here.

IV. EXPERIMENTAL RESULTS

The following part of this chapter will be devoted to the description of experimental results.

The first experiment concerns two-photon absorption spectroscopy of the polariton dispersion around the $n_{\mathrm{B}} = 1$ resonance. In contrast to TPA to the biexciton resonance, one has $\hbar\omega_{exc}$ and $\hbar\omega_{abs}$ around one-half of the exciton energy. This means that the resonance denominator in second-order perturbation theory is rather large and the effect is weak. The exciton spectra of many semiconductors have been investigated by TPA, e.g. in CuCl, CuBr, ZnO, CdS, ZnSe, ZnTe, Cu_2O, SnO_2 and CsBr. Data are compiled, e.g., in the reviews by Fröhlich (1981) and by Hönerlage et al. (1985).

In CuCl, the dispersion of the longitudinal exciton branch and of the UPB have been investigated by varying the angle between $\vec{k}_{exc}$ and $\vec{k}_{abs}$. The LPB around the bottleneck is not accessible by this technique. Due to the increase of n toward the exciton resonance, one has for the LPB always

$$\left|\vec{k}_{exc} + \vec{k}_{abs}\right| \leq \left|\vec{k}_{ex}\right| + \left|\vec{k}_{abs}\right| < \left|\vec{K}_{final}\right| \tag{16}$$

with $\hbar\omega_{exc} + \hbar\omega_{abs} = \hbar\omega_{final}$.

The results are shown in Fig. 9 for CuCl as full circles. It should be mentioned that there is also a rich variety of linear spectroscopic techniques to detect the dispersion of exciton-polaritons, e.g., resonant Brillouin scattering or the investigation of Fabry–Perot modes. These techniques are beyond the scope of this chapter and the reader is referred, e.g., to the review of Hönerlage et al. (1985). For TPA measurements detected by TPA-induced photoconductivity, see, e.g., the work of Koren (1975) or Seiler et al. (1982 and 1983).

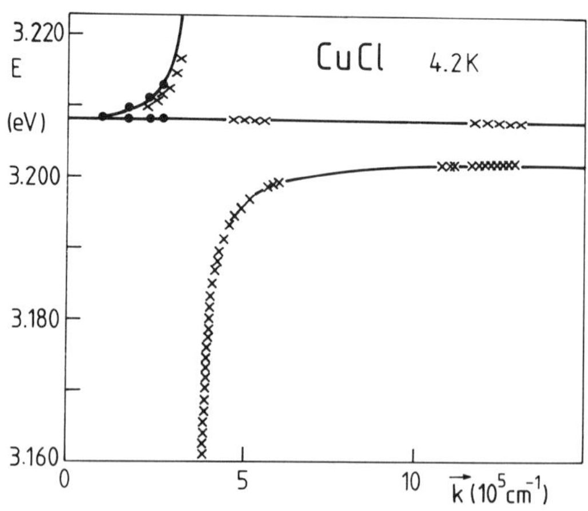

Figure 9. Calculated and measured dispersion of the polariton around the lowest free-exciton resonance (Γ_5) in CuCl. Solid lines: theory; full circles: results from two-photon absorption measurements; crosses: results from hyper-Raman scattering. From Hönerlage, B. et al. (1985).

A very efficient technique of nonlinear optical spectroscopy for the investigation of the polariton dispersion is the hyper-Raman scattering (HRS). As mentioned in section II, one excites with two quanta virtually a biexciton which then decays into two other quasi-particles. By selecting the direction of observation, one determines the direction of $\vec{k}_R$ in Eq. 9 and one measures $\hbar\omega_R$. In a mathematical procedure, one determines the polariton dispersion in a self-consistent way taking properly into account energy and momentum conservation and the refraction at the sample surfaces as explained in detail in the review by Hönerlage et al. (1985). The crosses in Fig. 9 are polariton-eigenstates determined experimentally by HRS. The determination of the polariton dispersion and of the set of parameters describing it has successfully been accomplished apart from CuCl, e.g., in CuBr, ZnO, CdS, ZnSe, ZnTe, PbI_2 or HgI_2. (For a review and extensive references, see Hönerlage et al., 1985). In some of the II–VI compounds, the HRS measurements have been complemented by classical reflection and transmission spectroscopy. In CuBr, CdS, ZnO and ZnTe, the investigations have also been carried out in magnetic fields up to 25 T, thus revealing the relevant parameters such as g-factors or diamagnet shifts.

The scattering processes between excitons have been investigated mainly by means of luminescence spectroscopy and line-shape analysis. (See, e.g., the review by Klingshirn and Haug, 1981, and the literature cited therein). There are some results of gain-spectroscopy by VSL described, e.g., by Hvam (1978) and by Hvam and Bivas (1980). A density-dependent collision broadening of the free-exciton resonance has also been observed in CdS by LIG-excitation spectroscopy for $I_{exc} \approx 1\ \text{kW/cm}^2$ (Fig. 10)

The next large group of optical nonlinearities that we shall discuss is connected with two-polariton transitions and induced absorption to the biexciton.

The first experimental information about the existence of biexcitons came from luninescence spectroscopy by Mysyrowicz et al. (1968). Again, we shall concentrate on the nonlinear optical aspects.

The absorption dip at $\hbar\omega_{abs} = E_{biex} - \hbar\omega_{exc}$ in Fig. 5 has been found by TPA and LATS, e.g., in CuCl, CuBr, ZnO, CdS or PbI_2 as summarized by Hönerlage et al. (1985). In Fig. 11, we present LATS spectra for CdS and ZnO. If $\hbar\omega_{exc}$ is increased, the dips shift to the red. A plot $\hbar\omega_{dip} = f(\hbar\omega_{exc})$ yields a straight line with a slope of -1 allowing us to determine the eigenenergies of the biexciton states with high precision. For ZnO, one found from the data in Fig. 11b three biexciton levels containing either two holes from the upmost valence band A or one from the A- and one from the B-valence bands or two B-holes. In CuBr, it was found that the biexciton

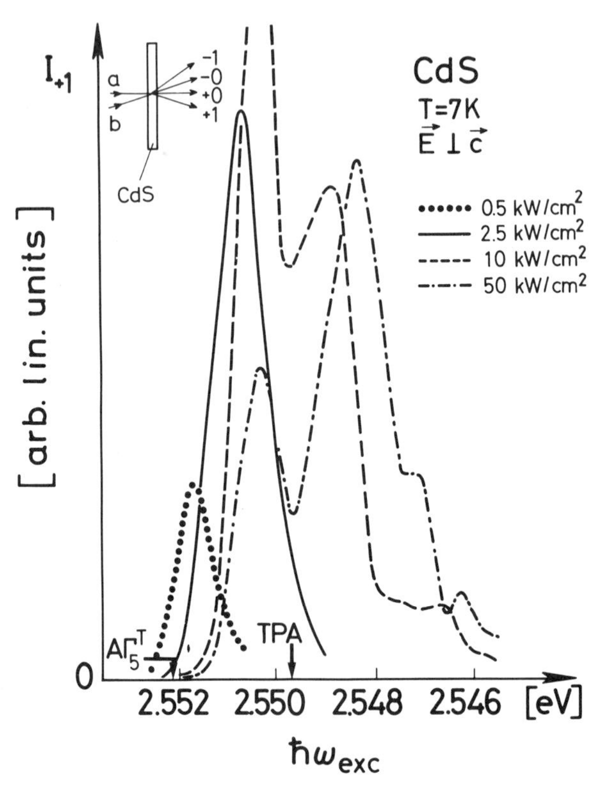

Figure 10. LIG-excitation spectra of CdS at low temperature and for a polarization $\vec{\mathbf{E}} \perp \vec{\mathbf{c}}$. The curves for $I_{exc} = 0.5$ and $2.5\ \text{kW/cm}^2$ are mainly due to an excitation-induced collision broadening of the $A\Gamma_5$ exciton resonance. For $I_{exc} = 10$ and $50\ \text{kW/cm}^2$ also two-photon transitions to the biexciton are involved. From Kalt, H. et al. (1985).

ground state is split into three sublevels with symmetries Γ_1, Γ_3 and Γ_5. By changing the geometry either between pump and probe beam in EPB experiments or between pump and luminescence in LATS experiments, it was possible to determine the dispersion of biexcitons in CdS, CuCl or CuBr. The renormalization of the real part of $\tilde{n}$ or of $\vec{k}$ around $\hbar\omega_{abs}$ can be detected, e.g., by HRS experiments, if one of the polaritons involved falls on the anomaly. Corresponding experiments have been performed, e.g., for CuCl or CdS. In Fig. 12, we give an example for CdS. One clearly sees the deviation of the relation $\hbar\omega = f(\hbar\omega_{exc})$ from the smooth curve that one would have without renormalization. Lasing connected with HRS in a resonator configuration has been used in CdS by Baumert and Broser (1981) in connection with the dependence of the emission on the scattering

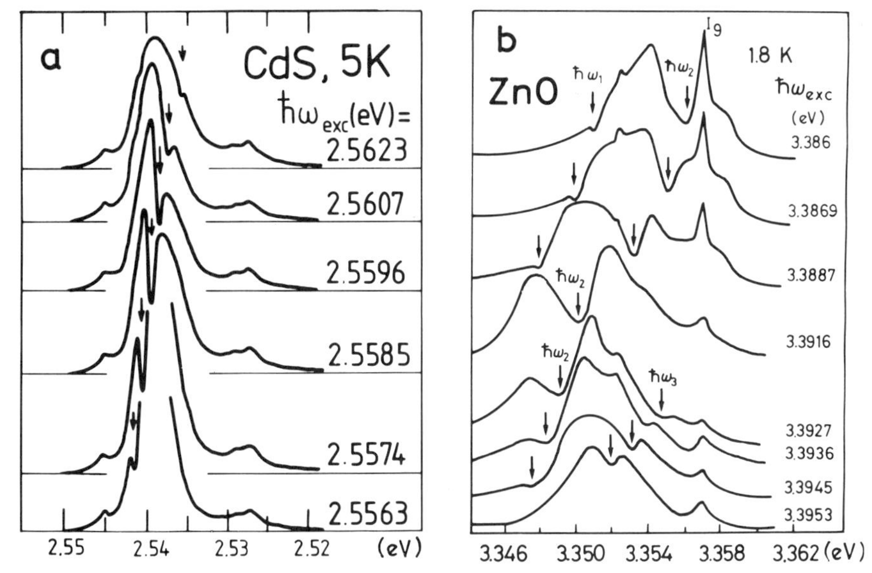

Figure 11. Luminescence spectra of CdS and ZnO showing dips (↓) which shift with varying energy of the exciting photons $\hbar\omega_{exc}$ as $\hbar\omega_{dip} + \hbar\omega_{exc}$ = constant. From Schrey, H. et al. (1979) and from Hvam, J. M. et al. (1983).

geometry to construct a tunable "four-polariton laser." Second-order HRS has been described by Schrey et al. (1979) in CdS.

While TPA, LATS and HRS experiments are sensitive either to the renormalization of the imaginary or of the real part of $\tilde{n}$, respectively, both aspects enter in LIG measurements or in excitation-induced circular dichroism (ECD). The principle of ECD has been presented already in Section II. Experiments have been performed successfully in CuCl, e.g., by Kuwata et al. (1981) and Itoh and Katono (1982), which has cubic T_d symmetry so that circularly polarized waves are good eigenstates. In hexagonal materials such as CdS or ZnO, this would be true only for waves propagating parallel to the $\vec{c}$-axis. Unfortunately, these materials grow preferentially as thin platelets containing the $\vec{c}$-axis and this is eventually the reason why attempts to find ECD failed.

LIG experiments using the nonlinearities connected with the biexciton to our knowledge were first performed in CuCl by Maruani et al. (1978). Actually, these measurements gave a strong impetus to LIG experiments with tunable lasers close to the electronic resonances in semiconductors. Later on, LIG was also used to investigate biexcitons in CdS and ZnO; a

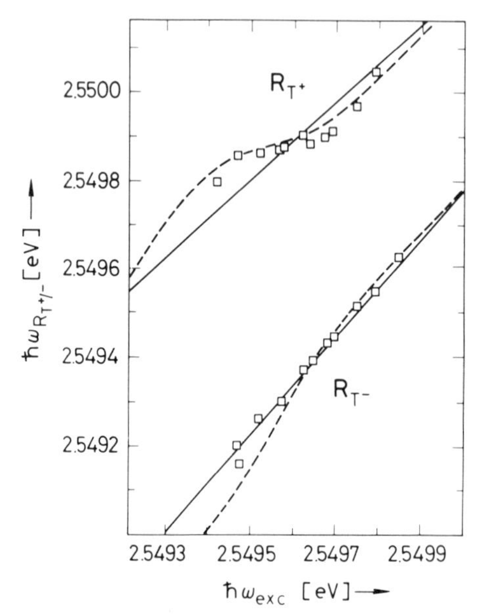

Figure 12. Hyper-Raman scattering experiment in CdS in near forward scattering geometry. The spectral position of the hyper-Raman scattering lines R_T^+ and R_T^- is plotted as a function of $\hbar\omega_{exc}$. Squares: experimental data; solid lines: theory without renormalization; dashed lines: theory with renormalization. From Lyssenko, V. G. et al. (1982).

recent review is given by Kalt et al. (1986). In Fig. 10, we give LIG excitation spectra for CdS showing the biexciton resonance for $I \gtrsim 10$ kW/cm^2. Especially on the low-energy wing, the LIG is mainly a phase grating, while amplitude effects dominate at the position of TPA around 2.55 eV and on the low-energy flank of the free exciton. In ZnO, the three biexciton levels known from LATS (Fig. 11) are also seen in LIG by Kalt et al. (1986). The induced absorption from really excited excitonlike polaritons to the biexciton has been investigated with psec time resolution in CuCl, e.g., by Lévy et al. (1979).

The decay time of the biexcitons has been determined, e.g. by Gale and Mysyrowicz (1975) and by Ostertag and Grun (1977), to be of the order of 1 nsec. Time- and polarization-resolved measurements of the luminescence allowed in CuCl to distinguish under resonant excitation $\hbar\omega_{exc} = \frac{1}{2}E_{biexc}$ between the coherent process HRS and the luminescence from a cold gas of biexcitons as reported, e.g., by Lévy et al. (1976), by Hönerlage et al. (1978) and by Mita et al. (1981). Luminescence-decay measurements of the M-band in CdS gave different results concerning the biexciton lifetime. Under the assumption that the M-band is due to biexciton decay (see the discussion of this point by Klingshirn and Haug, 1981), a lifetime of the biexciton in case of resonant excitation of only 20 psec has been found by Baumert (1983), while nonresonant excitation resulted in decay times of about 600 psec. The short lifetime is attributed to the giant oscillator strength of TPA from the crystal ground state to the biexciton level, which should also show up in the decay processes. The 600 psec are interpreted as connected with the decay time of excitons under the assumption that biexcitons are rapidly formed from excitons under nonresonant excitation.

LIG experiments also showed in CdS and ZnO some features that are connected with induced absorption from the exciton to the biexciton level. Furthermore, these experiments also gave some indications for (AB) biexcitons in CdS. A recent review is given by Kalt et al. (1986).

Now we shall concentrate on the optical nonlinearities connected with the formation of an electron–hole plasma. Luminescence experiments are here very informative for indirect-gap materials such as Ge, Si or GaP. In direct-gap semiconductors, the emission spectra are usually distorted by stimulated emission. Only a few measurements are known, where the spontaneous spectra have been observed, e.g., in CdS or GaAlAs e.g., by Lyssenko et al. (1975) and by Capizzi et al. (1985). In both cases, the main point was to reduce the dimensions of the excited volume to some $(\mu m)^3$. More reliable results are usually obtained by reflection and transmission measurements with the EPB method.

As a first experimental result, we show in Fig. 13 transmission spectra of weak, broad-band dye-laser pulses without and with simultaneous excitation for CdS, $CdSe_{1-x}Se_x$ and CdSe from Majumder et al. (1986). For the pure materials one finds a red shift of the absorption edge and optical amplification below the quasi-chemical potential in agreement with the predictions of Fig. 7. Furthermore, one sees that the Fabry–Perot modes are shifting to the blue, indicating a decrease of the refractive index. From these experiments it is possible to deduce the renormalization of the optical properties induced by the creation of an EHP pumped at $\hbar\omega_{exc}$ with I_{exc}. By extrapolation it is also possible to deduce the self-renormalization. Corresponding data are given, e.g., for CdS by Bohnert et al. (1984). The alloy semiconductor $CdS_{1-x}Se_x$ shows a different behaviour. The absorption edge is unchanged above the mobility edge; a bleaching of absorption and optical gain appear below. This is the spectral region of excitons localized by fluctuations of the width of the forbidden gap which result from composition fluctuations; see, e.g., the work by Permogorov et al. (1982) and by Cohen et al. (1982). At present, it is not fully understood why

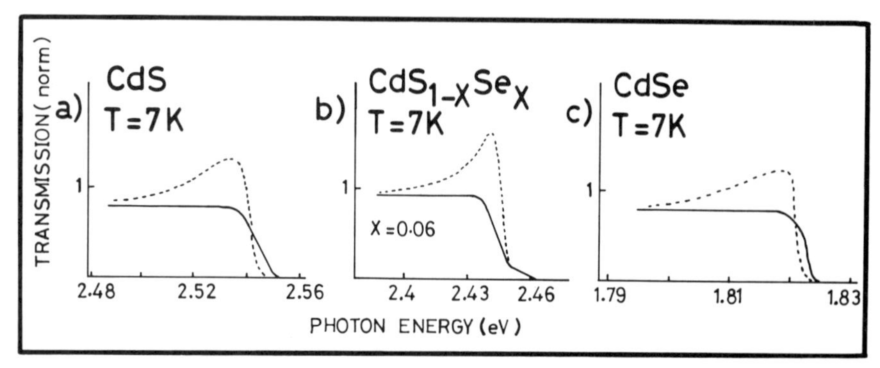

Figure 13. The transmission spectra of weak, spectrally broad dye-laser pulses through samples of CdS, $CdS_{1-x}Se_x$ and CdSe. Solid line: without additional excitation; dashed lines: with additional excitation in the band-to-band transition region. $I_{exc} = 2\ MW/cm^2$, $T = 7$ K. From Majumder, F. A. et al. (1986).

the red shift of the absorption edge is not as it is for pure constituents. Localized electron–hole pairs are certainly less efficient in screening, which results in reduced many-particle effects and band-gap shrinkage.

Now we return to pure materials. It has been found, e.g., by Bohnert et al. (1981), that the excitonic reflection spectra disappear when an EHP is created. In Fig. 14, we show excitation spectra of the integral plasma gain G and the modulation of the reflectivity ΔR around the exciton resonance, i.e. $\Delta R = R_{max} - R_{min}$ (see, e.g., Fig. 2). In agreement with theory it is found that G tends to zero when the excitonic features appear and vice versa. The transition occurs around 2.54 eV. This is the value of the chemical potential, i.e. the crossover from gain to absorption in CdS at low temperature for a rather large range of plasma densities n_p as seen from Fig. 20 in the work just cited. For $\hbar\omega_{exc}$ below 2.54 eV, it is no longer possible to pump an EHP efficiently. By studying the gain spectra as a function of excitation intensity and temperature, it has been found that the EHP does not reach a liquidlike equilibrium state predicted by the theory of Röseler and Zimmermann (1977) for $T < T_c$ with $T_c \approx 65$ K for CdS. This behaviour is in contrast to indirect-gap materials, where plasma droplets have been detected by a large variety of experimental techniques as reviewed by Rice (1977) and by Hensel et al. (1977). The reason for this difference is the short carrier lifetime τ_p in the direct-gap materials ($\tau_p \approx 100$ psec), which prevents a "segregation" in liquid drops and a surrounding gas. Only strong density fluctuations occur in direct-gap materials. In some experiments, the excitation independence of the crossover from gain to absorption has been

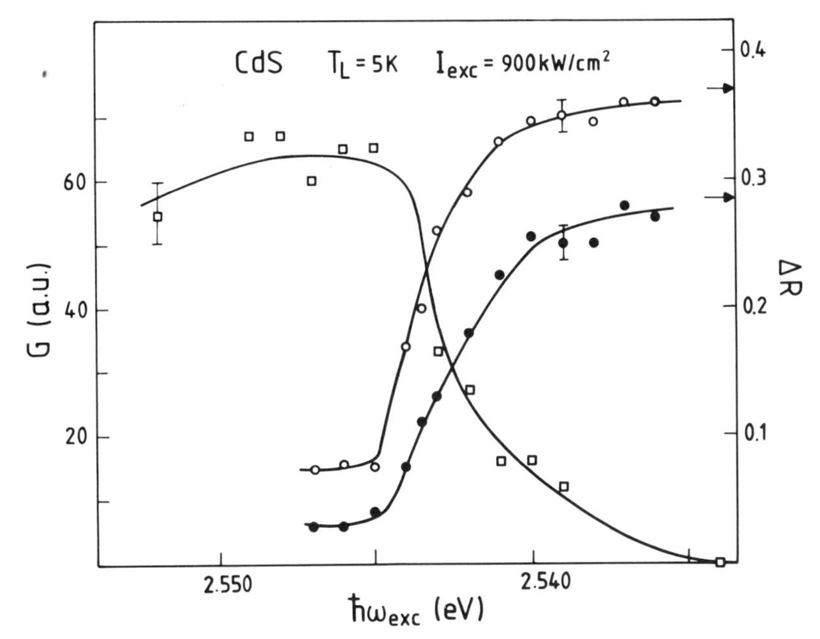

Figure 14. Excitation spectra of the integrated gain G of the electron–hole plasma and of the modulation ΔR of the $A\Gamma_5$ and $B\Gamma_5$ exciton resonance in CdS. From Bohnert, K. et al. (1981).

used as an argument for the existence of a liquid in various II–VI materials such as ZnSe, ZnTe or CdS. This is, however, a misinterpretation of the data. Actually, the crossover remains almost constant for a broad range of plasma densities n_p, as just mentioned. On the other hand, it can happen that n_p may be pinned by the onset of stimulated emission. Obviously, a rather constant crossover energy is not conclusive for the existence of a liquid.

The temperature dependence of the shift of the absorption edge is shown for CdS in Fig. 15. As explained in Section II, one finds a red shift at low temperature and a blue shift at room temperature. The latter phenomenon has also been described by Henneberger et al. (1985). The transition occurs around 160 K. There one finds almost no change of the absorption edge. The blue shift of the Fabry–Perot modes, i.e. the decrease of the refractive index, indicates that there is a decrease of absorption at higher photon energies connected, e.g., with a screening of the attractive e–h interaction.

Under inhomogeneous excitation conditions, the EHP has a tendency to expand. To get some insight in the drift length, spatially resolved EPB experiments have been performed, e.g., in CdS by Majumder et al. (1985).

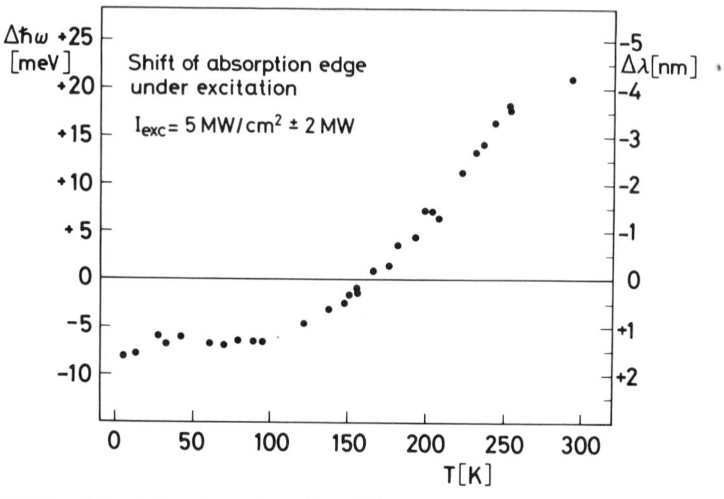

Figure 15. The shift of the absorption edge of CdS as a function of the temperature with additional excitation, according to Swoboda, H.-E. et al. (1986).

Under ns-pulsed excitation, it has been found that the drift length in CdS is about 5 μm while the corresponding value for CdSe is about 12 μm. In both cases, "slow" drift is present, i.e. the drift velocity is small as compared to the Fermi velocity of electrons or holes. The same result has been found by various independent measurements on the EHP in $GaAl_{1-x}As_x$ by Capizzi (1985).

A large number of time-resolved experiments with ps and sub-ps time resolution have been performed. Due to the spectral range of the emission of available lasers, the experiments used predominantly materials such as CdSe, GaAs or Si. Often the luminescence decay has been measured. In the following, we shall outline two EPB experiments. In the first experiment by Shank et al. (1978), the carriers were excited high into the band of CdSe or GaAs by UV (3075 Å) sub-ps pulses. The reflectivity was monitored in the red (6150 A) as a function of time after the excitation pulse. From the rate of the change in the reflectivity, an energy-relaxation speed of the excited carriers of 0.4 eV/ps has been deduced. In even more advanced experiments, the lifetime of carriers in states 100 meV above the band-gap of $Al_{0.34}Ga_{0.66}$. As has been determined by Tang and Erskine (1983) to be about 50 fs at room temperature. In another experiment by Fujimoto et al. (1984), the excitonic reflection of the $A\Gamma_5$ exciton in CdSe has been monitored during and after excitation with a pulse of about 80 fs duration.

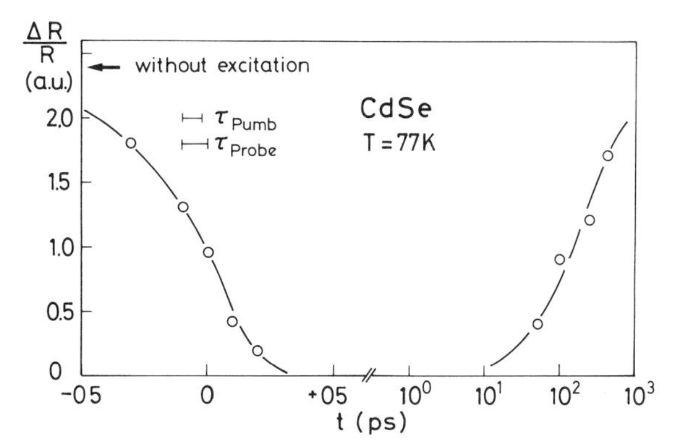

Figure 16. The temporal evolution of the excitonic reflection $\Delta R/R$ in CdSe at 77 K during and after excitation with an intense sub-ps pulse. Note the transition from a linear to a logarithmic time scale. $\hbar\omega_{exc}$ is in the band-to-band transition region. Curves deduced from experimental data given by Fujimoto, J. G. et al. (1984).

The probe pulse from a white-light continuum was about 100 fs long. In Fig. 16, we show a curve deduced from the experimental data of Fujimoto et al. It is concluded that the excitons are screened by the formation of an EHP in times shorter than 100 fs, the decay time of the plasma can be deduced from the reappearance of the exciton structure to be around 200 psec. Fehrenbach et al. (1985) claimed that excitons survive to much higher densities ($20 \cdot n_{Mott}$) if they are resonantly excited in contrast to band-to-band excitation. Though excitons as neutral entities with finite binding energy are less efficient for screening in three-dimensional materials as compared to the free carriers created by band-to-band excitation, the quantitative evaluation of the data of Fehrenbach et al. is to a certain extent the subject of the discussion. One point is that no detailed line-shape analysis of the absorption spectrum under excitation has been performed. The appearance of an absorption peak at the exciton energy is not a conclusive proof for the existence of excitons as bound states. Such a peak can appear also in an EHP due to the excitonic enhancement. Consequently, the density up to which excitons are still present might have been overestimated.

LIG spectroscopy has also been used successfully to investigate the properties of an EHP. CdS crystal has been pumped by two interfering UV-pulses (third harmonic of Nd–AG laser) of about 25-ps duration in an experiment by Saito and Göbel (1985), while the resulting grating was monitored by the diffraction of the second harmonic for which CdS is transparent. The plasma decay time was found to be about (150 ± 50) ps. In agreement with earlier work, it has been found that the EHP does not reach its equilibrium liquidlike state. For longer times, the LIG is governed by exciton and biexciton properties. Under ns-excitation, both the transmitted beams and the diffracted orders show a significant nonlinear dynamic behaviour in self-diffraction experiments performed in CdS at low temperature as shown by Kalt et al. (1986). The reason can be seen in induced-absorption optical bistability which will shortly be mentioned.

In the indirect-gap semiconductors, Si or Ge EPB experiments revealed renormalization effects around both the direct- and the indirect-gap excitons, e.g., by Balslev (1984) and by Schweizer et al. (1983). In the latter case, the screening of the direct exciton in Ge by an EHP has been investigated both experimentally and theoretically. Variations of the absorption spectra near the indirect gap due to plasma effects and free carrier absorption have been reported by Balslev (1984). Time-resolved LIG experiments by Smirl et al. (1978) and (1982a) in Ge showed the band filling in an EHP and the subsequent decay of the electron–hole pairs. On a much shorter time scale, anisotropic state filling has been detected in Ge by Bogges (1982) and by Smirl et al. (1982b) and in GaAs by Oudar et al. (1985) by LIG experiments. The intervalence-band transitions that become possible in the presence of a plasma (or under heavy p-doping, see Fig. 4) have been analyzed, e.g., for Ge by Aldrich and Silver (1980).

In the narrow-gap materials, one gets usually already for low temperatures a blue shift of the absorption edge due to the formation of a plasma in contrast to materials such as CdS or CdSe (Fig. 15). For the narrow-gap materials, the filling of the states overcompensates the reduction of the gap. This can be partly attributed to the small effective masses of the electrons, e.g., in InSb. A theoretical analysis of the renormalization of the optical properties of this material is found, e.g., in the review by Haug and Schmitt-Rink (1984). Experimental results for narrow-gap materials are reviewed, e.g., by Miller et al. (1981). The plasma expansion has been investigated by LIG experiments in InSb by Hagan et al. (1985). Varying the lattice constant, a drift length of the plasma of about 50 μm has been found. Narrow-gap materials usually have rather large optical nonlinearities which scale as $E_g^{-\alpha}$ where α is a number around four. Self-induced transparency on the basis of band filling has been reported as well as optical limiting as compiled by Miller et al. (1981).

To conclude this section, we shall present some photothermal nonlinearities. Fig. 17 shows transmission spectra of CdS with the EPB method. The sample is kept in air of room temperature. It is illuminated with the 514 nm-line (2.410 eV) of a cw Ar^+ laser at $I_{exc} = 2.3$ kW/cm^2 one observes a

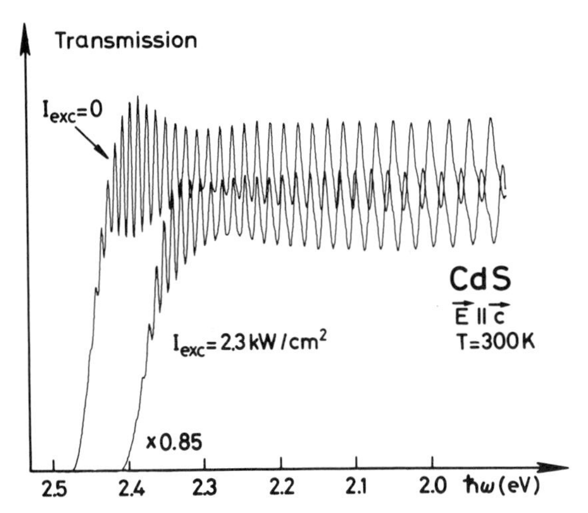

Figure 17. The transmission of a CdS sample kept in air at room temperature without and with excitation at 2.41 eV with an Ar^+ laser. From Lambsdorff, M. (1986).

red shift of the absorption edge, which is monitored by a cw-incandescent-lamp spectrum. The exponential absorption tail follows a Urbach–Martienssen rule with a temperature, which is determined by laser heating. In direct-gap materials with a gap larger than 1 eV, the electronic and thermal processes can rather easily be separated by their different time constants and typical excitation intensities (see, e.g., the work by Lambsdorff et al. 1986). In narrow-gap materials, the carrier lifetimes get longer with decreasing E_g resulting partly in comparable time constants for thermal and electronic effects. Since some of the narrow-gap materials show an increase of the gap with temperature (see Section II), this effect may overlap with band-filling phenomena. In other materials, thermal and electronic optical nonlinearities tend to compensate each other.

If a sufficient amount of energy is deposited in materials such as Si or Ge, one may locally melt a thin surface layer, which resolidifies after the pulse. This phenomenon is known as laser annealing and leads also to a rich variety of optical nonlinearities. In Si, e.g., one finds an increase of the reflectivity R due to the phase transition from solid to liquid and at even higher fluences a reduction of R due to evaporation of the material as observed by Shank (1985). These effects or the appearance of periodic structures after resolidification are beyond the scope of this contribution. The reader is referred, e.g., to the work by Smirl et al. (1985) or Van Driel et al. (1985).

V. CONCLUSION AND OUTLOOK

We hope that the discussion of the selected set of the experiments on nonlinear optics in semiconductors has shown that in the past two decades a rather consistent picture of the nonlinearities has been gradually developed.

Further activities will aim, among other things, at fs time-resolved investigation of relaxation processes in the bands at phase-decay processes and at the investigation of new materials among which superlattices or quantum-well structures will be most exciting. On the other hand, applied research activities will appear. The application potential of LIG and of the phase-conjugated waves reflected from induced gratings can be summarized by key words such as image amplification, real-time holography, reconstruction of distorted wavefronts, focusing of powerful laser beams to the deuterium–tritium pellets in laser fusion, compensation of dispersion in fiber communication networks, etc. The other main branch in the application of optical nonlinearities concerns optical bistability and its possible use in digital, optical data handling. An optical bistable element has under the same illumination conditions two (or more) stable and reversible states of different transmission and/orreflection, the occurrence of which depends on the history. These two states can be identified with the logic zero and one. So a bistable element acts as an optical memory. Various contributions to this book will be devoted to this subject. The rich variety of bistabilities in CdS has been recently reviewed, e.g., by Wegener et al. (1986a, b) and by Wegener and Klingshirn (1987).

REFERENCES

Aldrich, C. H., and Silver, R. N. (1980). *Phys. Rev. B* **21,** 600.

Balslev, I. (1984). *Phys. Rev. B* **30,** 3203.

Baumert, R. (1983). Private communication.

Baumert, R., and Broser, I. (1981). *Solid State Commun.* **38,** 31.

Benoit à la Guillaume, C., Debever, J. M., and Salvan, F. (1969). *Phys. Rev.* **177,** 567.

Bille, J. (1973). *Festkörperprobleme* **XIII**. (Advances in Solid State Physics **13**) 111.

Bogges, T. F., Smirl, A., and Wherrett, B. S. (1982). *Opt. Commun.* **43,** 128.

Bohnert, K., Anselment, M., Kobbe, G., Klingshirn, C., Haug, H., Koch, S. W., Schmitt-Rink, S., and Abraham, F. F. (1981). *Z. Phys. B* **42,** 1.

Bohnert, K., Fidorra, F., and Klingshirn, C. (1984). *Z. Phys. B* **57,** 263.

Broser, I., and Gutowski, J. (1985). *J. Crys. Growth* **72,** 313.

Brossat, T., and Raymond, F. (1985). *J. Crys. Growth* **72,** 280.

Capizzi, M., Frova, A., Modesti, S., Selloni, A., Staehli, J. L., and Guzzi, M. (1985). *Helv. Phys. Acta* **58,** 272.
Chelikowski, J. R., and Cohen, M. L. (1976). *Phys. Rev. B* **14,** 556.
Chemla, D. S., and Miller, D. A. B. (1985). *J. Opt. Soc. Am. B* **2,** 1155.
Cohen, E., and Sturge, M. D. (1982). *Phys. Rev. B* **25,** 3828.
Dean, P. J., and Herbert, D. C. (1979). In "Excitons," (K. Cho, ed.) *Topics in Current Physics* **14,** 55. Springer-Verlag.
Eichler, H. J. (1985). In *Proc. of Intern. School on Atomic and Molecular Spectroscopy*, 7th Course "Spectroscopy of Solid State Laser-Type Materials." Erice, Italy to be published by B. DiBortolo, ed. (1987).
Elkomoss, S. G., and Munschy, G. (1977). *J. Phys. Chem. Sol.* **38,** 557.
Fehrenbach, G. W., Schäfer, W., and Ulbrich, R. G. (1985). *J. Lumin.* **30,** 154.
Fröhlich, D. (1981). *Festkörperprobleme* **XXI** (Advances in Solid State Physics **21**) 363.
Fujimoto, J. G., Shevel, S. G., and Ippen, E. P. (1984). *Solid State Commun.* **49,** 605.
Gale, G. M., and Mysyrowicz, A. (1975). *Phys. Lett.* **54A,** 321.
Hagan, D. J., MacKenzie, H. A., Al Attar, H. A., and Firth, W. J. (1985). *Opt. Lett.* **10,** 187.
Haug, H. (1982). *Festkörperprobleme* **XXII** (Advances in Solid State Physics **22**) 149.
Haug, H. (1985). *J. Lumin* **30,** 171.
Haug, H., and Schmitt-Rink, S. (1984). *Progr. Quantum Electron.* **9,** 3.
Hegarty, J., and Sturge, M. D. (1985). *J. Opt. Soc. Am. B* **2,** 1143.
Henneberger, F., Puls, J., and Rossmann, H. (1985). *J. Lumin.* **30,** 204.
Hensel, J. C., Philips, T. G., and Thomas, G. A. (1977). *Solid State Phys.* **32,** 88.
Hönerlage, B., Vu Duy Pach, and Grun, J. B. (1978). *Phys. Status Solidi B* **88,** 545.
Hönerlage, B., Lévy, R., and Grun, J. B. (1982). *Opt. Commun.* **43,** 443.
Hönerlage, B., Lévy, R., Grun, J. B., Klingshirn, C., and Bohnert, K. (1985). *Phys. Rep.* **124,** 162.
Hvam, J. M. (1978). *J. Appl. Phys.* **49,** 3124.
Hvam, J. M., and Bivas, A. (1980). *Phys. Status Solidi B* **101,** 363.
Hvam, J. M., Blattner, G., Reuscher, M., and Klingshirn, C. (1983). *Phys. Status Solidi B* **118,** 179.
Itoh, T., and Katono, T. (1982). *J. Phys. Soc. Japan* **51,** 107.
Kalt, H., Lyssenko, V. G., Bohnert, K., and Klingshirn, C. (1985a). *J. Lumin.* **31 / 32,** 861.
Kalt, H., Lyssenko, V. G., Renner, R., and Klingshirn, C. (1985b). *J. Opt. Soc. Am. B* **2,** 1188.
Kalt, H., Renner, R., and Klingshirn, C. (1986). *IEEE J. Quantum Electron.*, in press.
Klingshirn, C. (1984). In *Energy Transfer Processes in Condensed Matter*, B. DiBartolo, ed. NATO ASI Series **114,** 285. Plenum Press.
Klingshirn, C. (1985a). In *AGARD Conference Proceedings* **No. 362** on Digital Optical Circuit Technology.
Klingshirn, C. (1985b). In *Proc. of Intern. School on Atomic and Molecular Spectroscopy*, 7th Course "Spectroscopy of Solid State Laser-Type Materials." Erice, Italy to be published by B. DiBortolo, ed. (1987).
Klingshirn, C., and Haug, H. (1981). *Phys. Rep.* **70,** 316.

Koren, G. (1975). *Phys. Rev. B* **11,** 802.
Kranz, H., and Haug, H. (1986). *J. Lumin.* **34,** 337.
Kulakovskii, V. D., Kukushkin, E. V., and Timofeev, V. B. (1981). *Sov. Phys. JETP* **54,** 366.
Kurvata, M., Mita, T., and Nagasawa, N. (1981). *Solid State Commun.* **40,** 911.
Lambsdorff, M. (1986). Diploma thesis, Frankfurt University, Frankfurt ain Main.
Lambsdorff, M., Dörnfeld, C., and Klingshirn, C. (1986). *Z. Phys. B*, **64**, 409.
Lévy, R., Klingshirn, C., Ostertag, E., Vu Duy Pach, and Grun, J. B. (1976). *Phys. Status Solidi B* **77,** 381.
Lévy, R., Hönerlage, B., and Grun, J. B. (1979). *Phys. Rev. B* **19,** 2326.
Liebler, J. G., Schmitt-Rink, S., and Haug, H. (1985). *J. Lumin.* **34,** 1.
Lindwurm, R., and Haug, H. (1983). *Z. Phys. B* **53,** 281.
Lyssenko, V. G., Revenko, V. I., Tartas, T. G., and Timofeev, V. B. (1975). *Sov. Phys. JETP* **41,** 163.
Lyssenko, V. G., Kempf, K., Bohnert, K., Schmieder, G., Klingshirn, C., and Schmitt-Rink, S. (1982). *Solid State Commun.* **42,** 401.
Madelung, O. (1970). *Grundlagen der Halbleiterphysik*, Heidelberger Taschenbücher, vol. 71. Springer-Verlag.
Madelung, O., Schulz, M., and Weiss, H., eds. (1982). *Landolt-Bornstein New Series, Group III*, vol. 17. Springer-Verlag.
Majumder, F. A., Swoboda, H.-E., Kempf, K., and Klingshirn, C. (1985). *Phys. Rev. B* **32,** 2407.
Majumder, F. A., Shevel, S., Lyssenko, V. G., Swoboda, H.-E., and Klingshirn, C. (1987). *Z. Phys. B* **66,** 409.
Maruani, A., and Chemla, D. S. (1981). *Phys. Rev. B* **23,** 841.
Maruani, A., Oudar, J. L., Batifol, E., and Chemla, D. S. (1978). *Phys. Rev. Lett.* **41,** 1372.
Masumoto, Y., and Shionoya, S. (1980). *J. Phys. Soc. Japan* **49,** 2236.
Miller, A., Miller, D. A. B., and Smith, S. D. (1981). *Adv. Phys.* **30,** 697.
Mita, T., and Nagasawa, N. (1978). *Opt. Commun.* **24,** 345.
Mita, T., Sotome, K., and Ueta, M. (1981). *J. Phys. Soc. Japan* **50,** 134.
Mysyrowicz, A., Grun, J. B., Lévy, R., Bivas, A., and Nikitine, S. (1968). *Phys. Lett.* **26A,** 615.
Ostertag, E., and Grun, J. B. (1977). *Phys. Status Solidi B* **82,** 335.
Oudar, J. L., Abram, I., Migus, A., Hulin, D., and Etchepore, J. (1985). *J. Lumin.* **30,** 340.
Permogorov, S., Reznitsky, A., Verbin, S., Müller, G. O., Flögel, P., and Nikiforova, N. (1982). *Phys. Status Solidi B* **113,** 589.
Razbirin, B. S., Uralt'sev, I. N., and Mikhailov (1978). *Solid State Commun.* **25,** 799.
Rice, T. M. (1977). *Solid State Phys.* **32,** 1.
Röseler, M., and Zimmermann, R. (1977). *Phys. Status Solidi B* **83,** 85.
Saito, H. (1985). *J. Lumin.* **30,** 303.
Saito, H., and Göbel, E. O. (1985). *Phys. Rev. B* **31,** 2360.
Schrey, H., Lyssenko, V. G., and Klingshirn, C. (1979a). *Solid State Commun.* **32,** 897.
Schrey, H., Lyssenko, V. G., Klingshirn, C., and Hönerlage, B. (1979b). *Phys. Rev. B* **20,** 5267.

Schweizer, J., Forchel, A., Hangleiter, A., Schmitt-Rink, S., Löwenau, J. P., and Haug, H. (1983). *Phys. Rev. Lett.* **51,** 698.
Segall, B. (1967). In *Physics and Chemistry of II–VI Compounds*, M. Aven and J. S. Prener, eds. North-Holland, Amsterdam, p. 1.
Seiler, D. G., Heiman, D., Feigenblatt, R., Aggarwal, R. L., and Lax, B. (1982). *Phys. Rev. B* **25,** 7666.
Seiler, D. G., Heiman, D., and Wherrett, B. S. (1983). *Phys. Rev. B* **27,** 2355.
Shank, C. V., Auston, D. H., Ippen, E. P., and Teschke, O. (1978). *Solid State Commun.* **26,** 567.
Shen, Y. R. (1984). *The Principles of Nonlinear Optics*. J. Wiley, New York.
Shevel, S., and Kunz, M. (1986). Private communication.
Smirl, A. L., Lindle, J. R., and Moss, S. C. (1978). *Phys. Rev. B* **18,** 5489.
Smirl, A. L., Moss, S. C., and Lindle, J. R. (1982a). *Phys. Rev. B* **25,** 2645.
Smirl, A. L., Boggess, T. F., Wherrett, B. S., Perryman, G. P., and Miller, A. (1982b). *Phys. Rev. Lett.* **49,** 933.
Smirl, A. L., Boggess, T. F., Moss, S. C., and Boyd, I. W. (1985). *J. Lumin.* **30,** 272.
Swoboda, H.-E. (1986). Private communication.
Tang, C. L., and Erskine, D. J. (1983). *Phys. Rev. Lett.* **51,** 840.
Unuma, Y., Abe, Y., Masumoto, Y., and Shionoya, S. (1984). *Phys. Status Solidi B* **125,** 735.
Van Driel, H. M., Sipe, J. E., and Young, J. F. (1985). *J. Lumin.* **30,** 446.
Wegener, M., Dörnfeld, C., Lambsdorff, M., Fidorra, F., and Klingshirn, C. (1986a). *SPIE Intern. Symposium on Optical Chaos*, Quebec, **667,** 102.
Wegener, M., Klingshirn, C. Koch, S. W., and Banyai, L. Semicond. Science and Technology 1, 366 (1986b).
Wegener, M., and Klingshirn, C. *Phys. Rev. A* **35,** 1740 and 4247 (1987).
Wolfe, J. P. (1985). *J. Lumin.* **30,** 82.
Yoshida, H., and Shionoya, S. (1983). *Phys. Stat. Sol. B* **115,** 203.
Yoshikumi, Y., Saito, H., and Shionoya, S. (1979). *Solid State Commun.* **32,** 665.
Zimmermann, R., Kilimann, K., Kraeft, W. D., Kremp, D., and Röpke, R. (1978). *Phys. Status Solidi B* **90,** 175.

3 MICROSCOPIC THEORY OF THE OPTICAL BAND EDGE NONLINEARITIES

Hartmut Haug

INSTITUT FÜR THEORETISCHE PHYSIK
UNIVERSITÄT FRANKFURT, ROBERT-MAYER-STR. 8
D-6000 FRANKFURT, FEDERAL REPUBLIC OF GERMANY

I. INTRODUCTION

In the introductory chapter, we mentioned already that the origin of the observed large optical nonlinearities of semiconductors is—except for some high-field effects such as the ac-Stark shift—not the direct intensity-depen-

OPTICAL NONLINEARITIES
AND INSTABILITIES IN
SEMICONDUCTORS

ISBN 0-12-332915-9

dence of the optical coefficients, but rather their dependence on the density of the electronic excitations (or the density of thermal excitations that result from the nonradiative decay of the electronic excitations). The density of these excitations has to be determined self-consistently from rate equations in which the density-dependent optical coefficients and the light intensity enter. We shall use in Section III.B nonequilibrium Green's functions to derive the relevant equations that describe the optical properties. We limit ourselves to the screened Hartree–Fock approximation. Schäfer and Treusch (1986) (see also Schäfer, Chapter 6) discussed in a similar approach attempts to go beyond this level in order to treat particularly dense exciton systems. But even on the level of the screened Hartree–Fock approximation, one obtains a rather intricate system of equations for the intra- and interband retarded Green's functions and particle propagators, which allow a self-consistent calculation of all spectral properties and excitation densities. This system of equations has not yet been exhaustively analyzed, but it allows us to obtain as limiting cases various formulations that have been treated recently: for situations that might be realized in low-temperature subpicosecond experiments as carried out first by Fehrenbach et al. (1982) and later other groups (see e.g. Chemla et al. Chapter 4, Peyghambarian and Gibbs, Chapter 12, and Ulbrich, Chapter 5), one obtains pair equations. The formal analogy of these equations with those of a superconductor has been exploited in coherent amplitude approximations by many authors. See e.g. Keldysh and Kozlov (1968), Zimmermann (1976), Silin (1977), Comte and Nozieres (1982), and Nozieres and Schmitt-Rink (1985). However, these investigations do not include the influence of the applied laser field. Only recently, Comte and Mahler (1986) included the laser field in a pair theory. If one takes into account only the real part of the complex self-energies, our theory contains their field-dependent gap equation which describes a light-enhanced excitonic pairing. However, a consistent physical picture can only be obtained by including also the lifetime effects, i.e. the imaginary part of the self-energies. Very recently, Schmitt-Rink and Chemla (1986) also included the laser field into a pair theory for nonresonantly excited excitons in order to explain the experimentally observed ac-Stark effect. The nonequilibrium Green's function theory is the natural framework for the treatment of such high-field effects. In this book, we leave the discussion of highly excited exciton systems (in which no plasma is present) mainly to Schäfer, Chapter 6, and to Chemla, Miller and Schmitt-Rink, Chapter 4. In situations where the electronic excitations are mainly free-carriers, e.g. due to rapid thermal ionization of the optically generated bound electron–hole pairs, the situation simplifies somewhat and we obtain under certain approximations for the two-point interband polarization the coherent wave equation that reduces in the low-excitation limit to the equation of Stahl (1979) (see also Stahl and Balslev 1987). This equation has been used mainly in a real-space representation to calculate the influence of finite geometries on the optical properties of semiconductors in the low-excitation limit. For the interband susceptibility we obtain the Bethe–Salpeter equation in the screened ladder approximation (Zimmermann et al., 1978). Simultaneously one obtains rate equations for the population of the renormalized states. Thus the present nonequilibrium theory justifies in quasi-equilibrium situations the use of one- and two-particle temperature-independent Green's functions together with rate equations for the excitation densities, which were used by Löwenau et al. (1982) and by Haug and Schmitt-Rink (1984 and 1985) in order to calculate the nonlinear optical response of laser-excited semiconductors. This approach implies the adiabatic elimination of the polarization as a dynamic variable, which is justified in a plasma because of the frequent carrier collisions.

In Section III.C we shall discuss briefly how according to Haug and Schmitt-Rink (1984) rather accurate numerical solutions of the Bethe–Salpeter equation can be obtained by a numerical matrix inversion. The resulting nonlinear optical spectra will be discussed for various semiconductors.

If band-filling effects are partly neglected, one can get an approximate but analytical expression for the density-dependence of the absorption spectrum according to a calculation of Banyai and Koch (1986). This approach will be discussed in Section III.D and it will be shown that the resulting expression allows us to describe the transition from an excitonic spectrum with sharp resonances at low light intensities to a broad-band plasma spectrum at high intensities, at least qualitatively.

Readers who are not interested in the theoretical foundation of the calculations of the optical nonlinearities may skip the following section and turn directly to the Sections III.C and D.

II. NONEQUILIBRIUM MANY-BODY THEORY

Laser-excited semiconductors are examples of externally perturbed systems. Due to the large number of strongly interacting electronic excitations, simple equations for the reduced single-particle density matrix elements which are often used in nonlinear optics are not so appropriate for semiconductors. The many-body theory that is most suited for the strongly interacting nonequilibrium system of electronic excitations is the nonequilibrium Green's function theory, which has been introduced in its final form by Keldysh (1965). Earlier versions are due to Schwinger (1961), Kadanoff and Baym (1962), and Koreman (1966). This technique has been applied to laser-excited semiconductors by Ivanov and Keldysh (1983), by Haug (1985) and by Schäfer and Treusch (1986).

In the presence of a laser field, one has to generalize the single-particle electron Green's functions in two ways:

a. In addition to the normal intraband Green's functions, the coherent light field induces interband Green's functions. Thus one has to treat a Green's function matrix with respect to the band index ρ.
b. The external perturbation causes not only a renormalization of the single-particle spectra but also finite populations of these states. Therefore, it is no longer possible to treat only one type of Green's functions, say e.g. the time-ordered ones. As Keldysh (1965) has shown, it is necessary to introduce with respect to the time-ordering index (upper index) a two-by-two matrix.

A. Definitions

We use the nomenclature of Haug (1985) and correct at the same time the misprints that unfortunately occurred in this paper. We use the following definitions for the Green's functions (with $\hbar = 1$ throughout this paper):

$$\begin{aligned} G_{\rho\nu}^{++} &= -i\langle \mathbf{T} a_{\rho k}(t) a_{\nu k}^{\dagger}(t')\rangle; \qquad & G_{\rho\nu}^{+-} &= +i\langle a_{\nu k}^{\dagger}(t') a_{\rho k}(t)\rangle; \\ G_{\rho\nu}^{-+} &= -i\langle a_{\rho k}(t) a_{\nu k}^{\dagger}(t')\rangle; \qquad & G_{\rho\nu}^{--} &= -i\langle \mathbf{T}' a_{\rho k}(t) a_{\nu k}^{\dagger}(t')\rangle. \end{aligned} \tag{1}$$

Here, $a_{\rho k}$ is the annihilation operator of an electron in the band ρ with the wavevector k. For simplicity we assume spatial homogeneity, but the formalism also allows us to study transport problems in a locally inhomogenous system (Hänsch and Mahan, 1983).

$G_{\rho\nu}^{++}$ is the usual time-ordered Green's function; $G_{\rho\nu}^{+-}$ is the particle propagator, because $G_{\rho\rho}^{+-}(k, t, t) = in_{\rho k}(t)$. Similarly, $G_{\rho\nu}^{-+}$ is the hole propagator, because $G_{\rho\rho}^{-+}(k, t, t) = -i[1 - n_{\rho k}(t)]$. Finally, $G_{\rho\nu}^{--}$ is the anti–time-ordered Green's function, the anti–time-ordering operator $\mathbf{T}'$ orders the operator with an earlier time argument to the left. The matrix $G_{\rho\nu}^{ij}$ obeys the Dyson equation

$$G_{\rho\nu}^{ij} = G_{\rho}^{0ij}\delta_{\rho\nu} + G_{\rho}^{0ik}\Sigma_{\rho\mu}^{kl}G_{\mu\nu}^{lj} \tag{2}$$

and

$$G_{\rho\nu}^{ij} = G_{\rho}^{0ij}\delta_{\rho\nu} + G_{\rho\mu}^{ik}\Sigma_{\mu\nu}^{kl}G_{\nu}^{0lj}, \tag{3}$$

where the summation convention is used. The self-energy matrix can be evaluated by using the familiar Feynman diagram techniques, summing at internal vertices also over the Keldysh index. For a two-band model the Dyson eq. 2 has the following graphical representation where $\nu \neq \rho$. Note the formal similarity with superconductivity, which becomes even more evident in the electron–hole picture. The nondiagonal or interband Green's function corresponds to the anomalous Green's function in the theory of superconductivity.

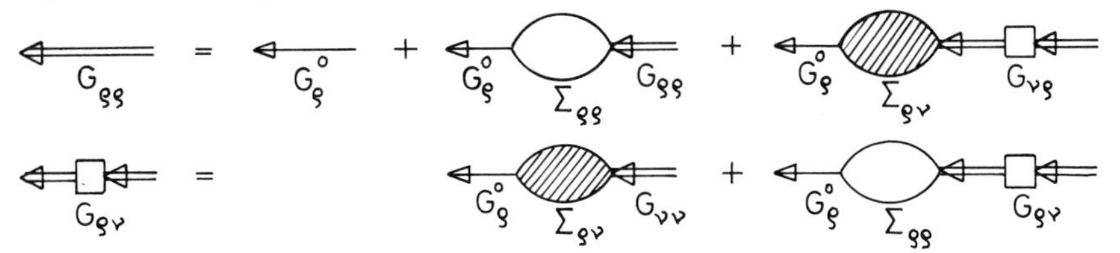

The retarded and advanced Green's functions are defined by

$$G_{\rho\nu}^{r}(k,t,t') = -i\Theta(t-t')\left\{\left[a_{\rho k}(t);\, a_{\nu k}^{\dagger}(t')\right]\right\}. \tag{4}$$

$$G_{\rho\nu}^{a}(k,t,t') = i\Theta(t'-t)\left\{\left[a_{\rho k}(t);\, a_{\nu k}^{\dagger}(t')\right]\right\}. \tag{5}$$

Here, [;] is the Fermi anticommutator. From the definitions (1), (4) and (5) it follows that

$$G^{r} = G^{++} - G^{+-} = G^{-+} - G^{--}, \qquad G^{a} = G^{++} - G^{-+} = G^{+-} - G^{--}. \tag{6}$$

Here, we omitted the band indices.

B. Equations for the Retarded Green's Functions

Using eq. 6 one gets from the matrix Dyson eq. 2 the Dyson equation for the retarded Green's function

$$G^{r} = G^{0r} + G^{0r}\Sigma^{r}G^{r}, \tag{7}$$

and from 3

$$G^{r} = G^{0r} + G^{r}\Sigma^{r}G^{0r}, \tag{8}$$

where

$$\begin{aligned}\Sigma^{r} &= \Sigma^{++} + \Sigma^{+-} = -\Sigma^{--} - \Sigma^{-+},\\ \Sigma^{a} &= \Sigma^{++} + \Sigma^{-+} = -\Sigma^{--} - \Sigma^{+-}.\end{aligned} \tag{9}$$

Next we apply the differential operator $G_\rho^0(t)^{-1} = \dfrac{i\partial}{\partial t} - \epsilon_\rho$ to the Dyson eq. 7 of $G_{\rho\nu}^r$ and $G_\nu^0(t')^{-1*} = -\dfrac{i\partial}{\partial t'} - \epsilon_\nu$ to eq. 8. We obtain

$$G_\rho^0(t)^{-1}G_{\rho\nu}^r(t,t') = \delta(t-t')\,\delta_{\rho\nu} + \int dt_1\,\Sigma_{\rho\mu}^r(t,t_1)G_{\mu\nu}^r(t_1,t') \quad (10a)$$

and

$$G_\nu^0(t')^{-1*}G_{\rho\nu}^r(t,t') = \delta(t-t')\,\delta_{\rho\nu} + \int dt_1\,G_{\rho\mu}^r(t,t_1)\Sigma_{\mu\nu}^r(t_1,t'). \quad (10b)$$

It is convenient to introduce $\sigma = t - t'$ and $\tau = (t + t')/2$. The two-point functions vary on a microscopic time scale with respect to σ, but only on a macroscopic scale with respect to τ. As a next step the Fourier transform is taken with respect to the relative time coordinate σ. Thus one gets, e.g., $G^r(k, \omega, \tau)$. One sees that these functions allow us quite naturally to calculate spectra that change in time. Similar arguments also hold for the space coordinates, if local homogeneity is not assumed. Then one would use operators that create and annihilate particles at the positions r and r', and introduce again difference and center of mass coordinates ρ and R, respectively, followed by a Fourier transformation with respect to ρ to obtain in general, e.g., $G^r(k, \omega, R, \tau)$.

Adding eqs. 10a and b, one obtains (suppressing the wave-number dependence)

$$\begin{aligned}&\left[G_\rho^r(\omega,\tau)^{-1} + G_\nu^r(\omega,\tau)^{-1}\right]G_{\rho\nu}^r \\ &\quad = 2\delta_{\rho\nu} + \left[\Sigma_{\rho\mu}^r(\omega,\tau)G_{\mu\nu}^r(\omega,\tau) + G_{\rho\mu}^r(\omega,\tau)\Sigma_{\mu\nu}^r(\omega,\tau)\right],\end{aligned} \quad (11)$$

where

$$G_\rho^r(\omega,\tau)^{-1} = \omega - \epsilon_\rho - \Sigma_{\rho\rho}^r(\omega,\tau) \quad (12)$$

which is renormalized only by the intraband self-energy. In the second line of eq. 12 the diagonal elements of the self-energy are not included.

In a two-band model the formal solution for the intra- and interband retarded Green's functions are (with $\rho \neq \nu$)

$$G_{\rho\rho}^r(\omega,\tau) = \frac{(G_\nu^r)^{-1}}{Z}, \quad (13)$$

where

$$Z = \left(G_\rho^r\right)^{-1}\left(G_\nu^r\right)^{-1} - \Sigma_{\rho\nu}^r \Sigma_{\nu\rho}^r,$$

and (14)

$$G_{\rho\nu}^r(\omega, \tau) = \frac{\Sigma_{\rho\nu}^r}{Z} = G_{\rho\rho}^r \Sigma_{\rho\nu}^r G_\nu^r.$$

In these results the effects of a strong coherent laser field and of the attractive electron–hole interaction, both contained in $\Sigma_{\rho\nu}^r$, are thus calculated starting with the dressed function G_ρ^r, which contains the renormalization due to such important physical processes as intraband Coulomb collisions as will be shown below. The fact is a basic advantage of the present theory as compared to other theoretical treatments in which the field renormalizations are introduced into the unrenormalized Green's functions.

C. Equations for the Particle Propagators

In addition to the retarded Green's functions, we need the particle propagators. The other Green's functions, namely G^{--} and G^{-+}, can be expressed via eqs. 6 in terms of G^r and G^{+-}. From the matrix Dyson eqs. 1 and 2 we get

$$\begin{aligned} &G_\rho^0(t)^{-1} G_{\rho\nu}^{+-}(t, t') \\ &\quad = \int dt_1 \left[\Sigma_{\rho\mu}^{++}(t, t_1) G_{\mu\nu}^{+-}(t_1, t') + \Sigma_{\rho\mu}^{+-}(t, t_1) G_{\mu\nu}^{--}(t_1, t')\right] \end{aligned} \tag{15}$$

and

$$\begin{aligned} &G_\nu^0(t')^{-1*} G_{\rho\nu}^{+-}(t, t') \\ &\quad = -\int dt_1 \left[G_{\rho\mu}^{++}(t, t_1) \Sigma_{\mu\nu}^{+-}(t_1, t') + G_{\rho\mu}^{+-}(t, t_1) \Sigma_{\mu\nu}^{--}(t_1, t')\right], \end{aligned} \tag{16}$$

where we used

$$G_\rho^0(t)^{-1} G_{\rho\rho}^{0++}(t, t') = \delta(t - t'),$$

$$G_\rho^0(t)^{-1} G_{\rho\rho}^{0+-}(t, t') = 0,$$

and

$$G_\nu^0(t')^{-1*} G_{\nu\nu}^{0--}(t, t') = -\delta(t - t').$$

From the difference of eq. 15 and eq. 16, we get after a Fourier transform with respect to the difference time coordinate σ:

$$\begin{aligned}\left(i\frac{\partial}{\partial\tau}+\epsilon_{\nu}-\epsilon_{\rho}\right)&G^{+-}_{\rho\nu}(\omega,\tau)\\ &=\Sigma^{r}_{\rho\mu}(\omega,\tau)G^{+-}_{\mu\nu}(\omega,\tau)-G^{+-}_{\rho\mu}(\omega,\tau)\Sigma^{a}_{\mu\nu}(\omega,\tau)\\ &\quad-\Sigma^{+-}_{\rho\mu}(\omega,\tau)G^{a}_{\mu\nu}(\omega,\tau)+G^{r}_{\rho\mu}(\omega,\tau)\Sigma^{+-}_{\mu\nu}(\omega,\tau).\end{aligned} \tag{17}$$

From the diagonal elements $\rho=\nu$ of this equation, one gets the rate equations, using the identity $n_{\rho k}=i\int G^{+-}_{\rho\rho}\,d\omega/2\pi$. The equation for the diagonal elements can be put into the form:

$$\begin{aligned}\left(\frac{i\partial}{\partial\tau}\right)G^{+-}_{\rho\rho}(\omega,\tau)&=\Sigma^{+-}_{\rho\rho}(\omega,\tau)G^{-+}_{\rho\rho}(\omega,\tau)-\Sigma^{-+}_{\rho\rho}(\omega,\tau)G^{+-}_{\rho\rho}(\omega,\tau)\\ &\quad+G^{r}_{\rho\nu}(\omega,\tau)\Sigma^{+-}_{\nu\rho}(\omega,\tau)-\Sigma^{+-}_{\rho\nu}(\omega,\tau)G^{a}_{\nu\rho}(\omega,\tau)\\ &\quad-G^{+-}_{\rho\nu}(\omega,\tau)\Sigma^{a}_{\nu\rho}(\omega,\tau)+\Sigma^{r}_{\rho\nu}(\omega,\tau)G^{+-}_{\nu\rho}(\omega,\tau),\end{aligned} \tag{18}$$

with $\nu\neq\rho$. From the off-diagonal elements $\rho\neq\nu$, one obtains the interband polarization $\langle P^{(+)}\rangle=\sum_k d_k\langle a^{+}_{2k}a_{1k}\rangle=i\sum_k\int\frac{d\omega}{2\pi}d_kG^{+-}_{12}$, which contains all information about the optical properties of the system. $d_k=er_{21}(k)$ is the dipole matrix element. The equation for the interband particle propagator is $(\rho\neq\nu)$

$$\begin{aligned}\left(\frac{i\partial}{\partial\tau}+G^{r}_{\rho}(\omega,\tau)^{-1}-G^{a}_{\nu}(\omega,\tau)^{-1}\right)&G^{+-}_{\rho\nu}(\omega,\tau)\\ &=+\Sigma^{r}_{\rho\nu}(\omega,\tau)G^{+-}_{\nu\nu}(\omega,\tau)-G^{+-}_{\rho\rho}(\omega,\tau)\Sigma^{a}_{\rho\nu}(\omega,\tau)\\ &\quad-\Sigma^{+-}_{\rho\nu}(\omega,\tau)G^{a}_{\nu\nu}(\omega,\tau)+G^{r}_{\rho\rho}(\omega,\tau)\Sigma^{+-}_{\rho\nu}(\omega,\tau)\\ &\quad-\Sigma^{+-}_{\rho\rho}(\omega,\tau)G^{a}_{\rho\nu}(\omega,\tau)+G^{r}_{\rho\nu}(\omega,\tau)\Sigma^{+-}_{\nu\nu}(\omega,\tau).\end{aligned} \tag{19}$$

D. Hamiltonian

In order to evaluate the equations for G^{r} and G^{+-}, we have to specify which interactions will be taken into account. We consider within a two-band model the interactions with the external coherent light field and with the empty continuum of all light modes in order to describe the influence of spontaneous emission. Furthermore, all direct Coulomb interactions will be considered, but the interband exchange interactions will not be included, which is justified if the exciton binding energy is small compared to the band gap.

The Hamiltonian of the system is:

$$H = \sum_{\rho, k} E_{\rho k} a^{\dagger}_{\rho k} a_{\rho k} + \sum_{\lambda} \Omega_{\lambda} b^{\dagger}_{\lambda} b_{\lambda} - \left[E^{(-)} P^{(+)} \exp(i\omega_0 t) + \text{h.c.}\right]$$
$$+ \sum_{\lambda, k} g_{\lambda k}\left(b^{\dagger}_{\lambda} \; a^{\dagger}_{2k} a_{1k} + \text{h.c.}\right) \tag{20}$$
$$+ \frac{1}{2} \sum_{\rho, \nu, q, k, k'} V_q a^{\dagger}_{\rho k+q} a^{\dagger}_{\nu k'-q} a_{\nu k'} a_{\rho k},$$

where the unperturbed band $\rho = 1$ corresponds to the conduction band and $\rho = 2$ to the valence band:

$$E_{1k} = E_g + \frac{k^2}{2m_e} \qquad E_{2k} = -\frac{k^2}{2m_h},$$

and (21)

$$V_q = \frac{4\pi e^2}{\epsilon_0 q^2}.$$

ϵ_0 is the background dielectric constant of the unexcited crystal. The interaction with the classical field $E^{(+)}$ and $E^{(-)}$ is written in the rotating wave approximation, d_k and $g_{\lambda k}$ are related optical matrix elements. Ω_λ is the energy of the spontaneously emitted photons, which are treated as a bath. The laser is resonant at the energy $\omega_0 = k^2/2m + E_g$, where m is the reduced electron-hole mass. The resonance energies in both bands are $\epsilon^r_1 = (\omega_0 - E_g)m/m_e + E_g$ and $\epsilon^r_2 = -(\omega_0 - E_g)m/m_h$. Adding and subtracting $H_0 = \sum \epsilon^r_\rho a^{\dagger}_{\rho k} a_{\rho k} + \sum \omega_0 b^{\dagger}_{\lambda} b_{\lambda}$ one gets $H = H_0 + (H - H_0) = H_0 + H'$. In the interaction representation $H = e^{iH_0 t} H' e^{-iH_0 t}$ one can eliminate the fast time-dependence:

$$H = \sum_{\rho,k} \epsilon_{\rho k} a^{\dagger}_{\rho k} a_{\rho k} + \sum_{\lambda} \omega_{\lambda} b^{\dagger}_{\lambda} b_{\lambda} - \left[E^{(-)} P^{(+)} + \text{h.c.}\right]$$
$$+ \sum_{\lambda,k} g_{\lambda k}\left(b^{\dagger}_{\lambda} a^{\dagger}_{2k} a_{1k} + \text{h.c.}\right) + \frac{1}{2} \sum_{\rho,\nu,q,k,k'} V_q a^{\dagger}_{\rho k+q} a^{\dagger}_{\nu k'-q} a_{\nu k'} a_{\rho k}, \tag{22}$$

where the shifted energies are given by $\epsilon_{1k} = \Delta_k \dfrac{m}{m_e}$, $\epsilon_{2k} = -\Delta_k \dfrac{m}{m_h}$ with $\Delta_k = E_g + \dfrac{k^2}{2m} - \omega_0$, and $\omega_\lambda = \Omega_\lambda - \omega_0$. For simplicity we assume $E^{(+)} = E^{(-)} = E$ (real).

E. Calculation of the Self-Energies

Now we can evaluate the diagonal and off-diagonal self-energy matrix elements for the radiative interactions and for the Coulomb interactions. The latter will be taken in the screened Hartree–Fock approximation:

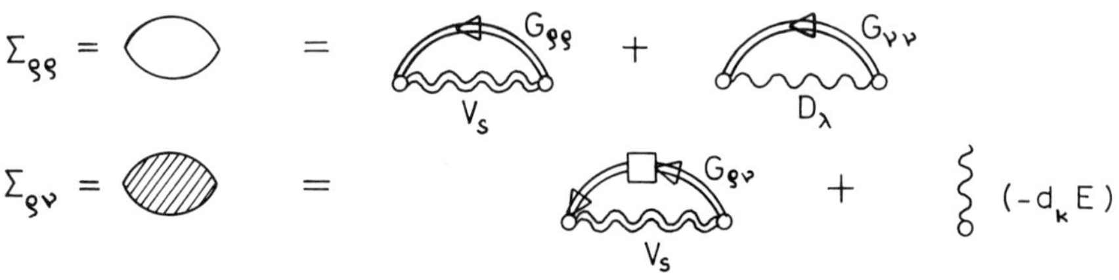

$\Sigma_{\varrho\varrho}$ = = $G_{\varrho\varrho}$ + $G_{\nu\nu}$

V_s D_λ

$\Sigma_{\varrho\nu}$ = = $G_{\varrho\nu}$ + $(-d_k E)$

V_s

These diagrams also show the necessity of treating the Coulomb and radiative effects consistently, because both stem from the basic electromagnetic interaction. In the language of quantum electrodynamics, the Coulomb interaction is mediated by the exchange of a longitudinal photon, while the radiative interaction is due to the exchange of a transverse photon.

F. Retarded Intraband Self-Energies

Let us start with the evaluation of the retarded self-energies $\Sigma^r = \Sigma^{++} + \Sigma^{+-}$. The intraband RPA–Coulomb self-energy $\Sigma^{rC}_{\rho\rho}(k, \omega)$ is given by:

$$i\sum_{k'}\int\frac{d\omega'}{2\pi}\left[V_S^{++}(k',\omega')G^{++}_{\rho\rho}(k-k',\omega-\omega')\right.$$
$$\left.-V_S^{+-}(k',\omega')G^{+-}_{\rho\rho}(k-k',\omega-\omega')\right] \tag{23}$$

With $V_S^{++} = V_S^r + V_S^{+-}$ and $G^{++} = G^r + G^{+-}$ we get

$$\Sigma^{rC}_{\rho\rho} = i\sum_{k'}\int\frac{d\omega'}{2\pi}\left[V_S^r\left(G^r_{\rho\rho} + G^{+-}_{\rho\rho}\right) + V_S^{+-}G^r_{\rho\rho}\right] \tag{24}$$

In order to proceed we assume that the collisions between the electronic excitations are so rapid that they establish a thermal distribution within each band, i.e. we assume a quasi equilibrium. Furthermore, we make use of the spectral representation for the screened Coulomb potential:

$$V_S^r(k,\omega) = V_k - \int \frac{d\omega'}{2\pi} \frac{2\,\mathrm{Im}\, V_S^r(k,\omega')}{\omega - \omega' + i\delta} \tag{25}$$

$$V_S^{+-}(k,\omega) = ig(\omega)\, 2\mathrm{Im}\, V_S^r(k,\omega),$$

where $g(\omega) = \dfrac{1}{\exp(\omega\beta) - 1}$ is the Bose distribution.

For the Green's functions we use self-consistently renormalized energies $e_{\rho k} = \epsilon_{\rho k} + \Sigma_{\rho\rho}^r(k, e_{\rho k})$, and make the approximations:

$$G_{\rho\rho}^r(k,\omega) \simeq G_\rho^r \simeq \frac{1}{\omega - e_{\rho k} + i\delta} \tag{26}$$

and

$$G_{\rho\rho}^{+-}(k,\omega) = if_\rho(\omega) A_{\rho\rho}(k,\omega),$$

where $f_\rho(\omega) = \dfrac{1}{\exp[(\omega - \mu_\rho)\beta] + 1}$ is the Fermi function with the quasi-chemical potential μ_ρ in the band ρ, and $-iA_{\rho\rho} = G_{\rho\rho}^r - G_{\rho\rho}^a$ is the spectral function, which reduces in the quasi-particle approximation to $A_{\rho\rho}(k,\omega) \simeq 2\pi\,\delta(\omega - e_{\rho k})$.

The resulting retarded intraband Coulomb self-energy is

$$\Sigma_{\rho\rho}^{rC}(k,\omega) = -\sum_{k'} V_{k'} f_\rho(e_{\rho,k-k'}) + \sum_{k'} \int \frac{d\omega'}{2\pi} \left[g(-\omega') + f_\rho(e_{\rho,k-k'})\right] \frac{2\mathrm{Im}\, V_S^r(k',\omega')}{\omega - \omega' - e_{\rho,k-k'} + i\delta} \tag{27}$$

This is the well-known RPA exchange and correlation energy. As pointed out by Haug and Schmitt-Rink (1984), it is advantageous to reorganize eq. 27 into a screened exchange and a Coulomb-hole term. In the dominant frequency approximation one gets by neglecting the recoil energies $e_{\rho,k-k'} - e_{\rho k}$ the quasi-static approximation for the self-energy

$$\Sigma_\rho^r(k, e_{\rho k}) = -\sum_{k'} V_S(k',0) f_\rho(e_{\rho,k-k'}) + \frac{1}{2}\sum_{k'} \left[V_S(k',0) - V_{k'}\right] \tag{28}$$

The last term is nothing but the self-energy of a point-charge in a plasma. In the low-density limit where all band-filling effects can be neglected, the last term of eq. 28 gives rise to a band-gap shift of $\Delta E_g = \Sigma_1^r + \Sigma_2^r = V_S(r) - V(r)$ for $r \to 0$. With a Debye screening wave-number k_S, one gets the approximate result that the band gap shrinks in a photo-excited semiconductor due to the screening of the Coulomb potential by $\Delta E_g = -e^2 k_S/\epsilon_0$, called the Debye shift.

The imaginary part of eq. 27 yields a dynamic collision broadening:

$$\Gamma_\rho^C(k,\omega) = -\pi \sum_{k'} \int \frac{d\omega'}{2\pi} \left[g(-\omega') + f_\rho(e_{\rho,k-k'}) \right] \times 2\,\mathrm{Im}\, V_S^r(k',\omega')\, \delta(\omega - \omega' - e_{\rho,k-k'}) \tag{29}$$

This collision broadening has been evaluated by Haug and Tran Thoai (1980). It is just equal to the sum of the intraband scattering rates in and out of state ρ, $k - k'$ under emission and absorption of a plasmon, respectively. These rates are proportional to $[1 + g(\omega)](1 - f_\rho) + g(\omega) f_\rho = 1 + g(\omega) - f_\rho = -[g(-\omega) + f_\rho]$. This dynamic collision broadening is important for the line-shape calculation of the band-tail absorption. It consists of a combination of the usual band tail due to the density-dependent collision broadening and of a smeared out side-band due to plasmon-assisted transitions. These plasmon-assisted transitions are a combination of an optical transition and an Auger process, i.e. a radiative recombination process in which additionally an electron is excited to higher energies (see Müller and Haug, 1987).

Next we evaluate the retarded radiative self-energy due to spontaneous emission. According to Haug (1984), one gets a result that has the structure of eq. 24 if one replaces the screened potential by the unrenormalized bath photons $g^2 D_\lambda^0$ and $G_{\rho\rho}$ by $G_{\nu\nu}$ with $\nu \neq \rho$. Using the formulae

$$D_\lambda^{0r} = \frac{1}{\omega - \omega_\lambda + i\delta} + \frac{1}{\omega + \omega_\lambda + i\delta} \quad \text{and} \quad D_\lambda^{0+-} = -i2\pi\, \delta(\omega + \omega_\lambda) \tag{30}$$

one gets the result

$$\Sigma_{\rho\rho}^{r,R}(k, e_\rho) = -\sum_\lambda \frac{g^2[1 - f_\nu(e_{\nu k})]}{e_{\rho k} - e_{\nu k} - \omega_\lambda + i\delta} - \sum_\lambda \frac{g^2[f_\nu(e_{\nu k})]}{e_{\rho k} - e_{\nu k} + \omega_\lambda + i\delta} \tag{31}$$

In this formula only the resonant term has to be taken into account in order to be consistent with the original Hamiltonian. The radiation broadening, e.g., of the conduction band is thus in the weak-field limit, where $e_{1k} = e_{ck}$

and $e_{2k} = e_{vk}$, given by

$$\Gamma_c^R(k, e_{ck}) = \pi \sum_{\lambda} g^2[1 - f_2(e_{vk})]\, \delta(e_{ck} - e_{vk} - \omega_\lambda)[1 - f_2(e_{vk})] \quad (32)$$

in accordance with Fermi's golden rule. The radiative lifetime (32) in direct-gap semiconductors is of the order of nanoseconds, while intraband scattering relaxation times resulting from (29) are in an optically generated plasma less than picoseconds. Thus the retarded radiative self-energies can normally be neglected compared to the corresponding Coulomb self-energies.

G. Retarded Interband Self-Energies

Let us now turn to the evaluation of the retarded interband self-energy matrix elements $\Sigma_{\rho\nu}^r = \Sigma_{\rho\nu}^{rR} + \Sigma_{\rho\nu}^{rC}$ according to the self-energy diagrams just given. The radiative self-energies due to the interaction with the coherent laser field are particularly simple:

$$\Sigma_{\rho\nu}^{++R} = -d_k E \quad \text{and} \quad \Sigma_{\rho\nu}^{+-R} = 0, \quad \text{thus} \quad \Sigma_{\rho\nu}^{rR} = -d_k E \qquad (33)$$

The retarded interband Coulomb self-energy is

$$\begin{aligned}\Sigma_{\rho\nu}^{rC}(k, \omega) = i \sum_{k'} \int \frac{d\omega'}{2\pi} \Big\{ V_S^r(k', \omega')\big[G_{\rho\nu}^r(k - k', \omega - \omega') \\ + G_{\rho\nu}^{+-}(k - k', \omega - \omega')\big] \\ + V_S^{+-}(k', \omega') G_{\rho\nu}^r(k - k', \omega - \omega')\Big\}.\end{aligned} \qquad (34)$$

Using again eqs. 25 for the screened potential, eq. 14 for $G_{\rho\nu}^r$ and the stationary solution of eq. 19 for $G_{\rho\nu}^{+-}$ we get an integral equation for the interband self-energy, which resembles the gap equation in superconductivity. That $\Sigma_{\rho\nu}^r$ plays indeed the role of the BCS-gap function can be seen from the poles of $G_{\rho\rho}^r$ in eq. 13. Taking $\Sigma_{\rho\nu}^r$ in a dominant frequency approximation and introducing the renormalized single-particle energies $e_\rho = \epsilon_\rho + \Sigma_{\rho\rho}^r(e_\rho)$, the eigenfrequencies of eqs. 13 and 14, which are given by the zeros of the denominator Z, are:

$$\omega_{1,2} = \tfrac{1}{2}\left[e_1 + e_2 \pm \sqrt{(e_1 - e_2)^2 + 4\Sigma_{12}^2}\right]. \qquad (35)$$

In contrast to superconductivity, the gap is not only due to the pair-forming attractive e–h interaction but in addition is induced by the coherent light field. If we disregard all Coulomb effects, the coherent field alone introduces a gap $\Delta\epsilon = \Sigma_{12}^R = d_k E$ at the resonance frequencies in the conduction

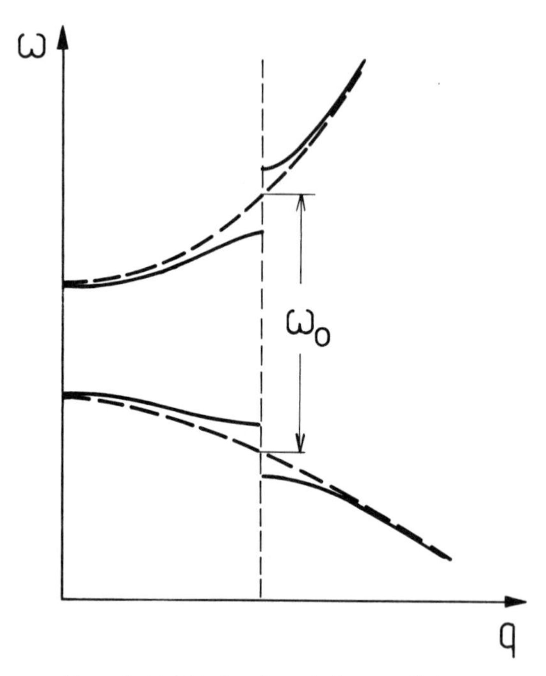

Figure 1. Light-induced gap (schematically).

and valence bands, as is shown in Fig. 1 (see also Elesin, 1971 and Harbich and Mahler, 1981). However, this gap could only be observed if its size is larger than the energy broadening due to the Coulomb and radiative transition rates. Recently, Comte and Mahler (1986) discussed an intensity-dependent gap equation derived within a pair-formalism, which can indeed be obtained from eqs. 33 and 34 in the high-field limit. For this purpose we take into account only the first leading term of eq. 19 and thus approximate the interband polarization by

$$\begin{aligned} G_{\rho\nu}^{+-} &= \Sigma_{\rho\nu} \frac{G_{\nu\nu}^{+-} - G_{\rho\rho}^{+-}}{\left(G_\rho^r\right)^{-1} - \left(G_\nu^a\right)^{-1}} \\ &= i\Sigma_{\rho\nu} \frac{f_\nu A_{\nu\nu} - f_\rho A_{\rho\rho}}{e_\nu - e_\rho}. \end{aligned} \tag{36}$$

Using the spectral decomposition

$$A_{\rho\rho} = u_\rho^2\,\delta(\omega - \omega_1) + v_\rho^2\,\delta(\omega - \omega_2)$$

with

$$u_\rho^2 = \frac{\omega_1 - e_\nu}{\omega_1 - \omega_2} \quad \text{and} \quad v_\rho^2 = -\frac{\omega_2 - e_\nu}{\omega_1 - \omega_2}$$

we find by disregarding all screening effects the inhomogenous gap equation

$$\Sigma_{\rho\nu} = -d_k E + \sum_{k'} \frac{V_{k'}\Sigma_{\rho\nu,k-k'}}{\omega_1 - \omega_2}\left[f(\omega_2) - f(\omega_1)\right], \tag{37}$$

which is just the temperature-dependent generalization of the gap-equation discussed by Comte and Mahler (1986). We leave the treatment of the complete gap equation with complex self-energies and with screening to a future investigation.

H. Particle Self-Energies

Finally, we evaluate the particle self-energy Σ^{+-}. The intraband matrix elements are given by:

$$\Sigma_{\rho\rho}^{+-R}(k,\omega) = ig^2 \sum_{\lambda} f_\nu(\omega + \omega_\lambda) A_{\nu\nu}(k, \omega + \omega_\lambda) \tag{38}$$

similarly

$$\Sigma_{\rho\rho}^{-+R}(k,\omega) = -ig^2 \sum_{\lambda} \left[1 - f_\nu(\omega - \omega_\lambda)\right] A_{\nu\nu}(k, \omega - \omega_\lambda) \tag{39}$$

and

$$\Sigma_{\rho\rho}^{+-C}(k,\omega) = i\sum_{k'} \int \frac{d\omega'}{2\pi} g(\omega') 2\,\mathrm{Im}\, V_S^r(k',\omega') \times f_\rho(\omega - \omega') A_{\rho\rho}(k - k', \omega - \omega') \tag{40}$$

The interband matrix elements are:

$$\Sigma_{\rho\nu}^{+-R} = 0, \tag{41}$$

and

$$\Sigma_{\rho\nu}^{+-C}(k,\omega) = -\sum_{k'} \int \frac{d\omega'}{2\pi} g(\omega') 2\,\mathrm{Im}\, V_S^r(k',\omega') G_{\rho\nu}^{+-}(k - k', \omega - \omega') \tag{42}$$

In quasi equilibrium, the intraband self-energies can also be expressed in terms of a spectral function

$$\Sigma_{\rho\rho}^{+-} = -if_\rho \Gamma_{\rho\rho} \quad \text{and} \quad \Sigma_{\rho\rho}^{-+} = i(1 - f_\rho)\Gamma_{\rho\rho}, \tag{43}$$

where

$$
\begin{aligned}
i\Gamma_{\rho\rho} &= \Sigma^{+-}_{\rho\rho} = \Sigma^{+-}_{\rho\rho} - \Sigma^{-+}_{\rho\rho} = \Sigma^{r}_{\rho\rho} - \Sigma^{a}_{\rho\rho} \\
&= -i\left(\left(G^{a}_{\rho\rho}\right)^{-1} - \left(G^{r}_{\rho\rho}\right)^{-1}\right) = -\frac{A_{\rho\rho}}{G^{r}_{\rho\rho}G^{a}_{\rho\rho}}.
\end{aligned}
\tag{44}
$$

I. Rate Equations

We are now in the position to calculate the rate equations from eq. 18 of the intraband particle propagators. Under quasi-equilibrium conditions, which are established by fast intraband scattering processes, we need only to calculate the total number of electrons within one band

$$
N_{\rho}(\tau) = i\sum_{k}\int\frac{d\omega}{2\pi}G^{+-}_{\rho\rho}(k,\omega,\tau)
\tag{45}
$$

The Coulomb scattering processes do not alter the total number of electrons within one band, thus they cancel completely in the rate equation for N_{ρ}.

From (18) we get

$$
\frac{\partial N_{\rho}}{\partial\tau} = \sum_{k}\int\frac{d\omega}{2\pi}\left(\Sigma^{+-}_{\rho\rho}G^{-+}_{\rho\rho} - \Sigma^{-+}_{\rho\rho}G^{+-}_{\rho\rho}\right) - iE\left(P_{\rho\nu} - P^{*}_{\rho\nu}\right),
\tag{46}
$$

where

$$
P_{\rho\nu} = i\sum_{k}\int\frac{d\omega}{2\pi}d_{k}G^{+-}_{\rho\nu}.
$$

The last term is simply the absorption rate. Using the relations

$$
-i(P_{12} - P^{*}_{12}) = \mathrm{Im}\,P = \frac{\epsilon''(\omega_0)E}{4\pi} = \frac{n(\omega_0)c\alpha(\omega_0)E}{4\pi\omega_0}
$$

and

$$
I = \frac{n(\omega_0)cE^2}{4\pi}
$$

one gets the usual rate equation

$$
\frac{\partial N_1}{\partial\tau} = \frac{\alpha(\omega_0)I}{\omega_0} - R,
\tag{47}
$$

where R is the rate of spontaneous emission:

$$
\begin{aligned}
R = \sum_{k,\lambda}\int g^2\frac{d\omega}{2\pi}A_{11}(\omega)\{&f_1(\omega)[1 - f_2(\omega - \omega_\lambda)]A_{22}(\omega - \omega_\lambda) \\
&- [1 - f_1(\omega)]f_2(\omega + \omega_\lambda)A_{22}(\omega + \omega_\lambda)\}.
\end{aligned}
\tag{48}
$$

In the low-field limit where $A_{11}(\omega) = 2\pi\delta(\omega - e_1)$ and $A_{22}(\omega) = 2\pi\delta(\omega - e_2)$, only the first term of eq. 48 contributes and gives the usual spontaneous emission rate. In general, both terms give a finite contribution due to the band mixing by the field and the electron–hole interaction.

J. Optical Polarization

The interband particle propagator given by eq. 19 contains all information about the nonlinear optical properties of the system. In order to understand the content of eq. 19, we shall make some simplifying assumptions through which we shall derive various successfully applied theoretical approaches.

1. Transitions without Interband Coulomb Interaction

If we disregard all interband Coulomb self-energies, we get in the stationary case the solution (see Haug 1984)

$$P_{12} = i\sum_{k'}\int\frac{d\omega'}{2\pi}\frac{d_k^2E(G_{22}^{+-} - G_{11}^{+-}) + G_{12}^r\Sigma_{22}^{+-} - \Sigma_{11}^{+-}G_{12}^a}{(G_{11}^r)^{-1} - (G_{22}^a)^{-1}} \tag{49}$$

Following Haug (1984), one can show that eq. 49 contains the known low- and high-field limits.

LOW-FIELD LIMIT. From eq. 13, one gets without interband Coulomb effects in the low-field limit

$$G_{12}^r = G_{11}^r d_k E G_2^r \simeq G_1^r d_k E G_2^r \tag{50}$$

Here again, all Green's functions with one index only contain no interband self-energy renormalizations. In quasi equilibrium one gets

$$P_{12} = i\sum_{k'}\int\frac{d\omega'}{2\pi}d_{k'}^2E\frac{f_2(A_2 - G_1^rG_2^r\Gamma_2) - f_1(A_1 - G_1^aG_2^a\Gamma_1)}{(G_{11}^r)^{-1} - (G_{22}^a)^{-1}} \tag{51}$$

Using the relation 44, one gets

$$P_{12} = -\sum_k d_k^2E\int\frac{d\omega}{2\pi}\int\frac{d\omega'}{2\pi}\frac{f_2(\omega) - f_1(\omega')}{\omega - \omega' + i\delta}A_2(\omega)A_1(\omega'), \tag{52}$$

which gives the usual linear response susceptibility $\chi(\omega_0) = P_{12}/E$. With sharp spectral distributions one would obtain

$$\chi(\omega_0) = -\sum_k d_k^2\frac{f_2(e_2) - f_1(e_1)}{e_2 - e_1 + i\delta}, \tag{53}$$

with $e_2 - e_1 = e_{vk} + \omega_0 - e_{ck}$.

HIGH-FIELD LIMIT. In this limit the intraband self-energies including that of spontaneous emission can be neglected in comparison with $d_k E$. Thus, the polarization is given by

$$P_{12} = -\sum_k d_k^2 E \int \frac{d\omega}{2\pi} f(\omega) \frac{A_{22} - A_{11}}{\epsilon_2 - \epsilon_1 + i\delta}. \tag{54}$$

The stationarity of the rate equations requires due to the absence of spontaneous emission $\mu_1 = \mu_2 = 0$, i.e. the bands are filled with electrons up to energies that are resonant with the coherent laser radiation (total bleaching). Using the eigenfrequencies $\omega_{1,2}$ (35), we find

$$P_{12} = -\sum_k d_k^2 E \frac{f(\omega_2) - f(\omega_1)}{\omega_2 - \omega_1 + i\delta}, \tag{55}$$

which is of the same form as the linear response result (53) except that e_1, e_2 are replaced by the renormalized energies ω_1, ω_2.

2. Transitions with Coulomb Interband Interactions

LOW-FIELD LIMIT. Now we consider the low-field limit, but we will take the most important effect of the Coulomb interband interaction into account, namely the formation of bound pairs, provided the screened Coulomb potential is strong enough. In order to allow for the possibility of exciton formation, the attractive Coulomb potential has to be included to infinite order, i.e. the interband polarization function $G_{\rho\nu}^{+-}$ must obey an integral equation. Thus we want again to select those terms on the r.h.s. of eq. 19 that contain $G_{\rho\nu}^{+-}$. These terms are contained in the first line. Retaining only these terms, we get under stationary conditions:

$$\left(G_\rho^{r-1} - G_\nu^{a-1}\right) G_{\rho\nu}^{+-} = -d_k E\left(G_{\rho\rho}^{+-} - G_{\nu\nu}^{+-}\right) - \Sigma_{\rho\nu}^{rC} G_{\rho\rho}^{+-} + G_{\nu\nu}^{+-} \Sigma_{\rho\nu}^{aC} \tag{56}$$

Selecting from the Coulomb interband energies (eq. 34) those terms that contain $G_{\rho\nu}^{+-}$, we find

$$\begin{aligned}\left(G_\rho^{r-1} - G_\nu^{a-1}\right) G_{\rho\nu}^{+-} = {} & -d_k E\left(G_{\rho\rho}^{+-} - G_{\nu\nu}^{+-}\right) \\ & + i \sum_{k'} \int \frac{d\omega'}{2\pi}\left[G_{\rho\rho}^{+-}(k,\omega) V_s^r(k',\omega')\right. \\ & \left. - G_{\nu\nu}^{+-}(k,\omega) V_s^a(k',\omega')\right] G_{\rho\nu}^{+-}(k-k',\omega-\omega')\end{aligned} \tag{57}$$

This is the desired integral equation for the intraband Green's function. It can be shown (see Müller et al., 1987) that the derivation of eq. 57 from eq. 19 becomes exact in the weak-field limit for a statically screened Coulomb potential. Anyhow, the dynamic screening complicates the solution of eq. 57 considerably. Normally (see, e.g., Zimmermann et al., 1978, and Haug and Schmitt-Rink, 1984) the frequency dependences of V_s and of the self-energies are treated only approximately according to a method of Shindo (1970). Here, we shall consider only a statically screened potential. Then we can immediately integrate eq. 57 over ω and obtain an integral equation for the interband polarization function

$$P_{cv}(k) = i\int \frac{d\omega}{2\pi} G_{cv}^{+-}.$$

With the quasi-particle approximation of eq. 26 we find:

$$(e_{vk} - e_{ck} + i\delta)P_{cv}(k) = (f_v - f_c)\Big(d_k E - \sum_{k'} V_s(k')P_{cv}(k-k')\Big), \quad (58)$$

where $f_c = f_c(e_{ck})$ depends on the renormalized shifted energy. Eq. 58 is closely related to the band-edge equation proposed by Stahl (1979) and by Stahl and Balslev (1986). Eq. 58 is indeed the momentum representation a band-edge equation for stationary, homogeneous situations. Using the effective mass approximation and a Fourier transform with respect to k, eq. 58 reduces in the limit of vanishing excitation density exactly to Stahl's band-edge equation:

$$\left(\omega_0 - E_g + \frac{\nabla^2}{2m} + V_s(r)\right)P_{cv}(r) = d(r)E, \quad (59)$$

where m is the reduced electron-hole mass and $d(r)$ the dipole matrix element in the space representation $d(r) = \sum_k d_k \exp(ikr)$, with $r = r_e - r_h$. Within the effective mass approximation the dipole matrix element is given by

$$d_k = e\sqrt{\frac{E_g}{4m}}\left(\frac{1}{E_g + \frac{k^2}{2m}}\right).$$

The k-dependence of d_k thus causes the smearing-out of $d(r)$, which has been suggested by Stahl (1979). It is easy to generalize the derivation by including variations in the center coordinates τ and R (see Müller et al., 1987) to get the full variation of the two-point function $P_{cv}(r_e, r_h, \tau)$ also in the case of finite excitation densities.

Finally, we introduce an interband susceptibility function $\chi(k) = P_{cv}(k)/E$ for which we find the Bethe–Salpeter equation in the screened ladder approximation:

$$\chi(k) = \chi_0(k)\left(1 - \frac{1}{d_k}\sum_q V_s(k-q)\chi(q)\right), \tag{60}$$

where

$$\chi_0(k) = \frac{f_v - f_c}{e_{vk} - e_{ck} + i\delta}. \tag{61}$$

This equation has been investigated for plasma screening by Zimmermann et al. (1978) and was used by Löwenau et al. (1982) to calculate plasma-density dependence of the band-edge spectra.

HIGH-FIELD LIMIT. In the high-field limit with interband Coulomb interaction again the result in eq. 58 remains valid, the only difference being that the population factors $f_v - f_c$ have to be replaced by those of the renormalized states $f(\omega_2) - f(\omega_1)$. This equation has to be used if one calculates the ac-Stark shift for nonresonantly excited excitons (see Schmitt-Rink and Chemla, 1986, and Müller et al., 1987).

III. PLASMA-DENSITY DEPENDENCE OF THE BAND-EDGE SPECTRA

In this section we shall describe how one can solve the integral equation 60 for the low-field limit by a numerical matrix inversion following a method that has been used by Löwenau et al. (1982) and described in detail by Haug and Schmitt-Rink (1984). Note first that the knowledge of the complex interband susceptibility function allows us to determine the optical susceptibility and the complex optical dielectric function:

$$\chi(\omega_0) = \sum_k d_k \chi(k) \quad \text{and} \quad \epsilon(\omega_0) = \epsilon_0[1 + 4\pi\chi(\omega_0)] \tag{62}$$

As described above, it is assumed that the screening is due to an electron–hole plasma, which may originate from a rapid ionization of optically excited excitons or which may have been excited directly by the laser. In order to simplify the calculation, we use (instead of the full RPA screening) the plasmon pole approximation (Lundquist, 1967; Zimmermann and Rösler, 1976; Rice, 1974; Overhauser, 1971; Haug and Schmitt-Rink, 1984). The statically screened Coulomb potential is in this approximation given by

$$V_s(k) = V_k \frac{1}{1 + \frac{\kappa^2}{k^2} \frac{1}{1 + C\frac{\kappa^2 k^2}{\omega_{pl}^2}}}, \tag{63}$$

where C is within the single plasmon pole approximation given by $C = 1/(16m^2)$. It has been found that the screening effects are somewhat overestimated by this formula, better results are obtained if C is enlarged by a factor of 2 to 4.

As a first step, we reformulate eq. 60 by introducing the vertex function $\Gamma(k, \omega_0)$:

$$\chi(k, \omega_0) = \Gamma(k, \omega_0)\chi_0(k, \omega_0). \tag{64}$$

The Bethe–Salpeter equation for $\Gamma(\text{p}, \omega)$ which follows from eq. 60 is

$$\Gamma(k, \omega_0) = 1 - \frac{1}{d_k} \sum_q V_s(k - q)\chi_0(q, \omega_0)\Gamma(q, \omega_0). \tag{65}$$

Considering only s-wave scattering in eq. 65, the screened Coulomb potential is replaced by its angle-averaged value:

$$I(k, q) = \frac{1}{2}\int_{-1}^{+1} d\cos\theta V_s(k - q), \tag{66}$$

where θ is the angle between k and q. In the low-density limit, the averaged potential has a logarithmic singularity at $k = q$. In order to remove this singularity we add and subtract on the r.h.s. of eq. 65 the term:

$$\frac{1}{d_k} \sum_q \frac{2k^4 I(k, q)}{q^2(k^2 + q^2)} \chi_0(k, \omega)\Gamma(k, \omega).$$

The q-integration in this compensation term can be done explicitly. The remaining difference term in the BSE:

$$\frac{1}{d_k} \sum_q I(k, q)\left(\frac{2k^4}{q^2(k^2 + q^2)} \chi_0(k, \omega_0)\Gamma(k, \omega_0) - \chi_0(q, \omega_0)\Gamma(q, \omega_0)\right).$$

is now a smooth function of k and q, because the bracket vanishes for $k \to q$.

Performing the remaining one-dimensional integration by means of a Gaussian quadrature, we obtain a system of linear equations which can be solved by matrix inversion. In order to obtain a good overall accuracy, we use a mesh with 80 to 180 points.

Finite lifetime effects can be taken into account by replacing $i\delta$ in eq. 61 by a phenomenological broadening $i\gamma$. In order to retain the correct crossover from gain to absorption ($\omega = \mu$) in the high-density limit, eq. 61 should be replaced by (see previous section):

$$\chi_0(k, \omega_0) = \int_{-\infty}^{+\infty} \frac{d\omega'}{2\pi} \frac{2\gamma}{(\omega' - e_{ck} + e_{vk})^2 + \gamma^2} \frac{f_v(e_{vk}) - f_c(\omega' + e_{vk})}{\omega - \omega' + i\delta} d_k . \tag{67}$$

In Fig. 2 we have plotted the resulting absorption and refraction spectra $\alpha(\omega)$ and $n(\omega)$, respectively, for GaAs for an electronic temperature of 10 K and for various free-carrier concentrations n according to Löwenau et al. (1982).

For vanishing free-carrier concentration ($n = 0$), one sees the 1s- and 2s-exciton absorption lines resolved with $\gamma = 0.05E_0$, where E_0 is the exciton Rydberg. The ionization continuum agrees well with the Elliot formula. With increasing plasma concentration, the edge of the ionization continuum shifts rapidly to lower energies, whereas the lowest exciton line stays almost constant, due to the high degree of compensation between the decrease of the e–h binding energy and the band-gap reduction. The curve with $n = 5 \times 10^{15}$ cm^{-3} is slightly above the ionization threshold, but still a large excitonic enhancement is present. Finally, at very high densities, the band-filling effects give rise to a large optical gain. The crossover from gain to absorption is at $\omega = \mu = \mu_e + \mu_h$. Naturally gain can never be realized in a single-beam experiment where the quasi-chemical potential μ remains always above the laser frequency $\mu \geq \omega_l$.

The associate refraction spectra show that around the 1s-exciton level, large dispersive changes occur when the exciton is ionized with increasing plasma concentration. Particularly, for $\omega < \omega_{1s}$ the refraction index decreases with increasing free-carrier concentration. This optical nonlinearity due to the ionization of the exciton is the mechanism responsible for the dispersive optical bistability in GaAs which was observed first by Gibbs et al. (1979).

The situation is somewhat different in InSb. For this material Miller et al. (1979) reported the first observation of a dispersive optical bistability. InSb is a narrow-gap semiconductor. Thus, the exciton has a very small binding energy and it is practically always ionized. This can be described by using the rather large broadening $\gamma = 2E_0 = 1$ meV. The corresponding absorption and refraction spectra are shown in Fig. 3 for an electronic tempera-

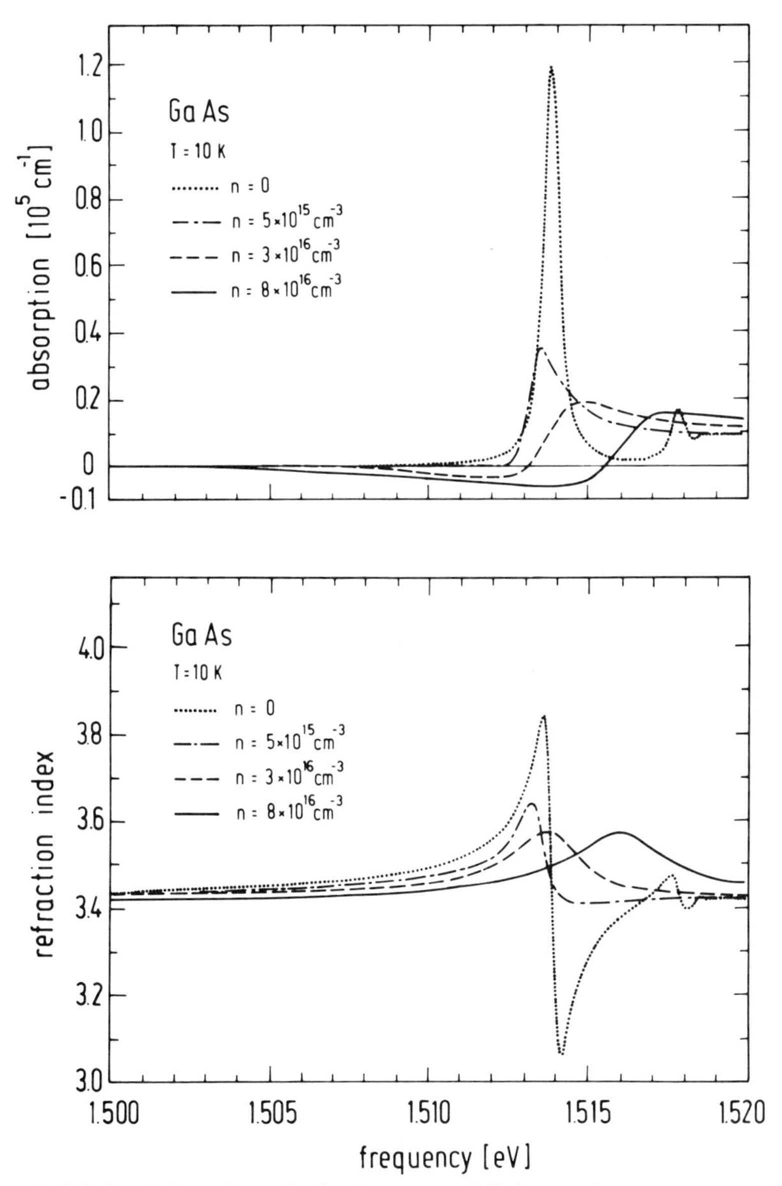

Figure 2. Calculated absorption and refraction spectra of GaAs at a plasma temperature of 10 K and various plasma densities according to Löwenau et al. (1982).

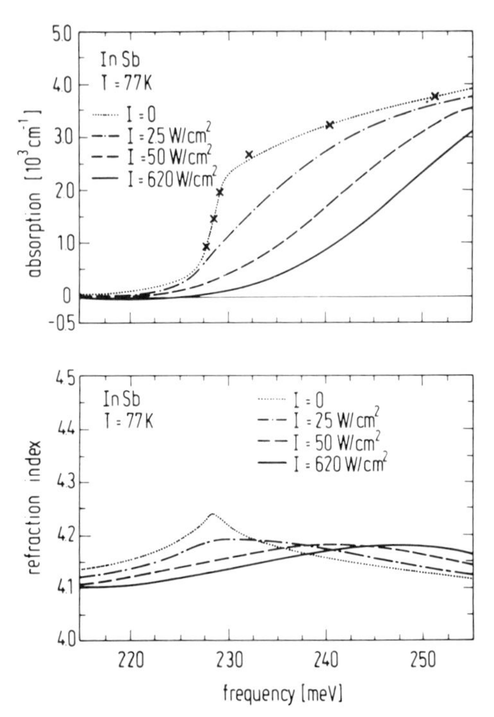

Figure 3. Calculated absorption and refraction spectra of InSb at a plasma temperature of 77 K and various plasma densities according to Löwenau et al. (1982).

ture of 77 K and various pump intensities I again according to Löwenau et al. (1982). The spectrum of the unexcited sample shows a large deviation from the parabolic density of states due to the influence of excitonic enhancement in agreement with the experimental observations of Gobeli and Fan (1960). In this spectral region, the effect of excitonic enhancement is much larger than that of the nonparabolicity of the conduction band. The frequency of the pump beam was assumed to be $\omega = 225$ meV. In the stationary case, the absorption rate equals the rate of spontaneous recombination: $\alpha(\omega)I/\omega = R_{\text{spon}} \approx \mathrm{B}n^2$, where B is the bimolecular recombination constant. This relation was used to calculate the free-carrier concentration as a function of the incident intensity.

The high-intensity absorption spectra show pronounced band-filling effects. For the intensity $I = 620\ \mathrm{W/cm^2}$, the free-carrier concentration is saturated, because the chemical potential has nearly reached the excitation frequency. The crystal is now transparent for the pump beam.

The associated refraction spectra show again large dispersive changes, which are obviously due to the band-filling effects. Thus, the origin of the dispersive optical nonlinearity in InSb is quite different from that in GaAs. However, both nonlinearities can be calculated in the same way by solving the Bethe–Salpeter equation 60.

At the end of this section we shall present a comparison of experimental and theoretical results. The most reliable measurements are due to Schweizer et al. (1983) for the absorption spectrum of the optical transitions at the direct gap of Ge. Ge is an indirect-gap semiconductor. Its band structure consists of one valence band maximum at the Γ-point and two conduction band minima at the L- and Γ-points. Optically created electrons relax rapidly to the indirect-gap minimum at the L-point, from where they can only decay by phonon-assisted transitions. Therefore, one can achieve with rather moderate cw-excitation a plasma concentration sufficient to ionize the exciton at the direct gap.

For the theoretical description of the direct transitions in Ge, the Γ-electron Fermi distribution has been replaced by zero in all relevant equations. The screened Coulomb potential is calculated by taking into

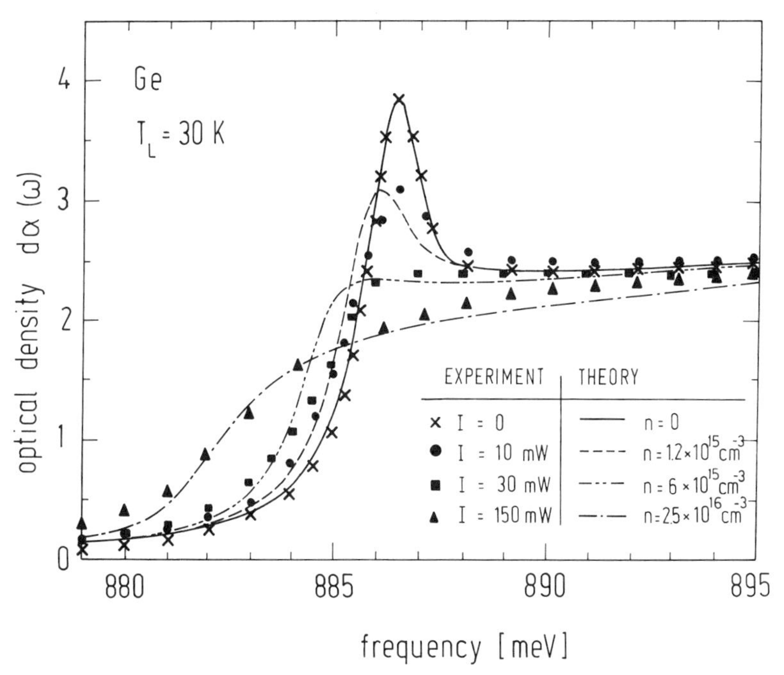

Figure 4. Measured and calculated absorption spectra at the direct gap of Ge according to Schweizer et al. (1983).

account the contribution of the free L-electrons and of the holes. Fig. 4 shows the measured and the calculated optical densities for a crystal thickness of $d = 10\ \mu m$ and various excitation powers P. The lattice temperature was 30 K. One sees that the theoretical and the experimental lineshapes are in good agreement. Only near the ionization threshold there is a small discrepancy in form of a slight red shift of the calculated spectra.

The direct transitions in Ge have also been investigated below the critical temperature for droplet formation, where a phase separation occurs. It has been shown by Zimmermann et al. (1981) that the corresponding experimental spectra are also well described by the quasi-static approximation.

IV. ANALYTICAL SOLUTION FOR A SIMPLIFIED MODEL

Recently Banyai and Koch (1986) attempted an approximate, but analytical solution for the excitonic absorption spectrum. They begin with linear response theory, i.e. the low-field limit as it was done in the last part of Section III.C. They then slightly reformulate the Kubo formula for Im $\chi(\omega)$ by extracting a frequency-dependent factor $\tanh[(\omega - \mu)\beta]$. This has the advantage that the transition from gain to absorption at $\omega = \mu$ is always correct, irrespective of any further approximations. They suggested that after this rearrangement, all band-filling effects can be neglected. In this approximation the Kubo formula can be expressed in terms of a generalized Elliot formula

$$\chi''(\omega) \approx 2\pi \tanh[\beta(\omega - \mu)]\, d_{cv}^2 \sum_n |\psi_n(r = 0)|^2\, \delta(\omega - E_n), \tag{68}$$

where ψ_n and E_n are the solutions of the Wannier equation with a screened Coulomb potential. Obviously in such an approximation the considered electron–hole pair is not aware which states are no longer available because they are already filled by free carriers. However, in situations where the concentration of the free carriers is relatively low and its distribution is sufficiently smeared out, this approximation yields good and very useful results.

The crucial final step of Banyai and Koch (1986) is to replace the screened Coulomb potential by the related Hulthen potential

$$V_{\mathrm{H}} = \frac{2e^2 k_s}{\epsilon_0} \frac{1}{e^{2rk_s} - 1}. \tag{69}$$

For this potential the needed eigenfunctions $\psi_n(r = 0)$ can be obtained analytically via the Jost functions. The energies of the bound states are

$$E_n^b = -E_0\left(\frac{1}{n} - k_s a_0\right)^2. \tag{70}$$

Particularly the ground state energy shifts in lowest order with

$$E_1^b - E_0 \simeq -2E_0 k_s a_0, \tag{71}$$

which compensates the Debye shift of the band gap $\Delta E_g = \dfrac{-e^2 k_s}{\epsilon_0} = -2E_0 k_s a_0$, which has been discussed in connection with eq. 28. Thus, the exciton resonance does not shift in lowest order. Note that the choice of parameters in the Hulthen potential, which was also made by Haug et al. (1986) deviates somewhat from that of Banyai and Koch (1986). There, the parameters of the Hulthen potential were chosen with the help of Bargman's theorem, which yields the same number of bound states for V_{H} and V_s, while the present choice guarantees the asymptotic compensation. A further consequence of this choice is that the exciton resonance vanishes when

$$k_s(n) a_0 = 1, \tag{72}$$

which is at least qualitatively the correct Mott criterion. The formula for this resulting absorption spectrum is

$$\alpha(\omega) = a \tanh[(\omega - \mu)\beta][b(\omega) + c(\omega)], \tag{73}$$

where $b(\omega)$ and $c(\omega)$ are the contributions of the bound states and of the continuum, respectively.

$$b(\omega) = \sum_l \pi\delta_\gamma\left\{\frac{\omega - E_g(N)}{E_0} + \left(\frac{1}{l} - \frac{l}{g}\right)^2\right\} \times \frac{2(g - l^2)(2l^3 - g)}{l^3 g^2} \prod_{n=1,\infty;\, n\neq l} \frac{n^2\left(n^2 l^2 - (g - l^2)^2\right)}{(n^2 - l^2)(n^2 l^2 - g^2)}, \tag{74}$$

where the sum runs over all bound states. The parameter g is $g = (k_s a_0)^{-1}$. The contribution of the ionization continuum is

$$c(\omega) = \int dx\sqrt{x}\, \delta_\gamma\left(\frac{\omega - E_g(N)}{E_0} - x\right) \prod_{n=1,\infty}\left(1 + \frac{2gn^2 - g^2}{(n^2 - g^2) + n^2 g^2 x}\right) \tag{75}$$

The lineshape function $\delta_\gamma(x)$ can be chosen as a Lorentzian. More realistic

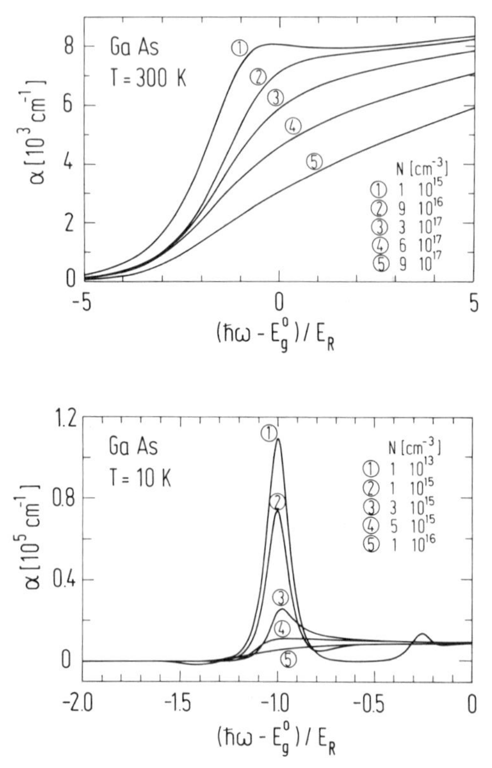

Figure 5. Calculated absorption spectra of GaAs for a) low and b) high plasma temperatures according to eq. 73 by Banyai and Koch (1986).

band-edge spectra are obtained by the choice

$$\delta_\gamma(x) = \frac{E_0}{\pi\gamma\cosh\left(\dfrac{xE_0}{\gamma}\right)}, \tag{76}$$

which approximates the exponential Urbach tail. In Fig. 5 the absorption spectrum calculated with formulae 73 to 76 is plotted for GaAs at a low and a high plasma temperature. A comparison with the numerical solution of the Bethe–Salpeter equation shows that at least qualitatively, all features of the density-dependent changes of the spectrum around the exciton ionization are retained in this model. Recently Lee et al. (1986) have shown that the experimentally observed room-temperature optical nonlinearities of GaAs are well described by this model (see also Chapter 12).

The existence of an analytical expression for the optical nonlinearity is crucial for the investigations of optical instabilities that occur in optical feedback systems (see Koch, Chapter 11). Examples of such instabilities are the optical bistability, self-oscillations and chaotic behaviour of passive or active semiconductor devices, which are analyzed in the second part of this book.

ACKNOWLEDGMENTS

I acknowledge stimulating discussions with S. Schmitt-Rink, S. W. Koch, L. Banyai and J. Müller. They all gave valuable contributions to various parts of this work. I thank A. Stahl, J. Treusch, C. Comte and S. Schmitt-Rink for sending me their papers prior to publication. This work has been supported by the Deutsche Forschungsgemeinschaft through the Sonderforschungsbereich Frankfurt-Darmstadt and by the Council of the European Community through the stimulation program "EJOB."

REFERENCES

Banyai, L., and Koch, S. W. (1986). *Z. Phys. B.* **63,** 283.
Comte, C., and Mahler, G. (1986). *Phys. Rev. B* **34**, 7164.
Comte, C., and Nozieres, P. (1982). *J. Physique* **43,** 1069; (1982) ibid. **43,** 1083.
Elesin, V. F. (1971). *Sov. Phys. JETP* **32,** 328.
Fehrenbach, G. W., Schäfer, W., Treusch, J., and Ulbrich, R. G. (1982). *Phys. Rev. Lett.* **49,** 1281.
Gibbs, H. M., McCall, S. L., Venkatesan, T. N. C., Gossard, A. C., Passner, A., and Wiegmann, W. (1979). *Appl. Phys. Lett.* **35,** 451.
Gobeli, G. W., and Fan, H. Y. (1960). *Phys. Rev.* **119**, 613.
Hänsch, W., and Mahan, G. D. (1983). *Phys. Rev. B* **28,** 1902.
Harbich, T., and Mahler, G. (1981). *Phys. Status Solidi B* **104,** 185.
Haug, H. (1985). *J. Lumin.* **30,** 171.
Haug, H., and Schmitt-Rink, S. (1984). *Prog. Quantum Electron.* **9,** 3.
Haug, H., and Schmitt-Rink, S. (1985). *J. Opt. Soc. Am. B* **2,** 1135.
Haug, H., and Tran Thoai, D. B. (1980). *Phys. Status Solidi B* **98,** 581.
Haug, H., Koch, S. W., and Lindberg, M. (1986). *Physica Scripta*, **13**, 187.
Ivanov, A. L. and Keldysh, L. V. (1983) *Sov. Phys. JETP* **57**, 234.
Kadanoff, L. P., and Baym, G. (1962). *Quantum Statistical Mechanics*. W. A. Benjamin, New York.
Keldysh, L. V. (1965). *Sov. Phys. JETP* **20,** 1018. For an introduction in this technique, see Landau-Lifshitz (1975). *Theoretical Physics*, vol. 10, *Kinetiques*. Pergamon Press, New York.
Keldysh, L. V., and Kozlov, A. N. (1968). *Sov. Phys. JETP* **27**, 521.
Keldysh, L. V., and Silin, A. P. (1975). *Sov. Phys. JETP* **42,** 535.
Koreman, V. (1966). *Ann. Phys.* **39,** 72.
Lee, Y. H., Chavez-Pirson, A., Koch, S. W., Gibbs, H. M., Park, S. H., Morhange, J., Jeffery, A., Peyghambarian, N., Banyai, L., Gossard, A. C., and Wiegmann, W. (1986). *Phys. Rev. Lett.* **57,** 2446.

Löwenau, J. P., Schmitt-Rink, S., and Haug, H. (1982). *Phys. Rev. Lett.* **49,** 1511.
Lundquist, B. I. (1967). *Phys. Kondens. Mater.* **6,** 193; (1967). Ibid. **6,** 206.
Miller, D. A. B., Smith, S. D. and Johnston, A. (1979). *Appl. Phys. Lett.* **35,** 658.
Müller, J., and Haug, H. (1987). *J. Lumin.*, **37**, 97.
Müller, J., Mevis, R., and Haug, H. (1987). *Z. Physik B* **69,** 179.
Nozieres, P., and Schmitt-Rink, S. (1985). *J. Low Temp. Phys.* **59,** 195.
Overhauser, A. W. (1971). *Phys. Rev. B* **3,** 1888.
Rice, T. M. (1974). *Nuovo Cimento* **23B,** 226.
Schäfer, W., and Treusch, J. (1986). *Z. Phys. B* **63,** 407.
Schmitt-Rink, S., and Chemla, D. S. (1986). *Phys. Rev. Lett.* **57,** 2752.
Schweizer, H., Forchel, A., Hangleiter, A., Schmitt-Rink, S., Löwenau, J. P., and Haug, H. (1983). *Phys. Rev. Lett.* **51,** 698.
Schwinger, J. (1961). *J. Math. Phys.* **2,** 407.
Shindo, K. (1970). *J. Phys. Japan* **29,** 287.
Silin, A. P. (1977). *Soc. Phys. Solid State* **19,** 77.
Stahl, A. (1979). *Phys. Status Solidi* **94,** 221.
Stahl, A., and Balslev, I. (1987). *Electrodynamics of the Semiconductor Band Edge*, Springer Tracts in Modern Physics vol. 110. Springer-Verlag, Berlin.
Zimmermann, R. (1976). *Phys. Status Solidi B* **76,** 191.
Zimmermann, R., and Rösler, M. (1976). *Phys. Status Solidi B* **75,** 633.
Zimmermann, R., Kilimann, K., Kräft, W. D., Kremp, D., and Röpke, G. (1978). *Phys. Status Solidi B* **90,** 175.
Zimmermann, R., Rösler, M., and Asnin, V. M. (1981). *Phys. Status Solidi B* **107,** 579.

4 NONLINEAR OPTICAL PROPERTIES OF SEMICONDUCTOR QUANTUM WELLS

D. S. Chemla and D. A. B. Miller

AT & T BELL LABORATORIES
HOLMDEL, NEW JERSEY

S. Schmitt-Rink

AT & T BELL LABORATORIES
MURRAY HILL, NEW JERSEY

I. INTRODUCTION

The concept of excitons has been in existence for over fifty years, and it is a key concept in the physics of the optical properties of semiconductors. Two developments over the past few decades have opened up new areas of the

OPTICAL NONLINEARITIES
AND INSTABILITIES IN
SEMICONDUCTORS

ISBN 0-12-332915-9

physics of excitons in semiconductors. The technology of lasers enables us to generate controlled densities of excitons and electron–hole pairs to see the many-body effects of these particles on each other through the changes that they induce in the optical properties. Furthermore, the short-pulse optical techniques that have recently become available allow the dynamics of these particles to be observed with temporal resolution comparable to the elementary time scales of the particles, such as the phonon-scattering times and the orbit times of the excitons themselves. The second development has been the technology of layered semiconductor growth, which is now able to grow structures at will with atomic monolayer precision. Among the many uses of this growth technology is the ability to generate structures that can confine excitons in dimensions less than the usual Bohr radius of the exciton. Although one might expect at first sight that this gross interference with the exciton would destroy it, in fact, the opposite is true. The resulting excitons are actually so robust that clear excitonic absorption resonances can be seen at room temperature, a fact that actually makes them easier to study than bulk excitons.

The combination of lasers and layered structures for studying exciton physics is barely five years old, and it is primarily this rapidly evolving field that is the subject of this chapter. We shall discuss the physics of excitons and the resulting linear optical absorption in quantum wells (QWs) before going on to the changes induced by optically generated exciton and electron–hole pair populations. Finally, we shall consider the effects of virtual populations of excitons. The emphasis will be on the physics of these processes, presented in an intuitive fashion as far as possible, rather than on complete, rigorous treatments or historical reviews of the literature. The article is self-contained as far as possible, although we have attempted to make it complementary in emphasis to the chapter by Miller et al. For a larger perspective on active research in optical properties of quantum wells, the reader is referred to the recent collection of articles (Chemla and Pinczuk, 1986).

II. EXCITON STATES IN SEMICONDUCTOR QUANTUM WELLS

In semiconductor QWs, the electrons (e) and holes (h) are free to move in the plane of the layers (x, y) and their motion normal (z) to the layers is controlled by the potential discontinuities at the interfaces (Dingle et al., 1974). It has been shown that the single-particle states are well described within the effective mass approximation (EMA). The wave functions are plane waves for the in-plane motion. The quantization normal to the layers gives "particle in a box" wave functions, which are sinusoidal in the well and exponential outside. The low-energy states form a discrete set of two-dimensional (2D) valence (v) and conduction (c) subbands with a steplike density of states. These states can be labeled by a "normal motion" quantum number $n_z = 1, 2, \ldots$. The states with energy above the band discontinuities form three-dimensional (3D) continua with a density of states presenting resonances at specific energies (Bastard, 1984). The nature of these single-particle states is discussed in more detail in chapter 13. In order to discuss the absorption spectra, we need to know how e–h pair states are constructed from these single-particle states, and this is briefly discussed in this section.

QWs have properties intermediate between 2D and 3D material; it is therefore instructive first to compare excitonic effects in pure 2D and 3D before discussing the influence of the finite well thickness. The theory of excitons (x) in 3D semiconductors is extremely well documented (Elliot, 1957). It turns out that, within the EMA, it can be easily extended to Wannier excitons with a pure 2D motion and $1/r$ e–h attraction (Shinada and Sugano, 1966).

Let us consider first the ideal case of a semiconductor of band gap E_g with two isotropic valence and conduction bands with masses $m_v = -m_h$ and $m_c = m_e$, respectively. Throughout this chapter, we shall use as units of energy and length the 3D Rydberg and Bohr radius, respectively, $R_0 = e^4\mu/2\epsilon_0^2\hbar^2$ and $a_0 = \epsilon_0\hbar^2/\mu e^2$, where $\mu^{-1} = m_e^{-1} + m_h^{-1}$ is the reduced e–h mass and ϵ_0 the dielectric constant. The solution of the 2D Schroedinger equation gives a series of bound states with energies $E_n^{2D} = E_g - R_0/(n - 1/2)^2$ as compared to the 3D result $E_n^{3D} = E_g - R_0/n^2$. The wave function of the 1s state in real space is

$$U_{1s}^{2D}(r) = \left(\frac{2}{\pi}\right)^{1/2} \frac{2}{a_0} \exp\left(-\frac{2r}{a_0}\right) \tag{1a}$$

and in momentum space

$$U_{1s}^{2D}(k) = \frac{(2\pi)^{1/2} a_0}{\left[1 + (a_0 k/2)^2\right]^{3/2}} \tag{1b}$$

Let us recall for comparison that in 3D, these functions are

$$U_{1s}^{3D}(r) = \frac{1}{(\pi)^{1/2} a_0^{3/2}} \exp\left(-\frac{r}{a_0}\right) \tag{2a}$$

and

$$U_{1s}^{3D}(k) = \frac{(8\pi)^{1/2} a_0^{3/2}}{\left[1 + (a_0 k)^2\right]^2} \tag{2b}$$

In 2D, the 1s exciton binding energy and Bohr radius are $R_{2D} = 4R_0$ and $a_0/2$, respectively, and the maximum radial charge density occurs at $a_{2D} = a_0/4$. Furthermore, the probability of finding the e and h in the same unit cell varies like $|U_n^{2D}(r = 0)|^2 = (2n - 1)^{-3}|U_{1s}^{2D}(r = 0)|^2$, whereas in 3D, $|U_n^{3D}(r = 0)|^2 = (n)^{-3}|U_{1s}^{3D}(r = 0)|^2$, which shows that the e–h correlation in excited bound states decreases more rapidly in 2D than in 3D. Much the same can be said for the scattering states (i.e. the unbound electron–hole pair states). In 2D, one finds

$$|U_k^{2D}(r = 0)|^2 = \frac{2}{[1 + \exp(-2\pi/a_0 k)]} \tag{3a}$$

where $a_0 k = [(E - E_g)/R_0]^{1/2}$ is the dimensionless e–h relative momentum. In 3D,

$$|U_k^{3D}(r = 0)|^2 = \frac{2\pi/a_0 k}{[1 - \exp(-2\pi/a_0 k)]} \tag{3b}$$

Real semiconductor QWs have a finite thickness and depth, and one faces many complications. The motion normal and in the plane of the layer are coupled by the Coulomb interaction. Because of the penetration of the wave functions into the barrier medium, the differences of the valence and conduction band effective masses inside and outside the well have to be accounted for. Finally, in III–V compounds, the hole motion in the plane is complex because the degeneracy of the light (lh) and heavy hole (hh) bands at the Γ point is lifted giving rise to two series of subbands. Away from $k = 0$, these subbands couple (Chang and Schulman, 1983) and acquire a highly nonparabolic dispersion so that a constant effective mass can no longer be defined for the holes (Broido and Sham, 1986). Several approximations must be made to make the problem tractable (Lee and Lin, 1979; Miller et al., 1981; Bastard et al., 1982; Greene et al., 1984).

First, the coupling among the various subbands and light and heavy holes is neglected. This is justified as long as the energy separation between the subbands is much larger than the exciton binding energy. Since excitons are built up from a linear combination of free e–h states distributed according to U_{1s}, the valence subband dispersion is averaged to give a "mean effective mass" for the holes. The value of this mass in the literature is quite often a matter of personal taste. Finally, variational procedures are used, starting from well-behaved wave functions, to solve the exciton Schroedinger equation. The main results are the following. For a trial wave function inseparable in $z_{e,h}$ and $(x, y)_{e,h}$ and infinitely deep wells, a smooth variation of the 1s exciton binding energy from $4R_0$ to R_0 is found for L_z/a_0 varying from 0 to ∞, in good agreement with physical intuition (Miller et al., 1981; Bastard et al., 1982). Unfortunately, this situation does not correspond to real QWs because in the limit of very thin wells, the wave function penetrates more and more in the barrier medium as a result of the finite confining potential, and the exciton becomes less and less confined. As $L_z \to 0$, the exciton tends toward the 3D exciton of this material and the binding energy toward the corresponding Rydberg (Greene et al., 1984). Thus, the enhancement of the binding energy shows an optimum $< 4R_0$ for narrow but finite QWs.

In order to gain physical insight into real QW excitons, it is desirable to have a reasonably good exciton ground state wave function in a closed form. For narrow QWs, the confinement energy is much larger than the binding energy and one can assume that the motion along z is governed by the QW potential discontinuity only. It is then possible in the variational calculation to use as a trial wave function the separable form $\phi_e(z_e)\phi_h(z_h)U_{1s}^{2D}(r)$, where $\phi_{e,h}(z_{e,h})$ are the $n_z = 1$ single-particle wave functions in the z direction and $U_{1s}^{2D}(r)$ has the same functional form as in Eq. 1a, with a variational radius parameter a_{2D} being used instead of $a_0/4$. This approach gives results comparable to more sophisticated theories (Miller et al., 1985). In the case of 100 Å GaAs/AlGaAs QWs it is found that $a_{2D} \sim 65$ Å and $R_{2D} \sim 9$ meV, as compared to $a_0 \sim 140$ Å and $R_0 \sim 4.2$ meV. If one accounts for the small penetration of the wave function into the barrier material, approximately 20 Å, the charge distribution is like a slightly oblate spheroid with nevertheless a substantial reduction in the average e–h separation (as measured by a_{2D}), which is responsible for the increased binding energy.

One can even go one step further and replace ϕ_e and ϕ_h by the single-particle wave functions of an infinite well of somewhat larger adjusted thickness. This accounts approximately for the penetration into the barrier medium and is mathematically the simplest trial wave function that can be chosen while still preserving the principle features of the actual exciton state (Miller et al., 1985).

In summary, in the pure 2D case, the exciton ground state binding energy is four times larger than in 3D and the relative motion wave function forms a flat disk with a radial charge density maximum at $a_{2D} = a_0/4$ rather than at a_0 as in 3D. The excited states are more separated from the ground state than in 3D and, correspondingly, the probability of finding the e and h in the same unit cell decreases more rapidly. In real QWs, the situation is intermediate between the 2D and 3D limits. Although the exciton conserves a rather isotropic charge distribution, the relative separation between the electron and hole in the 1s state is substantially reduced as compared to 3D. This results in a binding energy and a probability of finding the e and h in the same unit cell that is considerably larger than in 3D. In addition, in III–V materials, the light and heavy holes can give distinct exciton states.

III. LINEAR ABSORPTION IN SEMICONDUCTOR QUANTUM WELLS

In QWs, as in 3D semiconductors, the optical spectra near the band gap energy are profoundly influenced by the e–h correlation that induces excitonic resonances below the gap and enhancement above. In fact, it has been found experimentally that excitonic effects are strongly enhanced in the absorption spectra of QWs (Dingle et al., 1974; Chemla, 1983, 1985; Chemla and Miller, 1985; Miller and Kleinman, 1985). This is due to the combination of several factors discussed in the previous section. In this section we shall briefly present the main results obtained on the linear excitonic optical absorption in semiconductor QWs.

In bulk semiconductors, the optical absorption spectrum is proportional to the atomic transition probability, as modified by the probability of finding the e and h in the same unit cell. The imaginary part of the optical susceptibility χ_{3D} is formally given by

$$\operatorname{Im} \chi_{3D} = 2\pi |er_{cv}|^2 \sum_n |U_n^{3D}(r=0)|^2 \, \delta\left(\hbar\omega - E_n^{3D}\right) \tag{4a}$$

where er_{cv} is the atomic dipole matrix element, and $U_n^{3D}(r)$ the envelope wave function of the e–h relative motion, E_n^{3D} being the corresponding pair energy. Here, n runs both over bound and scattering states.

In QWs, χ is strictly a nonlocal function in real space, but for L_z much less than the wavelength of the light, a suitably averaged susceptibility can be defined (Miller et al., 1986a). For infinitely deep wells, the e and h single-particle wave functions are orthonormal, and the spatial average of χ over the QW yields the selection rule that optical transitions can only take place between valence and conduction subbands having the same quantum number n_z. The Coulomb interaction couples the various transitions, which among other things gives rise to a Fano lineshape of excitons associated with excited subbands. In narrow QWs, in which the wave functions are consequently approximately separable, one can ignore this subtlety; each transition has its own exciton series

$$\operatorname{Im}\chi_{2D} = 2\pi|er_{cv}|^2 \sum_n |U_n^{2D}(r=0)|^2 \,\delta\big(\hbar\omega - E_n^{2D}\big)/L_z \tag{4b}$$

where $U_n^{2D}(r)$ is the associated in-plane relative motion wave function.

In reality, the e and h tunnel by different amounts into the barriers, which leads to small but finite transition probabilities for the "forbidden transitions," i.e. those transitions for which the wave functions would otherwise be orthogonal. If an asymmetry is introduced, for example by applying an electric field, all of these transitions become partly allowed and are clearly observed (Miller et al., 1986a, b). (See also chapter 13 for a discussion).

In III–V compounds, because of the heavy and light hole mixing away from $k = 0$, one can even observe transitions that are normally parity forbidden at the microscope level (Chang and Schulman, 1983). The angular momentum selection rules for the interband matrix element show that for transitions from $k = 0$ heavy and light hole states to $k = 0$ conduction band states, the $|r_{cv}|^2$ are in a ratio 3/4 and 1/4 for optical fields polarized parallel to the layers, and 0 and 1 for orthogonal polarization. In doped or laser-excited QWs, the Coulomb interaction couples states with different momentum k, so that this rule is lifted (Sooryakumar et al., 1985; Ruckenstein et al., 1986).

At low temperatures, exciton resonances are clearly observed at the onset of each interband transition, the light and heavy hole doublet being well resolved at least in the lowest transitions. At the onset of continuum absorption, above the band gap discontinuity, delocalized exciton resonances were resolved by resonant Raman scattering and a weak structure was also seen in absorption (Zucker et al., 1984b).

The exciton lines in QWs are broad compared to other exciton lines at low temperature. The majority of the broadening has been identified as due to the unavoidable fluctations in the QW thickness (Weisbuch et al., 1981) and is therefore an inhomogeneous broadening. The best samples have interfaces with an islandlike structure with steps one monolayer high. The lateral correlation length depends on the details of the growth conditions; it is usually of the order of several hundred Å. This results in potential fluctuations seen by the excitons that produce very interesting localization effects both in linear and in nonlinear optics (Hegarty and Sturge, 1985;

Takagahara and Hanamura, 1986). The inhomogeneous line width increases as the thickness of the QW decreases; a typical value of the line width at half maximum is ~ 2 meV for 100 Å GaAs–AlGaAs QWs. Information on the homogeneous line width that underlies the dominant inhomogeneous broadening has been obtained by resonant Rayleigh (Hegarty and Sturge, 1985) and Raman (Zucker et al., 1987) scattering; it can be as narrow as 0.1 meV in GaAs–AlGaAs QWs. The line broadening can be accounted for in Eq. 4b by replacing the Dirac functions by phenomenological profiles with a finite width.

Below-gap absorption has been investigated in GaAs–AlGaAs QWs using a sensitive nonlinear optical technique that will be discussed in Section IV.B. It was found that just below the $n_z = 1$ hh-exciton peak, the absorption drops exponentially in agreement with Urbach's rule (Von Lehmen et al., 1986a). The steepness is comparable to that obtained for bulk GaAs. At low temperatures, LO phonon-assisted transitions are observed (Von Lehmen et al., 1987). About 30 meV below the $n_z = 1$ hh-exciton, the residual absorption can be as low as 10 cm^{-1} at room temperature and 0.1 cm^{-1} at 77 K.

A most remarkable property of III–V QW structures is the persistence in the absorption spectrum of well-resolved exciton resonances at room temperature and even higher (Miller et al., 1982a, b; Weiner et al., 1985a). An example of the GaAs–AlGaAs QW room temperature absorption spectrum is shown in Fig. 1. The spectrum is remarkably clear. The three $n_z = 1$, 2 and 3 plateaus are well resolved with exciton resonances at the onset of each. Similar spectra have been observed in wave guides containing a single QW (Weiner et al., 1985b). Using these structures, selection rules for light polarized parallel and perpendicular to the layers have been investigated. It was found that, within the experimental accuracy, the measured ratios of the transition strengths are in good agreement with those predicted by angular momentum theory.

The enhanced exciton oscillator strength that accompanies the larger binding energy plays a very important role in these results. However, alone it is not sufficient to explain the observations; other semiconductors have excitons more tightly bound than III–V QWs and yet their resonances disappear long before room temperature. In 3D, the exciton binding energy, through its dependence on the dielectric constant and the effective masses, scales roughly as the band gap. The dominant exciton–phonon interaction in polar semiconductors is that with LO phonons (Liebler et al., 1985) and its strength, through the dependence on the ionicity, also scales approxi-

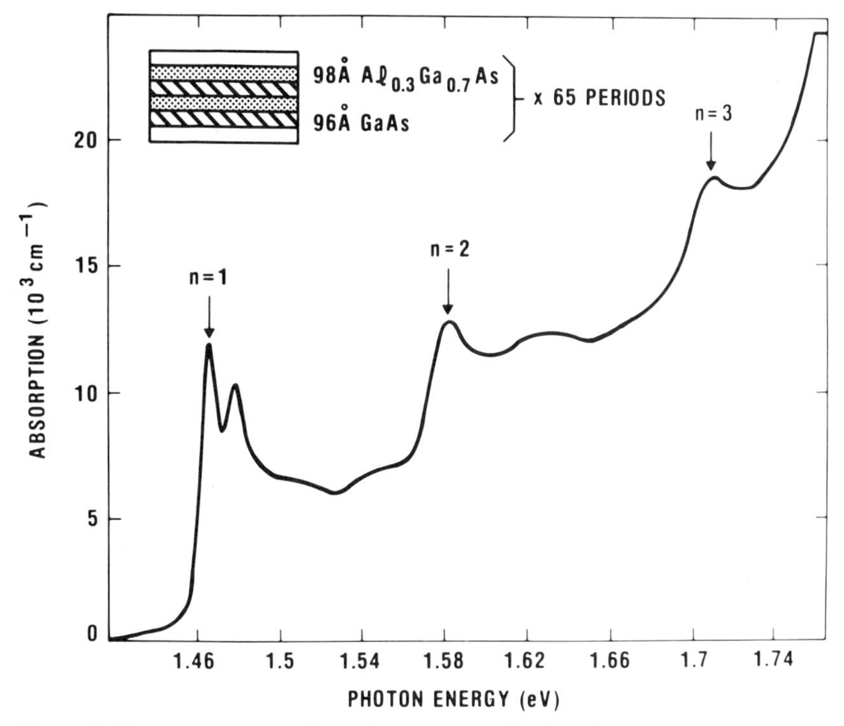

Figure 1. Room temperature absorption spectrum showing the $n_z = 1$, 2 and 3 exciton resonances and flat continua. At the fundamental edge, the heavy and light hole excitons are resolved. Between the $n_z = 1$ and $n_z = 2$ resonances, a weak structure due to a "forbidden transition" can be seen.

mately with the band gap. Thus, in 3D materials, increasing the exciton binding energy also increases the exciton sensitivity to thermal broadening and exciton resonances are only seen at low temperatures almost regardless of band gap energy. In QWs, the increased binding energy results from the artificial reduction of the e–h separation by the confinement in the QW, without any significant change of the coupling to LO phonons, as shown by resonant Raman-scattering studies (Zucker et al., 1983, 1984a). Thus, for QW excitons, the dominant mechanism for thermal broadening is essentially not modified. This was confirmed by studies of the temperature broadening of the exciton peak in GaAs–AlGaAs (Miller et al., 1982b) and GaInAs–AlInAs (Weiner et al., 1985a) QWs. It was found that the line width varies as the sum of a constant inhomogeneous term Γ_{inh} and a term proportional to the number of LO phonons

$$\Gamma(T) = \Gamma_{inh} + \frac{\Gamma_{\mathrm{LO}}}{\left[\exp\left(\frac{\hbar\omega_{\mathrm{LO}}}{k_B T}\right) - 1\right]} \tag{5}$$

The parameter Γ_{LO}, which describes the strength of the interaction with LO phonons in GaAs–AlGaAs QWs ($\Gamma_{\mathrm{LO}} \sim 5$ meV), is smaller than but comparable to that in the bulk ($\Gamma_{\mathrm{LO}} \sim 8$ meV). This shows that the efficiency of collisions with LO phonons is hardly changed. Absorption profiles such as that of Fig. 1 are very well fitted around the lowest transitions by two Gaussian lines at the heavy and light hole resonances and a phenomenologically broadened continuum (Chemla et al., 1984).

At high temperatures, excitons are unstable against collisions with thermal phonons. In fact, in GaAs–AlGaAs QWs the LO phonon energy (~ 36 meV) is much larger than the exciton binding energy (~ 9 meV). Thus, a single exciton–LO-phonon collision can ionize the exciton and release a free e–h pair with a substantial excess energy. If the temperature broadening of the exciton resonance is interpreted as a lifetime reduction due to thermal LO-phonon scattering, it is found that at room temperature, excitons live only ~ 0.4 ps in GaAs–AlGaAs and ~ 0.3 ps in GaInAs–AlInAs QWs. Simple thermodynamic arguments show that the excitons have a very small probability of recovering. Consequently, at room temperature, a photon absorption process can be pictured as

$$\hbar\omega \rightarrow x \rightarrow e - h \tag{6}$$

As we shall see later, this sequence includes interesting dynamics in the nonlinear optical response of QWs.

IV. EXPERIMENTAL AND THEORETICAL INVESTIGATIONS OF NONLINEAR OPTICAL EFFECTS

The basic mechanisms responsible for the nonlinear response of semiconductors are different depending on whether virtual or real excited-state populations are generated. Excitations below and above the absorption edge have to be clearly distinguished. Before presenting particular results, it is useful to discuss these mechanisms in very broad terms.

In the case of virtual transitions induced by excitation well below the absorption edge, the electric field of the coherent laser beam induces a coherent polarization that persists only as long as the field is applied. The nonlinear polarization can couple various optical fields, which exchange photons via the material, but no energy is deposited in the latter. During the application of the field, the state of the sample is a coherent superposition of excited states. The coherently driven components have exactly the same properties as if they were actually populated although they do not participate in any relaxation process; if they did relax by any collisions, their quantum-mechanical phase would be disturbed, they would no longer be coherent with the other components, and as far as the driving field is concerned, the associated photons would have been absorbed in real transitions. By a simple application of the uncertainty principle, the amount of time the photons spend in the excited state is $\sim \hbar/\Delta E$, where ΔE is the difference between the photon energy and the excited state energy. As long as this time is short compared to the scattering time τ_s of the excited state, the transitions will be predominantly virtual. Virtual transitions are therefore only favored under off-resonance excitation (i.e. large ΔE). The field-induced optical nonlinearities are usually small because of the large differences ΔE between the photon energies and the absorption edge; however, they are extremely fast ($\sim \hbar/\Delta E$).

Excitation above the absorption edge generates predominantly real excited-state populations with no particular quantum-mechanical coherence; this is because there are many states within ΔE of the photon energy that are consequently strongly absorbing and also strongly scattered. Because of the strong absorption, however, such above-gap excitation produces large excited-state populations for a given excitation intensity, and hence large changes in the optical properties of the material although these changes are incoherent with each other and with the driving field. These populations have a finite lifetime and participate in a number of relaxation processes. The changes they induce in the optical spectra persist as long as the populations themselves. For the specific case of semiconductors, the excited e–h states can form bound (excitons, trions, biexcitons) or unbound (e–h plasma) states, which produce different effects (Haug and Schmitt-Rink, 1984). Furthermore, ultrafast photogeneration can produce nonthermal distributions of these states. Two relaxation times should be distinguished. The first is the time it takes the excited e and h to reach a thermodynamic equilibrium among themselves (usually a fraction of a ps), and the second one is the time it takes them to equilibrate with the lattice (in the ps range). Optical transients can occur if excited species transform into one another during relaxation. For example, excitons that are generated by resonant excitation within the absorption peak are only stable at low temperatures, where they can last as long as several nanoseconds. In the case of QWs at room temperature, they are unstable and very quickly transform into free e–h pairs. Because the two species have different effects on the optical spectra, it is necessary to distinguish among them experimentally and theoretically.

In this section, we shall review the experimental and theoretical investigations of excitonic nonlinearities in QWs. We shall then distinguish between plasma, exciton and field-induced effects. Experimentally, field effects are produced by excitation well below the absorption edge and plasma effects by excitation above it. At room temperature, exciton effects are more difficult to select; not only resonance excitation must be used but they must be observed in times short compared to the exciton ionization time. This requires femtosecond spectroscopic techniques.

The experimental method most widely used is the "pump and probe" technique, in which a strong pump beam is applied to the sample and a weak test beam is used to probe the excited sample. Measuring the transmitted test beam provides information on the absorption and gain of the sample. It is also possible to measure the photons diffracted by interference between the pump and test beam in directions different from that of the incident beams. The diffraction efficiency provides information both on the change of refractive index and absorption. These techniques have been extensively discussed in the literature (Chemla et al., 1984).

A. Nonlinear Optical Effects Induced by Thermalized Plasmas

We consider first the experiments in which an optically created real and thermalized e–h plasma is responsible for the changes in optical properties. The plasma can be generated either directly above the gap (Chemla et al., 1984) or indirectly following exciton creation by resonant absorption (Miller et al., 1982b) or LO phonon-assisted absorption (Von Lehmen et al., 1986a, 1987). For thermalization to be completed, observation must take place a few ps after excitation.

It is found experimentally that the changes in the absorption spectrum do not depend on the excitation wavelength or on its duration. They depend only on the density N of excited pairs. A typical example of the modification of the absorption spectrum as a function of the pair density is presented in Fig. 2 for GaAs–AlGaAs QWs and N between ~ 0 and $\sim 2 \times 10^{12}$ cm^{-2}. In the particular case of this set of spectra, the sample was excited with a 100 fs pump pulse about 60 meV above the hh-exciton resonance and was probed several ps later by a broadband test continuum.

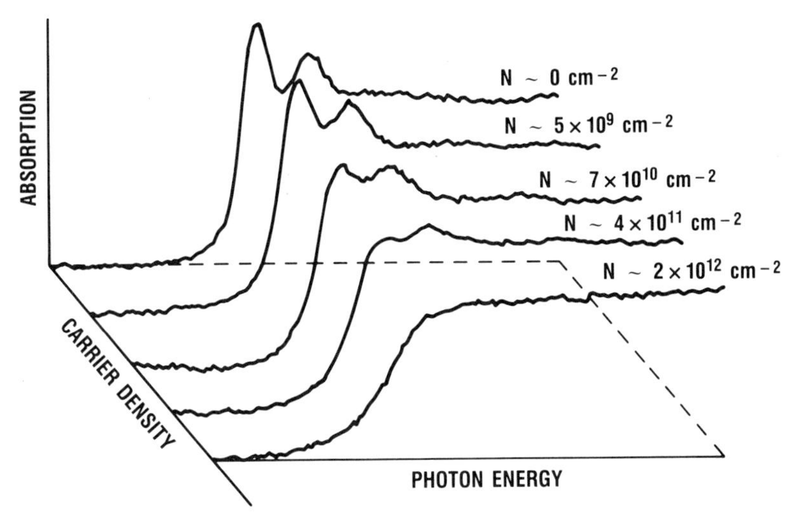

Figure 2. Absorption spectra for free e–h pair densities N between ~ 0 and $\sim 2 \times 10^{12}$ cm^{-2}, measured after thermalization of the e–h plasma.

A very smooth evolution is seen; the two exciton resonances weaken and are progressively replaced by a steplike edge. The independence of the nonlinear absorption on the excitation wavelength and thus on the way the e–h plasma is created is clearly demonstrated in Fig. 3. Here, two differential transmission spectra (absorption without pump minus absorption with pump) are compared. They were measured about 2 ps after selective exciton generation by resonant pumping (full line) and direct e–h plasma generation by nonresonant pumping (dotted line). The pump intensities were adjusted to produce the same density of pairs. Within the experimental accuracy the two curves are not distinguishable.

The changes of the optical spectra are interpreted as follows (see also Section IV.B) (Löwenau et al., 1982; Haug and Schmitt-Rink, 1984, 1985; Schmitt-Rink et al., 1984, 1985, 1986; Schmitt-Rink and Ell, 1985; Schmitt-Rink, 1986). As the plasma density increases, the energy of free e–h pairs becomes renormalized (band gap renormalization), whereas that of the bound states hardly changes because of their charge neutrality. The binding energy of the bound states measured from the renormalized continuum decreases. The resonances lose oscillator strength, both because of the occupation of states out of which excitons are constructed (phase space filling) and because of the loss of e–h correlation (i.e. the excitons are becoming larger). The width of the peaks increases because of collisional

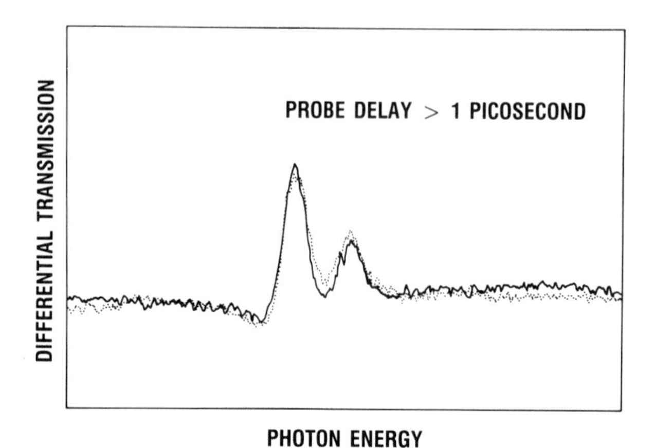

Figure 3. Differential transmission spectra for selective generation of hh-excitons (full line) and free e–h pairs (dotted line), after exciton ionization and thermalization of the e–h plasma.

broadening. At low densities, when the continuum is still far from the resonances, only the loss of oscillator strength and the broadening are apparent. However, if the density continues to increase, the resonances merge in the continuum, and eventually this continuum edge is the only remaining spectral feature. It is remarkable that in the case of high-quality GaAs–AlGaAs QWs, the profile of the final edge is exactly that of the broadened 2D continuum used to fit the unperturbed spectrum. This is shown in Fig. 4. The dots are obtained from the first ($N \sim 0$) and last ($N \sim 2 \times 10^{12}$ cm^{-2}) digitized spectra of Fig. 2, and the full lines correspond to the semi-empirical fit just described. In the upper part of the figure, the broadened continuum is also shown. A rigid red shift of ~ 17 meV of this continuum suffices to give an almost perfect coincidence with the high-density spectrum as shown in the lower part of the figure.

An interpretation similar to that given in the previous paragraph explains the saturation measurements performed under cw resonant pumping shown in Fig. 5 (Miller et al., 1982b). The experimental points (open circles) are very well fitted by the sum of two Lorentzian saturation expressions (full line). One corresponds to a species with very low saturation intensity ($I_s \sim 0.58$ kW cm^{-2}) and accounts for the hh-exciton bleaching. The other species is much harder to saturate ($I_s \sim 4.4$ kW cm^{-2}); it corresponds to the bleaching of the band-to-band transitions when the renormalized edge has shifted below the former position of the hh-exciton.

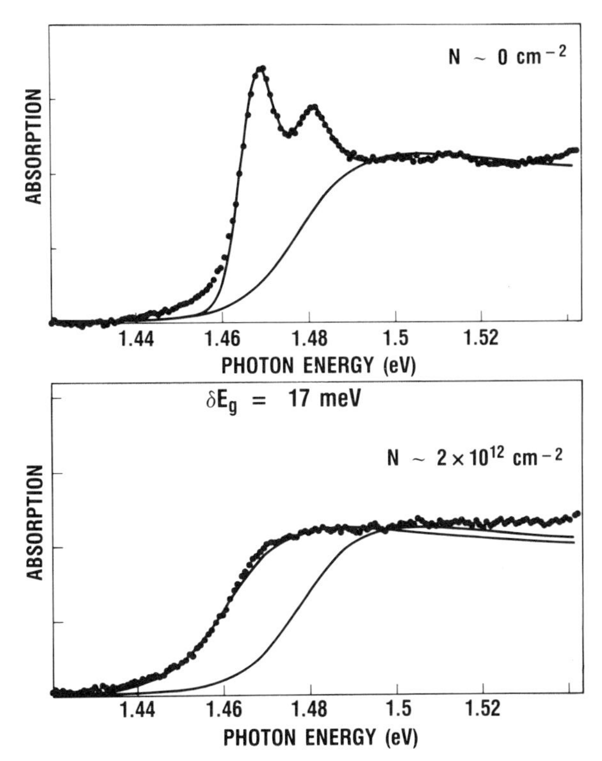

Figure 4. Semi-empirical fits of absorption before (upper part) and after (lower part) excitation. The first fit comprises two Gaussian resonances and a broadened continuum (also shown). The second uses only the 2D continuum rigidly shifted by $\delta E_g \sim 17$ meV.

Additional information on the refractive index effects were obtained from degenerate four-wave mixing (DFWM) experiments using tunable ps pump and probe lasers (Miller et al., 1983a; Chemla et al., 1984). In the low-density region, the bleaching of the exciton can be characterized by a nonlinear absorption cross section σ and a nonlinear refractive index η, which measure, respectively, the change of the absorption coefficient and refractive index induced by one e–h pair

$$\alpha = \alpha_0 - \sigma \overline{N} \tag{7a}$$

$$n = n_0 - \eta \overline{N} \tag{7b}$$

In these equations, in order to be consistent with the usual definitions, $\overline{N}$ is the pair density per unit volume. σ and η are related by the Kramers–Kronig

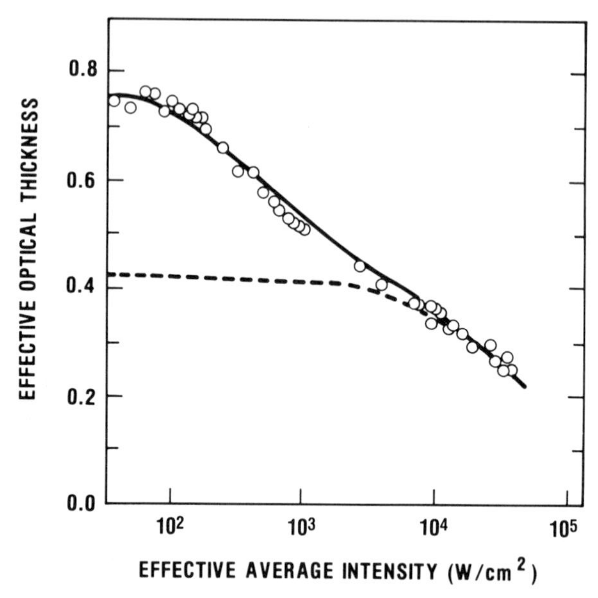

Figure 5. Saturation of the hh-exciton resonance measured under cw excitation (open circles). The full line corresponds to two Lorentzian saturable absorbers. One is very easy to saturate and corresponds to the hh-exciton bleaching. The other is much harder to saturate and corresponds to the bleaching of the renormalized continuum.

relations; together they determine the DFWM diffraction efficiency. Fig. 7 shows the spectra of σ and η, as obtained from a semi-empirical fit of the measured change in transmission shown in Fig. 6. The nonlinearities are extremely large and peak at $\sigma \sim 7 \times 10^{-14}$ cm^2 and $\eta \sim 3.7 \times 10^{-19}$ cm^3. Fig. 6c shows a comparison of the measured diffraction efficiency with the theoretical prediction based on the spectra of σ and η. The agreement is excellent.

The impressive magnitude of the room temperature nonlinear effects in GaAs–AlGaAs QWs was demonstrated by degenerate four-wave mixing using as the sole light source a cw laser diode. A diffraction efficiency of $\sim 0.5 \times 10^{-4}$ was observed in a 1.25 μm sample, using only a ~ 17 W cm^{-2} pump intensity (Miller et al., 1983b). Optical (bistable) devices using GaAs–AlGaAs and InGaAsP–InP QWs as the active medium in nonlinear Fabry–Perot etalons have been operated and have raised much interest for applications in all-optical logic (Peyghambarian and Gibbs, 1985; Tai et al., 1987a). It is likely however that these devices considerably exceed the densities at which the excitons cease to exist and that they rely on interband

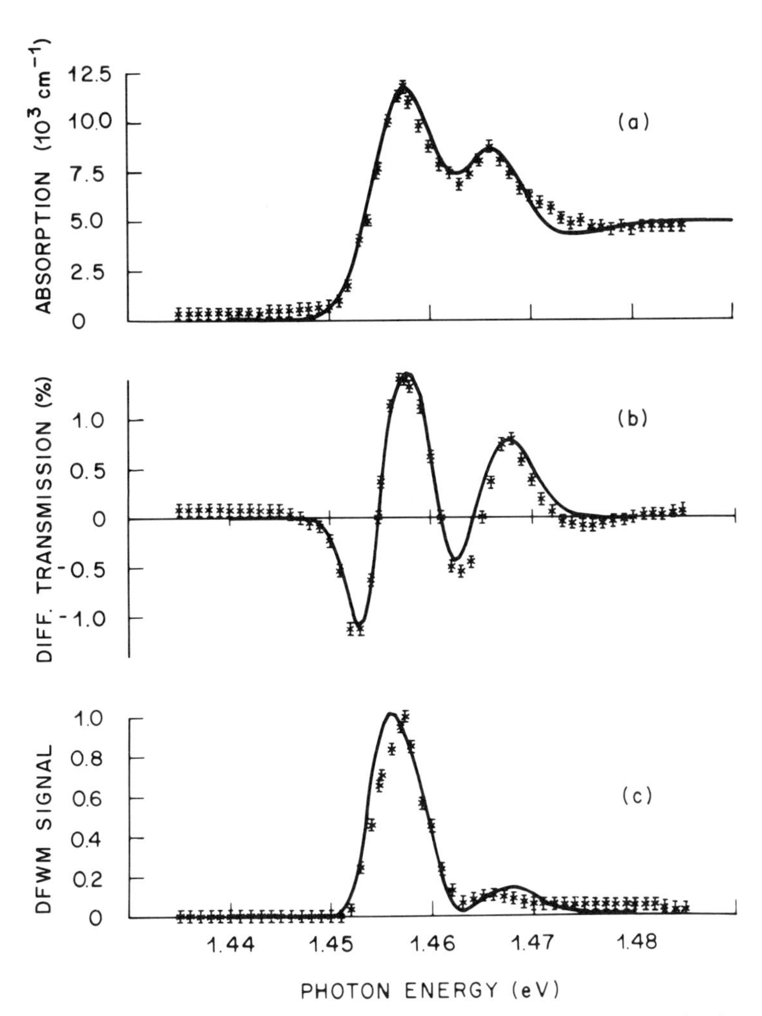

Figure 6. Semi-empirical fits of absorption (a) and differential transmission (b). The theoretical DFWM spectrum, shown by the full line in (c), is obtained from the nonlinear cross sections σ and η shown in Fig. 7.

saturation to complete the switching. GaAs–AlGaAs QWs have been used as saturable absorbers to passively mode-lock GaAs diode lasers. Stable operation producing pulses as short as 1.6 ps has been demonstrated (Smith et al., 1985).

As mentioned in Section III, recently the extreme sensitivity of excitons to the presence of an e–h plasma has been exploited to investigate the very low residual absorption at photon energies well below the band gap in QW

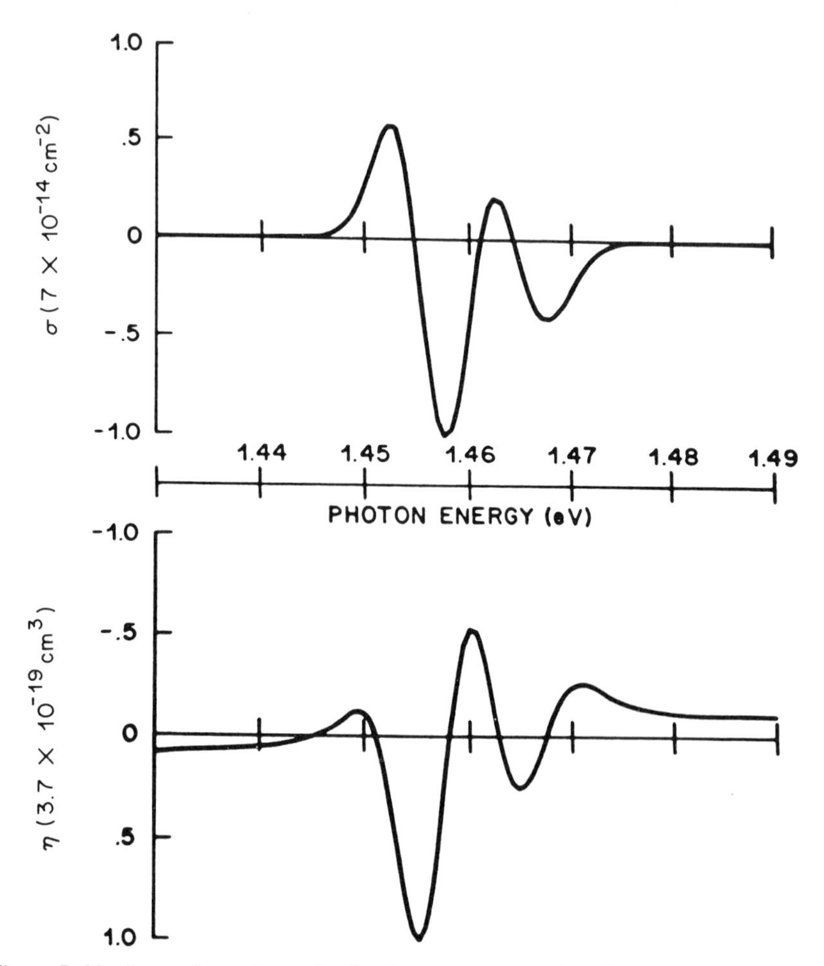

Figure 7. Nonlinear absorption and refraction spectra. σ and η describe respectively the change of absorption and refractive index induced by one e–h pair and are obtained from the fits shown in Fig. 6a and b.

structures. This includes the investigation of the so-called Urbach tail and of phonon side bands, processes in which a photon and some LO phonons are simultaneously absorbed (Liebler et al., 1985). The cross sections for these processes are very small. Until recently, they had been observed only in bulk materials (Sturge, 1962). It is extremely difficult to measure by classical methods such a small absorption in a micron thick QW and new nonlinear optical methods have to be devised to perform these investigations. Since the final result of absorption is the generation of real e–h pairs,

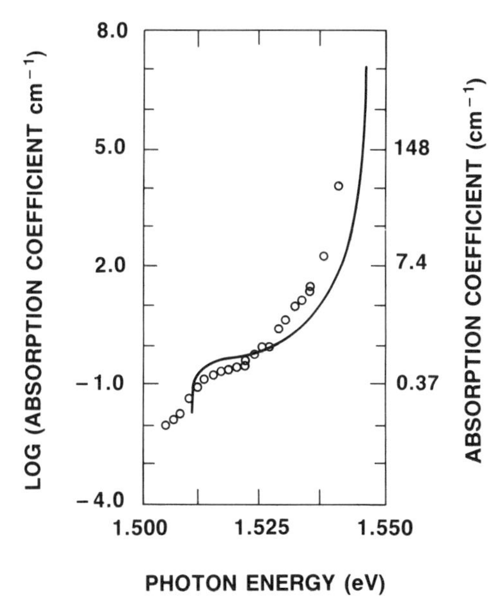

Figure 8. Below-gap LO phonon-assisted absorption. The open circles are measurements using a new nonlinear technique discussed in the text. The full line shows the one-LO–phonon contribution calculated from perturbation theory.

a very weak test beam can be used to probe the transmission of a sample at the peak of the hh-exciton resonance as a very intense pump beam excites the QW structure well below the gap. As the pump frequency is tuned, the test beam measures the directly or indirectly produced e–h plasma density and, hence, in the small signal regime, it maps out the pump absorption (Von Lehmen et al., 1986a, 1987). At high temperatures, when the phonon population is large, the below-gap absorption is found to follow the usual exponential Urbach's rule. However, at low temperatures, a kink appears approximately one LO–phonon energy below the hh-exciton resonance, as clearly shown by the experimental data (open circles) in Fig. 8 for a GaAs–AlGaAs QW at 77 K. This spectral feature corresponds to the one-LO–phonon edge; its profile and position are in remarkable agreement with an extension of the 3D second-order perturbation theory (Segall, 1966) to the 2D case, retaining as the final state the 1s state only. The theoretical line shape (full line) calculated without adjustable parameters agrees very well with the data. The agreement can be improved further if the zero-LO–phonon process on the high-energy side and many-phonon processes on the low-energy side are accounted for. Below the one-LO

-phonon edge, the absorption, although nonzero, is extremely small ($\sim 0.1\,\mathrm{cm}^{-1}$).

B. Nonlinear Optical Effects Induced by Excitons and Nonthermalized Plasmas

Exciton populations can be generated selectively by resonant excitation. At low temperatures, the lowest-energy excitons are stable and only disappear by recombination, a rather long process with characteristic times ranging from nanoseconds or hundreds of picoseconds for dipole-allowed recombination to seconds for the forbidden ones. At high temperatures, the excitons are unstable against collisions with energetic phonons and on the average live only for a time of the order of the mean collision time. As we discussed in Section III, in QWs this corresponds to a fraction of a ps; thus, ultrafast fs spectroscopic techniques are necessary to investigate exciton-induced nonlinear optical effects (Peyghambarian et al., 1984; Knox et al., 1985). In these experiments, a very weak broad-band fs continuum is used to probe over a large spectral range the transmission of a sample at tunable delay from a narrow-band fs pump.

In bulk materials, it was found that excitons produce much weaker nonlinear optical effects than plasmas (Fehrenbach et al., 1982). These observations have been interpreted as follows. In 3D, the dominant mechanism governing the bleaching of exciton resonances is the screening of the e–h interaction (i.e. the exciton state can disappear due to a Mott transition before phase space filling becomes strong). The screening by a plasma is extremely effective compared to screening by excitons because of the absence of a gap in the plasma's excitation spectrum. In the static long-wavelength limit, the presence of a plasma transforms the Coulomb potential into a Yukawa potential and the condition for the disappearance of bound states is the famous Mott criterion $\kappa_{3D} a_0 \sim 1$, where κ_{3D} is the inverse 3D screening length. Actually, the plasma cannot respond instantaneously to the sudden appearance of an e–h pair during the absorption process (charge density fluctuations can only develop in times of the order of the inverse plasma frequency) so that the screening is somewhat weaker. Nevertheless, intraband transitions in a plasma are not energy consuming, whereas excitons can only screen by transitions from their ground state to some excited state. Because of the gap in the excitation spectrum, this

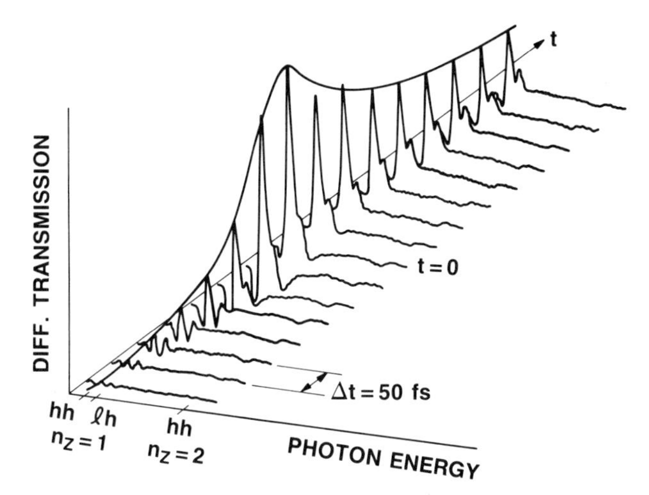

Figure 9. Differential transmission spectra measured with a broad-band 50-fs probe pulse in 50-fs intervals before and after resonant hh-exciton excitation with a 100-fs pump pulse. Initially, the absorption is reduced efficiently as excitons are created in the sample. It then recovers in ~ 300 fs as the excitons are ionized by collisions with thermal LO phonons. The absorption then settles at the same value as it would under cw excitation.

requires some energy so that screening by excitons is much weaker than plasma screening.

According to this picture, resonant exciton creation at room temperature should produce a weak bleaching of the exciton resonances that would increase substantially as the inefficient excitons transform into an efficient plasma. Experiments designed to observe this effect in QWs lead to different and surprising results that we discuss now (Knox et al., 1985). Fig. 9 shows differential transmission spectra measured with a broad-band 50 fs probe pulse in 50 fs intervals before and after resonant hh-excitation by a 100 fs pump pulse. Each spectrum corresponds to the integration of 2×10^6 shots and the resulting signal-to-noise ratio is $\sim 10^4$. The pump intensity is moderate and generates approximately $N \sim 2 \times 10^{10}$ cm^{-2} excitons. Instead of the expected delayed bleaching, one sees during the first half ps a very strong reduction of the absorption at the hh-exciton resonance, which then recovers in about one ps. The spectrum settles at the same level as in the case of thermalized plasma bleaching discussed in Section IV.A. The evolution of the hh-exciton bleaching is well described by the generation of excitons (the total number of excitons generated following the integral of the pump pulse), with subsequent exciton transformation into free e–h pairs in a 300 fs ionization time. The time constant of the process is in remarkable agreement with that evaluated from the thermal broadening (Section III). However, the surprising result is that the transient population of excitons produces a bleaching about twice as efficient as that of the thermalized plasma. The comparison of the differential spectra lineshape under resonant and above-gap nonresonant pumping (discussed shortly) also indicates that the bleaching mechanism is quite different from screening (Chemla and Miller, 1985). These results force us to reconsider the extension of the picture given above to QWs and room temperature.

The 2D nature of the effects was checked by investigation of single QW waveguides. Indeed, in multiple QW structures, it is possible that carriers generated in different layers interact via long-range Coulomb forces. Measurements of the exciton saturation using ps excitation show that, up to densities at which the exciton peak bleaches significantly, single-QW and multiple-QW structures behave similarly (Weiner et al., 1985b). This suggests that long-range Coulomb forces are effective within the same layer only or that they are not very important with respect to the exciton bleaching, in both single and multiple QWs.

In order to understand these results, one has to reconsider the basic mechanisms that can contribute to the bleaching of the exciton resonance. (We shall give a somewhat simplistic discussion here. See also Schmitt-Rink et al., 1985.) A somewhat deeper discussion of what is going on in 2D has been given elsewhere by one of us (Schmitt-Rink, 1986). In addition to the aforementioned direct screening (S), the absorption strength of exciton resonances can be diminished by the effects of the exclusion principle, phase space filling (PSF) (often referred to as "blocking") and fermion exchange (E); states in phase space that are already occupied are no longer accessible in optical transitions or available for exciton formation. Put in simple terms, one cannot optically create two e–h pairs (bound or free) on top of each other if they require the same $\underline{k}$-states. These effects involve only fermions of the same spin and are, of course, most efficient at short distances, whereas S is a long-range Coulomb effect. In 1D and 2D, particles see "more" of their near neighbors than in 3D, which suggests already at this point that "overlap" and "statistical" effects are more pronounced.

The effects of Pauli exclusion in an e–h plasma are well known. (For example, in the case of band-to-band transitions, PSF leads to the Burstein–Moss shift.) Their description requires the knowledge of the fermion distribution functions $f_e(k)$ and $f_h(k)$. In the case of an e–h plasma and depending on the temperature, $f_{e,h}(k)$ are given by Fermi or

Boltzmann distributions. To extend this principle to an exciton gas, it is only necessary to note that the resonant creation of one 1s exciton (at the bottom of its band) corresponds to an occupation probability $|U_{1s}(k)|^2$ in fermion space, which is equally shared between spin up and down particles, so that

$$f_e(k) = f_h(k) = N|U_{1s}(k)|^2/2 \quad (8)$$

where N in this case is the exciton density. This very intuitive result is, of course, theoretically substantiated (Comte and Nozieres, 1982; Haug and Schmitt-Rink, 1984; Schäfer and Treusch, 1986).

As for the effects of screening in 2D, various arguments can be given that they are much weaker than in 3D. Plasma S is determined by the number of carriers that can undergo intraband transitions and thus is proportional to the density of states (DOS). In 2D, the DOS is constant and hence the screening is weaker than in 3D. In particular, in the degenerate limit, the 2D Thomas–Fermi screening length reduces to a constant (Ando et al., 1982). This leads an extension of the 3D Mott criterion to 2D ad absurdum and shows that it is also rather meaningless to extend the random phase approximation (RPA) to 2D. All results will depend on the behavior of the screening at short distances, the proper description of which is outside the range of mean field theories. Much the same can be said for the nondegenerate limit, in which the breakdown of the RPA is signaled by the logarithmic divergence of the Debye shift (Totsuji, 1975; Haug and Schmitt-Rink, 1985; Schmitt-Rink, 1986). Obviously, what has to be accounted for is the exchange-correlation hole surrounding each particle (Schmitt-Rink, 1986). Particles of identical spin or charge try to avoid each other, which hinders the piling up of a screening charge. This reduces the screening beyond its mean field value. One further phenomenon worth mentioning is the important fact that in 1D and 2D, an arbitrarily small attractive potential always supports at least one bound state (similarly, all states are localized in 1D and 2D disordered systems). Plasma screening is strongest in the static long-wavelength limit, but even in that limit, it cannot ionize an exciton, in contrast to 3D systems.

The consequences of these characteristic properties of 2D systems for excitonic nonlinear optical effects were not considered until recently (Schmitt-Rink et al., 1985; Schmitt-Rink, 1986). If the effects of S are weak in QWs, then the bleaching of the exciton resonances by PSF and E will be related to the overlap of the electron and hole distributions in a plasma or in a resonantly excited exciton gas with the portion of the Brillouin zone that participates in the exciton wave function. Thus, a cold plasma, for

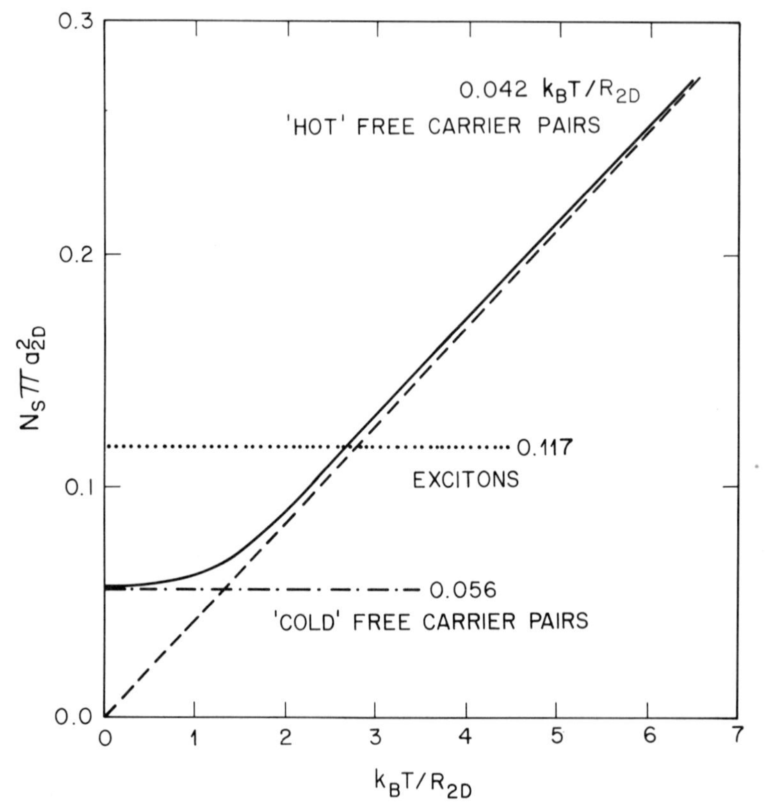

Figure 10. Comparison of the effects of Pauli exclusion (phase space filling and fermion exchange) induced by an exciton gas (dotted line), a "hot" e–h plasma (dashed line), and a "cold" e–h plasma (dashed-dotted line). The full line is a sketch of the transition from one limit to the other and is not calculated.

which $f_{e,h}(k)$ are Fermi distributions located at the bottom of the bands, should bleach more efficiently than the "zero temperature" exciton gas, for which $f_{e,h}(k)$ are given by Eq. 8. In turn, the exciton gas will be more efficient than a hot plasma as soon as the thermal spread of the Boltzmann distributions in k-space exceeds the inverse exciton Bohr radius. The correct theoretical description of PSF and E effects indeed confirms this intuitive analysis as shown in Fig. 10, where the respective 2D saturation densities N_s are plotted. This accounts then for the experimental observations, which are nothing but a comparison on a fs time scale of the effects of a resonantly excited exciton gas at short times to those of a hot (room temperature) e–h plasma at long times, the former effect being larger than the latter.

The theoretical results for the effects of PSF and E on 2D excitons shown in Fig. 10 have been obtained from an approximate calculation of the 1s exciton orbital wave function U_{1s}, as modified by the presence of other e–h pairs (Schmitt-Rink et al., 1985). The difference due to Pauli exclusion between the perturbed and unperturbed e–h relative motion Hamiltonians in momentum space is

$$\langle \underline{k}|\delta H|\underline{k}'\rangle_E = -\delta_{\underline{k},\underline{k}'} \sum_{\underline{k}''} V(\underline{k}-\underline{k}'')[f_e(k'') + f_h(k'')] + V(\underline{k}-\underline{k}')[f_e(k) + f_h(k)] \quad (9)$$

where $V(q)$ is the bare Coulomb interaction. The first term in Eq. 9 is a self-energy correction that comprises the electron and hole exchange self-energies, i.e. the energies that parallel spin particles gain by avoiding each other. The second term is the corresponding vertex correction, which describes the weakening of the attractive e–h interaction due to the exclusion principle. Screening would give rise to further corrections, except that for 2D systems we do not quite know how to treat it rigorously. In any case, as just discussed, we expect Eq. 9 to contain the major physics and the corrections due to screening to be relatively unimportant for the present discussion.

According to Eq. 4, the 1s exciton oscillator strength f_{1s} is proportional to the probability $|U_{1s}(r=0)|^2$ to find the e and h in the same unit cell. Through the first-order modification of U_{1s}, the perturbation Eq. 9 gives rise to a relative change in oscillator strength

$$\left.\frac{\delta f_{1s}}{f_{1s}}\right|_E = \sum_{n\neq 1s}\left[\frac{\langle n|\delta H|1s\rangle_E}{E_{1s}-E_n}\frac{U_n(r=0)}{U_{1s}(r=0)} + \frac{\langle 1s|\delta H|n\rangle_E}{E_{1s}-E_n}\frac{U_n^*(r=0)}{U_{1s}^*(r=0)}\right] \quad (10)$$

Consistent with Eq. 10, the change due to the blocking of the final 1s state is

$$\left.\frac{\delta f_{1s}}{f_{1s}}\right|_{PSF} = -\sum_{\underline{k}}[f_e(k) + f_h(k)]\frac{U_{1s}(k)}{U_{1s}(r=0)} \quad (11)$$

Both changes are, of course, fundamentally related; Eq. 11 is nothing but the correction to f_{1s} due to the proper normalization of U_{1s} in the presence of other pairs. Remembering that $U_{1s}(r=0) = \Sigma_{\underline{k}} U_{1s}(k)$, Eq. 11 can be simply interpreted as the average occupation of the components of the wave function in $\underline{k}$-space.

In leading order in the e–h pair density N, the relative change of the optical susceptibility χ in the spectral vicinity of the 1s exciton resonance can be written as

$$\begin{aligned}\frac{\delta\chi}{\chi} = -\frac{N}{N_s} &= -N\left[\frac{1}{N_s^E} + \frac{1}{N_s^{PSF}}\right] \\ &= \lim_{N\to 0}\left[\left.\frac{\delta f_{1s}}{f_{1s}}\right|_E + \left.\frac{\delta f_{1s}}{f_{1s}}\right|_{PSF}\right]\end{aligned} \tag{12}$$

where N_s is the overall saturation density and N_s^E and N_s^{PSF} are the saturation densities associated with exchange and phase-space filling respectively. This accounts automatically for the saturation behavior observed under cw pumping (Section IV.A). To evaluate Eq. 12, we need to perform the summation over all excited exciton states in Eq. 10. When the summation is carried out over the complete set of uncorrelated pair (plane wave) states, the following analytical results are obtained:

$$N_s\pi a_{2D}^2 \sim 0.117 \tag{13}$$

for the selective generation of 2D excitons and

$$N_s\pi a_{2D}^2 \sim 0.056, \qquad k_BT \ll R_{2D} \tag{14a}$$

$$\sim 0.042\frac{k_BT}{R_{2D}}, \qquad k_BT \gg R_{2D} \tag{14b}$$

for the generation of a thermalized 2D e–h plasma. These results have the following intuitive interpretation. When an exciton is created, the area occupied by it cannot sustain more excitons (Eq. 13). If we assign to the exciton some effective "hard disk" radius of the order a_{2D}, two disks of this radius cannot share the same space. Much the same can be said for cold free carriers (Eq. 14a). A single carrier cannot occupy the space occupied by an exciton and vice versa. Even in the case of hot free carriers (Eq. 14b), the same notion can be retained, only now we must note that only a fraction of the order $(\lambda/a_{2D})^2 \sim R_{2D}/k_BT$ occupy the phase space sampled by an exciton (λ is the thermal wavelength); i.e. a carrier with thermal energy greater than the exciton binding energy can occupy the same space as an exciton without violating Pauli exclusion.

Equations 13 and 14b explain all experimental results quantitatively (within a factor of 2), without any adjustable parameters, which is quite remarkable in view of the various approximations made.

The simple theory outlined in the previous paragraphs assumed that in 2D, screening is weak as compared to the effects of the Pauli principle. In order to test this hypothesis, a direct experimental comparison of PSF and E with S has been performed (Knox et al., 1986). The principle of these investigations is the following. Using fs laser pulses, it is possible to generate nonthermal carrier distributions in the continuum between the $n_z = 1$ and 2 resonances and to observe their effects on the absorption spectrum before and during their relaxation to the bottom of the band. If the excess energy is less than one LO phonon energy, exchange of energy with the lattice will take a rather long time and the carriers will first thermalize among themselves before reaching an equilibrium with the lattice. Immediately after excitation, the carriers do not occupy states out of which the $n_z = 1$ and 2 excitons are constructed and, thus, the PSF and E are not effective, whereas screening is right away effective. As the carriers equilibrate among themselves, their distribution evolves toward a thermal one and they start to fill up states at the bottom of the $n_z = 1$ subbands. Therefore, they produce PSF and E on the corresponding excitons. Only the high-energy tail of the thermalized distribution will extend up to the bottom of the $n_z = 2$ subbands, which in any case are almost orthogonal to the $n_z = 1$ subbands, so that for these resonances, PSF and E remain ineffective. Therefore, the evolution of the absorption at the $n_z = 1$ excitons during thermalization will first show the effects of S and then how the PSF and E are turned on, whereas at the $n_z = 2$ resonance there should be only little (if any) change during thermalization. Then, on a much longer time scale, the carriers will equilibrate with the lattice and eventually recombine.

This process is shown in Fig. 11, where differential transmission spectra in 50 fs intervals from 100 fs before to 200 fs after arrival of a 100 fs pump pulse are presented. The spectral distribution of the pump photons is shown at the bottom of the figure, the pump pulse produces approximately $N \sim 2 \times 10^{10}\ \mathrm{cm}^{-2}$ e–h pairs. Immediately upon arrival of the pump pulse, one sees absorption changes at the $n_z = 1$ and 2 resonances as well as a spectral hole burning in the continuum, corresponding to state filling by the nonthermal carrier distributions. The $n_z = 1$ excitons lose some oscillator strength and broaden slightly, whereas the $n_z = 2$ exciton experiences also a slight red shift. These changes are consistent with those due to the S of a plasma discussed in Section IV.A. Note that such small effects can only be seen because of the enormous improvements of fs spectroscopic techniques (Fork et al., 1981; Knox et al., 1984). As the carriers start to thermalize, the spectral hole in the continuum changes shape and shifts to lower energies. Accompanying this thermalization, the bleaching of the $n_z = 1$ resonances increases and finally settles after 200 fs at the same level as in the case of thermalized plasma generation discussed previously. The changes at the $n_z = 2$ resonance remain essentially the same, showing that the S has not changed significantly. Comparison of the $n_z = 1$ bleaching at $t = 0$ and

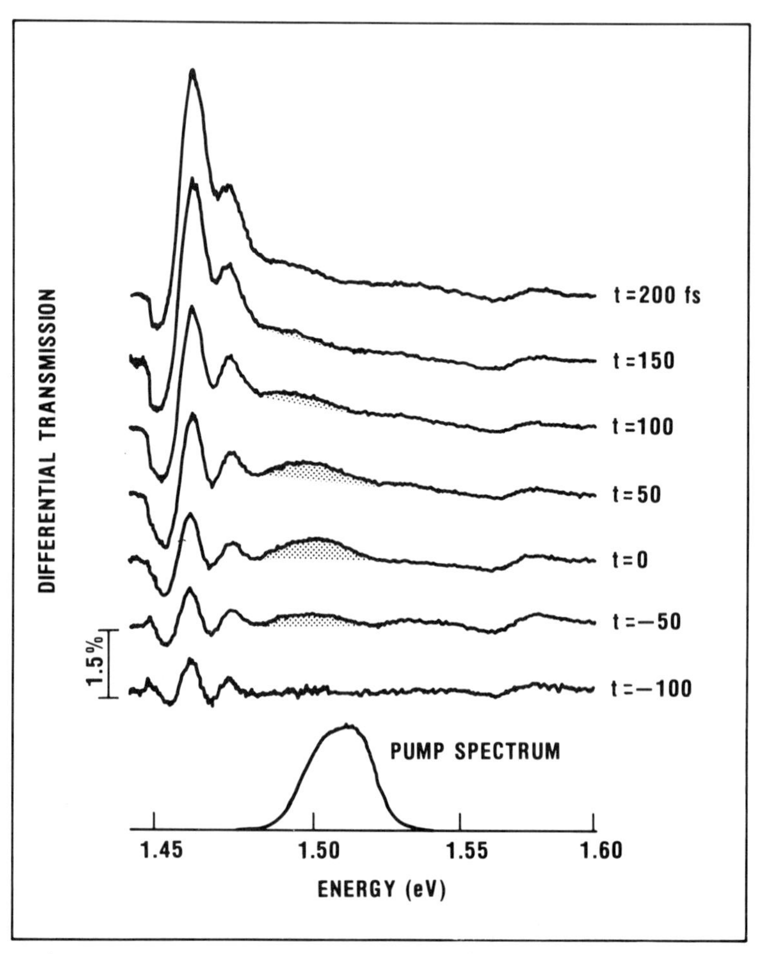

Figure 11. Differential transmission spectra measured with a broad-band 50-fs probe pulse in 50-fs intervals before and after excitation of nonthermal e and h distributions by a 100-fs pump pulse. Initially, the pump burns a spectral hole in the absorption spectrum, which then thermalizes within ~ 200 fs. During thermalization, the effects of the Pauli principle at the $n_z = 1$ resonances are "turned on."

after $t = 200$ fs shows that the combined effects of PSF and E are at least six times larger than those of S (the measurements can only give a lower bound for the ratio [PSF + E]/S, because it is always possible for some carriers to be generated by the very front edge of the pulse and thus to thermalize early). These experiments confirm that the effects of plasma S on excitons are strongly reduced in 2D as compared to 3D. They also contain a wealth of information about carrier relaxation in these important microstructures that is not directly relevant to our present discussion.

Yet another, although more indirect, confirmation of the theory was obtained from the study of the exciton shift in QWs at low temperatures (Peyghambarian et al., 1984; Hulin et al., 1986a; Masumoto et al., 1986). In 3D, the exciton resonances hardly change their position as the density of excitons increases. This can be explained by an almost perfect cancellation of the blue shift due to Pauli exclusion (short-range hard core repulsion) (Comte and Noziéres, 1982) and the red shift due to screening (long-range Van der Waals attraction) (Noziéres and Comte, 1982). In 2D QWs, the blue shift due to Pauli exclusion is no longer balanced and thus should be measurable (Schmitt-Rink et al., 1985; Schmitt-Rink, 1986). At low temperatures, the Hartree contribution to the shift of the exciton (expectation value of Eq. 9 in the 1s state) is given by (Schmitt-Rink et al., 1985; Bobrysheva et al., 1980)

$$\delta E_{1s}^{2D} = 16\pi N a_{2D}^2 (1 - 315\pi^2/2^{12}) E_{1s}^{2D} \tag{15}$$

Experiments performed on QWs of various thicknesses have shown this blue shift when excitons are selectively generated or when they are formed after a few ps from nonresonantly excited e and h. The shift is rather difficult to observe in thick layers but is very pronounced in the narrowest QW, where the dimensionality approaches the pure 2D limit. Its magnitude has been found to be in surprisingly good agreement with the theory (Hulin et al., 1986a). A blue shift was also observed in layered semiconductors, such as BiI_3 (Watanabe et al., 1986).

C. Field-induced Nonlinear Optical Effects

Coherent nonlinear optical processes are produced by the excitation of semiconductors in the transparency region well below the absorption edge. These effects are often referred to as "nonresonant" (Bloembergen, 1965), in the sense that the excitation does not coincide with an absorption resonance; they do have resonant enhancements as such absorption resonances are approached. Such nonlinearities have been widely investigated in the near and medium infrared (Chemla, 1980). The physical mechanisms that produce them can be described as follows. The valence electrons are held in equilibrium by the large atomic fields, which are of the order of $E_{at} = 10^8$–10^9 V/cm; the application of optical fields $E(\omega)$ not negligible as compared to E_{at} induces anharmonic charge density flucations that in turn can interact with other optical fields $E(\omega')$. The mutual coupling of the

$E(\omega)$ and $E(\omega')$ via the anharmonic charge density fluctuations causes exchange of photons among the optical fields. In the process, however, no energy is deposited in the material. The nonresonant effects are characterized by the fact that below-gap optical fields induce only virtual populations, which last as long as the fields are applied.

In the (textbook) quantum mechanical description of these processes, the "ground state" of the semiconductor in the presence of the optical fields is represented by a coherent superposition of excited (e–h) states. The expectation value of the polarization is calculated (at least formally) using the density matrix formalism. In most cases, spatial dispersion is negligible and only dipolar effects are retained. They are described by an interaction Hamiltonian of the form $-P \cdot E$, where P is the polarization of the medium. In the classical discussion of nonlinear optics, the expectation value of P is expanded in a power series of the $E(\omega)$'s, the coefficients being the nonresonant nonlinear susceptibilities $\chi^{(3)}$, $\chi^{(5)}, \ldots$. Finally, the field interaction is analyzed by solving the corresponding Maxwell equations coupled by the nonlinear polarizations (Shen, 1984).

The formal expressions of the nonlinear susceptibilities are written in terms of excited state wave functions and energies, which are usually not known. Most of the time they are treated as phenomenological parameters that are determined experimentally. Specific experimental investigations of excitonic effects in nonresonant nonlinear optical processes have been rather scarce and theoretical models have been limited to semi-empirical descriptions, where the exciton levels are characterized by ad hoc matrix elements and spectroscopically determined energies.

It is obvious that e–h, e–e and h–h correlation effects must play a crucial role in nonresonant nonlinear optical processes because of the strong coupling induced between the charge density fluctuations by the Coulomb forces. It is also clear that the effects of the Coulomb correlation must be treated in a self-consistent manner with the effects of the applied fields. Therefore, the correct description of nonresonant excitonic nonlinearities in semiconductors cannot follow a simple transposition of the description of atomic nonlinear optical processes. In this section, we review recent experimental investigations of these effects in GaAs QWs (Mysyrowicz et al., 1986; Von Lehmen et al., 1986b) and a theory that, for the first time, correctly accounts for exciton correlation effects on "nonresonant" nonlinear susceptibilities (Schmitt-Rink and Chemla, 1986).

How can excitonic effects be resolved in below-gap nonlinear optical processes? Obviously, they influence the absolute magnitude of the susceptibilities $\chi^{(n)}$. In practice, however, the determination of these quantities is not very useful because (1) they are rather difficult to measure accurately and (2) the $\chi^{(n)}$ are global parameters involving infinite sums over complete sets of states, in which the behavior of particular states is difficult to isolate. It is thus more interesting to turn to spectroscopic measurements.

Clearly, the most natural way to probe excitonic effects is to probe the absorption spectrum while the sample is strongly excited well below the gap and to look for transient changes that persist only as long as the excitation. It is fair to say that these effects were first observed accidentally. In time-resolved studies of the below-gap absorption in QWs (Section IV.A), strong and transient changes of the probe transmission, as shown in Fig. 12, were observed (Von Lehmen et al., 1986b). These changes can be positive or negative depending on the position of the test photon energy with respect to the peak of the hh-exciton. They last only as long as the pump pulse. If the detuning from the exciton resonance is not too large, as in Fig. 12, much weaker effects due to a photo-generated plasma are also seen. This process is easily distinguished from the field effects. It persists for the lifetime and completely disappears for large detuning. The fast component of the change in probe absorption was studied as a function of the pump parameters. It was found that it is linear in the inverse pump detuning $(E_{1s} - \hbar\omega_p)^{-1}$ for a fixed pump intensity (Fig. 13a), and linear in the pump intensity for a fixed pump detuning (Fig. 13b). Furthermore, a lineshape analysis of the fast and slow components as a function of the probe frequency around the $n_z = 1$ resonances evidences their very different spectral profiles (Von Lehmen et al., 1986b). The slow component has all the characteristics of the plasma-induced bleaching and broadening discussed in Section IV.A. The fast component is much larger and its differential transmission spectrum corresponds to a transient blue shift of the exciton resonances. For a pump detuning $\sim$ 30 meV and a pump intensity $\sim$ 8 MW/cm^2 the magnitude of the shift is $\sim$ 0.2 meV for the hh-exciton peak and $\sim$ 0.05 meV for the lh-exciton peak. The same effects have also been observed using fs pump and continuum probe pulses, although at much higher pump intensity (Mysyrowicz et al., 1986). Again, they were found accidentally in the study of the blue shift induced by real exciton populations (Section IV.B). In addition, in these experiments there is clear evidence of a loss of excitonic oscillator strength. In conclusion, the main effects of the strong below-gap excitation on the transmission of a weak test beam are a shift and bleaching of the resonances, henceforth referred to as the AC Stark effect.

How does this semiconductor AC Stark effect relate to the one studied extensively in atomic systems (Mollow, 1969, 1972)? Clearly, in the absence of Coulomb interactions (and thus of excitons), it would not be very

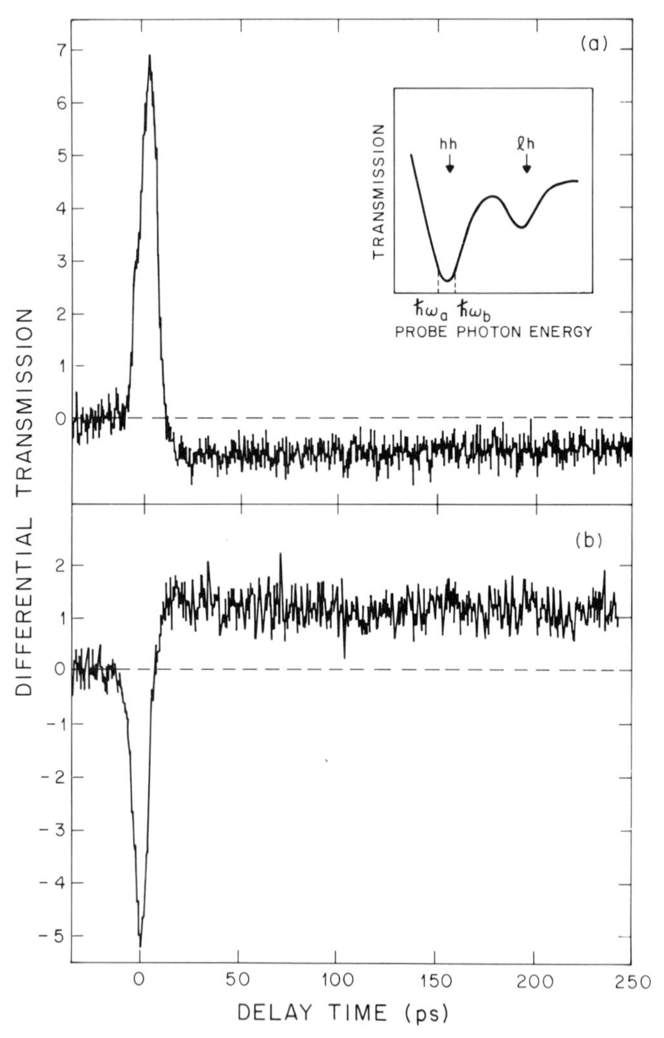

Figure 12. Differential transmission as a function of the pump-probe delay. The pump is tuned ~ 25 meV below the hh-exciton resonance. The probe is tuned ~ 1 meV below the hh-exciton resonance in (a) and ~ 1 meV above the hh-exciton resonance in (b).

different. Interband transitions with different e–h relative momenta $\underline{k}$ would then decouple, and each individual transition would behave in much the same fashion as a two-level atom coherently driven below resonance. The band-to-band transitions would be bleached and blue shifted and, at high pump intensities, optical gain would develop below $2\hbar\omega_p - E_g$. (In the latter process, two pump photons are destroyed and a test photon and free e–h pair are created.) In second order in the pump field E_p, the change in

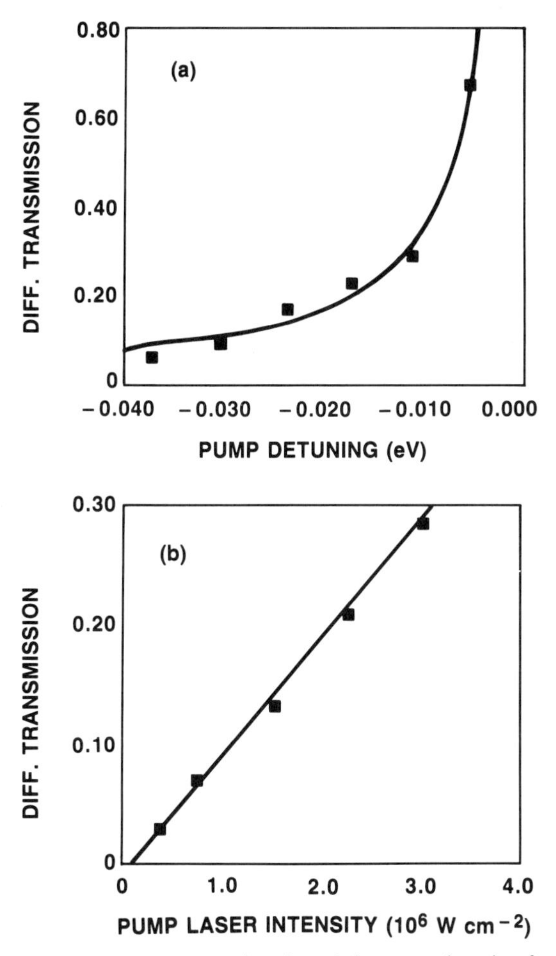

Figure 13. Differential transmission as a function of the pump detuning from the hh-exciton resonance (a) and pump intensity for a fixed pump detuning ~ 18 meV (b).

band gap would be

$$\delta E_g = 2\frac{|er_{cv}E_p|^2}{E_g - \hbar\omega_p} \tag{16}$$

Unperturbed exciton states are a superposition of free e and h states, the coefficients being the relative motion orbital wave functions $U_n(\underline{k})$, subject to

$$\left(E_n - E_g - \frac{\hbar^2 k^2}{2m}\right)U_n(\underline{k}) + \sum_{\underline{k}'} V(\underline{k} - \underline{k}')U_n(\underline{k}') = 0 \tag{17}$$

This simple Schroedinger equation serves to demonstrate how this free e–h picture has to be modified to account for the e–h interaction. Inside an exciton, e–h pair states (and thus interband transitions) with different relative momenta $\underline{k}$ are strongly Coulomb-correlated, so that a pair with given $\underline{k}$ does not only experience the external field $E_{ext} \equiv E_p$ but also a significant internal one, the "molecular field" E_{mol} associated with virtual pairs created at $\underline{k}'$. Much the same can be said for the effects of the e–e and h–h interaction (see Eq. 9). At each $\underline{k}$, external and Coulomb fields combine to give an effective self-consistent "local field," $E_{loc} = E_{ext} + E_{mol}$ (Schmitt-Rink and Chemla, 1986). The problem is thus similar to that of a paramagnet in an external magnetic field.

For very small external fields, the "molecular fields" or "local field corrections" are large. They transform the free e–h pairs into excitons. (In a paramagnet, this corresponds to the transition from itinerant to local moment behavior.) As the pump field strength increases, so does the number of coherently excited excitons. The virtual exciton states start to interact, in much the same way as if they were actually populated. In fact, we recover Eqs. 9 through 11 if we identify $f_{e,h}(k)$ with virtual e and h distributions. Considering the leading 1s state contribution only, we can recover Eqs. 8 and 15, where N is now the number of virtually excited 1s excitons

$$N = 2\frac{|er_{cv}E_p|^2}{(E_{1s} - \hbar\omega_p)^2}|U_{1s}(r = 0)|^2 \tag{18}$$

Thus, for not-too-large pump intensities, we can carry over our previous results. The interaction of virtually excited excitons is exactly the same as that of real ones (Schmitt-Rink and Chemla, 1986).

The coherent "ground state" of a semiconductor in the presence of a pump beam is, of course, not what is being measured by a test beam. A weak test beam E_t induces a polarization, which has to be clearly distinguished from that induced by the pump beam. It measures the "excitation spectrum" (Schmitt-Rink and Chemla, 1986).

The Coulomb fields induced by the external perturbation E_t combine to give an effective field $\delta E_{ext} = E_t + \delta E_{mol} + \delta E_s$ at each $\underline{k}$. The first two terms $(E_t + \delta E_{mol})$ account again for the fact that the system responds to the exact local field, the third term describes the screening of the perturbation due to the induced charge density fluctuations. For not-too-large pump intensities, i.e. in the excitonic limit, the evaluation of the excitation spectrum, which determines the linear response to δE_{ext}, is very tedious.

Some of the results, such as the occurrence of excitonic gain below $\hbar\omega_p$, can be guessed. Of special interest here is the AC Stark shift of the exciton resonance. Keeping only the virtual 1s state contribution and neglecting virtual exciton–exciton interactions (which can produce an additional shift), one finds in leading order in the pump intensity (Schmitt-Rink and Chemla, 1986)

$$\delta E_{1s} = 2\frac{|er_{cv}E_p|^2}{E_{1s} - \hbar\omega_p}\frac{|U_{1s}(r = 0)|^2}{N_s^{PSF}} \tag{19}$$

where N_s^{PSF} is the saturation density due to excitonic phase space filling, defined in Section IV.B.

First of all, Eq. 19 explains the experimental results, without any adjustable parameters. For the experimental parameters given above, the theoretical hh-exciton blue shift is 0.15 meV and the ratio of the hh-exciton and lh-exciton shifts is 4. The agreement is excellent.

If we identify Eq. 19 with the AC Stark shift in a two-level system, we find that 1 exciton behaves like N_s^{PSF} two-level atoms, a result that we could have also obtained from the discussion given in Section IV.B. Vice versa, the first factor in Eq. 19 expresses the AC Stark shift of the atomic s and p states that form the conduction and valence bands. The second factor describes the renormalization of this atomic shift due to excitonic effects. Its numerator reflects the fact than an exciton is built up from a linear combination of Bloch states that themselves originate from the atomic states. The denominator contains the saturation density N_s^{PSF}, above which the concept of excitons becomes invalid.

Very recently, the AC Stark effect has also been observed in InGaAs–InP QWs (Tai et al., 1987b) and has been exploited to operate a GaAs–AlGaAs optical gate with subpicosecond switch on and off times (Hulin et al., 1986b).

V. CONCLUSIONS

In its few years of existence, the nonlinear optics of quantum wells has already proved to be a fertile ground for novel physics. It has, for example, increased our understanding of the linear optical properties of quantum wells, helping to explain the remarkable persistence of exciton resonances at room temperature. It has enhanced our knowledge of many-body effects in two and three dimensions. It has enabled us to measure the dynamics of particle interactions in confined systems. It has elucidated the many-body interactions of virtual populations. More generally, however, it has contributed to a growing awareness of the possibilities for new nonlinear optical processes for physics and devices through quantum engineering of small structures. This broader field is only just starting to grow, and it promises to be an exciting one in the years ahead.

REFERENCES

Ando, T., Fowler, A. B., and Stern, F. (1982). *Rev. Mod. Phys.* **54,** 437–672.

Bastard, G. (1984). *Phys. Rev. B* **30,** 3547–3549.

Bastard, G., Mendez, E. E., Chang, L. L., and Esaki, L. (1982). *Phys. Rev. B* **26,** 1974–1979.

Bloembergen, N. (1965). *Nonlinear Optics.* W. A. Benjamin, New York.

Bobrysheva, A. I., Zyukov, V. T., and Beryl, S. I. (1980). *Phys. Status Solidi B* **101,** 69–76.

Broido, D. A., and Sham, L. J. (1986). *Phys. Rev. B* **34,** 3917–3923.

Chang, Y. C., and Schulman, J. N. (1983). *Appl. Phys. Lett.* **43,** 536–538.

Chemla, D. S. (1980). *Rep. Prog. Phys.* **43,** 1191–1262.

Chemla, D. S. (1983). *Helv. Phys. Acta* **56,** 607–637.

Chemla, D. S. (1985). *J. Lumin.* **30,** 502–519.

Chemla, D. S., and Miller, D. A. B. (1985). *J. Opt. Soc. Am. B* **2,** 1155–1173.

Chemla, D. S., and Pinczuk, A., eds.(1986). *IEEE J. Quantum Electron.* **QE-22,** 1609–1921.

Chemla, D. S., Miller, D. A. B., Smith, P. W., Gossard, A. C., and Wiegmann, W. (1984). *IEEE J. Quantum Electron.* **QE-20,** 265–275.

Comte, C., and Nozieres, P. (1982). *J. Physique* **43,** 1069–1081.

Dingle, R., Wiegmann, W., and Henry, C. H. (1974). *Phys. Rev. Lett.* **33,** 827–830.

Elliot, R. J. (1957). *Phys. Rev.* **108,** 1384–1389.

Fehrenbach, G. W., Schaefer, W., Treusch, J., and Ulbrich, R. G. (1982). *Phys. Rev. Lett.* **49,** 1281–1284.

Fork, R. L., Greene, B. I., and Shank, C. V. (1981). *Appl. Phys. Lett.* **38,** 671–672.

Greene, R. L., Bajaj, K. K., and Phelps, D. E. (1984). *Phys. Rev. B* **29,** 1807–1812.

Haug, H., and Schmitt-Rink, S. (1984). *Prog. Quantum Electron.* **9,** 3–100.

Haug, H., and Schmitt-Rink, S. (1985). *J. Opt. Soc. Am. B* **2,** 1135–1142.

Hegarty, J., and Sturge, M. D. (1985). *J. Opt. Soc. Am. B* **2,** 1143–1154.

Hulin, D., Mysyrowicz, A., Antonetti, A., Migus, A., Masselink, W. T., Morkoc, H., Gibbs, H. M., and Peyghambarian, N. (1986a). *Phys. Rev. B* **33,** 4389–4391.

Hulin, D., Mysyrowicz, A., Antonetti, A., Migus, A., Masselink, W. T., Morkoc, H., Gibbs, H. M., and Peyghambarian, N. (1986b). *Appl. Phys. Lett.* **49,** 749–751.

Knox, W. H., Downer, M. C., Fork, R. L., and Shank, C. V. (1984). *Opt. Lett.* **9,** 552–554.

Knox, W. H., Fork, R. L., Downer, M. C., Miller, D. A. B., Chemla, D. S., Shank, C. V., Gossard, A. C., and Wiegmann, W. (1985). *Phys. Rev. Lett.* **54,** 1306–1309.

Knox, W. H., Hirlimann, C., Miller, D. A. B., Shah, J., Chemla, D. S., and Shank, C. V. (1986). *Phys. Rev. Lett.* **56,** 1191–1193.

Lee, Y. C., and Lin, D. L. (1979). *Phys. Rev. B* **19,** 1982–1989.
Liebler, J. G., Schmitt-Rink, S., and Haug, H. (1985). *J. Lumin.* **34,** 1–7.
Löwenau, J. P., Schmitt-Rink, S., and Haug, H. (1982). *Phys. Rev. Lett.* **49,** 1511–1514.
Masumoto, Y., Tarucha, S., and Okamoto, H. (1986). *J. Phys. Soc. Japan* **55,** 57–60.
Miller, D. A. B., Chemla, D. S., Smith, P. W., Gossard, A. C., and Wiegmann, W. (1982a). *Appl. Phys. B* **28,** 96–97.
Miller, D. A. B., Chemla, D. S., Eilenberger, D. J., Smith, P. W., Gossard, A. C., and Tsang, W. T. (1982b). *Appl. Phys. Lett.* **41,** 679–681.
Miller, D. A. B., Chemla, D. S., Eilenberger, D. J., Smith, P. W., Gossard, A. C., and Wiegmann, W. (1983a). *Appl. Phys. Lett.* **42,** 925–927.
Miller, D. A. B., Chemla, D. S., Smith, P. W., Gossard, A. C., and Wiegmann, W. (1983b). *Opt. Lett.* **8,** 477–479.
Miller, D. A. B., Chemla, D. S., Damen, T. C., Gossard, A. C., Wiegmann, W., Wood, T. H., and Burrus, C. A. (1985). *Phys. Rev. B* **32,** 1043–1060.
Miller, D. A. B., Chemla, D. S., and Schmitt-Rink, S. (1986a). *Phys. Rev. B* **33,** 6976–6982.
Miller, D. A. B., Weiner, J. S., and Chemla, D. S. (1986b). *IEEE J. Quantum Electron.* **QE-22,** 1816–1830.
Miller, R. C., and Kleinman, D. A. (1985). *J. Lumin.* **30,** 520–540.
Miller, R. C., Kleinman, D. A., Tsang, W. T., and Gossard, A. C. (1981). *Phys. Rev. B* **24,** 1134–1136.
Mollow, B. R. (1969). *Phys. Rev.* **188,** 1969–1975.
Mollow, B. R. (1972). *Phys. Rev. A* **5,** 2217–2222.
Mysyrowicz, A., Hulin, D., Antonetti, A., Migus, A., Masselink, W. T., and Morkoc, H. (1986). *Phys. Rev. Lett.* **56,** 2748–2751.
Nozieres, P., and Comte, C. (1982). *J. Physique* **43,** 1083–1098.
Peyghambarian, N., and Gibbs, H. M. (1985). *J. Opt. Soc. Am. B* **2,** 1215–1227.
Peyghambarian, N., Gibbs, H. M., Jewell, J. L., Antonetti, A., Migus, A., Hulin, D., and Mysyrowicz, A. (1984). *Phys. Rev. Lett.* **53,** 2433–2436.
Ruckenstein, A. E., Schmitt-Rink, S., and Miller, R. C. (1986). *Phys. Rev. Lett.* **56,** 504–507.
Schaefer, W., and Treusch, J. (1986). *Z. Phys. B* **63,** 407–426.
Schmitt-Rink, S. (1986). "Theory of Transient Excitonic Nonlinearities in Quantum Wells." *Proc. NSF Workshop on Optical Nonlinearities, Fast Phenomena and Signal Processing,* Tucson.
Schmitt-Rink, S., and Chemla, D. S. (1986). *Phys. Rev. Lett.* **57,** 2752–2755.
Schmitt-Rink, S., and Ell, C. (1985). *J. Lumin.* **30,** 585–596.
Schmitt-Rink, S., Chemla, D. S., and Miller, D. A. B. (1985). *Phys. Rev. B* **32,** 6601–6609.
Schmitt-Rink, S., Ell, C., Koch, S. W., Schmidt, H. E., and Haug, H. (1984). *Solid State Commun.* **52,** 123–125.
Schmitt-Rink, S., Ell, C., and Haug, H. (1986). *Phys. Rev. B* **33,** 1183–1189.
Segall, B. (1966). *Phys. Rev.* **150,** 734–747.
Shen, Y. R. (1984). *The Principle of Nonlinear Optics.* J. Wiley, New York.
Shinada, S., and Sugano, S. (1966). *J. Phys. Soc. Japan* **21,** 1936–1946.

Smith, P. W., Silberberg, Y., and Miller, D. A. B. (1985). *J. Opt. Soc. Am. B* **2,** 1228–1236.

Sooryakumar, R., Chemla, D. S., Pinczuk, A., Gossard, A., Wiegmann, W., and Sham, L. J. (1985). *Solid State Commun.* **54,** 859–862.

Sturge, M. D. (1962). *Phys. Rev.* **127,** 768–773.

Tai, K., Jewell, J. L., Tsang, W. T., Temkin, H., Panish, M., and Twu, Y. (1987a). *Appl. Phys. Lett.* **51,** 152 (1987).

Tai, K., Tsang, W. T., and Hegarty, J. (1987b). *Appl. Phys. Lett.* **51,** 86 (1987).

Takagahara, T., and Hanamura, E. (1986). *Phys. Rev. Lett.* **56,** 2533–2536.

Totsuji, H. (1975). *J. Phys. Soc. Japan* **39,** 253–254.

Von Lehmen, A., Zucker, J. E., Heritage, J. P., Chemla, D. S., and Gossard, A. C. (1986a). *Appl. Phys. Lett.* **48,** 1479–1481.

Von Lehmen, A., Chemla, D. S., Zucker, J. E., and Heritage, J. P. (1986b). *Opt. Lett.* **11,** 609–611.

Von Lehmen, A., Zucker, J. E., Heritage, J. P., and Chemla, D. S. (1987). *Phys. Rev. B* **35,** 6479–6482.

Watanabe, K., Karasawa, T., Komatsu, T., and Kaifu, Y. (1986). *J. Phys. Soc. Japan* **55,** 897–907.

Weiner, J. S., Chemla, D. S., Miller, D. A. B., Wood, T. H., Sivco, D., and Cho, A. Y. (1985a). *Appl. Phys. Lett.* **46,** 619–621.

Weiner, J. S., Chemla, D. S., Miller, D. A. B., Haus, H. A., Gossard, A. C., Wiegmann, W., and Burrus, C. A. (1985b). *Appl. Phys. Lett.* **47,** 664–667.

Weisbuch, C., Dingle, R., Gossard, A. C., and Wiegmann, W. (1981). *Solid State Commun.* **38,** 709–712.

Zucker, J. E., Pinczuk, A., Chemla, D. S., Gossard, A. C., and Wiegmann, W. (1983). *Phys. Rev. Lett.* **51,** 1293–1296.

Zucker, J. E., Pinczuk, A., Chemla, D. S., Gossard, A. C., and Wiegmann, W. (1984a). *Phys. Rev. Lett.* **53,** 1280–1283.

Zucker, J. E., Pinczuk, A., Chemla, D. S., Gossard, A. C., and Weigmann, W. (1984b). *Phys. Rev. B* **29,** 7065–7068.

Zucker, J. E., Pinczuk, A., Chemla, D. S., and Gossard, A. C. (1987). *Phys. Rev. B* **35,** 2892–2895.

5 DENSE NONEQUILIBRIUM EXCITONS: BAND EDGE ABSORPTION SPECTRA OF HIGHLY EXCITED GALLIUM ARSENIDE

R. G. Ulbrich

INSTITUT FÜR PHYSIK
UNIVERSITÄT DORTMUND
4600 DORTMUND 50
FEDERAL REPUBLIC OF GERMANY

I. INTRODUCTION

The band edge spectrum of a semiconductor depends critically on the degree of excitation in the form of electron–hole pairs and phonons. At low temperatures, near equilibrium, the edge is dominated by excitonic line spectra. Absorption of light or heat and also the presence of impurities change the energy and the occupation of electronic states in the crystal and modify its optical properties. This dependence is one aspect of nonlinear

OPTICAL NONLINEARITIES
AND INSTABILITIES IN
SEMICONDUCTORS

ISBN 0-12-332915-9

optics in semiconductors (Haug and Schmitt-Rink, 1985). In bulk geometry, with all sample dimensions much larger than the exciton Bohr radius a_B, the natural unit for e–h pair density is the reciprocal exciton volume. A typical a_B is of the order of 100 Å, so that the pair density for complete volume filling with 1s excitons is $\sim 2 \cdot 10^{17}\,\text{cm}^{-3}$. Considering a low-density pair recombination lifetime $\tau_R = 0.3$ ns, one needs at least $2 \cdot 10^4$ W/cm^2 cw pump power into a 1 μm thick crystal to maintain such a density or a deposit of more than 6 μJ/cm^2 pulse energy in a transient experiment.

With the currently available picosecond pulses from mode-locked lasers, the instantaneous density of mobile e–h pairs can actually be controlled and varied over a much larger range, from intrinsic values ($\sim 10^{10}\,\text{cm}^{-3}$) to more than $10^{21}\,\text{cm}^{-3}$, which marks the threshold of melting. Excitation of the crystal and diagnostics of its optical properties are in general performed with laser light pulses in single- or two-beam arrangements (Pilkuhn, 1985). Early experiments in Ge with nonresonant psec excitation far above E_G and probing reflectivity changes (Auston and Shank, 1974) were followed by work on absorption transients at the band gap of GaAs (Shank et al., 1978, 1979). Gradually the temporal and spectral resolution was pushed to the limits (Fleming and Siegman, 1986). At the same time, the basic pump-and-probe concept was refined into the more general description in terms of nonlinear optical four-wave mixing (Eichler et al., 1986). Most recently, ultrafast phenomena such as Coulomb scattering in dense continuum e–h pair distributions (Lin et al., 1986) and exciton ionization via LO phonon absorption at room temperature were studied experimentally in the 100-fs domain (Chemla and Miller, 1985).

Apart from the practical importance of this work for optoelectronic devices, there is considerable fundamental interest in optically excited e–h pair ensembles with adjustable density (Comte and Noziéres, 1982). They are ideal objects for empirical studies of quantum phenomena in many-particle systems, and one can choose the excitation conditions such that different kinds of nonequilibrium prevail. Of central importance for the edge spectra is the interplay of the screening of the long-range Coulomb interaction between both types of mobile charge carriers, electrons and holes, and their mutual binding (or "pairing") to neutral excitons. There is a close similarity with Mott's concept of the insulator-metal transition (Mott, 1974). The essential difference is the additional translational degree of freedom ascribed to the holes. Proper descriptions of the screening of the effective e–h interaction for nonequilibrium carrier distributions are difficult; the reader is referred to the treatments in this volume and the literature (Haug, 1985; Haug and Schmitt-Rink, 1985; Schäfer, 1986).

This chapter will discuss empirical band edge spectra of the model semiconductor GaAs under conditions of resonant picosecond pulse excitation. We shall point at the nonequilibrium nature of the e–h pair distribution and its connection with the screening properties of a dense exciton ensemble.

II. EXCITONS AT LOW DENSITIES: ABSORPTION SPECTRA AND E–H CORRELATION

At sufficiently low pair densities, the edge spectrum is dominated by sharp exciton resonances. Discrete bound pair states below the gap energy E_G and a smooth continuum above E_G give clear evidence for the long-range attractive Coulomb interaction between e and h. Fig. 1 shows the absorption spectrum of a 4.2 μm thick high-purity GaAs crystal at $T = 1.2$ K. The vapor-phase–grown material has residual shallow impurity concentration $N_A \lesssim N_D = 2 \cdot 10^{14}$ cm^{-3}. The spectrum was measured at low excitation ($n_{\text{pair}} < 10^{10}$cm^{-3}) with a filtered white light source, and the spectrometer behind the sample. We have performed a lineshape analysis of the two lowest resonances, 1s and 2s, in the polariton framework and find a broadening parameter $\Gamma = 40$ μeV. This is quite close to the intrinsic value due to exciton scattering on piezoelectrically active TA and LA phonon modes (Weisbuch and Ulbrich, 1982). The corresponding scattering time of 50 ps and the mean free path ~ 0.5 μm are in good agreement with cyclotron resonance linewidth data and low-field mobility measurements of photoexcited electrons taken on the same sample. All measured parameters of the spectrum of Fig. 1—exciton Rydberg Ry = 4.2 meV, the oscillator strength for $n = 1, 2, 3$ lines, the absolute value of the edge absorption coefficient $\alpha(E_G) = (8.5 \pm 0.5) \cdot 10^3$ cm^{-1}, and the spectral shape above the gap—are consistent with simple exciton theory in effective-mass-approximation (Elliott, 1957; Dow, 1976). Despite the actual Γ_8 valence band degeneracy, which leads to remarkable fine-structure in magneto-optical spectra (both below and above E_G), a simple two-band description with average masses $m_e^* = 0.067m_0$ and $m_h^* = 0.32m_0$ is a good starting point for further exploration of the spectra under conditions of high excitation.

In this simplified description, where we assume simple parabolic, spherical bands for e and h and a static dielectric screening constant $\epsilon_0 = 12.5$,

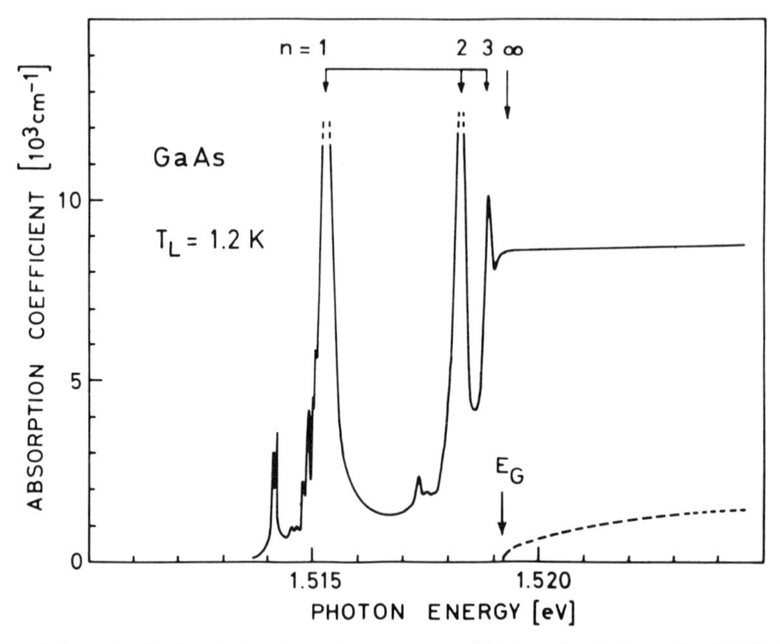

Figure 1. Low-density band edge absorption spectrum of high-purity GaAs at $T = 1.2$ K. The spectrum shows a Rydberg series of discrete bound pair states $n = 1, 2, 3$ and a smooth continuum above E_G. Additional sharp lines below $n = 1$ and $n = 2$ are due to residual impurities and their localized bound pair states. A fictitious independent-particle absorption spectrum (dashed curve, see text) is shown for comparison.

the relative motion of a single e–h pair is described by hydrogenic wavefunctions with $a_B = 110$ Å in GaAs. Fig. 2 shows the s-like wavefunctions $\Psi_{k,0}(r)$ (i.e. angular momentum zero) for four different pair energies relative to E_G, with $k = (E - E_G)/\text{Ry}$ in units of the exciton Rydberg, and proper normalization for both discrete and continuum states (Landau and Lifschitz, 1966). Note that there is strong correlation for small relative separation $r = |\vec{r}_e - \vec{r}_h|$, not only in all bound states, $k < 0$, but also in the continuum, $k \geq 0$. This leads to strong enhancement of oscillator strength, which is proportional to $|\Psi(r = 0)|^2$, in the continuum: the actual absorption at E_G and up to several Ry above E_G is much larger than in a fictitious independent-particle description, where the interband transition matrix element is taken between plane wave states, i.e. multiple e–h scattering is neglected (see Fig. 1, dashed curve). For $k > 0$ and large r, the correlation function becomes asymptotically a free spherical wave with wavelength $2\pi \cdot a_B/k$.

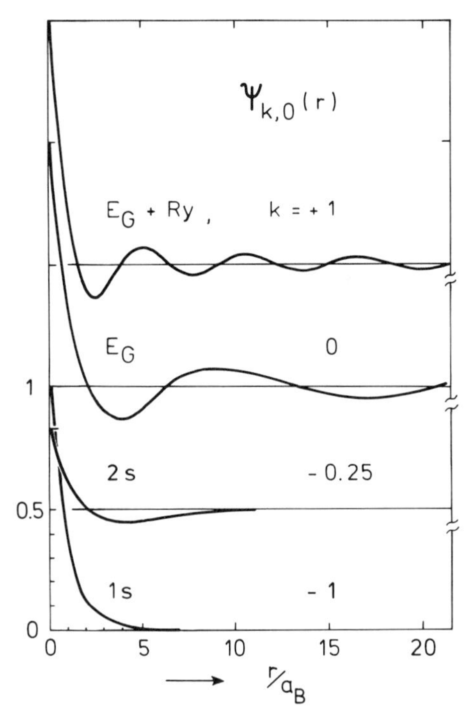

Figure 2. Wave functions for the relative motion of a single electron–hole pair. The parameter k indicates the pair energy E relative to E_G, in units of the exciton Rydberg Ry. Note that there is strong correlation for small $r = |\vec{r}_e - \vec{r}_h|$ not only in all bound states, $k < 0$, but also in the continuum, $k \geq 0$.

The lowest exciton resonance, $n = 1$, has the largest peak absorption—above 10^5 cm^{-1} in the 4.2 μm thick sample. This corresponds to less than 1000 Å penetration depth for narrow-band excitation. To assure homogeneous excitation, one has to use thinner samples. Provided that the confinement of the relative (and the center-of-mass) exciton motion by reflection at the surface boundaries is ideal, new resonances will occur in thinner samples and change the former 3-d hydrogenic spectrum gradually into its quasi–2-d "squeezed" counterpart. Real surfaces, however, cause diffuse scattering so that considerable broadening of the discrete resonances will be observed. A series of thin samples was prepared from the high-quality material by ion-milling and chemical etching from 6 μm down to less than 0.5 μm. The absorption spectra showed a successive increase in exciton linewidths, and the higher resonances $n = 3, 2$ disappeared gradually (Fehrenbach et al., 1985a). At 5000 Å thickness (roughly twice the abscissa

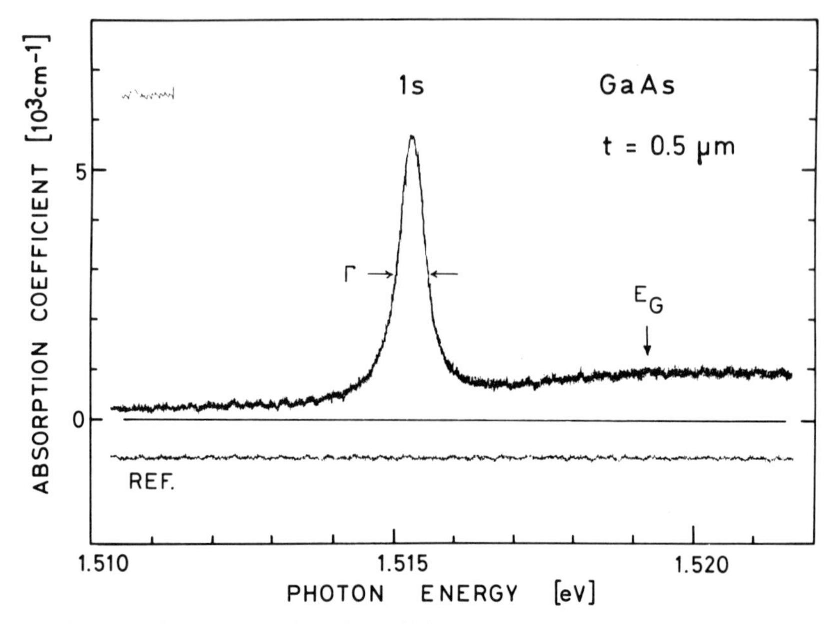

Figure 3. Absorption spectrum of a 0.5 μm thick GaAs sample at 1.2 K. Diffuse scattering at surface imperfections broadens excited exciton states with larger orbits ($n > 1$) and leaves only the lowest $n = 1$ resonance with halfwidth 0.6 meV.

scale of Fig. 2), only the lowest exciton resonance remained, as shown in Fig. 3. The 1s exciton halfwidth of 0.6 meV is now 15 times larger than in the 4.2 μm thick sample, and the pair lifetime has shortened to ~ 100 ps because of surface recombination.

III. TRANSIENT ABSORPTION SPECTRA AT HIGH EXCITATION

The pump-and-probe configuration has been applied in many variants. Depending on the choice of pump photon energy $\hbar\omega$, pump pulse intensity J and duration τ_P, and probe pulse delay τ, one can select widely different excitation conditions and focus at different kinds of nonequilibrium. If τ_M and τ_E are the average carrier momentum and energy relaxation times, we may classify the possible configurations as follows:

(i) Quasi-equilibrium, where electrons and holes form two thermal distributions with distinct quasi Fermi-levels and a common temperature $T_e = T_h$ not far from lattice temperature. This is the situation after excitation of e–h pairs at or above E_G and observation such that $\tau \gg \tau_E > \tau_M$.

(ii) Pulse excitation and observation with $\tau_E \gtrsim \tau > \tau_M$, so that the energy transfer within the carrier systems and from the carriers into the phonon system can be followed in experiment.

(iii) Ultrashort pulse excitation with $\tau_P < \tau_M$ and observation during or shortly after the pulse. This covers the evolution from the initial dielectric polarization pulse (with long-range coherence of e–h pairs and wavevector $\mathbf{K}_{pair} \simeq 0$) toward momentum relaxation with loss of coherence and subsequent energy spread.

(iv) Finally one can drive the crystal with intense light pulses under off-resonance conditions, i.e. below (but still close to) the lowest exciton resonance. The resulting shifts in edge spectra may then be described in the context of the dynamic Stark effect, with more or less preserved phase coherence in all virtually excited discrete and continuum e–h pair states (Mysyrowicz et al., 1986).

In GaAs at low temperature, and for injection close to E_G (i.e. kT and $\hbar\omega - E_G$ much smaller than $\hbar\omega_{LO}$), the time parameters are $\tau_M \simeq 10^{-11}$ s, $\tau_E \simeq 10^{-10}$ s, $\tau_R \simeq 10^{-9}$ s. τ_M depends on pair density roughtly as $n^{-0.5}$ above $n = 10^{13}$ cm^{-3} (Platzman and Wolff, 1973). It follows that electrons and holes can relax toward local equilibrium (with respect to energy) within their relatively short lifetime τ_R only on a μm-scale. In the case of a "cold" energy distribution with sufficient density, the inevitable shortening of τ_R due to stimulated emission—the reverse process of optical excitation—may worsen the situation. It is evident that optically excited e–h pairs constitute a rather "open" system, fed by energy input via creation, and drained by energy transfer into phonons or by radiative annihilation. Throughout the range of temperatures and pair densities discussed here, the vibronic excitations have a much larger specific heat than the e–h pairs, so that energy transfer into the phonons is essentially unidirectional. In the following we shall focus on case (ii) and discuss two pumping conditions: a) resonant excitation of 1s excitons (Fehrenbach et al., 1982) and b) continuum excitation above E_G (see Fig. 4) (Fehrenbach et al., 1985a; 1985b).

The deposited pair densities were determined in a direct way. For given laser configuration, a measurement of the excite pulse spectrum $J(\hbar\omega)$ before and after passage through the sample was performed. Integration of the difference between both spectra (corrected for reflectance at the sample surfaces) over $\hbar\omega$ gave the deposited pulse energy or the equivalent number of e–h pairs generated within the sample. The measured τ_R depended on

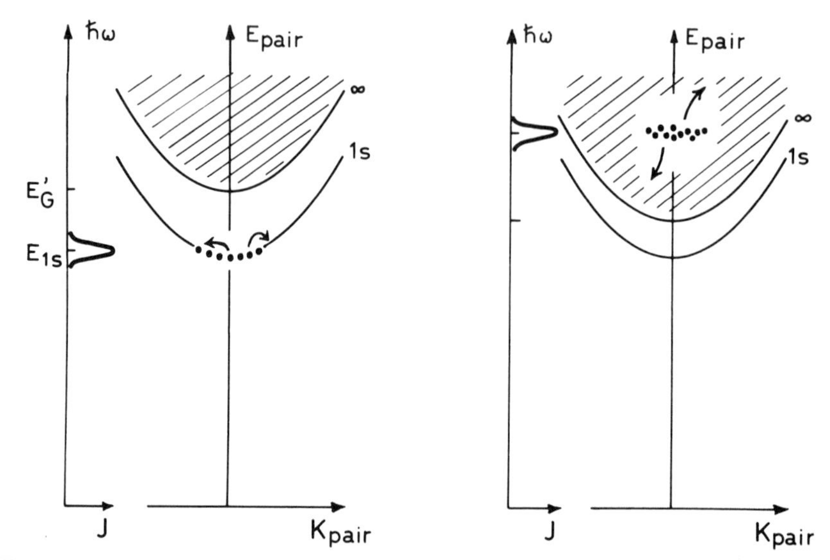

Figure 4. Left: Intense excitation with photon energy $\hbar\omega = E_{1s}$ creates a relatively long-lived, dense exciton distribution with dielectric screening properties and effective band gap E'_G.
Right: The same excitation above the gap generates continuum e–h pairs that randomize their momenta and energies within a few scattering times to form a quasi-equilibrium plasma state with metallic screening properties and much lower band edge.

sample thickness (τ_R = 100 ps at 0.5 μm, and $\gtrsim$ 1 ns at 4.2 μm). Probe delay times of ~ 10 ps were used to assure sufficient relaxation of pair density gradients in the depth of the sample via ballistic and/or diffusive transport. Propagation velocities measured under similar conditions are known to be larger than 1000 Å/ps (Ulbrich and Fehrenbach, 1979), so that reasonable homogeneity is achieved in the thin samples (< 1 μm) within 10 ps after excitation.

The results of both experiments are shown in Fig. 5. Observed exciton peak positions (circles) and line broadenings (vertical bars) are plotted as a function of deposited pair density. Measured 1s line shifts and broadenings in samples of different thickness (0.5 μm, 1.2 μm and 2 μm) were consistent. The 2s line could be followed only in the 2 μm sample. The experimental findings are:

(1) The 1*s* resonance remains constant in energy up to the onset of bleaching, and then its energy increases slightly.
(2) The 2*s* resonance shows a pronounced blue shift at much smaller densities.

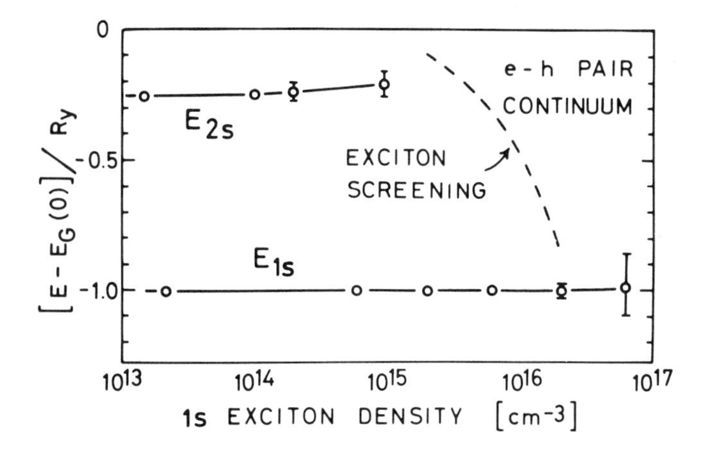

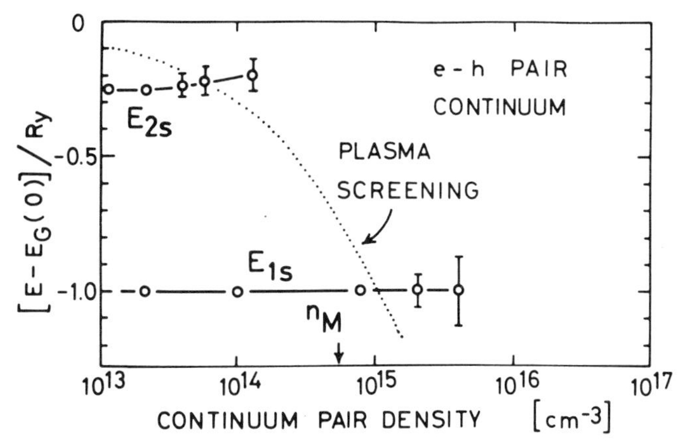

Figure 5. Top: Measured peak positions (circles) of 1s and 2s exciton absorption lines as a function of deposited 1s exciton density (under resonant pumping conditions). The vertical bars indicate excess linewidths. The dashed curve is a theoretical gap energy E'_G based on dielectric screening.

Bottom: As above, but for continuum excitation with $\hbar\omega = E_G + \text{Ry}$. The dotted curve is the effective band gap energy $E_{G'}$ for plasma screening. After Fehrenbach et al. (1985a).

(3) Bleaching densities for the 1s as well as for the 2s line under 1s resonant pumping are a factor of ~ 20 higher than with continuum pumping one Ry above E_G.

(4) Bleaching of 1*s* oscillator strength to one-half of its low-density value occurred at $8 \cdot 10^{16}$ cm^{-3} (1s pumping) and $4 \cdot 10^{15}$ cm^{-3} (continuum pumping).

Recent DFWM experiments by Schultheiss et al. on the transverse relaxation time T_2 of $1s$ excitons in thin GaAs quantum well samples of high quality confirmed finding (3). T_2 values scaled with the line broadenings just quoted, but at somewhat lower densities (Schultheiss et al., 1986).

We interpret the findings (1), (3) and (4) in the context of screening properties of the initially excited e–h pair distribution (Fehrenbach et al., 1982, 1985a, 1985b; Ulbrich, 1985; Schäfer, 1986). The key for qualitative understanding is the persistence of the dense exciton distribution against decay into the energetically more favourable plasma state. Excitation with a light pulse spectrum centered at the 1s resonance (see Fig. 4, left) prepares a *nonequilibrium* state of correlated e–h pairs that is *not* the ground state of the system at the prescribed density. The relaxation from this initial state with mean energy per pair $\langle E \rangle = E_{1s}$ toward the quasi-equilibrium plasma state is found to be rather slow (> 50 ps), according to the probe delays used in Fig. 5. The relative stability at densities high above the Mott criterion $n_M = 5 \cdot 10^{14}\ \mathrm{cm}^{-3}$ seems at first surprising. A consideration of the possible decay mechanisms of the dense exciton ensemble with initial energy width $\Delta E \ll \mathrm{Ry}$ shows, however, that auto-ionization of the discrete pairs via lowest-order pair–pair collisions is indeed slow because of kinematic restrictions. Breaking of an exciton requires at least one Ry energy. Only with larger ΔE of the order of Ry/2, and at larger density, will this mechanism ultimately trigger the transition into the plasma state. Another possible decay process, interband Auger recombination accompanied by free carrier generation, is improbable for densities below $10^{17}\,\mathrm{cm}^{-3}$.

Excitation with light pulses above E_G, on the other hand (see Fig. 4, right), prepares an initial state in which the electron and hole energies are quickly randomized, within a few Coulomb scattering times. Mean kinetic energy and temperature of the relaxed plasma will depend on the deposited pair density, through the dependence of single-particle energy shifts on n_{pair}. At given density, this plasma screens the long-range e–h interaction much more effectively than a 1s exciton distribution. This is the basic explanation for the findings (3) and (4).

IV. CONCLUSION

Edge absorption spectra of optically excited GaAs show at low temperature a strong dependence of exciton line shifts and broadenings on pump photon energy. Under conditions of resonant 1s exciton pumping, the existence of a metastable, long-lived (> 50 ps) nonequilibrium exciton distribution with densities ($> 2 \cdot 10^{16}\,cm^{-3}$) much above the Mott criterion has been demonstrated. Continuum excitation screens the e–h interaction much more strongly and gives $\sim$ 20 times smaller bleaching densities.

ACKNOWLEDGEMENTS

I am indebted to G. W. Fehrenbach for his cooperation in all stages of this work and to B. Gerlach, W. Schäfer and J. Treusch for numerous stimulating discussions.

REFERENCES

Auston, D. H., and Shank, C. V. (1974). *Phys. Rev. Lett.* **32,** 1120.
Chemla, D. S., and Miller, D. A. B. (1985). *J. Opt. Soc. Am. B* **2,** 1155.
Comte, C., and Noziéres, P. (1982). *J. Physique* **43,** 1069; 1083.
Dow, J. D. (1976). In *Optical Properties of Solids* (B. O. Seraphin, ed.), ch. 2. North-Holland, Amsterdam.
Eichler, H. J., Günther, P., and Pohl, D. W. (1986). *Laser-induced Dynamic Gratings*. Springer Series in Optical Sciences, vol. 50. Springer, Berlin.
Elliott, R. J. (1957). *Phys. Rev.* **108,** 1384.
Fehrenbach, G. W., Schäfer, W., Treusch, J., and Ulbrich, R. G. (1982). *Phys. Rev. Lett.* **49,** 1281.
Fehrenbach, G. W., Schäfer, W., and Ulbrich, R. G. (1985a). *J. Lumin.* **30,** 154.
Fehrenbach, G. W., Schäfer, W., Schuldt, K. H., Treusch, J., and Ulbrich, R. G. (1985b). *Proc. 17th Int. Conf. Phys. Semicond.* (J. D. Chadi and W. A. Harrison, eds.), p. 1251. Springer-Verlag, New York.
Fleming, G. R., and Siegman, A. E. (1986). In "Ultrafast Phenomena V" (*Proc. 5th OSA Topical Meeting,* 1986), Part V. Springer-Verlag, Berlin.
Haug, H. (1985). *J. Lumin.* **30,** 171. See also the contribution by W. Schäfer in this volume.
Haug, H., and Schmitt-Rink, S. (1985). *J. Opt. Soc. Am. B* **2,** 1135.
Landau, L. D., and Lifschitz, E. M. (1966). *Lehrbuch der Theoretischen Physik,* Vol. III, p. 122 ff. Akademie Verlag, Berlin.
Lin, W. Z., Fujimoto, J. G., Ippen, E. P., and Logan, R. A. (1986). In *Ultrafast Phenomena V* (*Proc. 5th OSA Topical Meeting, 1986*), p. 193. Springer-Verlag, Berlin.
Mott, N. F. (1974). *Metal-Insulator Transitions,* ch. 4.9. Taylor and Francis, London.
Mysyrowicz, A., Gibbs, H. M., Peyghambarian, N., Masselink, W. T., and Morkoc, H. (1986). In *Ultrafast Phenomena V* (*Proc. 5th OSA Topical Meeting, 1986*), Part V. Springer-Verlag, Berlin, p. 197.
Pilkuhn, M. H. (1985). *Proc. 3rd Trieste Symposium on High Excitation and Short Pulse Phenomena*. North-Holland, Amsterdam.

Platzman, P. M., and Wolff, P. A. (1973). In *Solid State Physics* (eds. H. Ehrenreich, F. Seitz, and D. Turnbull), Suppl. 13, ch. 1. Academic Press, New York.
Schäfer, W., and Treusch, J. (1986). *Z. Phys. B* **63,** 407.
Schultheiss, L., Kuhl, J., Honold, A., and Tu, C. W. (1986). *Phys. Rev. Lett.* **57,** 1635.
Shank, C. V., Auston, D. H., Ippen, E. P., and Teschke, O. (1978). *Solid State Commun.* **26,** 567.
Shank, C. V., Fork, R. L., Leheny, R. F., and Shah, Jagdeep (1979). *Phys. Rev. Lett.* **42,** 112.
Ulbrich, R. G. (1985). In *Festkörperprobleme XXV* (Advances in Solid State Physics 25) (ed. P. Grosse), p. 299. Vieweg Verlag, Braunschweig.
Ulbrich, R. G., and Fehrenbach, G. W. (1979). *Phys. Rev. Lett.* **43,** 963.
Weisbuch, C., and Ulbrich, R. G. (1982). In *Light Scattering III* (eds. M. Cardona, and G. Güntherodt). Springer Topics in Applied Physics vol. 51, p. 235 ff. Springer-Verlag, Berlin.

6 THEORY OF DENSE NONEQUILIBRIUM EXCITON SYSTEMS

W. Schäfer

INSTITUT FÜR PHYSIK
UNIVERSITÄT DORTMUND
4600 DORTMUND 50, FEDERAL REPUBLIC OF GERMANY

I. INTRODUCTION

Since the early work of Keldysh and Kopaev (1965), Hanamura (1970), and others, there has been a lot of theoretical investigation of dense exciton systems or, more generally, of two mutually interacting fermion systems. Besides the formal efforts to formulate a general theory for composite particles on the basis of the well-known many-body formalism for systems of simple particles (see, e.g., Girardeau 1978, Brittin 1980), most approaches end in a description of the properties of an exciton surrounded by

OPTICAL NONLINEARITIES
AND INSTABILITIES IN
SEMICONDUCTORS

ISBN 0-12-332915-9

others in low order in the density of excited pairs (Röpke et al. 1980). If an equilibrium situation is anticipated—in the relevant range of densities and temperatures—nearly all surrounding pairs are ionized and, thus, a theory of highly excited semiconductors can be formulated in terms of the influence of a two-component plasma on the properties of an additionally excited pair. But as is obvious for intuitive reasons, the transition from a weakly interacting exciton gas to a state, which can be comprehended as a two-component plasma, can be described neither by a low-density expansion for the interaction between excitons nor in terms of plasma physics. A perturbational treatment is only valid in low order in the density. There one can derive a first-order renormalization of the binding energy and of one-particle states, or e.g. in second order the occurrence of exciton–exciton bound states, i.e., biexcitons (Hanamura and Haug 1978). On the other hand, the plasma description neglects the correlation between excited electron–hole (e–h) pairs completely. To describe the region in which a transition from an exciton gas to a plasma occurs, a theory being equally valid in all orders of the density is necessary. The only approach fulfilling this claim, which is up to now worked out in the literature, is the $T = 0$ pair formalism, where all e–h pairs occupy a Bose condensed state (see, e.g., the review of Haug and Schmitt-Rink, 1984). This assumption, however, is not fulfilled in real experimental situations. Even when a monochromatic exciting electromagnetic wave produces a macroscopically filled state with nearly vanishing center-of-mass momentum, the width of the pair-distribution function remains finite in practice. Moreover, pair- and one-particle distribution functions are not restricted to those of thermal equilibrium, but are essentially determined by the details of the excitation mechanism. That means that the character of the initially prepared highly excited state strongly influences the resulting optical spectrum obtained by absorption of a weak pulse probing the linear response of the highly excited semiconductor. Corresponding experiments were performed in recent years in order to investigate the properties of dense exciton systems under various excitation conditions (Fehrenbach et al. 1982, Peyghambarian et al. 1984, Chemla 1985, Schultheis et al. 1986).

In this contribution we shall focus our interest on special excitation conditions preparing a state in which—on a ps-time scale—only the lowest excitonic bound states are occupied. To prepare such a state, the width of the resonantly exciting pulse must be small in comparison with the gap in the two-particle spectrum, which for low densities is determined by the exciton binding energy, in order to guarantee that only excitonic states and not plasmalike states too are occupied. In recent experiments (Fehrenbach et al. 1985, Schultheis et al. 1986) it was shown unambiguously that the optical properties of such states differ essentially from corresponding states characterized by the same density of excited pairs, but being dominated by plasmalike excitations above the continuum edge. For details we refer to the contributions of R. Ulbrich in this volume. On the first sight, such a state of dense excitons is expected to become unstable at about the same critical density, called Mott density (Mott 1967), at which bound states in an e–h plasma cease to be stable. For sufficiently long times after excitation, all excitons should be ionized due to exciton–exciton scattering and a description in terms of plasma physics should be valid. If however the width of the initial pair distribution is small in comparison with the binding energy, the ionization threshold cannot be reached by collisions of two pairs only, but multiexciton scattering processes are necessary to generate pairs above the continuum edge. Moreover, ionization by scattering with LO-phonons, which is extremely effective at room temperature (Knox et al. 1985), can be excluded for sufficiently low temperatures. Thus, the gap in the two-particle spectrum acts as a potential barrier between the excitonic state and the plasmalike state even at densities for which the energy of the latter is lower than in the excitonic case. Obviously, we have to deal with a system far from equilibrium; therefore, the application of nonequilibrium quantum statistical methods is necessary in order to treat such a system adequately. In the following we shall employ a nonequilibrium Green's function technique developed by Keldysh (1965) and by Kadanoff and Baym (1962) independently. At first we give a brief introduction into the general formalism and summarise the basic equations.

II. GREEN'S FUNCTIONS FOR NONEQUILIBRIUM SYSTEMS

We consider a system that develops in time according to the Hamiltonian $H = H_0 + H_{\text{ex}}$ where H_0 describes electrons in a rigid lattice coupled to the electromagnetic field

$$
\begin{aligned}
H_0 = {} & \frac{1}{2m}\int d^3r\, \Psi^+(\mathbf{r}t)\left[-i\hbar\nabla - \frac{e}{c}\mathbf{A}(\mathbf{r}t)\right]^2 + U(\mathbf{r}t)\Psi(\mathbf{r}t) \\
& + \frac{1}{2}\int d^3r\, d^3r'\, \Psi^+(\mathbf{r}t)\Psi^+(\mathbf{r}'t)V(\mathbf{r}-\mathbf{r}')\Psi(\mathbf{r}'t)\Psi(\mathbf{r}t) + H_{\text{f}}
\end{aligned}
\tag{2.1}
$$

The operators of the transverse electromagnetic field in Coulomb gauge and the particle operators fulfill the usual commutation and anticommutation relations, respectively; U is the rigid lattice potential, V the Coulomb potential. H_f denotes the Hamiltonian of the free electromagnetic field. The inclusion of lattice vibrations is straightforward using the methods reviewed by Lundquist and Hedin (1969) and we shall omit this point in our discussion. H_{ex} represents time-dependent external sources that drive the system out of equilibrium:

$$H_{ex} = \int d^3r \left[\phi_{ex}(\mathbf{r}t)\Psi^+(\mathbf{r}t)\Psi(\mathbf{r}t) - \frac{1}{c}\mathbf{J}_{ex}(\mathbf{r}t)\mathbf{A}(\mathbf{r}t) \right] \tag{2.2}$$

The time development of an operator, for example Ψ, under the action of H_{ex} is determined by

$$\Psi(t) = T_c \exp\left[-\frac{i}{\hbar} \int_c \overline{H}_{ex}(t')\, dt' \right] \overline{\Psi}(t), \tag{2.3}$$

where the bare on the operators indicates the interaction representation. In contrast to an equilibrium theory, the t'-integration has to be performed along a closed contour C beginning at a time $t_0 \to -\infty$ and go through t and back to t_0. T_c is the time-order operator along this contour, effecting that operators with time arguments occurring earlier on the contour are on the right from those with "later" arguments. In general, C may be a contour in the complex t-plane and several realizations are possible. (For a detailed discussion of this point see Langreth 1976.) For our problem, it is most convenient to use the version of the theory proposed by Keldysh. In his real-time formalism C goes from $-\infty$ to $+\infty$ on a positive branch and then back from $+\infty$ to $-\infty$ on a negative branch (Keldysh 1965). From the definition of this contour, one sees immediately that a calculation of chronological products in the nonequilibrium case leads to four different Green's functions corresponding to the four possibilities to arrange two time arguments on the contour C. We define

$$G_{++}(12) = -\frac{i}{\hbar}\langle T_+ \Psi(1)\Psi^+(2)\rangle \tag{2.4}$$

$$G_{--}(12) = \frac{i}{\hbar}\langle T_- \Psi(1)\Psi^+(2)\rangle \tag{2.5}$$

$$G_{+-}(12) = -\frac{i}{\hbar}\langle \Psi^+(2)\Psi(1)\rangle \tag{2.6}$$

$$G_{-+}(12) = -\frac{i}{\hbar}\langle \Psi(1)\Psi^+(2)\rangle \tag{2.7}$$

The argument 1, e.g., is as usual an abbreviation for $(r_1 t_1 \sigma_1)$ and $\langle\ \rangle$ means $Tr\rho_0$, where ρ_0 is the density matrix of the unperturbed system. The first two Green's functions are the usual time and antitime-ordered functions with both time arguments on the positive branch or on the negative branch, respectively, whereas G_{+-} is the particle and G_{-+} the hole propagator. Diagrammatic expansions for these nonequilibrium Green's functions are obviously identical in form to the usual equilibrium expansions. The essential difference is that integrations about internal time variables have to be performed along the contour C, rather than from $-\infty$ to $+\infty$ as in the $T = 0$ equilibrium theory or from 0 to $-i\beta$ as in $T \neq 0$ equilibrium theory. Here we shall not go into details of deriving equations of motion for the Green's function, but refer to the literature (Schwinger 1961, Kadanoff and Baym 1962, Du Bois 1967). The Dyson equation determining the one-particle Green's function can be written in the compact form

$$G(\bar{1}\bar{2}) = G_H^0(\bar{1}\bar{2}) + \int d\bar{3} G_H^0(\bar{1}\bar{3}) \left[i\frac{\hbar e}{mc} \langle \mathbf{A}(\bar{3}) \rangle \nabla + \frac{e^2}{2mc^2} \langle \mathbf{A}(\bar{3}) \rangle^2 \right] G(\bar{3}\bar{2})$$

$$+ \int d\bar{3}\bar{4} G_H^0(\bar{1}\bar{3}) \Sigma(\bar{3}\bar{4}) G(\bar{4}\bar{2}), \tag{2.8}$$

where we have introduced $\bar{1} = r_1 t_1 \sigma_1 b_1$ and $\int d\bar{1} = \int d^3 r_1 \int_{-\infty}^{\infty} dt_1 \Sigma_{\sigma_1 b_1}$. As a consequence of the integration along the contour C, we have in addition to each internal time integration a sum over a double-valued index b labelling the two branches. Thus the Dyson equation has the form of a matrix equation for the four matrix elements 2.4 through 2.7. Further, we have introduced G_{H}^0 as Green's function of the Hartree–Hamiltonian with the Hartree potential given by

$$V_{\mathrm{H}}(\bar{1}) = \phi(\bar{1}) - i\hbar b_1 \int d\bar{2} V(\bar{1} - \bar{2}) G(\bar{2}\bar{2}^{\pm}) \tag{2.9}$$

where $2^{\pm}$ implies t_2^+ or t_2^- for $b_1 = +1$ or $b_2 = -1$, respectively, and the Coulomb potential is defined as

$$V(\bar{1} - \bar{2}) = V(\mathbf{r}_1 - \mathbf{r}_2) \delta(t_1 - t_2) \delta_{b_1 b_2}. \tag{2.10}$$

If we neglect transverse interaction effects and fluctuations of the electromagnetic field, namely the difference between $\langle A^2 \rangle$ and $\langle A \rangle^2$, then the set of equations determining the self-energy Σ, the screened interaction $W = V + \Delta W$, the polarization propagator P, and the vertex-function Γ have the same form as in equilibrium theory (Lundquist and Hedin 1969):

$$\Sigma(\bar{1}\bar{2}) = -ihb_1\int d\bar{3}\bar{4}W(\bar{3}\bar{1})G(\bar{1}\bar{4})\Gamma(\bar{4}\bar{2}\bar{3}), \tag{2.11}$$

$$W(\bar{1}\bar{2}) = V(\bar{1}-\bar{2}) + \int d\bar{3}\bar{4}V(\bar{1}-\bar{3})P(\bar{3}\bar{4})W(\bar{4}\bar{2}), \tag{2.12}$$

$$P(\bar{1}\bar{2}) = ihb_1\int d\bar{3}\bar{4}G(\bar{1}\bar{3})G(\bar{4}\bar{1})\Gamma(\bar{3}\bar{4}\bar{2}), \tag{2.13}$$

$$\Gamma(\bar{1}\bar{2}\bar{3}) = -\delta(\bar{1}-\bar{2})\delta(\bar{1}-\bar{3}) + \int d\bar{4}\bar{5}\bar{6}\bar{7}\frac{\delta\Sigma(\bar{1}\bar{2})}{\delta G(\bar{4}\bar{5})}G(\bar{4}\bar{6})G(\bar{7}\bar{5})\Gamma(\bar{6}\bar{7}\bar{3}). \tag{2.14}$$

The coupling of particles to the electromagnetic field is described by

$$\Box\langle\mathbf{A}(\mathbf{r}t)\rangle = -\frac{4\pi}{c}\mathbf{J}(\mathbf{r}t) \tag{2.15}$$

where $\mathbf{J}$ is the total transverse current, which is given by

$$\mathbf{J}(1) = \mathbf{J}_{\text{ex}}(1) - i\frac{\hbar e}{2mc}\lim_{1\to 1'}(\nabla_1 - \nabla_{1'})G(\bar{1}\bar{1}'^{\pm}) - \frac{e^2}{mc^2}\langle\mathbf{A}(1)\rangle G(\bar{1},\bar{1}^{\pm}) \tag{2.16}$$

The calculation of the Green's function matrix elements in any particular approximation for the self-energy can fortunately be reduced to the calculation of only two independent contributions, as can be seen from the following relations, which are a consequence of the definitions 2.4 through 2.7.

$$G_{++} + G_{+-} = G_{--} + G_{-+} = G_r \tag{2.17}$$

$$G_{++} - G_{-+} = G_{--} - G_{+-} = G_a \tag{2.18}$$

The same relations are fulfilled by Σ, W and P. Even with vanishing external sources, the system of eqs. 2.8 and 2.11 through 2.14 constitutes an insolvable mathematical problem and we have at first to introduce an approximation for G_0 in the equilibrium case. In a quasi-particle approximation we have

$$G^0_{r,a}(\omega) = \sum_{nk}\frac{|nk\rangle\langle nk|}{\omega - \epsilon_{n\mathbf{k}} \pm i\delta} \tag{2.19}$$

$$G^0_{+-}(\omega) = -2\pi i\sum_{nk} f(\epsilon_{n\mathbf{k}})|\,nk\rangle\langle nk|\,\delta(\omega - \epsilon_{n\mathbf{k}}), \tag{2.20}$$

$$G^0_{-+}(\omega) = -2\pi i\sum_{nk}[1 - f(\epsilon_{n\mathbf{k}})]\,|nk\rangle\langle nk|\,\delta(\omega - \epsilon_{n\mathbf{k}}). \tag{2.21}$$

where f is the Fermi function $|n\mathbf{k}\rangle$ and $\epsilon_{n\mathbf{k}}$ are Bloch states and Bloch energies, respectively, which can be calculated for a semiconductor using, for example, a screened Hartree–Fock approximation for the self-energy (Strimati et al. 1982, Loui 1986). From eqs. 2.19 through 2.21, we see that the retarded Green's function contains only information about the particle spectrum, whereas the information about the particle distribution is contained in the particle- and hole-propagators. In equilibrium, both informations can be incorporated in only one Green's function, namely the time-ordered one. This is no longer possible in a general nonequilibrium situation, and we have to deal with G_r and G_{+-} functions. The set of equations determining these functions will be derived in the following section.

III. DECOMPOSITION OF THE DYSON EQUATION

The effect of external sources or external fields, respectively, consists in the excitation of the system from an equilibrium state to a state far from equilibrium. For sufficiently weak external fields, this process can be described in the framework of linear response theory and therefore all properties of the particle system are the same as in equilibrium. However, if we consider the excitation by a strong pulse, all propagators become complicated functionals of the external field. We shall restrict ourselves to an excitation for which the width of the exciting pulse is very small compared to the gap of the semiconductor. Analyzing the structure of a functional expansion of the Green's function in powers of the external field, one can show that the Dyson equation (2.8) can be separated into two equations (Schäfer and Treusch 1986): one equation that contains only even powers of the external field and another one for the contributions of odd order.

$$G^{\mathrm{g}} = G_0 + G_0 A G^{\mathrm{u}} + G_0 \Sigma^{\mathrm{g}} G^{\mathrm{g}} + G_0 \Sigma^{\mathrm{u}} G^{\mathrm{u}} \tag{3.1}$$

$$G^{\mathrm{u}} = G_0 A G^{\mathrm{g}} + G_0 \Sigma^{\mathrm{u}} G^{\mathrm{g}} + G_0 \Sigma^{\mathrm{g}} G^{\mathrm{u}}, \tag{3.2}$$

where "g" means the even and "u" the odd contributions to the Green's function matrix. Σ^{g} contains only field-dependent contributions to the self-energy, whereas the field-independent terms are included in G^0. The A^2-contribution to eq. 3.1 is already omitted as the diamagnetic intraband interaction as well as the interband interaction is negligible in the energy range of the interest. Further, we introduced the abbreviation

$$A(\bar{1}) = \frac{i\hbar e}{mc}\langle \mathbf{A}(\bar{1})\rangle \nabla \,. \tag{3.3}$$

A further simplification is obtained if we assume that the one-particle states can be described within a two-band model in the effective mass approximation and Coulomb vertices coupling valence and conduction states are negligible. Then only the exciting field couples valence and conduction states and G^{g} is reduced to the diagonal and G^{u} to the off-diagonal elements of the one-particle Green's function. Thus we obtain

$$G^{aa} = G_0^{aa} + G_0^{aa} A G^{ba} + G_0^{aa}\Sigma^{aa}G^{aa} + G_0^{aa}\Sigma^{ab}G^{ba} \tag{3.4}$$

$$G^{ab} = G_0^{aa} A G^{bb} + G_0^{aa}\Sigma^{ab}G^{bb} + G_0^{aa}\Sigma^{aa}G^{ab}, \tag{3.5}$$

where a and b are indices of e or h, respectively.

Formally we can solve eqs. 3.4 and 3.5 with respect to the diagonal element of the self-energy by introducing:

$$\tilde{G}^{aa} = G_0^{aa} + G_0^{aa}\Sigma^{aa}\tilde{G}^{aa} \tag{3.6}$$

$$G^{ab} = \tilde{G}^{aa}(A + \Sigma^{ab})G^{bb} \tag{3.7}$$

$$G^{aa} = G_0^{aa} + G_0^{aa}\overline{\Sigma}^{aa}G^{aa} \tag{3.8}$$

with

$$\overline{\Sigma}^{aa} = \Sigma^{aa} + (A + \Sigma^{ab})\tilde{G}^{bb}(A + \Sigma^{ba}). \tag{3.9}$$

As already mentioned, it is not necessary to treat the complete Green's function matrix, as only G_r and G_{+-}, for example, contain independent information. Using the relations 2.17 and 2.18, one obtains for the retarded Green's functions

$$\tilde{G}_r^{aa} = G_{r0}^{aa} + G_{r0}^{aa}\Sigma_r^{aa}\tilde{G}_r^{aa} \tag{3.10}$$

$$G_r^{aa} = G_{r0}^{aa} + G_{r0}^{aa}\overline{\Sigma}_r^{aa}G_r^{aa} \tag{3.11}$$

$$G_r^{ab} = \tilde{G}_r^{aa}\left(A + \Sigma_r^{ab}\right)G_r^{bb} \tag{3.12}$$

with

$$\overline{\Sigma}_r^{aa} = \Sigma_r^{aa} + \left(A + \Sigma_r^{ab}\right)\tilde{G}_r^{bb}\left(A + \Sigma_r^{ba}\right). \tag{3.13}$$

For the particle propagator it is more convenient to work with an inverse (differential) representation, as we can use $G_a^{0^{-1}}G_{+-}^0 = G_{+-}^0 G_a^{0^{-1}} = 0$ and thus the structure of the inhomogeneity is simplified. In a symmetrized form we obtain:

$$G_a^{0^{-1}} G_{+-}^{aa} - G_{+-}^{aa} G_a^{0^{-1}} = \overline{\Sigma}_{+-}^{aa} G_a^{aa} + \overline{\Sigma}_r^{aa} G_{+-}^{aa} - G_{+-}^{aa} \overline{\Sigma}_a^{aa} - G_r^{aa} \overline{\Sigma}_{+-}^{aa}, \tag{3.14}$$

where we defined

$$\begin{aligned} \overline{\Sigma}_{+-}^{aa} &= \Sigma_{+-}^{aa} + \Sigma_{+-}^{ab} \tilde{G}_a^{bb} \left(A + \Sigma_a^{ba} \right) \\ &\quad + \left(A + \Sigma_r^{ab} \right) \tilde{G}_{+-}^{bb} \left(A + \Sigma_a^{ba} \right) + \left(A + \Sigma_r^{ab} \right) \tilde{G}_r^{bb} \Sigma_{+-}^{ba} . \end{aligned} \tag{3.15}$$

The corresponding equation for $\tilde{G}$ is simply obtained by replacing G by $\tilde{G}$ and $\overline{\Sigma}$ by Σ in eq. 3.14. The off-diagonal elements of the particle propagator obey the equation

$$\begin{aligned} & G_a^{0^{-1}} G_{+-}^{ab} - G_{+-}^{ab} G_b^{0^{-1}} \\ &\quad = \Sigma_r^{aa} G_{+-}^{ab} + \Sigma_{+-}^{aa} G_a^{ab} + \Sigma_{+-}^{ab} G_a^{bb} + \left(A + \Sigma_r^{ab} \right) G_{+-}^{bb} \\ &\quad\quad - G_{+-}^{ab} \Sigma_a^{bb} - G_r^{ab} \Sigma_{+-}^{bb} - G_r^{aa} \Sigma_{+-}^{ab} - G_{+-}^{aa} \left(A + \Sigma_a^{ab} \right) \end{aligned} \tag{3.16}$$

Eqs. 3.14 and 3.16 are a general starting point for the derivation of kinetic equations for the one-particle distribution function and the two-particle distribution caused by the exciting field and scattering processes within the particle system. Even in the lowest nontrivial approximation for the self-energy, a general solution seems to be impossible, due to the fact that the time translation variance of the system is broken by the external field. Thus, all propagators depend separately on two times instead of depending only on the differences of time variables. The same holds for space variables, if the excitation yields a spatially inhomogeneous system.

The usual procedure (Kadanoff and Baym 1962) to deal with this problem is to express all quantities in terms of sum and difference variables, e.g., $G(x_1 x_2) \rightarrow G(x, X)$ with $x = x_1 - x_2 = (\mathbf{r}t)$ and $X = (x_1 + x_2)/2 = (\mathbf{R}T)$. Assuming that all quantities are slowly varying functions of the sum variables, one can perform a gradient expansion and finally take the Fourier transform of the equation with respect to difference variables. The basic assumption of slow variations on the scale of sum variables is not justified in the case of short-pulse excitation of semiconductors. We shall briefly discuss the conditions under which such a treatment is valid.

The two basic processes that take place after the excitation of a semiconductor by a strong laser pulse are renormalization of spectral properties and relaxation of the system to a quasi-equilibrium state. Both processes are governed by the induced contributions to the self-energy which itself depends not only implicitly on the induced time scale via the time dependence of the particle-distribution functions, but also explicitly. As will be seen later in more detail, the rapidly varying components cancel in the diagonal elements of the self-energy and the explicit time dependence consists in rapid oscillations on the induced time scale with frequencies being of the order of the pulse width. This behaviour in general prevents the validity of a gradient expansion, and therefore the introduction of a unique induced time-scale is only possible for times that are large in comparison with the duration of the pulse. If we restrict ourselves to the description of processes that take place sufficiently long after the excitation, we can assume as instantaneous all those processes that take place on a time scale of the order of the pulse duration. In the case of ps-excitation of excitons, this assumption concerns the renormalization of one- and two-particle energies as well as the relaxation within the particle system. Pair–pair collisions and emission or absorption of acoustic phonons can be described on the slow time-scale. For even larger times after the excitation, transverse interaction effects, being responsible for radiative recombination, become important. The onset of this process is the end of the induced time-scale. On the basis of the discussed assumptions, the evaluation of eq. 3.14 is completely analogous to the well-known treatment in transport theory; for details, we refer to the literature (see, e.g., Kadanoff and Baym 1962, Langreth 1976). Thus, we obtain from eq. 3.14

$$H_a(q, X)G^{aa}_{+-}(q, X) = \bar{\Sigma}^{aa}_{-+}(a, X)G^{aa}_{+-}(q, X) - \bar{\Sigma}^{aa}_{+-}(q, X)G^{aa}_{-+}(q, X) \tag{3.17}$$

with

$$H_a(q, X) = i\hbar\frac{\partial}{\partial T} + \frac{\hbar\mathbf{k}\nabla_R}{m_a} \tag{3.18}$$

and $q = (\mathbf{k}, \omega)$, where we have neglected gradient corrections and the induced contribution to the Hartree potential. The latter vanishes for zero temperature due to the charge neutrality of the system and is negligible, provided that the width of the pair distribution is sufficiently small. The corresponding equation for $\tilde{G}$ is obtained from eq. 3.17 by replacing G by $\tilde{G}$ and $\bar{\Sigma}$ by Σ. For the retarded (advanced) Green's functions we have

$$G^{ab}_{r,a}(q, X) = \left[h\omega - \frac{\hbar^2k^2}{2m_a} - \bar{\Sigma}^{aa}_{r,a}(q, X)\right]^{-1}. \tag{3.19}$$

A corresponding treatment of eq. 3.16 yields the following equation for the off-diagonal elements that are responsible for the coupling of the particles to the exciting field.

$$\int d^3q\, d\omega \left[i\hbar \frac{\partial}{\partial T} - \epsilon_a(\mathbf{k} + \alpha\mathbf{q}) + \epsilon_b(\mathbf{k} - \beta\mathbf{q}) \right] \cdot G^{ab}_{+-}(\omega_1\mathbf{k}, \omega\mathbf{q}) e^{i\mathbf{q}\mathbf{R}} e^{-i\omega T}$$

$$\begin{aligned} = \int d^3q\, d\omega \Big\{ \big[&\Sigma^{aa}_r(\omega_1 + \omega/2, \mathbf{k} + \alpha\mathbf{q}) \\ &- \Sigma^{bb}_a(\omega_1 - \omega/2, \mathbf{k} - \beta\mathbf{q}) \big] G^{ab}_{+-}(\omega_1\mathbf{k}\omega\mathbf{q}) \\ &+ \big[M(\mathbf{k}) A(\omega, \mathbf{q}) + \Sigma^{ab}_r(\omega_1\mathbf{k}\omega\mathbf{q}) \big] G^{bb}_{+-}(\omega_1 - \omega/2, \mathbf{k} - \beta\mathbf{q}) \\ &- \big[M(\mathbf{k}) A(\omega, \mathbf{q}) + \Sigma^{ab}_a(\omega_1\mathbf{k}\omega\mathbf{q}) \big] G^{aa}_{+-}(\omega_1 + \omega/2, \mathbf{k} + \alpha\mathbf{q}) \\ &+ \Sigma^{aa}_{+-}(\omega_1 + \omega/2, \mathbf{k} + \alpha\mathbf{q}) G^{ab}_a(\omega_1\mathbf{k}\omega\mathbf{q}) \\ &+ \Sigma^{bb}_{+-}(\omega_1 - \omega/2, \mathbf{k} - \beta\mathbf{q}) G^{ab}_r(\omega_1\mathbf{k}\omega\mathbf{q}) \\ &+ \Sigma^{ab}_{+-}(\omega_1\mathbf{k}, \omega\mathbf{q}) \big[G^{bb}_a(\omega_1 - \omega/2, \mathbf{k} - \beta\mathbf{q}) \\ &- G^{aa}_r(\omega_1 + \omega/2, \mathbf{k} + \alpha\mathbf{q}) \big] e^{i\mathbf{q}\mathbf{R}} e^{-i\omega T} \Big\} \end{aligned} \tag{3.20}$$

with $\mathbf{R} = \alpha\mathbf{r}_1 + \beta\mathbf{r}_2$, $\alpha = m_a/M$, $\beta = m_b/M$, $M = m_a + m_b$. $M(\mathbf{k})$ is the optical transition matrix element between valence and conduction states.

Eq. 3.20 represents a rather complicated equation for the pair states which are generated by the pulse and interact with the surrounding pairs. The physical meaning of the various contributions will become more evident in the next section.

IV. THE SCREENED HARTREE–FOCK APPROXIMATION

In the following, we shall restrict our discussion to the first nontrivial approximation for the self-energy, which is obtained from eq. 2.11 by neglecting vertex corrections:

$$\Sigma(\overline{1}\overline{2}) = i\hbar b_1 W(\overline{2}\overline{1}) G(\overline{1}\overline{2}) \tag{4.1}$$

For low pair-densities this approximation is not sufficient. E.g., interaction effects that lead to bound states of excitons are neglected (compare Schäfer and Treusch 1986). For intermediate densities, however, all relevant many-particle effects will be shown to be described within this approximation. From eq. 4.1 we obtain by using eqs. 2.17 and 2.18 the self-energies

$$\Sigma_r^{ab}(12) = i\hbar\left[W_{+-}(21)G_r^{ab}(12) + W_a(21)G_{+-}^{ab}(12) - \delta_{ah}\delta_{ab}V(1-2)G_{+-}^{0}(12)\right], \tag{4.2}$$

$$\Sigma_a^{ab}(12) = i\hbar\left[W_{+-}(21)G_a^{ab}(12) + W_r(21)G_{+-}^{ab}(12) - \delta_{ah}\delta_{ab}V(1-2)G_{+-}^{0}(12)\right], \tag{4.3}$$

$$\Sigma_{+-}^{ab}(12) = -i\hbar W_{-+}(21)G_{+-}^{ab}(12), \tag{4.4}$$

$$\Sigma_{-+}^{ab}(12) = i\hbar W_{+-}(21)G_{-+}^{ab}(12), \tag{4.5}$$

where we have taken into account that the field-independent contribution to the self-energy is already included in G^0 and has to be subtracted. The coupled system of integral equations, which is obtained from the general equations in the SHF-approximation can be surveyed most easily in a diagrammatic representation, which is shown in Fig. 1. The Green's function $\tilde{G}$ is renormalized by the screened exchange interaction between excited particles. The fully renormalized Green's function additionally depends on a further self-energy contribution. In lowest order in the density (i.e. second order in the field), this term describes the effects of phase-space filling due to the Pauli principle. The off-diagonal elements of the self-energy determining this contribution are represented by the ladder diagram, which describes excitonic states excited by the effective field. The effective field is connected with the external field via the sum of bubble diagrams (compare eq. 2.16), taking polariton effects into account. Finally, the induced longitudinal polarizability that determines the screened interaction W is shown in the last diagram of Fig. 1 and will be discussed in detail in the next section.

Next we evaluate eq. 3.20 in the SHF-approximation. As can be seen from the ladder diagrams in Fig. 1, the off-diagonal elements of the Green's function are given by the product of a transverse polarizability with the exciting field

$$G_{+-}^{ab}(\omega_1\mathbf{k}, \omega\mathbf{q}) = P_{+-}^{ab}(\omega_1\mathbf{k}, \omega\mathbf{q})A(\omega\mathbf{q}) \tag{4.6}$$

Using eqs. 4.2 through 4.6 in 3.20, an equation for P_{+-}^{ab} can be derived. As a consequence of the dynamic character of the induced screening, the ω_1

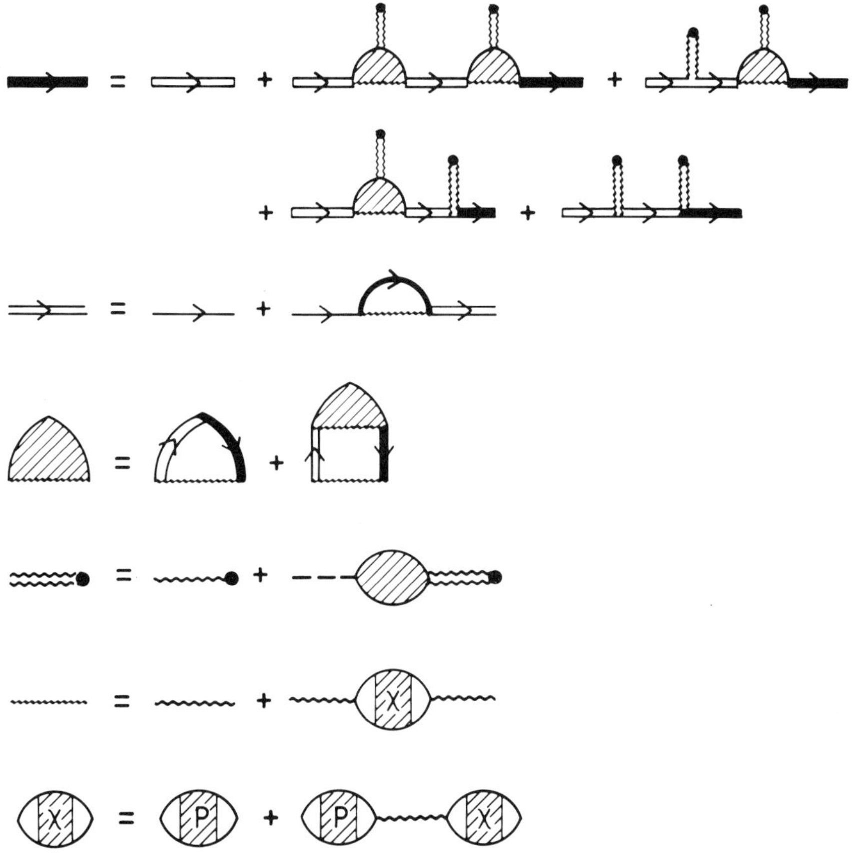

Figure 1. Diagrammatic representation of the set of integral equations, which determine the one-particle Green's function in SHF-approximation. The symbols have the following meanings: Green's functions G, and $\tilde{G}$, bare Coulomb-interaction V, screened interaction W, external vector-potential A_{ext}, effective vector-potential $\langle A \rangle$, transverse photon Green's function D^T — — — —, transverse polarizability P_{ch}. The longitudinal polarizability P is discussed in section V; see also Fig. 2.

frequency integration cannot be performed, because the pole structure of P with respect to ω_1 is not known. Such a problem is well known in exciton theory and usually treated in a first-order approximation with respect to the scattering with the longitudinal excitations that determine the screening. This so-called Shindo approximation (Shindo 1970, Haug and Tran Thoai 1978, Zimmermann et al. 1978) may be generalized to arbitrary order (compare Schäfer and Treusch 1986). In first order, one gets

$$P_{+-}^{ab}(\omega_1 \mathbf{k} \omega \mathbf{q}) = \frac{\left[G_{+-}^{bb}(\omega_1 - \omega/2, \mathbf{k} - \beta\mathbf{q}) - G_{+-}^{aa}(\omega_1 + \omega/2, \mathbf{k} + \alpha\mathbf{q})\right] P^{ab}(\omega \mathbf{k} \mathbf{q})}{\int \frac{d\omega_1}{2\pi}\left[G_{+-}^{bb}(\omega_1 - \omega/2, \mathbf{k} - \beta\mathbf{q}) - G_{+-}^{aa}(\omega_1 + \omega/2, \mathbf{k} + \alpha\mathbf{q})\right]} \tag{4.7}$$

Similar equations are valid for P_r^{ab} and P_a^{ab}, with G_{+-} in the denominator replaced by G_r or G_a, respectively. Within this approximation, we obtain from eq. 3.20, for P^{he}, e.g., the following equation (Schäfer and Treusch 1986):

$$\begin{aligned} &\left[i\hbar\frac{\partial}{\partial T} + \epsilon_{\mathrm{e}}(\mathbf{k} - \beta\mathbf{q}) - \epsilon_{\mathrm{h}}(\mathbf{k} + \alpha\mathbf{q}) + h\omega\right] P^{\mathrm{he}}(\omega\mathbf{k}\mathbf{q}) \\ &\quad -\hbar\sum_{k'} V(\mathbf{k} - \mathbf{k}')P^{\mathrm{he}}(\omega\mathbf{k}'\mathbf{q}) \\ &\quad -\hbar\sum_{k'}\{V(\mathbf{k} - \mathbf{k}')[1 - f_{\mathrm{eh}}(\mathbf{k}'\mathbf{q})]P^{\mathrm{he}}(\omega\mathbf{k}\mathbf{q}) - [1 - f_{\mathrm{eh}}(\mathbf{k}\mathbf{q})]P^{\mathrm{he}}(\omega\mathbf{k}'\mathbf{q}) \\ &\quad -\Delta W_{\mathrm{he}}(\omega\mathbf{k}\mathbf{k}'\mathbf{q})P^{\mathrm{he}}(\omega\mathbf{k}'\mathbf{q})/f_{\mathrm{eh}}(\mathbf{k}'\mathbf{q}) \\ &\quad +\Delta W_{\mathrm{he}}(\omega\mathbf{k}'\mathbf{k}\mathbf{q})P^{\mathrm{he}}(\omega\mathbf{k}\mathbf{q})/f_{\mathrm{eh}}(\mathbf{k}\mathbf{q})\} = f_{\mathrm{eh}}(\mathbf{k}\mathbf{q})M(\mathbf{k}), \end{aligned} \tag{4.8}$$

where we have introduced the following notation:

$$\begin{aligned} \Delta W_{\mathrm{he}}(\omega\mathbf{k}\mathbf{k}'\mathbf{q}) = \iint \frac{d\omega_1}{2\pi}\frac{d\omega_2}{2\pi} &\cdot\{[G_{+-}^{\mathrm{ee}}(\omega_1 - \omega/2, \mathbf{k} - \beta\mathbf{q})\,\Delta W_r(\omega_1 - \omega_2, \mathbf{k} - \mathbf{k}') \\ &- G_{+-}^{\mathrm{hh}}(\omega_1 + \omega/2, \mathbf{k} + \alpha\mathbf{q})\,\Delta W_a(\omega_1 - \omega_2, \mathbf{k} - \mathbf{k}') \\ &+ [G_a^{\mathrm{ee}}(\omega_1 - \omega/2, \mathbf{k} - \beta\mathbf{q}) - G_r^{\mathrm{hh}}(\omega_1 + \omega/2, \mathbf{k} + \alpha\mathbf{q})] \\ &\quad\cdot W_{+-}(\omega_1 - \omega_2, \mathbf{k} - \mathbf{k}')] \\ &\quad\cdot[G_{+-}^{\mathrm{ee}}(\omega_2 - \omega/2, \mathbf{k}' - \beta\mathbf{q}) - G_{+-}^{\mathrm{hh}}(\omega_2 + \omega/2, \mathbf{k}' + \alpha\mathbf{q})] \\ &- G_{+-}^{\mathrm{ee}}(\omega_1 - \omega/2, \mathbf{k} - \beta\mathbf{q})W_{-+}(\omega_1 - \omega_2, \mathbf{k} - \mathbf{k}') \\ &\quad\cdot[G_r^{\mathrm{ee}}(\omega_2 - \omega/2, \mathbf{k}' - \beta\mathbf{q}) - G_r^{\mathrm{hh}}(\omega_2 + \omega/2, \mathbf{k}' + \alpha\mathbf{q})] \\ &+ G_{+-}^{\mathrm{hh}}(\omega_1 + \omega/2, \mathbf{k} + \alpha\mathbf{q})W_{-+}(\omega_1 - \omega_2, \mathbf{k} - \mathbf{k}') \\ &\quad\cdot[G_a^{\mathrm{ee}}(\omega_2 - \omega/2, \mathbf{k}' - \beta\mathbf{q}) - G_a^{\mathrm{hh}}(\omega_2 + \omega/2, \mathbf{k}' + \alpha\mathbf{q})]\}. \end{aligned} \tag{4.9}$$

The distribution function f_{eh} is defined as

$$\begin{aligned} f_{eh}(\mathbf{kq}) &= 1 - f_e(\mathbf{kq}) - f_h(\mathbf{kq}) \\ &= i\int \frac{d\omega}{2\pi}\left[G^{ee}_{+-}(\omega,\mathbf{k}-\beta\mathbf{q}) - G^{hh}_{+-}(\omega,\mathbf{k}+\alpha\mathbf{q})\right]. \end{aligned} \tag{4.10}$$

Eq. 4.8 represents an effective exciton equation including phase-space filling, exchange and screening effects. These effects, however, are much more complicated here than in the case of plasma excitation, which is described by a formally equivalent equation (Zimmerman et al. 1978). In the plasma case, the distribution functions f are simply Fermi functions, while they have to be determined self-consistently for excitonic systems, in which excited and probed states are identical. This fact complicates also the screening properties, which are described by ΔW. To obtain insight into the self-consistency condition for the one-particle contribution, we shall calculate the self-energy contribution that describes the occupation of pair states. For simplicity, we shall only consider the case of static screening. The incorporation of dynamic effects is cumbersome but straightforward and not essential for the discussion of the basic physical effects. From eqs. 3.9 and 4.4, we obtain

$$\begin{aligned} &\Delta\Sigma^{bb}_{+-}(\omega,\mathbf{k},T,\mathbf{R}) \\ &\quad = (2\pi)^{-4} \sum_{\mathbf{q}_1\mathbf{q}_2\mathbf{Q}\mathbf{Q}'} \int d\omega_1\, d\omega_2\, d\Omega\, d\Omega' \\ &\qquad \cdot\{W(\mathbf{k}-\mathbf{q}_1)G^{ba}_{+-}[\omega_1+\Omega'/2,\mathbf{q}_1+\mathbf{Q}'/2-(\beta-\alpha)\mathbf{Q}/2,\Omega,\mathbf{Q}] \\ &\qquad \cdot\tilde{G}^{aa}_{+-}(\omega-\Omega/2+\Omega'/2,\mathbf{k}-\mathbf{Q}/2+\mathbf{Q}'/2) \\ &\qquad \cdot G^{ab}_{+-}[\omega_2-\Omega/2,\mathbf{q}_2-\mathbf{Q}/2-(\alpha-\beta)\mathbf{Q}'/2,\Omega',\mathbf{Q}'] \\ &\qquad \cdot W(\mathbf{k}-\mathbf{q}_2)e^{-i(\Omega+\Omega')T}e^{iR(\mathbf{Q}+\mathbf{Q}')}\}. \end{aligned} \tag{4.11}$$

Because dynamic screening effects are neglected, the ω_1- and ω_2-integrations can be performed without further assumptions. For the remaining frequency integrations, assumptions about the structure of the off-diagonal elements of G and the diagonal elements of $\tilde{G}$ are necessary. We assume a quasi-particle representation for the solution of eqs. 4.8 and 3.17, where all quantities may still depend on the induced variables R and T. This approximation is valid for slow variations in R and T.

$$P^{eh}(\Omega,\mathbf{k},\mathbf{Q}) = \sum_{nk'} \Psi^{eh}_n(\mathbf{k},\mathbf{Q})\Psi^{eh}_n(\mathbf{k}',\mathbf{Q})\big/[\Omega-\epsilon_n(\mathbf{Q})+i\delta]M(\mathbf{k}') \tag{4.12}$$

$$P^{he}(\Omega,\mathbf{k},\mathbf{Q}) = \sum_{nk'} \Psi^{he}_n(\mathbf{k},\mathbf{Q})\Psi^{he}_n(\mathbf{k}',Q)\big/[\Omega+\epsilon_n(\mathbf{Q})+i\delta]M(\mathbf{k}'). \tag{4.13}$$

Similarly, we approximate $\tilde{G}$ by

$$\tilde{G}^{ee}_{+-}(\omega, \mathbf{k}) = -2\pi i \tilde{f}_e(\mathbf{k})\delta[\omega - \tilde{\epsilon}_e(\mathbf{k})] \quad (4.14)$$

$$\tilde{G}^{hh}_{+-}(\omega, \mathbf{k}) = -2\pi i[1 - \tilde{f}_h(\mathbf{k})]\delta[\omega - \tilde{\epsilon}_h(\mathbf{k})], \quad (4.15)$$

where again the one-particle distribution $\tilde{f}$ and the quasi-particle energies $\tilde{\epsilon}$, renormalized by Σ, depend on R and T. These assumptions hold provided $\tilde{f}$ and $\tilde{\epsilon}$ are calculated consistently with eqs. 4.8 and 3.17. Having performed the frequency integrations, $\Delta\Sigma_{+-}$ can be written in the form

$$\Delta\Sigma^{ee}_{+-}(\omega \mathbf{k} T \mathbf{R}) = \sum_{nn'\mathbf{Q}\mathbf{Q}'} K(nn'\mathbf{Q}\mathbf{Q}'\omega k)\, e^{i[\epsilon_{n'}(\mathbf{Q}') - \epsilon_n(\mathbf{Q})]T} e^{i(\mathbf{Q}+\mathbf{Q}')\mathbf{R}} \sin xT/x \quad (4.16)$$

with

$$x = 2[\omega - \tilde{\epsilon}_h(\mathbf{k} - \mathbf{Q}/2 + \mathbf{Q}'/2)] - \epsilon_n(\mathbf{Q}) - \epsilon_{n'}(\mathbf{Q}'). \quad (4.17)$$

The quantity K is not given explicitly. The time dependence of $\Delta\Sigma_{+-}$ is characterized by the mentioned oscillations on the induced time-scale. Only in the limit of $T/\tau \gg 1$, where τ denotes the pulse duration, these oscillations become negligible. Formally we take the limit $T \to \infty$ and obtain as final results

$$\Delta\Sigma^{ee}_{+-}(\mathbf{k}, \omega) = 2\pi i \sum_{n\mathbf{Q}} \sigma^e_n(\mathbf{k}, \mathbf{Q}) n[\epsilon_n(\mathbf{Q})]\delta[\omega - \epsilon_h(\mathbf{k} - \mathbf{Q}) - \tilde{\epsilon}_n(\mathbf{Q})], \quad (4.18)$$

where we have introduced

$$\sigma^e_n(\mathbf{k}, \mathbf{Q}) = \sum_{\mathbf{q}\mathbf{q}'} W(\mathbf{k}, \mathbf{q})\Psi^{eh}_n(\mathbf{q} - \beta\mathbf{Q}, \mathbf{Q}) \cdot W(\mathbf{k}, \mathbf{q}')\Psi^{he}_n(\mathbf{q}' - \beta\mathbf{Q}, \mathbf{Q})[1 - \tilde{f}_h(\mathbf{K} - \mathbf{Q})] \quad (4.19)$$

The pair distribution function n is given by

$$n[\epsilon_n(\mathbf{Q})] = \tilde{\Psi}^{he}(0, \mathbf{Q})\tilde{\Psi}^{eh}(0, \mathbf{Q}) A_-[-\epsilon_n(\mathbf{Q}), -\mathbf{Q}] A_+[\epsilon_n(\mathbf{Q}), \mathbf{Q}] |M|^2 \quad (4.20)$$

with A_+ and A_- being the positive and negative frequency parts of A, respectively. $\tilde{\Psi}$ denotes the Fourier transform of Ψ.

In the last part of this section, we shall briefly discuss the quasi-equilibrium limit. From eq. 3.17, we see that in this limit the detailed-balance condition holds:

$$\bar{\Sigma}^{aa}_{+-} G^{aa}_{-+} = \bar{\Sigma}^{aa}_{-+} G^{aa}_{+-}. \quad (4.21)$$

From eqs. 4.21, 2.17 and 2.18, we obtain for G_{+-} the representation

$$G_{+-}^{aa}(\omega\mathbf{k}) = g(\omega\mathbf{k})\,\mathrm{Im}\,G_r^{aa}(\omega,\mathbf{k}) \tag{4.22}$$

with

$$g(\omega\mathbf{k}) = \frac{\overline{\Sigma}_{+-}(\omega\mathbf{k})}{\mathrm{Im}\,\overline{\Sigma}(\omega\mathbf{k})}. \tag{4.23}$$

In an equilibrium situation, g reduces to the Fermi function $\mathrm{f}(\omega)$. Due to the dynamic character of the self-energy $\Delta\Sigma_{+-}$, $\mathrm{Im}\,G_r$ possesses two poles: the usual quasi-particle pole and a pole induced by the exciting field, which corresponds to one-particle states contributing to the excitonic state. For a Bose-condensed state, these poles can be calculated analytically. For holes one obtains, e.g.,

$$\omega_{\pm}^{\mathrm{h}} = \tfrac{1}{2}(\tilde{\epsilon}_{\mathrm{h}} + \tilde{\epsilon}_{\mathrm{e}} - \epsilon) \pm \tfrac{1}{2}\left[(\tilde{\epsilon}_n - \tilde{\epsilon}_{\mathrm{e}} + \epsilon)^2 + 4\sigma^{\mathrm{h}}N\right]^{1/2} \tag{4.24}$$

with $\epsilon = \epsilon_{1s}(0)$.

From eqs. 4.8 and 4.22, one finds again the well-known $T = 0$ pair formalism (see, e.g., Zimmermann 1976, Comte and Nozieres 1982, Schäfer and Treusch 1986). If one is interested in the physically more relevant case of a pair distribution with a finite width, the pole structure of G_{+-} and G_{-+} functions have to be analyzed numerically. From such an analysis, one can construct again a simple representation:

$$G_{+-}^{\mathrm{hh}}(\omega,\mathbf{k}) = -2\pi i\left[1 - \mathrm{f}_{\mathrm{h}}(\mathbf{k})\right]\delta\left[\omega - E_{+}^{\mathrm{h}}(\mathbf{k})\right] \tag{4.25}$$

$$G_{-+}^{\mathrm{hh}}(\omega\mathbf{k}) = -2\pi i\mathrm{f}_{\mathrm{h}}(\mathbf{k})\,\delta\left[\omega - E_{-}^{\mathrm{h}}(\mathbf{k})\right] \tag{4.26}$$

$$G_{+-}^{\mathrm{ee}}(\omega,\mathbf{k}) = -2\pi i\mathrm{f}_{\mathrm{e}}(\mathbf{k})\,\delta\left[\omega - E_{-}^{\mathrm{e}}(\mathbf{k})\right] \tag{4.27}$$

$$G_{-+}^{\mathrm{ee}}(\omega,\mathbf{k}) = -2\pi i\left[1 - \mathrm{f}_{\mathrm{e}}(\mathbf{k})\right]\delta\left[\omega - E_{+}^{\mathrm{e}}(k)\right] \tag{4.28}$$

where $E_{\pm}$ are complex energies describing position and width of the poles in the spectral function, which are thus approximated by smeared out δ-functions. The one-particle distribution is given by the frequency integral over eq. 4.22. This representation permits an investigation of screening properties in arbitrary order in the density.

V. THE INDUCED POLARIZABILITY

Just as in the one-particle description discussed in the previous section, the exciting pulse modifies the two-particle properties by inducing additional many-particle effects. In the SHF-approximation, these effects are described as induced screening. Formally we can analyze the two-particle properties in the highly excited system by the same methods used for the one-particle Green's function. The technical details however are much more involved and even a brief outline is beyond the scope of this contribution. For a formal discussion we refer to Schäfer and Treusch (1986). Here we shall only sketch the main ideas and discuss the most important results. Starting point is eq. 2.13, which is given in a screened ladder approximation by

$$P(\bar{1}\bar{1}'\bar{2}\bar{2}') = -i\hbar b_1 G(\bar{1}\bar{2})G(\bar{2}'\bar{1}') + i\hbar b_1 \int d\bar{3}\bar{4} G(\bar{1}\bar{3})G(\bar{4}\bar{1}')W(\bar{4}\bar{3})P(\bar{3}\bar{4}\bar{2}\bar{2}') \tag{5.1}$$

As in the case of one-particle Green's functions, one can perform a decomposition in contributions being of even or odd order in the exciting field. The diagrammatic representation of the result for the contributions of even order is shown in Fig. 2. Obviously the physical meaning of these diagrams is twofold. Matrix elements of P that become resonant at energies of the order of the gap-energy, namely P^{eheh}, P^{hehe}, P^{heeh} and P^{ehhe}, describe the linear response of the highly excited system to a weak probe beam. The evaluation of the corresponding diagrams leads to an equation that is very similar to eq. 4.8, but contains additional phase-space and fermionic exchange contributions. The matrix elements P^{eeee}, P^{hhhh} and P^{eehh} are responsible for the induced screening effects, because they have resonances for energies in the vicinity of the gap of the two-particle spectrum. For the calculation of these contributions, we first have to perform the decomposition with respect to the branch indices. In general, this will lead to a system of sixteen coupled equations, for which up to now a general formal treatment is missing. However, in the static approximation for the screened interaction, a calculation of the diagrams in terms of G_a, G_r and G_{+-} functions is straightforward. Here we shall only present the final results for the sum of diagrams denoted as P_0, which are obtained using the same assumptions as those discussed in the previous section.

Using eq. 4.22, we obtain from the first diagram in Fig. 2:

$$P_r^{(1)}(\omega, \mathbf{q}) = 2 \sum_{\mathbf{k},\, a=\mathrm{e,h}} \left[\frac{f_a(\mathbf{k}) - f_a(\mathbf{k}+\mathbf{q})}{\omega - E_+^a(\mathbf{k}+\mathbf{q}) + E_-^a(\mathbf{k})} \right] \tag{5.2}$$

which has the usual form of a Lindhard polarizability for electrons and holes. The physical properties, however, are very different for densities for which the gap in the spectrum is still existent and the properties of the one-particle distribution function are essentially determined by the electron–hole correlation.

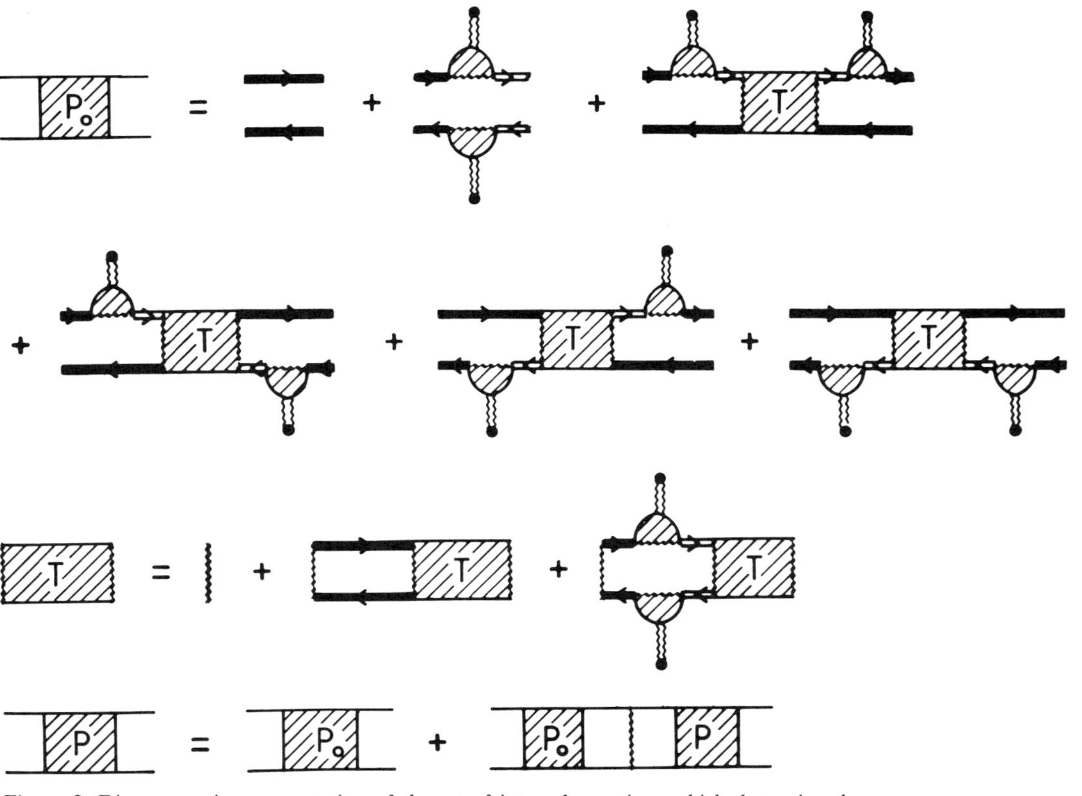

Figure 2. Diagrammatic representation of the set of integral equations which determine the longitudinal polarizability in SHF approximation.

The second diagram in Fig. 2 yields:

$$P_r^{(2)}(\omega, q) = -2 \sum_{\substack{n\mathbf{Q} \\ a \neq b}} \Psi_n^{ab}(\mathbf{k} + \mathbf{q} - \alpha\mathbf{Q}, \mathbf{Q})\Psi_n^{ba}(\mathbf{k} - \alpha\mathbf{Q}, \mathbf{Q}) n[\epsilon_n(\mathbf{Q})]$$

$$\cdot \left[\frac{1}{\omega - E_+^a(\mathbf{k} + \mathbf{q}) + E_-^a(\mathbf{k} - \mathbf{Q})} - \frac{1}{\omega + E_+^a(k + q) - E_-^a(\mathbf{k} - \mathbf{Q})} \right]. \quad (5.3)$$

In the Bose condensed limit, the sum of $P^{(1)}$ and $P^{(2)}$ corresponds to the form of excitonic screening which has been discussed by Nozieres and Comte (1982) and Haug and Schmitt-Rink (1984). The last four contributions to P_0 can be summarized as

$$P_r^{(3)}(\omega, q) = 2 \sum_{nn', \mathbf{Q}} \left| \langle \Psi_n | e^{i\alpha\mathbf{q}\mathbf{r}} - e^{-i\beta\mathbf{q}\mathbf{r}} | \overline{\Psi}_{n'} \rangle \right|^2 n[\epsilon_n(Q)]$$

$$\cdot \left[\frac{1}{\omega + \epsilon_n(Q) - \bar{\epsilon}_{n'}(\mathbf{q} + \mathbf{Q})} - \frac{1}{\omega - \epsilon_n(\mathbf{Q}) + \bar{\epsilon}_{n'}(\mathbf{Q} - \mathbf{q})} \right]. \quad (5.4)$$

The intermediate states $\overline{\Psi}_{n'}$ differ from the occupied states Ψ_n due to the fact that the T-matrix in the diagrams in Fig. 2 is completely renormalized with respect to the exciting field. However, the renormalization of the polarizability that couples to the exciting field is necessarily incomplete. In the low-density limit, eq. 5.4 corresponds to the excitonic polarizability (Röpke and Der 1979). For practical application of eq. 5.4 for arbitrary densities, it is suitable to separate the polarizability into the intraband ($n = n'$) and interband contributions. The latter can then be treated in an effective-gap approximation. To compare the relative importance of the different contributions, we have exactly evaluated eqs. 5.2 through 5.4 for $\omega = 0$ in the low-density limit, where wavefunctions are given by hydrogenic functions. (See, e.g., Landau and Lifschitz 1979.) Results are shown for $T = 0$ and $T = 15$ K in Figs. 3 and 4. The interband contribution to eq. 5.4 turns out to depend not very sensitively on the temperature, whereas the intraband contribution decreases rapidly with increasing temperature. Thus, we conclude that in the range of relevant temperatures, the intraband scattering is of minor importance, in contrast to other statements in the literature (Collet 1985). This can be understood quite easily if we consider the dependence on the wave-vector. An expansion for small q starts with q^4 for any pair distribution of finite width. If we take the limit of a Bose condensed state, the leading order is q^2. Thus, the rather artificial assump-

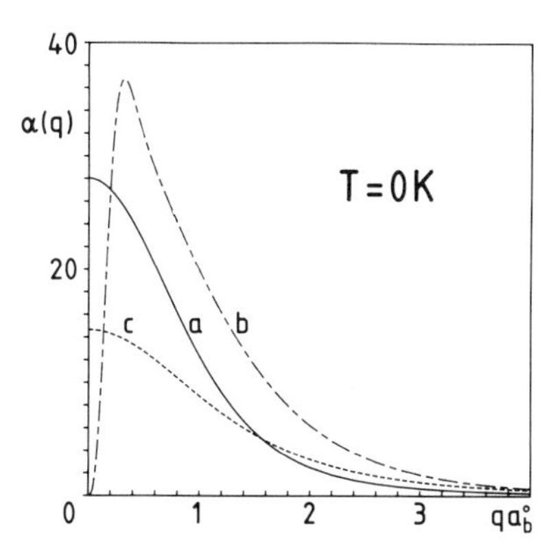

Figure 3. Different contributions to the static polarizability $\alpha(q) = v(q)P(q)$ in the low-density limit for $T = 0$: a: interband contribution to P^3; b: intraband contribution to P^3; c: sum of P^1 and P^2.

tion of vanishing width of the pair-distribution function converts a quadrupol contribution to the dielectric function into a dipole contribution. This leads to a drastic overestimation of the intraband scattering. It should be noted that the same reason leads to an even more extreme overestimation of the bosonic exchange contributions to the self-energy, which were recently calculated in the low-density limit (Stolz and Zimmermann 1984). The third curve in Figs. 3 and 4 shows that for very low densities, the contribution of $P^{(1)} + P^{(2)}$ is nearly negligible. In the nonlinear density region, the relative importance of $P^{(1)}$ increases, whereas the contributions of $P^{(2)}$ and $P^{(3)}$ decrease, due to the change of the matrix elements (see Fig.

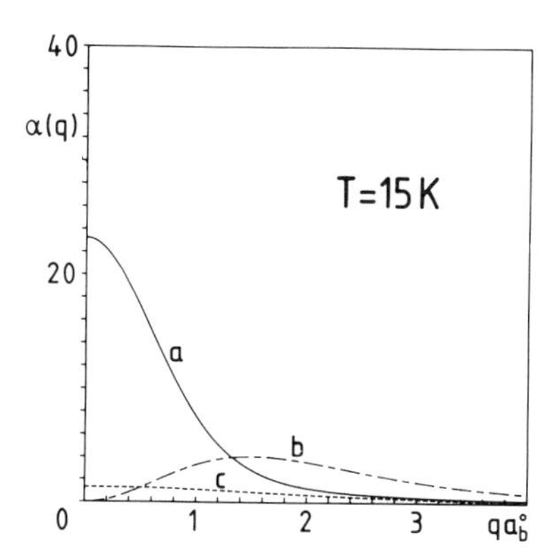

Figure 4. Different contributions to the excitonic polarizability in the low-density limit for $T = 15$ K.

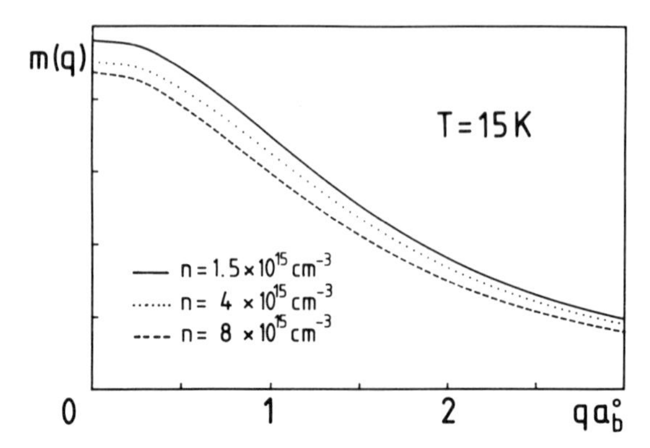

Figure 5. The matrix element determining the intraband polarizability in a mean-gap approximation $m(q) = 2v(q)\langle\Psi_0|1 - e^{iqr}|\Psi_0\rangle$ for different densities.

5) which overcompensates the effect of the gap shrinkage. At very high densities, $P^{(1)}$ reduces to the Lindhard polarizability of the independent electron and hole subsystems, while all other contributions vanish.

VI. RESULTS AND DISCUSSION

Up to now, numerical calculations have only been performed in the quasi-equilibrium limit, which was sketched at the end of section IV. Further basic assumptions are discussed in the following.

A realistic self-consistent computation of the two-particle distribution from the solution of eq. 3.20 would require that in eq. 4.8 also higher-order scattering processes would be incorporated. Probably, scattering processes beyond the SHF scheme are important. Further, the change of distribution with induced time becomes crucial. To perform such a program numerically without incisive simplifications is clearly out of the range of present computational possibilities. Thus, we assume that on the relevant time scale no scattering processes from the excitonic groundstate into the continuum will occur, and we take the pair distribution as a Boltzmann distribution. The second basic assumption concerns the treatment of the screening functions. The excitonic interband polarizability as well as the contributions $P^{(1)}$ and $P^{(2)}$ are treated in a mean-gap approximation. Assuming the mean gap to be given by the binding energy, this parameter is fixed by the numerical solution of the homogeneous part of eq. 4.8. Intraband contributions can be treated exactly. With these assumptions, it is possible to

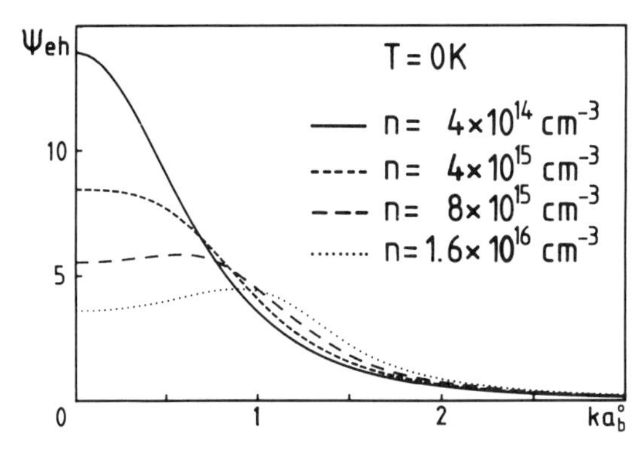

Figure 6. Pair wave-functions at $T = 0$ for different densities.

construct an analytical model for the screening function, which enters the evaluation of eq. 4.9, as well as eqs. 4.25 through 4.28. The rest of the work can be done using numerical standard methods.

We have solved eq. 4.8 self-consistently with respect to the pair-wave functions the one-particle distribution function and the mean gap for densities in the range of the Mott density and for different widths of the pair distribution function. Material parameters, namely electron- and hole-masses, binding energy and Bohr radius are chosen to describe approximately GaAs:

$$m_e = 0.07m_0, \qquad m_h = 0.48m_0,$$

$$a_B^0 = 110\text{ Å} \quad \text{and} \quad \epsilon_b^0 = 4.2\text{ meV}.$$

Results for pair-wave functions are shown in Figs. 6 and 7. For $T = 0$, we find a strong reduction of the amplitude, which decreases with increasing

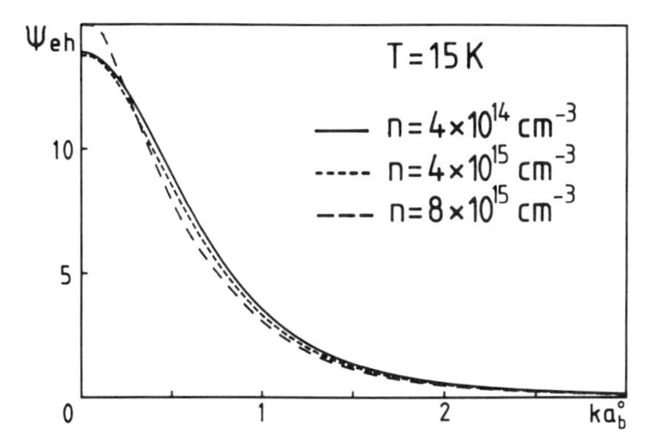

Figure 7. Pair wave-functions at $T = 15$ K for different densities.

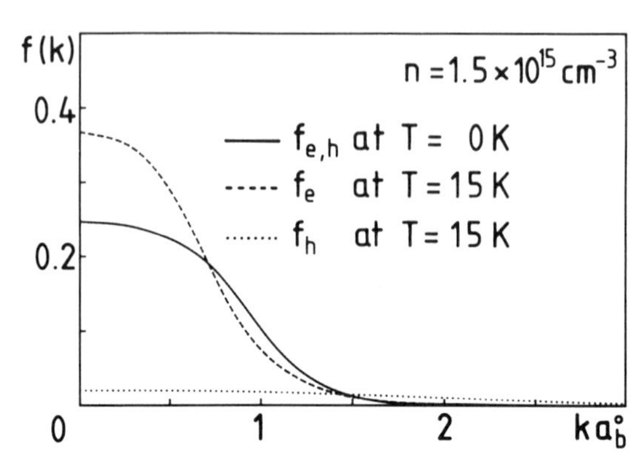

Figure 8. One-particle distribution functions for electrons and holes at $T = 15$ K in comparison to the one-particle distribution at $T = 0$.

density, reflecting the large phase-space effects in a "cold" exciton gas (compare also Schmitt-Rink et al. 1985). At a temperature of 15 K, which corresponds to a width of the pair-distribution of 0.3 ϵ_b^0, this effect has vanished completely. Up to densities of 5×10^{15} cm^{-3}, the pair-wave function is nearly unchanged, as well as the excitonic groundstate energy, indicating that also the oscillator strength and, thus, the optical spectrum in the vicinity of the excitonic peak remains the same as in the low-density limit. However, the binding energy is reduced due to the gap shrinkage to a value of 0.7 ϵ_b^0. At the same density and for $T = 10$ K, the height of the excitonic peak in the optical spectrum calculated for the case of plasma excitation (Haug and Schmitt-Rink 1984) is reduced to one-fourth of the low density value. This shows that in the excitonic phase at finite temperatures, highly correlated pair-states are considerably more stable than in the plasma phase. Phenomenologically, there are at least two reasons for this higher stability. At first, the excitonic screening is much less effective than plasma screening, as a consequence of the gap in the two-particle spectrum. The second reason is that for finite temperatures, excitonic phase space effects decrease rapidly with increasing temperatures due to the large hole mass in GaAs, as is illustrated in Fig. 8.

ACKNOWLEDGMENTS

I am grateful to J. Treusch and R. Ulbrich for stimulating discussions. For their cooperation during the present investigation, I thank R. Binder and

K. H. Schuldt. Furthermore, financial support from the Deutsche Forschungsgemeinschaft is acknowledged.

REFERENCES

Brittin, W. E., and Sakakura, A. Y. (1980). *Phys. Rev. A* **21**, 2050.
Chemla, D. S. (1985). *High Excitation and Short Pulse Phenomena* (M. H. Pilkhuhn, ed.). North-Holland, Amsterdam.
Collet, J. (1985). *J. Phys. Chem. Sol.* **46**, 417.
Comte, C., and Nozieres, P. (1982). *J. Physique* **43**, 1069.
Du Bois, D. F. (1967). *Lectures in Theoretical Physic*, **IXc**, W. E. Brittin (ed.). Gordon and Breach.
Fehrenbach, G. W., Schäfer, W., Treusch, J., and Ulbrich, R. G. (1982). *Phys. Rev. Lett.* **49**, 1281.
Fehrenbach, G. W., Schäfer, W., Schuldt, K. H., Treusch, J., and Ulbrich, R. G. (1985). *Proc. 17th Int. Conf. Phys. Semicond.* (J. D. Chadi and W. A. Harrison, eds.). Springer-Verlag, New York.
Girardeau, M. (1978). *J. Math. Phys.* **19**, 2605.
Hanamura, E. (1970). *J. Phys. Soc. Japan* **29**, 50.
Hanamura, E., and Haug, H. (1978). *Phys. Rev.* **43C**, 423.
Haug, H., and Schmitt-Rink, S. (1984). *Proc. Quantum Electron.*, **Vol 9**, 3.
Haug, H., and Tran Thoai, D. B. 91978). *Phys. Stat. Sol. B* **85**, 561.
Kadanoff, L. P., and Baym, G. (1962). *Quantum Statistical Mechanics*. W. A. Benjamin, New York.
Keldysh, L. V. (1965). *Sov. Phys. JETP* **20**, 4.
Keldysh, L. V., and Kopaev, Y. V. (1965). *Sov. Phys. Solid State* **6**, 2219.
Knox, W. H., Fork, R. L., Downer, M. C., Miller, D. A. B., Chemla, D. S., and Shank, C. V. (1985). *Phys. Rev. Lett.* **54**, 1306.
Landau, L. D., and Lifschitz, E. M. (1979). *Lehrbuch der Theoretischen Physik,* Bd. III. Akademie Verlag, Berlin.
Langreth, D. C. (1976). *Linear and Nonlinear Electron Transport in Solids*. (D. T. Devreese and V. E. Vàn Doren, eds.). Plenum Press.
Loui, S. (1986). *Proc. 18th Int. Conf. Phys. Semicond.* (Engström O. ed). Chalmers, Göteborg.
Lundquist, S., and Hedin, S. (1969). *Solid State Physics,* vol. 23, 1.
Mott, N. F. (1967). *Philos. Mag.* **6**, 287.
Nozieres, P., and Comte, C. (1982). *J. Physique* **43**, 1083.
Peyghambarian, N., Gibbs, H. M., Jewell, J. L., Antonetti, A., Migus, A., Hulin, D., and Mysyrowicz, A. (1984). *Phys. Rev. Lett.* **53**, 2433.
Röpke, G., and Der, R. (1979). *Phys. Status Solidi B* **92**, 501.
Röpke, G., Seifert, T., Stolz, H., and Zimmermann, R. (1980). *Phys. Status Solidi B* **100**, 215.
Schäfer, W., and Treusch, J. (1986). *Z. Phys. B* **63**, 407.

Schmitt-Rink, S., Chemla, D. S., and Miller, D. A. B. (1985). *Phys. Rev. B* **32**, 6601.
Schultheis, L., Kuhl, J., Honold, A., and Tu, C. W. (1986). *Proc. 18th Int. Conf. Phys. Semicond.* [Engström O. ed]. Chalmers, Göteborg.
Schwinger, J. (1961). *J. Math. Phys.* **2**, 407.
Shindo, K. (1970). *J. Phys. Soc. Japan* **29**, 287.
Stolz, H., and Zimmermann, R. (1984). *Phys. Status Solidi B* **124**, 201.
Strinati, G., Mattausch, H. J., and Hanke, W. (1980). *Phys. Rev. Lett.* **45**, 290.
Zimmermann, R., (1976). *Phys. Status Solidi B* **75**, 191.
Zimmermann, R., Kilimann, K., Kraeft, W. G., Kremp, D., and Röpke, D. (1978). *Phys. Status Solidi B* **90**, 175.

7 OPTICAL DECAY AND SPATIAL RELAXATION

G. Mahler and T. Kuhn

INSTITUT FÜR THEORETISCHE PHYSIK
UNIVERSITÄT STUTTGART, PFAFFENWALDRING 57
D-7000 STUTTGART 80, FEDERAL REPUBLIC OF GERMANY

A. Forchel and H. Hillmer

4 PHYSIKALISCHES INSTITUT
UNIVERSITÄT STUTTGART, PFAFFENWALDRING 57
D-7000 STUTTGART 80, FEDERAL REPUBLIC OF GERMANY

OPTICAL NONLINEARITIES
AND INSTABILITIES IN
SEMICONDUCTORS

ISBN 0-12-332915-9

I. INTRODUCTION

The following chapter describes theoretical and experimental results of a study of ambipolar transport in optically excited semiconductors. Laser-excited semiconductors generally have to be treated as nonequilibrium systems. Usually the nonequilibrium description is reduced to a quasi-equilibrium concept in which the local decay by recombination of carrier pairs is included. The typical experimental situation, however, includes strongly inhomogeneous spatial conditions such as excitation within a thin (about 1 μm) surface layer. The spatial inhomogeneities imply that in addition to recombination, transport effects have to be considered in order to describe the properties of the ambipolar carrier system.

Linear optics is concerned with the optical response exclusively due to intrinsic (i.e. material) properties of the respective system in thermal equilibrium. It follows that this linear response must break down when the perturbing electromagnetic field becomes large enough to modify the thermal reference state. It is a general trend of nonequilibrium states that they are less "universal" in the sense that more details—even on a macroscopic level—become relevant for their specification.

In all cases the most detailed description refers to the actual state of the coupled electron–phonon system. If this driven state is known, the radiative dissipation channels (spontaneous luminescence) can be analyzed. Supplementary, the linear response with respect to a second weak test beam can be calculated and compared with the experiment (Klingshirn *et al.*, 1981). In the "macrolimit" this response defines a local complex dielectric function depending on the parameters of the driving field, which thus bridges the gap to the more conventional phenomenological description (Shen, 1976). For inhomogeneous systems with complicated internal structure, however, such a description will, in general, no longer be adequate.

In many cases, the investigation has to cope with a number of serious problems:

1. Description of the nonequilibrium state in terms of single-particle distribution functions (Mahler *et al.*, 1985),
2. Influence of many-body effects on the electronic quasi-particle energies,
3. Influence of many-body effects (electron–hole correlations) on the coupling to the phonon field, (Schmitt-Rink *et al.*, 1980)
4. Influence of the finite size effects (more generally, of spatially inhomogeneous systems) on the kinetics (Kuhn *et al.*, 1987).

In addition, there may even be cross-effects, e.g., an influence of nonequilibrium on self-energy corrections.

The purpose of our present study is to investigate primarily problems (1) and (4). As we shall show, (2) can easily be incorporated to lowest order, while (3) is expected to be of minor importance, at least in the cases addressed in our theoretical approach.

The first part of this chapter addresses the theoretical description of carrier transport. In a second part, optical time-of-flight studies of ambipolar transport in Si are presented and compared with the theory.

II. TRANSPORT MODEL

A. Definition of the System

1. Structure Model

It is common practice in the discussion of the elementary excitations of solid state material to assume the structure (i.e., the positions of the nuclei) to be given. This is well justified and easily verified experimentally for macroscopic crystalline systems. In the present case, however, we shall allow for intentional structural changes (i.e., changes in local material parameters) with respect to one spatial direction, say *z*-direction. In such cases, the structural specification will typically be less well known and supplemented by model assumptions. If those changes are discontinuous on the length scale of the lattice constant a_0, the resulting structure is of layer-type composition. If, furthermore, the thickness of each layer is typically in the micron range, the electronic single particle may still adequately be described within the effective mass approximation (EMA). For length scales in the submicron region, size quantization with respect to the spectrum may come into play, leading to subband structures or continuum resonances (Kelly *et al.*, 1985).

Presently we shall consider the simplest structural model, an individual isolated layer of thickness $L \gg a_0$ (slab geometry). The influence of the two surfaces at $z = 0$ and $z = L$ on the particle kinetics will be modelled by fixed reflection coefficients R_0, R_L. This means that a particle arriving at the respective surface is specularly reflected with probability $R_{0,L}$ and recombines with probability $1 - R_{0,L}$. Reflection coefficients close to 1 correspond to "good-quality" surfaces in the sense that they contribute only a low density of surface recombination channels.

2. Quasi Particles

The electronic excitations are assumed to be given by the bulk band structure, reduced here to a two-band scheme with band gap E_g. We shall restrict ourselves to an indirect semiconductor. This restriction has consequences with respect to the time scale of the spontaneous decay, τ_{sp} (it is then the slowest process in our problem and can be neglected for times $\ll \tau_{sp}$) and with respect to the selection rules to be applied for the luminescence spectra. Electron- and hole-band, respectively, are assumed to be isotropic and nondegenerate, and specified by the EMA parameters m_e, m_h. This means that for Si, for example, properly averaged band parameters have to be used. The lattice excitations are also taken as the bulk phonon modes, reduced to a single effective acoustic mode of constant sound velocity v_s. The neglect of optical phonons is justified, provided the excess energy, E_{exc}, of the electronic particles is less than the optical frequency, and the phonon temperature T_{ph} is low enough.

This three-component quasi-particle system is still too complicated for a detailed kinetic study. In the excitonic limit the electronic subsystem is well described by a single component with effective mass $m^* = m_e + m_h$. In the plasma limit we apply the "pseudospin approximation" (Kuhn *et al.*, 1987). Here it is assumed that electrons and holes are indistinguishable particles except for their charge. This means that $m_e^* = m_h^* = m^* = (m_e + m_h)/2$ and all coupling constants are the same, which allows us to treat the electronic subsystem *kinetically* as a single-component system, as the charge does not couple to the translational degrees of freedom. It is thus the analogue of the so-called ambipolar approximation usually applied on the hydrodynamic level (Mahler *et al.*, 1985).

3. Excitation Model

The laser with excitation energy $\hbar\omega_L$ creates electrons and holes of excess energy

$$E_{exc} = \frac{\hbar^2 k_L^2}{2m^*} = \frac{1}{2}(\hbar\omega_L - E_g) \tag{1}$$

and isotropic in $\mathbf{k}$-space. The effect of interactions on the quasi-particle dispersion relation could easily be included to lowest order, if we replace E_g by $E_g[n(z)] = E_g + \Delta E_{xc}[n(z)]$, where ΔE_{xc} is the mean-field contribution of exchange and correlation.

It is assumed that the laser spot diameter is large compared to L, so that we can neglect any radial dependence. For the local generation rate g we thus write

$$g(\mathbf{k}, z, t) = g(t)\,\delta(k_L^2 - k^2)e^{-\lambda z} \tag{2}$$

The time dependence of a pulse excitation is modelled by a Gaussian of width τ_L. $\lambda(E_{exc})$ is the reciprocal penetration depth, which may also depend on the electronic state built up in the surface region; such a nonlinear effect, however, would only change the length scale and will not be considered in detail.

B. Boltzmann Equation

1. Phonon system

The phonons are supposed to remain in thermal equilibrium at temperature T_{ph}: their distribution function is therefore known, $f_{ph} = f_{ph0}$, and independent of the electronic state. This bath approximation considerably simplifies our problem, but excludes such phenomena as the so-called phonon-wind (Wolfe, 1983).

2. Electronic System

The quasi-classical Boltzmann equation is in the absence of external forces

$$\frac{\partial f_e}{\partial t} + \frac{\hbar k_z}{m^*}\frac{\partial}{\partial z} f_e(\mathbf{k}, z, t) = g(\mathbf{k}, z, t) + J_{ee} + J_{eph} \tag{3}$$

Here J_{ee} and J_{eph} are the electron–electron- and electron–phonon-collision integrals, respectively. Using the concept of self-scattering (Fawcett, 1973), we add a term $\Gamma \cdot f(\mathbf{k}, z, t)$ on both sides of eq. 3 and integrate to get the formal solution

$$f_e\left(\mathbf{k}, z, t + \frac{1}{\Gamma}\right) = f_e(\mathbf{k}, 0, t)e^{-\gamma z} + \frac{m^*}{\hbar k_z}\int_0^z e^{-\gamma(z-z')}S(\mathbf{k}, z', t)\,dz' \tag{4}$$

with

$$\gamma := \frac{m^*\Gamma}{\hbar k_z}, \tag{5}$$

$$S(\mathbf{k}, z, t) := g(\mathbf{k}, z, t) + J_{ee} + J_{eph} + \Gamma f_e \tag{6}$$

Unique solutions are enforced by the boundary conditions just discussed. The resulting integral equation has different forms for $k_z > 0$ and $k_z < 0$.

Taking $k_z > 0$ we introduce the vectors $\mathbf{k}_+ = (k_x, k_y, k_z)$ and $\mathbf{k}_- = (k_x, k_y, -k_z)$ and obtain

$$\mathrm{f_e}\left(\mathbf{k}_+, z, t + \frac{1}{\Gamma}\right) = \frac{m^*}{\hbar k_z}\left[\int_0^z e^{-\gamma z'} S(\mathbf{k}_+, z - z', t)\, dz' + R_0 e^{-\gamma z} \int_0^L e^{-\gamma z'} S(\mathbf{k}_-, z', t)\, dz'\right] \tag{7a}$$

$$\mathrm{f_e}\left(\mathbf{k}_-, z, t + \frac{1}{\Gamma}\right) = \frac{m^*}{\hbar k_z}\left[\int_0^{L-z} e^{-\gamma z'} S(\mathbf{k}_-, z + z', t)\, dz' + R_L e^{-\gamma(L-z)} \int_0^L e^{-\gamma z'} S(\mathbf{k}_+, L - z', t)\, dz'\right] \tag{7b}$$

Here terms containing the factor $\exp(-\gamma L)$ have been neglected. This method can easily be extended to more complicated layered systems containing interfaces, quantum wells, etc. Now the collision integrals are modelled by the relaxation-time approximation:

$$J_{\mathrm{ee}} = -\frac{\mathrm{f_e} - \mathrm{f_e'}}{\tau_{\mathrm{ee}}} \tag{8}$$

describes electron–electron scattering, the fastest process. Here

$$\mathrm{f_e'}(\mathbf{k}, z, t) = n(z, t)\left[\frac{\hbar^2}{2\pi m^* k_{\mathrm{B}} T(z, t)}\right]^{3/2} \exp\left\{-\frac{\hbar^2[\mathbf{k} - \mathbf{k}_{\mathrm{e}}(z, t)]}{2m^* k_{\mathrm{B}} T(z, t)}\right\}^2 \tag{9}$$

is a displaced Maxwellian, the parameters of which are

$$\begin{aligned} n(z, t) &= \int d^3k\, \mathrm{f_e}(\mathbf{k}, z, t), \\ n(z, t)\mathbf{k}_{\mathrm{e}}(z, t) &= \int d^3k\, \mathbf{k}\, \mathrm{f_e}(\mathbf{k}, z, t), \\ \frac{3}{2} n(z, t) k_{\mathrm{B}} T(z, t) &= \frac{\hbar^2}{2m^*} \int d^3k (\mathbf{k} - \mathbf{k}_{\mathrm{e}})^2 \mathrm{f_e}(\mathbf{k}, z, t). \end{aligned} \tag{10}$$

Electron–phonon collisions are introduced by (Nag, 1980)

$$J_{\mathrm{eph}} = -\frac{\mathrm{f_e} - \mathrm{f_{e0}}}{\tau_{\mathrm{eph}}(k)} \tag{11}$$

with

$$\frac{1}{\tau_{\mathrm{eph}}(k)} = \frac{m^* k_{\mathrm{B}} T_{\mathrm{ph}} (E_1^*)^2}{\pi \hbar^3 v_s^2 \rho} k \tag{12}$$

and

$$\mathrm{f}_{e0}(\mathbf{k}, z) = n_0(z, t)\left[\frac{\hbar^2}{2\pi m^* k_B T_{ph}}\right]^{3/2} \exp\left[-\frac{\hbar^2 k^2}{2m^* k_B T_{ph}}\right]. \quad (13)$$

$n_0(z, t)$ is adjusted to satisfy the current continuity equation, E_1^* is the effective acoustic deformation potential, and ρ the mass density. This ansatz allows for a local nonequilibrium between the two quasi-particle systems, $T \neq T_{ph}$ and $\mathbf{k}_e \neq \mathbf{k}_{ph} = 0$. According to the axial symmetry of our problem, f_e depends only on k_z and $k_{\parallel} = (k_x^2 + k_y^2)^{1/2}$. We approximate the $k_{\parallel}$ dependence by a local Maxwellian

$$\mathrm{f}_e(k_{\parallel}, k_z, z, t) := f_1(k_z, z, t)\frac{\hbar^2}{2\pi m^* k_B T(z, T)} \exp\left[-\frac{\hbar^2 k_{\parallel}^2}{2m^* k_B T(z, t)}\right] \quad (14)$$

and integrate eq. 3 over k_x and k_y to get an integral equation for the one-dimensional distribution function f_1:

$$\frac{\partial \mathrm{f}_1}{\partial t} + \frac{\hbar k_z}{m^*}\frac{\partial \mathrm{f}_1}{\partial z} = g_1(k_z, z, t) - \frac{\mathrm{f}_1 - \mathrm{f}_{10}}{\tau_1} - \frac{\mathrm{f}_1 - \mathrm{f}_1'}{\tau_{ee}} \quad (15)$$

where $\mathrm{f}_{10}, (\mathrm{f}_1')$ is the one-dimensional (shifted) Maxwellian. The original 3-dimensional distribution function is then recovered from eq. 14.

With Θ denoting the step function, the corresponding effective generation rate is

$$g_1(k_z, z, t) = g_0'(t)\Theta(k_L^2 - k_z^2)e^{-\lambda z} \quad (16)$$

and

$$\tau_1^{-1}(k_z, z, t) = \frac{\sqrt{2}\, k_B T_{ph}(E_1^*)^2 m^{*3/2}(k_B T)^{1/2}}{\pi \hbar^4 v_s^2 \rho}\left\{q + \frac{\sqrt{\pi}}{2}e^{q^2}[1 - \Phi(q)]\right\} \quad (17)$$

with $q = \hbar k_z (2m^* k_B T)^{-1/2}$ and Φ the error function.

C. Hydrodynamic Description

The hydrodynamic variables are obtained by averaging microscopic single-particle variables over $\boldsymbol{k}$-space. We thus find for the local particle density

$$n(z, t) = \int d^3k\, \mathrm{f}_e(\mathbf{k}, z, t) \quad (18)$$

and for the energy density

$$u(z, t) = \int d^3k \, E(\mathbf{k}) f_e(\mathbf{k}, z, t). \tag{19}$$

The respective local current densities are

$$\mathbf{j}_n(z, t) = \frac{\hbar}{m^*} \int d^3k \, \mathbf{k} f_e(\mathbf{k}, z, t) := n \mathbf{v}_e \tag{20}$$

$$\mathbf{j}_u(z, t) = \frac{\hbar}{m^*} \int d^3k \, \mathbf{k} E(\mathbf{k}) f_e(\mathbf{k}, z, t) \tag{21}$$

In a phenomenological description (linear transport) these currents are decomposed into separate contributions. Restricting ourselves to $\mathbf{j}_n$, we find (Mahler *et al.*, 1985)

$$\mathbf{j}_n(z, t) = -D \nabla n - D_T \frac{\nabla T}{T} + n \mathbf{w} \tag{22}$$

The first term is thus interpreted to arise from isothermal diffusion, the second term describes thermodiffusion, and the last term an additional drift ($\mathbf{w}$ has to be distinguished from $\mathbf{v}_e$). The density- and temperature-dependence of the transport coefficients D, D_T is known. $\mathbf{w}$ could be deduced from the proper balance equations for particle number, energy and momentum, provided the proper boundary conditions were known. This is not the case.

In the present kinetic study, we are in a better position. As the boundary conditions for the distribution function are easily specified from an analysis of the microscopic processes, we are able to find $f(\mathbf{k}, z, t)$ and thus infer $\mathbf{w}$ from the difference between our calculated j_n and the diffusive contributions. In the present case, all currents are in the z-direction.

III. STATIONARY SOLUTIONS: NUMERICAL RESULTS

The bulk materials used are $\tau_{ee} = 10$ ps, $v_s = 9 \times 10^5$ cm/s, and $\rho = 2.33$ g/cm^3. Numerical stability requires

$$\Gamma \geq \frac{1}{\tau_1} + \frac{1}{\tau_{ee}} \tag{23}$$

The values of m^* and E_1^* can be chosen to reproduce the ambipolar diffusion constant D for excitonic transport, D_{FE}, or ambipolar transport, D_{eh}. For nondegenerate systems, there is no qualitative difference. As ambipolar motion is consistent with the situation in the optical response discussed in this section, we choose here

$$m^* = 0.43m_0 \tag{24}$$

which implies $E_1^* = 8.8$ eV. In the experimental sections, on the other hand, we shall use the appropriate excitonic quantities. The following results are obtained for the control parameters

$$L = 20\ \mu m$$
$$\frac{1}{\lambda} = 1\ \mu m$$
$$R_0 = R_L = 0.5$$
$$E_{\text{exc}} = 8 \text{ meV } (60 \text{ meV})$$
$$T_{\text{ph}} = 10 \text{ K}$$

The resulting density, temperature, and velocity profiles v_e are shown in Fig. 1 (solid line: profiles for $E_{\text{exc}} = 8$ meV; broken line: profiles for $E_{\text{exc}} = 60$ meV). The local density $n(z)$ depends (for the nondegenerate case and without any many-body effects) linearly on g_0. The absolute value of n can therefore be scaled to any value below the density of degeneracy, $n_{\text{deg}}(T)$. Fig. 1a shows the density in units of $n_{\max}$, the maximum density reached at the spatial position $z = z_{\max}$.

We immediately see that the density profile has a pronounced maximum at $z_{\max} \geq 0$, i.e. there is a spatial region with $\nabla n \geq 0$. This is due to reverse diffusion, as first predicted by Mahler *et al.* (1983), and due to surface recombination ($R_0 \leq 1$). Reverse diffusion is possible only in the presence of a strong negative temperature gradient, as seen in Fig. 1b, which increases with E_{exc}. Surface recombination manifests itself in the negative hydrodynamic velocity v_e close to the excited boundary. v_e increases significantly close to the opposite boundary at $z = L$, as part of the particles with negative velocities are missing due to surface recombination.

Unfortunately, a direct local measurement of f_e or of the hydrodynamic profiles is not possible. The analysis of the *bulk*-optical decay channel shows that the observed luminescence intensity should be proportional to

$$I(E) := \frac{1}{N} \int \zeta(E, z)\, dz \tag{25}$$

where N is a normalization constant,

$$\zeta(E, z) := \int dE' D_e(E') f_e(E', z) D_h(E - E') f_h(E - E', z), \tag{26}$$

and $D_e(E)$ is the electronic density of states in energy space. The resulting

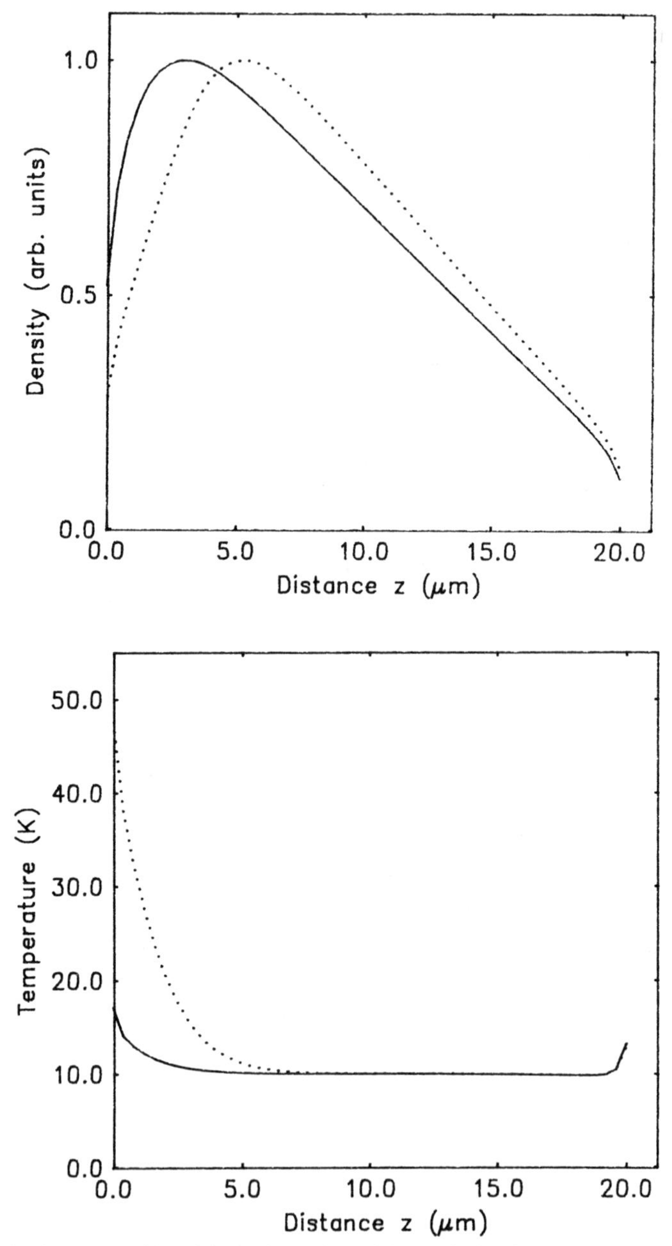

Figure 1. Stationary profiles of hydrodynamic variables calculated for a 20 μm sample. (a) density $n(z)/n_{\max}$; (b) temperature $T(z)$; (c) velocity $v_e(z)$.

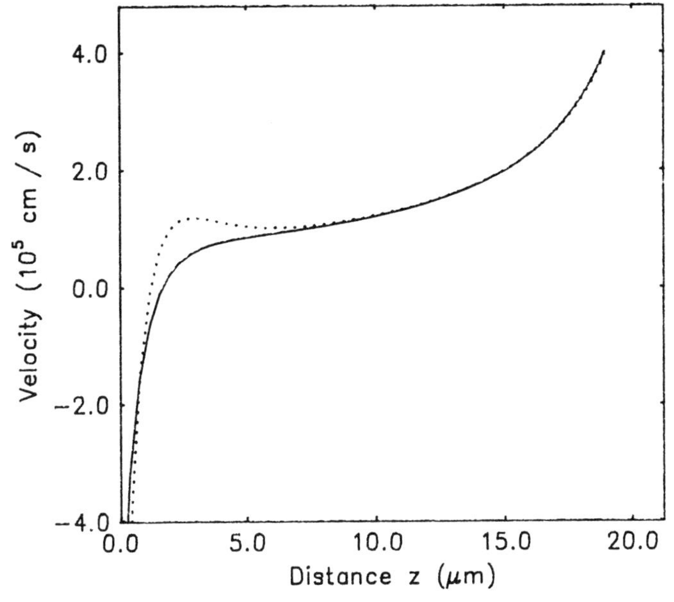

Figure 1. (*Continued*).

$I(E)$ for $E_{exc} = 60$ meV is shown in Fig. 2 (broken line). One may expect that $I(E)$ should be dominated by $\zeta(E, z_{max})$; the dotted line shows $\zeta(E, z_{max})$, normalized to the same maximum. The solid line, finally, depicts $\zeta'(E, z_{max})$, where the actual f_e has been replaced by a displaced Maxwellian with the parameter set $[n_{max}, T = T(z_{max}), w_z = v_e(z_{max})]$. We see that even the last step of approximation is well justified, i.e. despite spatial inhomogeneity, the bulk luminescence can be described by a single local distribution function. We find $w_z = 1.0 \cdot 10^5$ cm/s, which is just at the limit below which w_z no longer has a measurable effect on the spectrum. w_z is practically only of diffusive origin; for the case of nondegenerate plasmas drift contributions become important only on length scales below 5 μm (Kuhn *et al.*, 1987).

It should be noted, however, that if the luminescence was dominated by the spatial contribution at $z = L$ (or $z = 0$), a considerably higher velocity and thus a stronger line-shape distortion could result (see Fig. 1c).

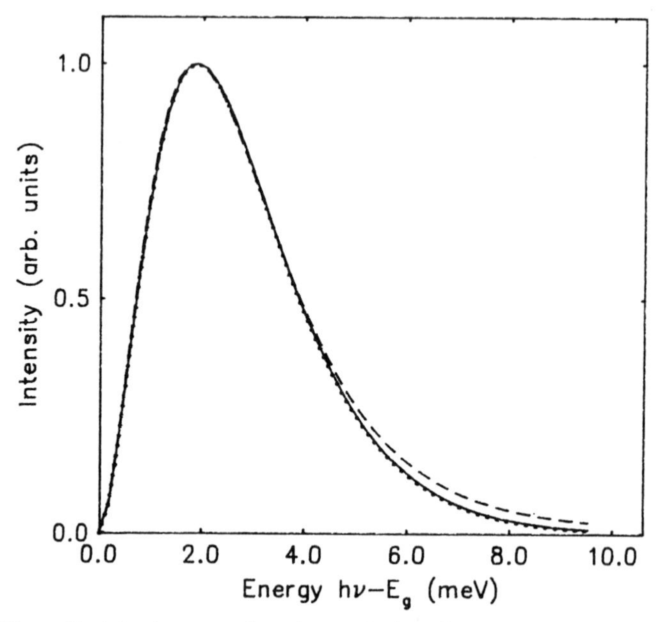

Figure 2. Normalized luminescence line shapes. Broken line: spatially averaged response. Dotted line: response exclusively from the spatial region with maximum density. Solid line: response from a shifted Maxwellian with parameter set $n = n_{\max}$, $T = T(z = z_{\max})$, $w = v(z = z_{\max})$.

IV. TIME-DEPENDENT BEHAVIOR

Time-dependent phenomena occur under various experimental conditions. A typical situation is encountered when the system relaxes into its stationary state, after the light field has been switched on or off. Recent experimental data refer to pulse excitations (Hillmer *et al.*, 1987).

A. Time-of-Flight Method

A time-of-flight (TOF) experiment generally is based on the measurement of the number of carrier pairs that have travelled over a given distance as a function of time and provides direct access to the transport quantities. In our TOF studies, carrier pairs are generated at the front surface of Si wafers by the exciting laser pulse and those carrier pairs that arrive at the back surface are detected in a thin layer with modified crystal properties. The method has very high spatial and temporal resolution, in particular com-

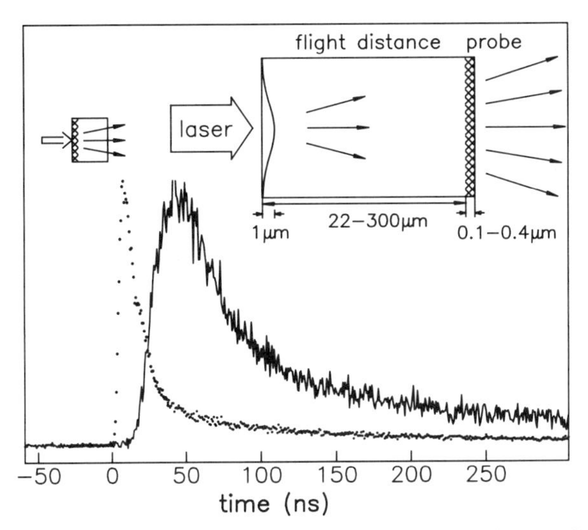

Figure 3. Sample geometry and the two different excitation conditions of the BE layer (cross-hatched). Time-resolved BE luminescence after direct excitation (dotted line) and excitation via ambipolar transport (solid line). For $t > 15$ ns, the decay observed in the case of direct excitation is caused by diffusion of carrier pairs that have been excited beyond the thickness of the implanted layer by the laser.

pared to so-called imaging experiments introduced by Greenstein and Wolfe (1981), and hence can be used for high-resolution investigations of ambipolar transport.

For the present studies, we use Si wafers with thicknesses L between 300 μm and 22 μm. The schematic design of our samples is depicted in the inset of Fig. 3. In one side of the wafers, ions are implanted which act as shallow impurities. These impurities are used as an optically active carrier sensor. Due to the characteristic bound exciton (BE) emission carrier pairs arriving at the implanted layer can be detected (cross-hatched area in the right inset of Fig. 3). The time-dependent luminescence intensity of the BE displays the carrier concentration $n(L, t)$ at the doped surface as a function of time. Because the penetration depth of the ions is only about 0.2 μm, the spatial resolution of the experiment is very high.

The carriers are generated at the undoped surface by short pulses of an Ar^+ laser ($\lambda = 514.5$ nm, pulse width $\Gamma_L = 150$ ps). The detecting photon-counting system consists of a fast S1 photomultiplier, amplifiers, discriminators and a multichannel analyzer (overall time resolution 0.5ns). For the experiments, the samples are mounted in a temperature-controlled dewar (1.8 K $< T <$ 40 K).

In order to serve as a useful carrier detector, the bound exciton luminescence has to fulfill different requirements. Most importantly, the BE lifetime has to be short compared to the transit time of the carrier pairs to display the number of arriving carrier pairs as a function of time without further noticeable broadening. Furthermore, thermal stability of the BE over a broad range of temperatures is required as well as a spectral position, which does not overlap with the intrinsic luminescence. Note that direct optical excitation of the sensors by the emission of carrier pairs can be neglected here. Si is an indirect gap semiconductor and, therefore, the intrinsic emissions in high-purity material are shifted by the phonon energy below the bandgap. Combined with the very small intrinsic luminescence intensity and the low density of impurities ($\sim 10^{17}$ cm^{-3}), this does not allow a significant optical excitation by intrinsic recombination luminescence. We have investigated predominantly indium and thallium bound excitons which fulfill all requirements reasonably well as demonstrated in publications by Forchel *et al.* (1985) and Hillmer *et al.* (1987). For details of BE properties in Si, e.g., see Dean *et al.* (1967).

For a sample with 95 μm thickness, Fig. 3 shows the typical response of the indium BE emission as a function of delay time when the excitation of the bound exciton is due to carrier transport (solid line) in comparison with direct optical excitation (dotted line). Due to the diffusive processes during the carrier transport, the carrier concentration profile at the probe layer is strongly broadened in the first case compared to that observed after direct excitation. In the following, for all experimental evaluations, the origin of the time axes has been determined by comparison with the BE response after direct excitation. This effectively cancels the influence of capture times of the excitons at the dopands and is essential for the quantitative evaluation of the data.

Depending on the concentration and the temperature of the carriers, different states of the ambipolar many-particle system are stable as reported by Hensel *et al.* (1977) and Forchel *et al.* (1982). For the interpretation of the experiments, it is therefore important to determine which species can be detected by the probe layers. The carrier pairs may exist in the form of free excitons (FE), electron-hole droplets (EHD) or above the critical temperature for droplet formation as electron-hole plasma (EHP). In Fig. 4, we depict the results of an intensity-dependent investigation of the sensor bound exciton luminescence in comparison with the FE and EHD luminescence. In the left part of Fig. 4, we show the intensity variation after direct optical excitation and, in the right part, after excitation by carrier transport

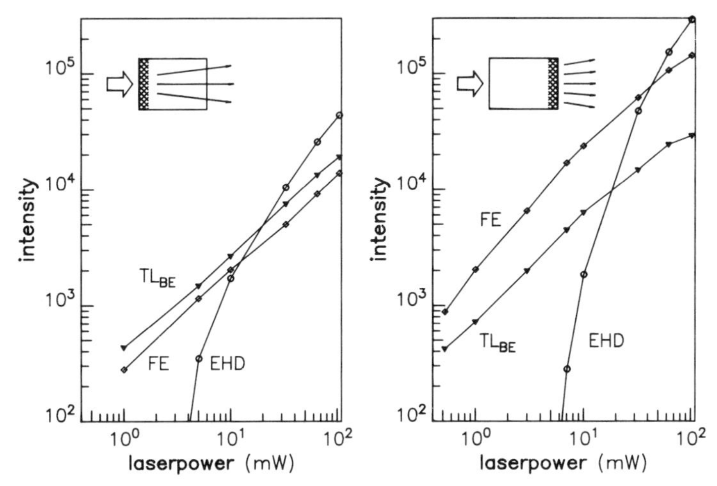

Figure 4. Intensity variation of the FE, EHD and BE emission as a function of the average laser power. The peak powers used in the experiments are higher by about a factor of 4×10^4 than the average values given on the abscissa.

as a function of average laser power. We observe the same intensity variation for the BE and the FE intensities over the entire range of pumping powers investigated. In contrast, the EHD intensity shows a distinctively different excitation power dependence. This indicates that we observe FE transport in our TOF experiments. The BE emission apparently is not influenced by EHD droplets even for direct optical excitation of the implanted surface (Fig. 3). This is most likely due to screening of the BE binding energy at the high carrier densities within the EHD.

B. Experimental Results

Fig. 5 depicts the time-resolved concentration, $n(L, t)$, in the sensor layer of a sample with $L = 53\ \mu m$ flight distance. Compared with the profile of the 95 μm sample in Fig. 3, we observe a much shorter delay of the onset of the emission with decreasing transport distance as well as a faster decay. With increasing temperature at a fixed distance ($L = 28\ \mu m$ in Fig. 6), the profiles shift to longer times with drastically flattened slopes of the rise and decay. Qualitatively, the observed behavior can be explained easily. With increasing distance, the carriers need more time to reach the sensors, which leads to the observed increase of the delays. For longer

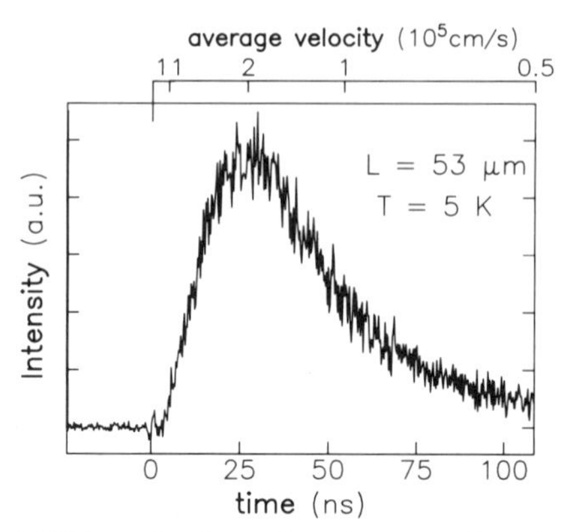

Figure 5. Time-resolved BE intensity $I \sim n(L, t)$ for a sample with a flight distance $L = 53$ μm.

transport distances, the influence of the diffusion is expected to broaden the profiles; the same effect comes also from higher thermal velocities with increasing temperature.

The bottom axis of Fig. 5 displays the response time which may be interpreted as an integral transport time for the given sample. By using the known thickness of the transport layer ($L = 53$ μm), we can convert the time axis into a velocity axis v^*. This velocity axis is displayed on top of Fig. 5 for nonzero positive delay times. The maximum of $n(L, t)$ corresponds to $v^* = 2 \cdot 10^5$ cm/s, though there are significant contributions to $n(L, t)$ at much higher velocities up to about $1 \cdot 10^6$ cm/s. It should be noted, however, that v^* cannot be transformed simply into the proper hydrodynamic velocity $v_e(z = L, t)$ (compare eq. 20). v^* represents therefore only a crude estimate of v_e.

Most likely different physical mechanisms are responsible for the observed distribution of v^*. First of all, it is well known that diffusive motion includes a wide range of velocities varying from very high to very low values. For example, by assuming diffusion only, one can easily calculate that for a flight distance of 20 μm and a diffusion coefficient $D = 100$ cm^2/s, the velocity distribution decreases only to 3.3% amplitude for $v = 1 \cdot 10^6$ cm/s. Note that this value exceeds the longitudinal acoustic phonon velocity significantly. This contradicts results by Wolfe (1983), who reported the existence of an upper limit for the electron–hole pair velocity in optically excited Si at the phonon velocity ("sound barrier"). In our

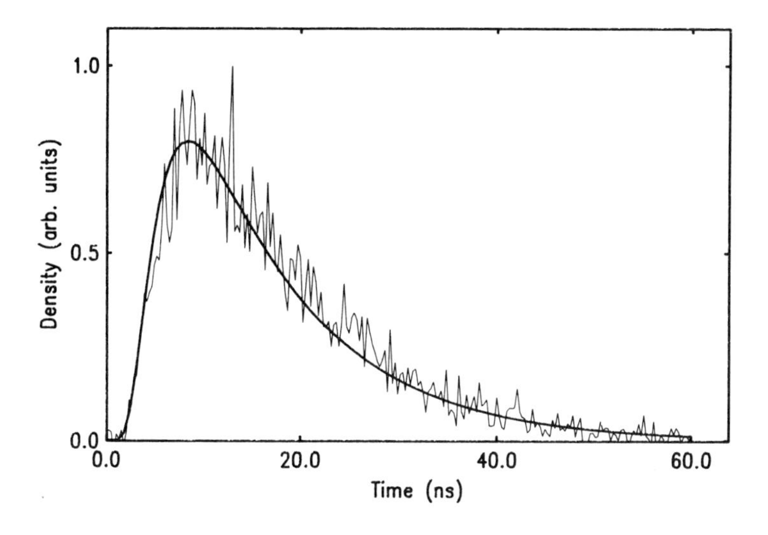

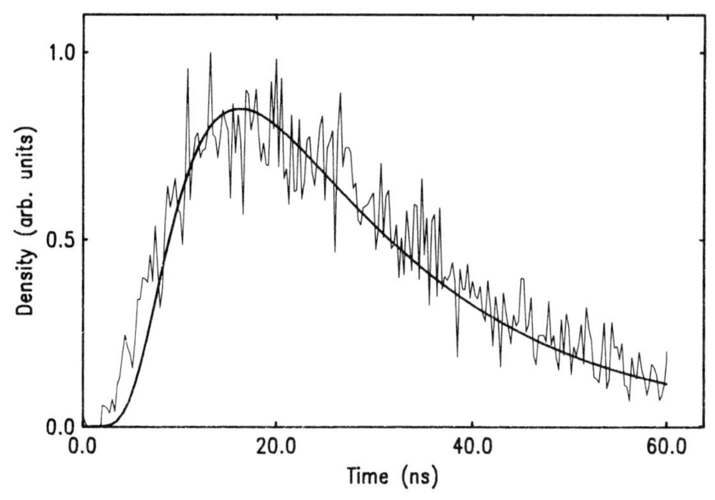

Figure 6. The measured profiles for a 28 μm sample together with the best fits using the hydrodynamic model with the parameters diffusivity and surface recombination velocity.

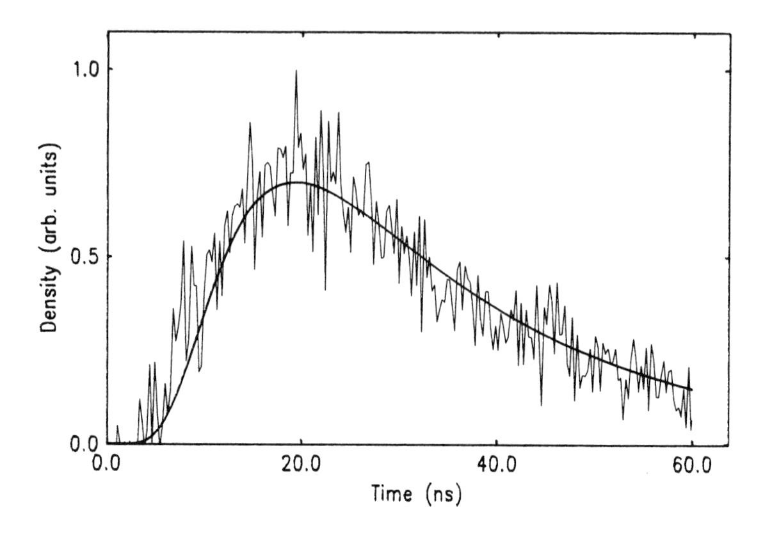

(a) $T_{ph} = 1.8\ K;\ D = 130\ cm^2/s,\ s = -v_e(z = 0) = v(z = L) = 2.26 \cdot 10^5\ cm/s$

(b) $T_{ph} = 10\ K;\ D = 50\ cm^2/s,\ s = 3.8 \cdot 10^5\ cm/s$

(c) $T_{ph} = 17\ K;\ D = 37\ cm^2/s,\ s = 21 \cdot 10^5\ cm/s$

experiments we observe a noticeable fraction of carriers with velocities above the speed of sound and we do not observe any discontinuous variation of the profile at the sound velocity.

In addition to the diffusive mechanism, the sample surfaces will have a strong influence on the carrier profiles. At the surface, reflection and recombination processes similarly influence the carrier transport. By reflection, carriers change the direction of motion, and by recombination, carriers are extracted, both yielding a preferential direction of motion. As is shown in detail in the analysis of our experimental data by the hydrodynamic model, a combination of appropriate diffusivities and surface effects is necessary to explain the data.

C. Numerical Results

It is straightforward to predict the response $n(L, t)$ from our kinetic analysis, provided we know all the actual material and control parameters. However, as the quality of the surfaces is not known, and as details of the band structure cannot be included, such a procedure would readily explain only the qualitative features. Alternatively, a simple fit procedure can be based on the hydrodynamic approach, in which the boundary conditions $R_0 = R_L \leq 1$ are modelled by a surface recombination velocity $s = -v_e(z=0) = v_e(z=L)$. Assuming an absorption length of 1 μm, we have calculated the response function $n(L = 28\ \mu\text{m}, t)$ for different sample temperatures. The resulting theoretical detector signals shown in Fig. 6 are in comparison with the experimental data. The values for the fit parameters s and D are given in the figure caption.

As can be seen, very satisfying fits are obtained. We point out that no additional drift w_z has to be included. Ignoring the boundary conditions, satisfactory fits can only be obtained if a drift velocity $w_z \leq 10^5$ cm/s of then unknown origin is included.

The temperature dependence of the diffusivities determined from the fits is shown in the left part of Fig. 7. The diffusivity reaches 130 cm^2/s at the lowest bath temperature and decreases to 37 cm^2/s at 17 K. Using the Einstein relation $D = \mu k_B T/q$ where μ denotes the mobility and q the elemental charge, we obtain the mobility–temperature relation shown in the right part of Fig. 7.

In a nonpolar semiconductor such as silicon, the following scattering mechanisms can affect the motion of free excitons: scattering with acoustic and optical phonons via the deformation potential and scattering with

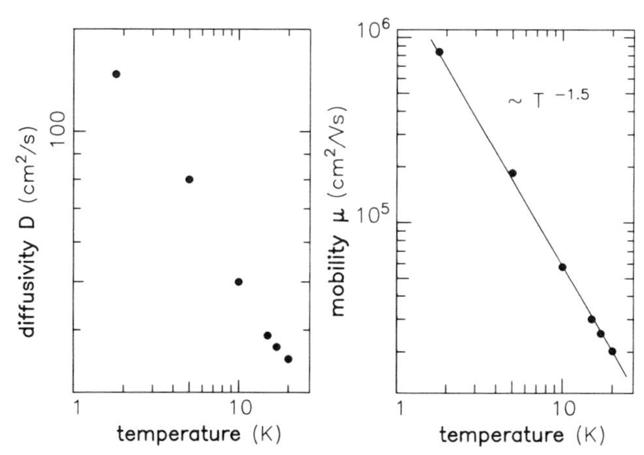

Figure 7. Results of the hydrodynamic profile fits for the diffusivity and the mobility as a function of temperature. For the mobility, a fit calculated from scattering by acoustic phonons via the deformation potential is included.

ionized and neutral impurities. The scattering rates of these different mechanisms can be added using the Mathiessen rule as shown, e.g., by Heywang and Pötzl (1982), giving the total scattering rate. This leads to a summation of different temperature dependences where an adjustable coefficient accounts for the importance of each specific mechanism. These parameters are varied in the fits to the measured mobility data. For the best fit displayed in Fig. 7, we only need the contribution of the acoustic phonons interacting with the FE via the deformation potential showing the temperature dependence $T^{-1.5}$. Since we use ultrapure samples (20,000 Ωcm), this strong dominance of the acoustic-phonon scattering is expected.

To evaluate the relaxation time $\tau(T)$, the following formula is used:

$$\tau(T) = 1.5 m_{\mathrm{op}} \frac{D}{k_{\mathrm{B}} T} \tag{27}$$

where m_{op} is the optical FE mass and energy averaging has been performed in $\langle \tau \rangle$. Because of the large mass difference between m_{hh} and m_{lh}, the FE are essentially heavy hole excitons. We use for the optical FE mass $m_{\mathrm{op}} = 3(2/m_t + 1/m_1)^{-1} + m_{\mathrm{hh}}$.

The fit of Fig. 7 is equivalent to $\tau(T) = 1.4 \text{ ns} \cdot T^{-1.5}$. To compare this result with the theoretically calculated relaxation time value, we use the formula reported by Nag (1980), Eq. 12, where E_1 denotes $E_1^* = E_{1\mathrm{e}} - E_{1\mathrm{h}}$ the FE deformation potential and m the FE density of states mass, calculated according to $m_d = (m_t^2 \cdot m_1)^{1/3} + (m_{\mathrm{hh}}^{3/2} + m_{\mathrm{lh}}^{3/2})^{2/3}$. Using literature values given by Landolt-Börnstein (1982), we obtain $\tau(T) = 1.2 \text{ ns} \cdot T^{-1.5}$ in good agreement with the measurement. Note that in order to calculate the deformation potential, we have assumed that the FE is scattered as a single particle (hydrostatic scattering limit). If the electron and hole in the FE were scattered independently, an average of the unipolar scattering rates would apply, in quantitative disagreement with the experimental data. Our measurements indicate therefore that the hydrostatic scattering limit applies for FE transport at low temperature in Si.

V. CONCLUSIONS

We have demonstrated that carrier kinetics sensitively depend on the injection conditions and surface conditions. In particular, the average velocity v_{e} has been found to depend on much more experimental details than expected and specified so far. It is therefore not surprising to find seemingly contradictory answers in the measurements published up to now.

Due to reflection and recombination, the surface can act as a source of net momentum. However, for a nondegenerate system and for the length scales discussed here ($L > 10\ \mu m$), the total $v_e(z > 10\ \mu m)$ does not exceed some 10^5 cm/s under stationary conditions. The time-dependent behavior after pulse excitation can be characterized by $v_e(z = L, t)$, which should become supersonic only for $t \leq 1$ ns; this is unfortunately beyond the reach of the experiment. Nevertheless, these limits determined by the length- and time-scales of the experiments do certainly not apply to individual carrier velocities.

In the present chapter, we have used a time-of-flight method to study the carrier transport in Si experimentally. In the experiment, suitable dopants were implanted within about 0.2 μm of the nonexcited sample surface and served as FE detectors. We observe average velocity distributions showing no influence of a "sound barrier," i.e., no discontinuous behavior of the $v^*(z = L, t)$ occurs at the sound velocity. Fitting the transport profiles with a hydrodynamic model, surface and volume effects can be separated. This yields a FE relaxation time $\tau(T) = 1.4\ \text{ns} \cdot T^{-1.5}$ for the temperature range of 1.8 K to 20 K. Interaction of the FE with longitudinal acoustic phonons via the deformation potential is the dominant scattering mechanism.

Note that the present experimental and theoretical evaluation concentrates on the behavior of ambipolar transport at low carrier densities (nondegenerate or excitonic limit). Previously, different authors (Forchel *et al.* 1983, Kempf and Klingshirn 1984, and Nather *et al.* 1985) have studied velocity-induced changes of the optical response under high-excitation conditions. In order to explain the high-excitation data, a rigidly shifted Fermi distribution has been proposed where the shift of the Fermi sphere with respect to the origin represented the expansion velocity. Typical expansion velocities are in the $2 \cdot 10^6$ cm/s to $5 \cdot 10^6$ cm/s range. For these conditions, our theoretical model should be extended to include the influence of degeneracy and phonon nonequilibrium: As a result, the length- and time-scales for relaxation may considerably increase.

With suitable modifications, our TOF method has a large potential for studies of the carrier transport over micro- and submicrometer distances. Recently we have used specially designed double quantum well structures to study the ambipolar transport in GaAlAs (Hillmer *et al.* 1986). In these structures, two quantum wells of different thicknesses are used as detectors for carrier pairs at the beginning and the end of the flight distance. Quantum wells are used because they allow the definition of the transport distance with an accuracy better than 0.1 μm. We observe in these structures average velocities v^* of about $2 \cdot 10^6$ cm/s and very high diffusivities. These high values are most likely due to surface reflection and recombination as in the case presented here. Due to the fast relaxation and recombination times in the quantum wells, these structures may be used to study transport over ultrasmall distances. Further work to investigate excitonic and plasma transport in similar structures is in progress.

ACKNOWLEDGEMENTS

We appreciate the collaboration of B. Laurich and G. Mayer during this work. Stimulating discussions with M. H. Pilkuhn and H. Schweizer are gratefully acknowledged. The financial support of this work by the Deutsche Forschungsgemeinschaft and the Stiftung Volkswagenwerk is greatly appreciated.

REFERENCES

Dean, P. J., Haynes, J. P., and Flood, W. F. (1967). *Phys. Rev.* **161**, 711–729.
Fawcett, W. (1973). In *Electrons in Crystalline Solids*, A. Salam, ed. International Atomic Energy Agency, Vienna, pp. 531–618.
Forchel, A., Laurich, B., Wagner, J., Schmid, W., and Reinecke, T. L. (1982). *Phys. Rev. B*, **25**, 2730–2747.
Forchel, A., Schweizer, H., and Mahler, G. (1983). *Phys. Rev. Lett.* **51**, 501–504.
Forchel, A., Laurich, B., Hillmer, H., Tränkle, G., and Pilkuhn, M. (1985). *J. Lumin.* **30**, 67–81.
Greenstein, M., and Wolfe, J. (1981). *Phys. Rev. B.* **24**, 3318–3348.
Hensel, J. C., Phillips, T. G., Rice, T. M., and Thomas, G. A. (1977). In *Solid State Physics*, (H. Ehrenreich, F. Seitz, and D. Turnbull, eds.,) vol. 32, ch. 2, pp. 88–314.
Heywang, W., and Pötzl, H. W. (1982). In *Bandstructure and Current Transport*, Semiconductor Physics, vol. 3, p. 196. Springer-Verlag, Berlin.
Hillmer, H., Kuhn, T., Laurich, B., Forchel, A., and Mahler, G. (1987). *Physica Scripta* **35**, 520–523.
Hillmer, H., Mayer, G., Forchel, A., Löchner, K., and Bauser, E., (1986). *Appl. Phys. Lett.* **49**, 948–950.
Kelly, M. J., and Nicholas, R. J. (1985). *Rep. Prog. Phys.* **48**, 1699–1741.
Kempf, K., and Klingshirn, C. (1984). *Solid State Commun.* **49**, 23–26.
Klingshirn, C., and Haug, H. (1981). *Phys. Rep.* **70**, 315–398.
Kuhn, T., and Mahler, G. (1987). *Phys. Rev. B*, **35**, 2827–2833.
Landolt-Börnstein (1982). *Numerical Data and Relationships*, (K. H. Hellwege, ed.), vol. 17a, O. Madelung, ed., pp. 45–47, 63.
Mahler, G., and Forchel, A. (1983). *Helv. Phys. Acta* **56**, 875–880.
Mahler, G. and Fourikis, A. (1985), *J. Lumin.* **30**, 18–36.
Nag, B. R., (1980). In *Solid State Science*, (H. J. Queisser, ed.), vol. 11, p. 174. Springer-Verlag, Berlin.
Nather, H. and Quagliano, L. G. (1985). *J. Lumin.* **30**, 50–64.
Schmitt-Rink, S., Tran Thoai, D. B., and Haug, H. (1980). *Z. Phys.* **B39**, 25–31.
Shen, Y. R., (1976). *Rev. Mod. Phys.* **48**, 1–32.
Wolfe, J. P., (1983). *Proc. 16th. Int. Conf. Phys. Semicond. 1982.* Physica 117B and 118B, vol. 1, pp. 321–326.

8 OPTICAL NONLINEARITIES DUE TO BIEXCITONS

R. Lévy, B. Hönerlage, and J. B. Grun

LABORATOIRE DE SPECTROSCOPIE ET D'OPTIQUE DU CORPS SOLIDE
UNITÉ ASSOCIÉE AU C.N.R.S. N° 232, UNIVERSITÉ LOUIS PASTEUR,
5, RUE DE L'UNIVERSITÉ, F-67084 STRASBOURG CEDEX, FRANCE

I. INTRODUCTION

In direct bandgap semiconductors, the dielectric function $\epsilon(\omega)$ shows resonance structures in the vicinity of the energy gap, if the optical transition between the uppermost valence band and the lowest lying conduction band is dipole-allowed. In large bandgap materials, some of these resonances are observed well below the fundamental gap: they are due to excitonic polaritons. Since these excitons govern the optical properties of

ISBN 0-12-332915-9

the material close to the band edge, the optical nonlinearities are influenced by their mutual interactions and by their density-dependent interaction with other quasi particles.

In large bandgap materials, it is well known that two excitons may couple together to form excitonic molecules or biexcitons (Grun *et al.*, 1982; Hönerlage *et al.*, 1985a; Chemla and Maruani, 1982). This coupled state has a binding energy that separates its resonance from the two-exciton continuum (Klingshirn and Haug, 1981). If the excitons that form the biexcitons are dipole active, the transition from the crystal ground state to the biexciton ground state is always allowed by absorption of two photons with the same linear polarization. Therefore, this two-photon resonance will show up in the dielectric properties of the material and in their nonlinearities.

Optical nonlinearities have been intensively studied in a situation where the light field is not resonant with the elementary excitations of the material. In this case, which covers classical nonlinear optics, the dielectric polarization may be developed into a power series of the light field E(t) in the form:

$$P = \epsilon_0 \chi \mathrm{E} = \epsilon_0 \big(\chi^{(1)} \cdot \mathrm{E} + \chi^{(2)} \cdot \mathrm{EE} + \chi^{(3)} \cdot \mathrm{EEE} + \cdots \big). \tag{1}$$

In relation 1, χ is the nonlinear dielectric susceptibility tensor, which is related to the dielectric function $\epsilon(\omega)$ by the relation:

$$\epsilon(\omega) = 1 + \chi(\omega) \tag{2}$$

The quantities $\chi^{(2)}, \chi^{(3)}, \cdots$ denote the different orders of the nonlinear susceptibility tensor and ϵ_0 is the permitivity of vacuum. Eqs. 1 and 2 govern the linear and nonlinear optical response of the medium. However, when studying its optical properties near the exciton and biexciton resonances, the development of P in a power series of E is no longer possible since it does not converge easily, close to resonances. Therefore, in this spectral region, a different approach has to be chosen. This has been made, e.g., by using Green's functions (Haug, 1982; Haug and Schmitt-Rink, 1984; Kranz and Haug, 1986; März *et al.*, 1980; May *et al.*, 1979), unitary transformations in connection with operator techniques (Abram, 1983; Abram and Maruani, 1982; Sung and Bowden, 1984; Bowden and Sung, 1979; Haus *et al.*, 1985) or by solving Heisenberg's equation of motion (Inoue, 1985). Using these formalisms, the resonant part of the optical response can be calculated by considering transverse excitons and biexcitons and taking into account the relevant selection rules (Cho and Itoh, 1984; Hanamura, 1981). Then, nonresonant contributions due to band-to-band transitions or other exciton states are neglected. This approximation is possible because of the large binding energies of excitons and biexcitons which separate well the resonances.

We will discuss here in more details an approach based on the density-matrix formalism. This method allows us to introduce phenomenologically finite lifetimes of the elementary excitations as well as coherence dephasing times (Yariv, 1975; Agrawal and Carmichael, 1979; Tokihiro and Hanamura, 1980). We shall apply this theory explicitly to CuCl, since it is the most widely studied material (Henneberger and May, 1982; Bigot and Hönerlage, 1983; Hönerlage and Bigot, 1984a) in which biexcitons lead to optical nonlinearities. This is probably due to the fact that CuCl has a particularly simple band structure and large binding energies of the quasi particles. This allows us to consider in the density matrix formalism only the crystal ground state $|1\rangle$, two transverse exciton states $|2\rangle$ and $|4\rangle$ and the biexciton state $|3\rangle$. The tensor character of χ can be neglected. If a single linearly polarized light field is considered, the selection rules simplify the system even further and we have to consider only one of the exciton states explicitly.

We shall first discuss the physical origin of the different nonlinearities in the dispersion and absorption of the material. We shall then present different experimental methods used with CuCl to study these anomalies as functions of the intensity, the photon energy and the polarization of the excitation in "pump and probe" experiments. As for the dynamics of the nonlinearities in CuCl, we shall discuss their importance for single-beam experiments, optical bistability and pulse shaping, using a quasi-stationary approximation. In two-beam experiments, the same dynamic effects show up and their evolution is important in the spectral region around the exciton–biexciton transition.

II. THE NONLINEAR DIELECTRIC FUNCTION IN PUMP AND PROBE EXPERIMENTS

A. Theoretical Description of Nonlinearities Due to Biexcitons

Let us consider the three-level system discussed in the introduction under steady state excitation of a pump beam (index p) with frequency ω_p. The dielectric properties of the system are then analyzed using a test beam (index t), which may have a different photon energy $\hbar\omega_t$ but the same (linear) polarization as the pump beam. The Hamiltonian of the system writes in the dipole approximation:

$$H = H_0 - \mu \mathrm{E}(\omega_t, \omega_p) \tag{3}$$

where H_0 describes the (diagonalized) three-level system consisting of the crystal ground state with energy $E_0 = 0$, the exciton with energy E_{ex} and the biexciton state (energy E_{Bi}). μ is the dipole operator, and only the transition matrix elements $\langle 2|\mu|1\rangle = \mu_{\mathrm{ex}}$ and $\langle 3|\mu|2\rangle = \mu_{\mathrm{Bi}}$ are different from zero as discussed in the introduction. Without lack of generality, μ_{ex} and μ_{Bi} are assumed to be real. The electromagnetic radiation field has the form:

$$\mathrm{E}(\omega_t, \omega_p) = \mathrm{E}_0^t \cos(\omega_t t) + \mathrm{E}_0^p \cos(\omega_p t) \tag{4}$$

where the field amplitudes E_0^p and E_0^t are time-independent. With these definitions and the selection rules, the Hamiltonian H has the matrix representation:

$$H = \begin{pmatrix} 0 & -\mu_{\mathrm{ex}}\mathrm{E} & 0 \\ -\mu_{\mathrm{ex}}\mathrm{E} & E_{\mathrm{ex}} & -\mu_{\mathrm{Bi}}\mathrm{E} \\ 0 & -\mu_{\mathrm{Bi}}\mathrm{E} & E_{\mathrm{Bi}} \end{pmatrix} \tag{5}$$

Using the Schrödinger equation (Yariv, 1975), H determines the time evolution of the density matrix elements $\rho_{ij}[i, j \in (1, 2, 3)]$ when solving the equation:

$$\frac{\delta \rho_{ij}}{\delta t} = \frac{i}{\hbar}[\rho, H]_{ij} - \Gamma_{ij}\rho_{ij} \tag{6}$$

Using this formalism, one can calculate the ensemble average of the expectation value of an operator B from:

$$\langle \overline{B} \rangle = \mathrm{tr}(\rho B). \tag{7}$$

Since the (macroscopic) polarization P is given by

$$P = N\langle \bar{\mu} \rangle, \tag{8}$$

we obtain with the selection rules:

$$P = N\,\mathrm{tr}(\rho\mu) = N\left[(\rho_{12} + \rho_{21})\mu_{\mathrm{ex}} + (\rho_{23} + \rho_{32})\mu_{Bi}\right]. \tag{9}$$

On the other hand, P is related to the electric field by the convolution product

$$P = \epsilon_0 \mathrm{Re}(\chi * \mathrm{E}) \tag{10}$$

where χ is the complex dielectric susceptibility discussed in the introduction.

We are now interested in the periodic solutions of Eq. 6 for the density matrix elements ρ_{ij} which have the time-dependence imposed by the test and the pump beams. We therefore develop ρ_{ij} in a Fourier series of the form:

$$\rho_{ij} = \sum_{nn'} \rho_{ij}^{nn'} e^{i(n\omega_t + n'\omega_p)t} \tag{11}$$

Introducing this definition into Eqs. 9 and 10, we find for $\chi(\omega_t)$, after performing the convolution product (Bigot and Hönerlage, 1983; Hönerlage and Bigot, 1984a):

$$\chi(\omega_t) = \frac{2N}{\epsilon_0 E_0^t}\left[\mu_{\mathrm{ex}}\left(\rho_{12}^{10} + \rho_{21}^{10}\right) + \mu_{\mathrm{Bi}}\left(\rho_{23}^{10} + \rho_{32}^{10}\right)\right] \tag{12}$$

When introducing Eq. 11 into Eq. 6, the latter equation leads to an infinite system of coupled equations for the Fourier coefficients $\rho_{ij}^{nn'}$. This system is truncated, by retaining only resonant and antiresonant terms provided that $|n| + |n'| \leq 2$. As Hönerlage and Bigot have shown (1984b) for the degenerate case, higher-order terms contribute only to less than 3‰ to the total susceptibility. In addition, the fact that we also keep antiresonant terms gives rise to the correct pole structure of $\chi(\omega_t)$. At low pump intensities, $\epsilon(\omega_t)$ coincides with the value calculated from the one-oscillator model. Since only one exciton state and the biexciton state are retained in the density matrix, the linear contributions of the other oscillators are taken globally into account by introducing a background dielectric constant ϵ_{b} in Eq. 2, which now can be written as:

$$\epsilon(\omega_t) = \epsilon_{\mathrm{b}} + \chi(\omega_t) = \epsilon' + i\epsilon'' \tag{2'}$$

From Eq. 2′, we calculate the wavevector Q_t and the absorption α_t of the test field in the presence of the pump beam to:

$$\begin{aligned} Q_t &= \frac{\omega_t}{\sqrt{2}\,c}\left[\epsilon' + \sqrt{(\epsilon')^2 + (\epsilon'')^2}\right]^{1/2} \\ \alpha_t &= \frac{\omega_t}{\sqrt{2}\,c}\left[-\epsilon' + \sqrt{(\epsilon')^2 + (\epsilon'')^2}\right]^{1/2} \end{aligned} \tag{13}$$

Fig. 1 gives the dispersion of the test beam if the polarization of the pump beam is parallel to that of the test beam for a photon density of $n_{\mathrm{p}} = 10^{15}$ photons/cm^3. The photon density is related to the electric field in this approximation by

$$n_{\mathrm{p}} = \frac{E_{\mathrm{p}}^2 \tilde{\epsilon}_{\mathrm{p}} \epsilon_0}{2\hbar\omega_{\mathrm{p}}} \tag{14}$$

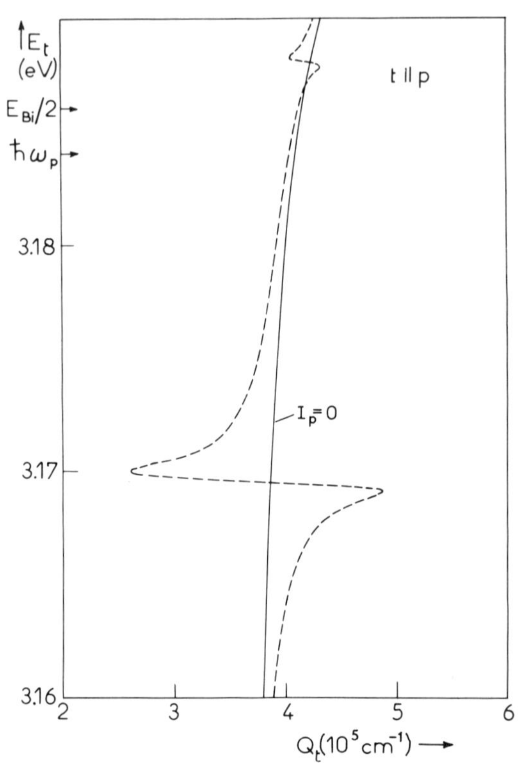

Figure 1. Dispersion $\hbar\omega_t(\mathbf{Q}_t)$ of a test beam in CuCl, when excited by a pump beam at $\hbar\omega_p = 3.184$ eV with $n_p = 0$ (full line) and $n_p = 10^{15}$ photons/cm³ (dashed line) for parallel polarization of the two beams (B. Hönerlage and F. Tomasini, 1986).

where $\tilde{\epsilon}_p$ is, for the sake of simplicity, the real part of the dielectric constant of the pump beam at low intensities.

The damping constants used are those approximately known at low intensities of excitation (Hönerlage and Tomasini, 1986):

$$\hbar\Gamma_{12} = \hbar\Gamma_{23} = 2 \times 10^{-4}\,\text{eV}$$

$$\hbar\Gamma_{13} = 4 \times 10^{-4}\,\text{eV}$$

$$\hbar\Gamma_{ii} = 1.5 \times 10^{-6}\,\text{eV}.$$

In Fig. 1, the photon energy of the pump beam is kept constant at $\hbar\omega_p = 3.184$ eV, i.e. outside the biexciton resonance ($E_{Bi}/2 = 3.186$ eV).

The exciton parameters are taken from Vu Duy Phach *et al.* (1977a and b), Bivas *et al.* (1977) and Hönerlage *et al.* (1978), but spatial dispersion is neglected in our model. We observe two anomalies in the polariton dispersion: the first one, around 3.1695 eV, is due to transitions between exciton and biexciton states. It is called "induced absorption resonance" (IA). The other one, around 3.188 eV, is due to the resonant creation of biexcitons by the simultaneous absorption of one photon coming from the pump and another one from the test. It is called "two-photon resonance" in the following. While the first one is at a fixed energy such as $E_{\mathrm{Bi}} = E_{\mathrm{ex}} + \hbar\omega_t$ and vanishes if the exciton population $\rho_{22} = 0$ is imposed, the latter resonance shifts if $\hbar\omega_{\mathrm{p}}$ is varied, fulfilling the relation $E_{\mathrm{Bi}} = \hbar\omega_{\mathrm{p}} + \hbar\omega_t$. It is worthwhile noticing that the first anomaly changes the dispersion in a wide spectral range and can be simulated in a model that neglects exciton populations by an intensity dependence of ϵ_b. At even higher intensities, as discussed by Hönerlage and Bigot (1984a), a resonance close to $\hbar\omega_{\mathrm{p}}$ shows up. It is due to a periodic variation of the population difference $(\rho_{22} - \rho_{33})^{\circ\circ}$ with the difference frequency $(\hbar\omega_{\mathrm{p}} - \hbar\omega_t)$. The same resonance structures show up when calculating α_t, but in this case, the nonlinearities are much more confined spectrally (Bigot and Hönerlage, 1983; Hönerlage and Bigot, 1984a; Hönerlage and Tomasini, 1986).

It is important to point out that single-beam interactions can be treated in our model by taking the test and pump beams as degenerate (Hönerlage and Bigot, 1984b). In this case, only the two-photon resonance appears in the dispersion and absorption of CuCl. As shown in Fig. 2, it gives rise to a resonance structure around half the biexciton energy, which vanishes at small photon densities, becomes Stark-shifted and broadens at higher photon densities.

Let us now discuss how the dielectric function depends on the relative polarization of the two beams. Fig. 3a represents the level schema and the selection rules for the different optical transitions for parallel polarizations of both beams. Simple (double) arrows indicate one-(two-)photon transitions. We notice that the transverse exciton state $|4\rangle$ does not show up in the optical transitions and the (spin-flip) scattering $|2\rangle \leftrightarrow |4\rangle$ can be accounted for by the lifetime $1/\Gamma_{22}$ of the exciton state.

In two-beam experiments, however, the linear polarization of both beams relative to one another is an additional degree of freedom, since different polarization components couple with the two different transverse exciton states ($|2\rangle$ and $|4\rangle$ in Fig. 3b). If a pump beam excites the crystal of CuCl close to the biexciton two-photon resonance, a test beam polarized parallel

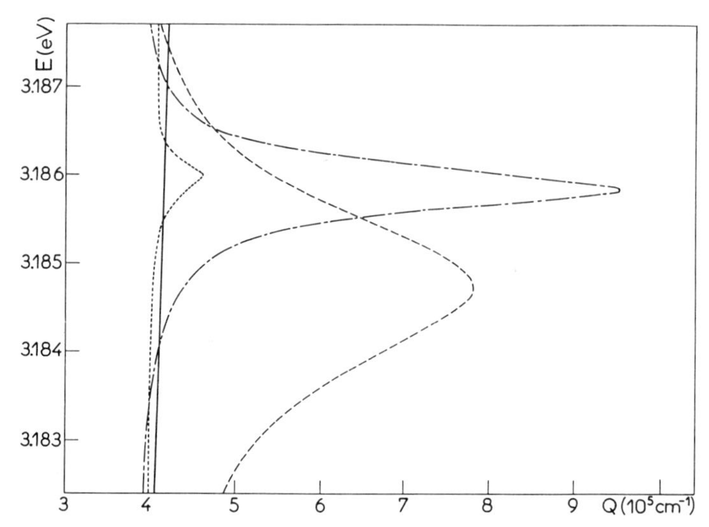

Figure 2. Dispersion $\hbar\omega(\mathbf{Q})$ for a single (pump) beam near half the biexciton energy for different intensities (full line: $n_p = 0$; dotted line: $n_p = 10^{15}$ photons/cm^3; dashed-dotted line: $n_p = 10^{16}$ photons/cm^3; dashed line: $n_p = 10^{17}$ photons/cm^3) (B. Hönerlage and J. Y. Bigot, 1984b).

to the pump beam will be absorbed by this two-photon absorption process (Haug, 1982; Haug and Schmitt-Rink, 1984), but will remain unabsorbed when being polarized perpendicularly (Fig. 3b). Of course, as indicated in Fig. 3, biexcitons can always be created by the absorption of two photons from the test beam. As we have discussed with Fig. 1, besides the simultaneous two-photon absorption, biexcitons can also be generated in a two-step process (induced absorption process), in which excitons are first created. In the case of pump and test beams with parallel polarizations, excitons are created, in the state $|2\rangle$ for instance, by the pump beam, the transition to the biexciton state being further on performed by a one-photon absorption; this process is forbidden for two-crossed polarized beams. However, an appreciable population of excitons can nevertheless be obtained in the state $|4\rangle$ via the creation of biexcitons and their radiative recombination ($|1\rangle \rightarrow |3\rangle \rightarrow |4\rangle$) or, directly, from transverse excitons $|2\rangle$ which scatter elastically to the state $|4\rangle$.

We have calculated the dielectric function in the case of crossed polarizations of both beams (Hönerlage and Tomasini, 1986; Lévy *et al.*, 1986). We have introduced explicitly a coupling matrix element R between the exciton

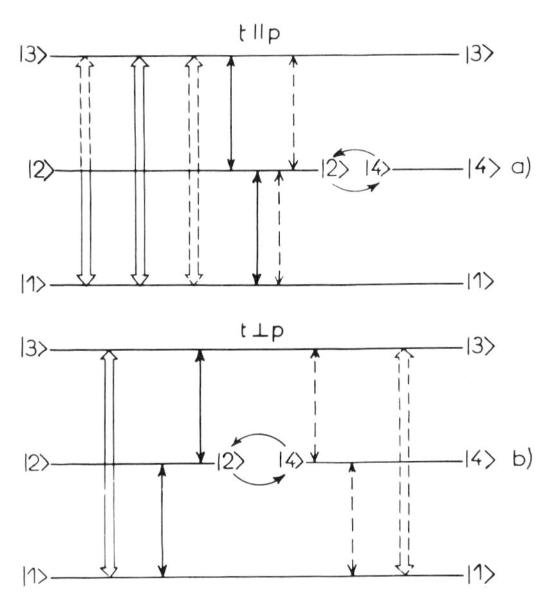

Figure 3. Selection rules in a four-level system under the excitation of two pump and test beams having a) parallel polarizations and b) crossed polarizations. Simple arrows: one-photon transitions; double arrows: two-photon transitions (B. Hönerlage and F. Tomasini, 1986).

states $|2\rangle$ and $|4\rangle$ and followed the same lines as in the parallel case, assuming the same dampings for states $|2\rangle$ and $|4\rangle$. The result for the dispersion is shown in Fig. 4. In this case, the resonance at $\hbar\omega_t = 3.188$ eV vanishes due to selection rules as does the resonance close to $\hbar\omega_{\mathrm{p}}$ since the latter is due to a coherent variation of populations which cannot be induced by beams with crossed polarizations. The induced absorption anomaly around 3.1695 eV, however, still shows up.

If we choose $R = 0$ (dashed-dotted line), we obtain a "Z"-type dispersion in this spectral region. This abnormal behaviour is due to the fact that excitons are mainly created from the recombination of biexcitons (Haug, 1985). If R is increased to $R = 10^{-4}$ eV, we find that the dispersion follows again the normal ("S"-type) resonance structure. If R is further increased, the dispersion does not change. This indicates that $R = 10^{-4}$ eV is sufficient to populate the exciton state $|4\rangle$ by spin-flip scattering.

In the following, we present time-integrated experimental results. We show that the dispersion changes due to exciton–biexciton transitions and we analyze these results in the framework of the stationary theory we have discussed.

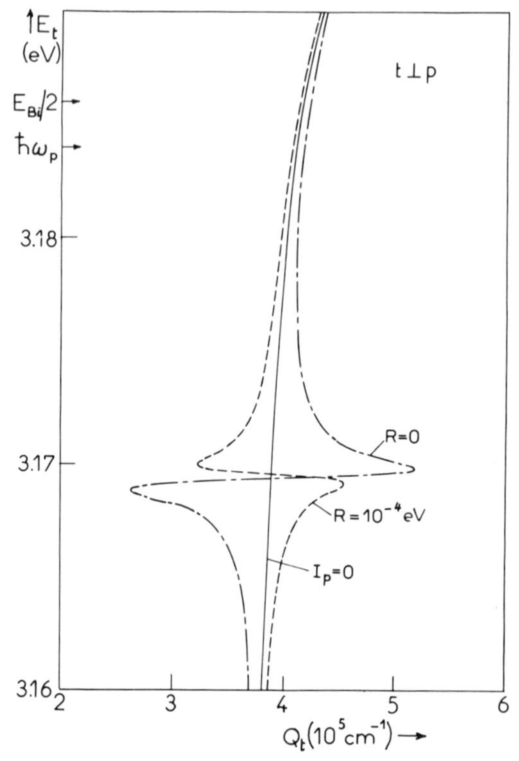

Figure 4. As Fig. 1 for crossed polarizations of the pump and test beams: $R = 10^{-4}$ eV (dotted line), $R = 0$ (dashed-dotted line) (B. Hönerlage and F. Tomasini, 1986).

B. Experimental Evidence for Optical Nonlinearities Due to Exciton–Biexciton Transitions

In CuCl, biexcitons may be created by the absorption of two photons (polaritons) of an intense light beam when their energy $\hbar\omega_p$ is equal to half the biexciton energy $E_{Bi}/2$. This process is resonantly enhanced since the photon energy is close to the exciton energy and also because of the giant oscillator strength of the exciton–biexciton transition (Grun *et al.*, 1982; Hönerlage *et al.*, 1985a; Chemla and Maruani, 1982; Gogolin, 1974; Gogolin and Rashba, 1973). If the photon energy of the exciting light field is detuned from the biexciton resonance, biexcitons are created only virtually. They recombine, obeying to energy and momentum conservation, into two polaritons (index R and C):

$$2\hbar\omega_p = \hbar\omega_R + \hbar\omega_C \text{ and } 2\mathbf{Q}_p = \mathbf{Q}_R + \mathbf{Q}_C \tag{15}$$

where $\mathbf{Q}_p$ is the wavevector of the exciting beam polaritons, $(\hbar\omega_R, \mathbf{Q}_R)$ and $(\hbar\omega_C, \mathbf{Q}_C)$ are the energies and wavevectors of the polaritons created throughout the scattering process. This process, called hyper-Raman (HR) or two-photon Raman scattering, has first been used to determine the polariton dispersion in CuCl at low intensities of excitation and, further on, more complicated polariton structures (Hönerlage *et al.*, 1985a; Grun *et al.*, 1982; Hönerlage *et al.*, 1980; Blattner *et al.*, 1982; Nozue *et al.*, 1981). At higher intensities, both polaritons have the dispersions sketched in Figs. 1 or 4, depending on the polarization of the emission. Since the density of the created polaritons is small, they play the role of the "test" beam described in our preceding discussion, and their dispersion depends on the photon density of the exciting beam. The dispersion of this last beam (index p) is that of Fig. 2 since it plays the role of the pump beam. Therefore, hyper-Raman scattering can be used to study renormalization effects of the dielectric function (Kurtze *et al.*, 1980; Itoh *et al.*, 1978; Itoh and Suzuki, 1978). In our case, the polaritons with photon energy $\hbar\omega_R$ are observed, and the two-photon anomaly of Fig. 1 gives rise to a doublet structure of the HR emission.

The experimental setup used (Lévy *et al.*, 1985a; Grun *et al.*, 1983) is shown in Fig. 5. The samples are excited by the light of a grazing incidence dye laser (αNND in ethanol or BiBuQ in toluene) pumped by a Lambda Physik excimer laser (EMG 101). The emission of the dye laser lasts 2.5 ns (FWHM) with a repetition rate of 13 Hz. The spectral width of the dye laser emission is measured with a Fabry–Perot interferometer to be about 0.03 meV (FWHM). The dye laser emission is tunable from 3.15 to 3.23 eV. The light is strongly polarized in the plane defined by the dye laser cavity. The intensity of the laser emission can be varied by a set of neutral density filters. The dye laser is focused under an angle of incidence α onto the surface of the sample, in a spot of about 100 μm in diameter. The maximum intensity I_M is of about 50 MW/cm^2.

The samples studied are copper chloride monocrystalline platelets of 30 μm thickness, grown by vapor phase transport in a reduced argon atmosphere. They are cooled down to pumped liquid helium temperature (2K). The emission from the rear surface of the crystal (forward scattering configuration) is detected under an angle β through a 3/4m Spex spectrograph working in the second order. Spectra are analyzed by a PAR optical multichannel analyzer, visualized on an oscilloscope and registered by an XY recorder.

Fig. 6a shows a typical hyper-Raman spectrum of CuCl samples. The intensity of the laser is kept at 10% of the maximum intensity I_M available

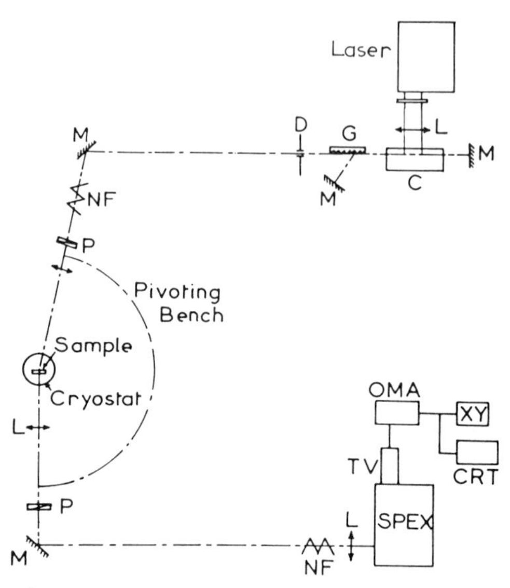

Figure 5. Set-up for hyper-Raman scattering experiments. M: mirror; C: dye cell; D: diaphragm; NF: neutral density filter; P: polarizer; G: grating; OMA: optical multichannel analyzer; TV: SIT picture tube; L: lens, Spex: Spetrograph; XY: plotter; CRT: oscilloscope (R. Lévy *et al*., 1985a).

in order to avoid the destruction of the samples. The photon energy is $\hbar\omega_p = 3.1837$ eV. The HR emission is detected in a forward configuration with an angle $\beta = 11°$. Beside the diffusion of the dye laser, two hyper-Raman lines R_T^+, R_T^- are observed. When $\hbar\omega_p = 3.1847$ eV, both lines split into a doublet structure as can be seen in Fig. 6b.

Figs. 7 and 8 represent the spectral positions of the HR emission lines R_T^+ and R_T^- as functions of the energy $\hbar\omega_p$ of the exciting photons in two different forward scattering configurations. The angle of incidence α is equal to zero in both cases, the angles of observation β are 11° and 23°, respectively.

We have fitted the experimental spectral positions of the HR lines R_T^+ and R_T^- with calculated ones represented by the continuous curves in Fig. 7. We have used the following parameters: a density of photons $n_p = 4.3 \times 10^{14}$ photons/cm^3, a dephasing constant $\hbar\Gamma_{12} = 3 \times 10^{-4}$ eV and an arbitrarily chosen value of $\hbar\Gamma_{ii} = 1.5 \times 10^{-6}$ eV. The global shift between the spectral positions of these HR emission lines and those obtained at low intensities of excitation (the dotted lines in Fig. 7) is due to the induced

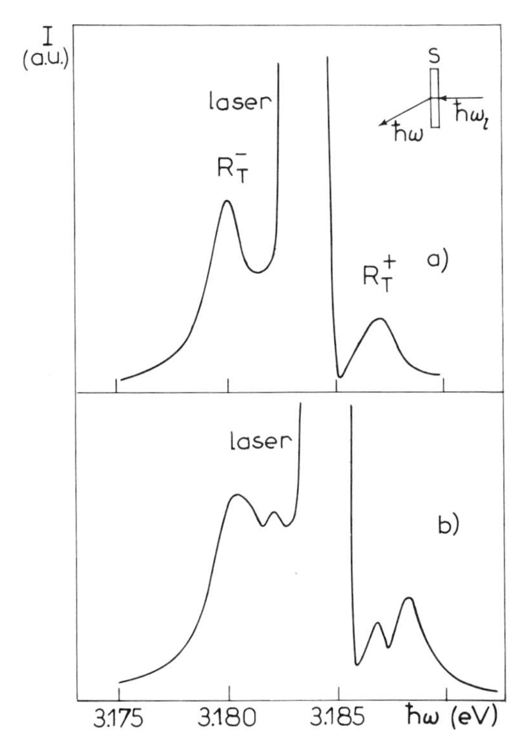

Figure 6. Typical hyper-Raman scattering spectra for $\beta = 11°$, $\alpha = 0°$ and $I = 0.1\ I_M$; and (a) $\hbar\omega_p = 3.1837$ eV, (b) $\hbar\omega_p = 3.1847$ eV (J. B. Grun *et al.*, 1983).

absorption anomaly at 3.1695 eV (see Fig. 1). It reflects the presence of a large number of excitons created through the nonlinear process.

The induced polariton branch due to the two-photon absorption of a photon $\hbar\omega_p$ of the exciting laser and a photon emitted ($\hbar\omega_R$ or $\hbar\omega_C$) is also well taken into account as shown by the good fit between experimental results and the calculated continuous curves. This induced branch, which corresponds to a splitting of the HR lines, is observed around the energies:

$$\hbar\omega_i = E_{Bi} - \hbar\omega_p$$

and (16)

$$\hbar\omega_i = 3\hbar\omega_p - E_{Bi}$$

as obtained from Eqs. 15. These energies are represented by the dashed-dotted lines drawn in Fig. 7.

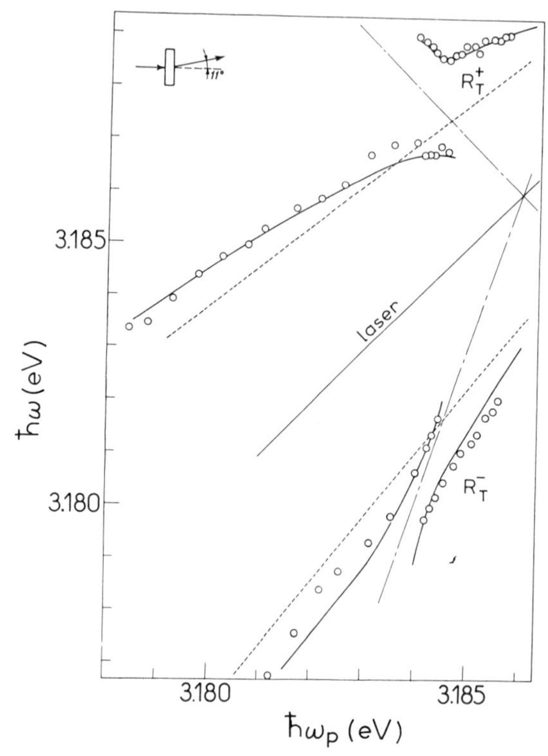

Figure 7. Experimental positions of hyper-Raman emission lines of CuCl when excited under an angle of incidence $\alpha = 0°$ and for a direction of observation $\beta = 11°$ (R. Lévy *et al.*, 1985a).

In Fig. 8, the calculated spectral positions of the HR emission lines represented by the continuous curve correspond to an excitation intensity of $n_p = 10^{15}$ photons/cm^3 and exciton and biexciton energy relaxation dampings Γ_{22} and Γ_{33} equal to $\hbar\Gamma_{ii} = 2 \times 10^{-5}$ eV. The overall shift of the HR lines is again satisfactorily explained.

Besides the shift and splitting of the HR lines, we also observe a nonlinear shift when the photon energy of the exciting laser is tuned across half the biexciton energy. It corresponds to the anomaly shown in Fig. 2, where the wavevector of the exciting beam varies considerably if the photon energy is changed slightly around the biexciton resonance.

The different renormalization phenomena have also been studied by HR scattering as functions of the excitation intensity. However, the intensity of the light emitted in HR scattering, which is a spontaneous process, was too

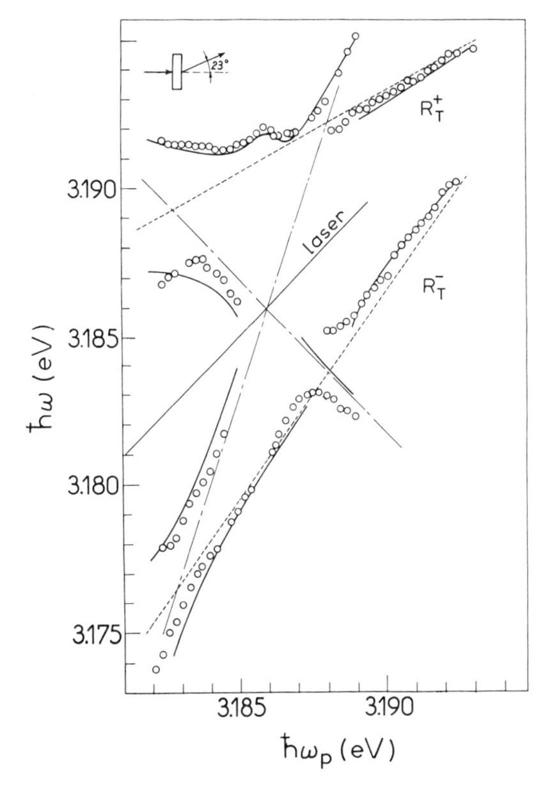

Figure 8. Experimental positions of hyper-Raman emission lines of CuCl when excited under an angle of incidence $\alpha = 0°$ and for a direction of observation $\beta = 23°$ (R. Lévy *et al.*, 1985a).

low to allow a systematic study to be made. We have, therefore, looked for a more efficient induced process: degenerate four-wave mixing (Chemla *et al.*, 1979; Maruani *et al.*, 1978; Chase *et al.*, 1983) would have permitted us to determine coherence times and wave functions of the intermediate states involved in the process. Nondegenerate, noncollinear four-wave mixing was, however, the adequate spectroscopic tool to perform momentum space spectroscopy as HR scattering does, but in a much more efficient way.

Similar to electronic CARS, nondegenerate four-wave mixing is the induced process corresponding to the HR scattering as the spontaneous one (Hönerlage *et al.*, 1982; Hönerlage *et al.*, 1983; Lévy *et al.*, 1984). The experimental configuration used is similar to that of HR scattering, as sketched in Fig. 9. Biexcitons are again created virtually by the absorption

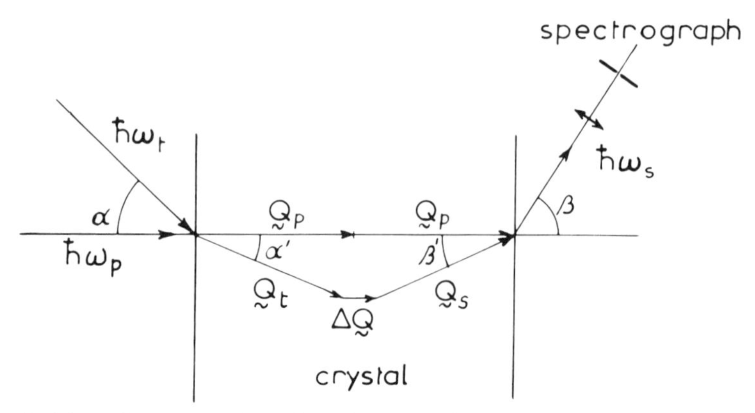

Figure 9. Schematic representation of nondegenerate four-wave mixing processes (R. Lévy *et al.*, 1986).

of two polaritons with energy $\hbar\omega_p$ and wavevector $\mathbf{Q}_p$ inside the crystal. They are provided by the tunable dye laser described already, called "pump." The biexciton recombination is now induced by a second tunable dye laser, called "test," which is similar to the first one and pumped by the same excimer laser. The energy and wavevector of the corresponding polaritons are $\hbar\omega_t(\mathbf{Q}_t)$. A parametric signal of photon energy $\hbar\omega_s$ and wavevector $\mathbf{Q}_s$ is generated, obeying to energy conservation:

$$\hbar\omega_s = 2\hbar\omega_p - \hbar\omega_t \tag{17}$$

However, there is a phase-mismatch $\Delta\mathbf{Q}$ in the process. It is equal to:

$$\Delta\mathbf{Q} = 2\mathbf{Q}_p - \mathbf{Q}_t - \mathbf{Q}_s \tag{18}$$

As shown by Hönerlage *et al.* (1982), the signal intensity I_s is maximum, when the phase-matching condition is fulfilled. As described by Lévy *et al.* (1986), this is achieved when the angle α between the two laser beams and the angle of observation β (as shown in Fig. 9) are small and if $\alpha = \beta$ holds. In the situation of phase matching, the photon energy of the parametric emission $\hbar\omega_s$ is equal to the energy $\hbar\omega_R$ of the HR emission.

We choose here $\alpha = \beta = 11°$ and a maximum intensity $I_p^{max} = 50$ MW/cm^2 of the pump beam. We keep the photon energy $\hbar\omega_p$ of the pump beam fixed and vary the photon energy of the test beam. Fig. 10 gives the intensity of the parametric emission I_s as a function of its photon energy $\hbar\omega_s$ (excitation spectrum). In this case, $\hbar\omega_p = 3.1845$ eV is chosen. The polarizations of the pump and test beams are taken perpendicular to one another. A well-defined maximum is found, which shifts to higher energies

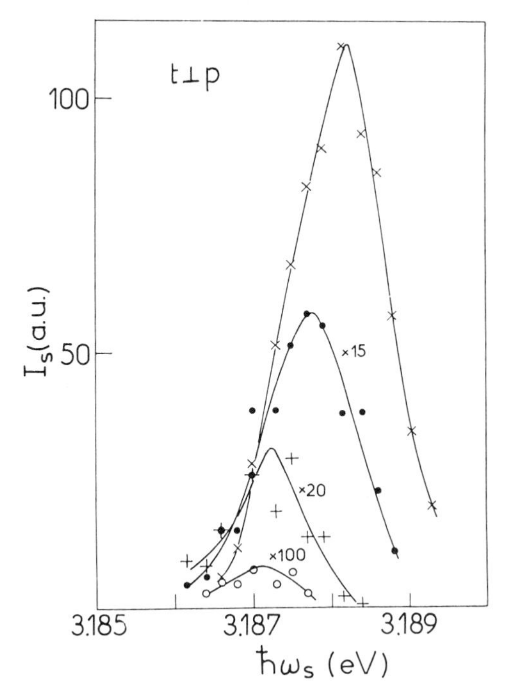

Figure 10. Excitation spectrum of the parametric emission for different intensities I_p of the pump beam. Test and pump beams are polarized perpendicularly to each other. The photon energy of the pump beam is fixed at 3.1845 eV [$I_p/I_p^{max} = 0.1(\times)$; 10^{-2}(●); $2.5 \times 10^{-3}(+)$; 5×10^{-4}(○)] (B. Hönerlage *et al.*, 1983).

with increasing intensities of the pump beam as already noticed in HR scattering (Lévy *et al.*, 1985a; Grun *et al.*, 1982). This shift can also be seen in Fig. 11, where we plot the spectral position of the maximum of the signal excitation spectrum as a function of the pump beam intensity I_p. The arrow denotes the spectral position at which the phase matching occurs at small intensities for 10° as calculated in the framework of the one-oscillator model with the parameters of Vu Duy Phach *et al.* (1977a), Bivas *et al.* (1977) and Hönerlage *et al.* (1978). This value of 10° is in agreement with the experimental value of $(11 \pm 2)°$. The photon density of the pump beam n_p is calculated from the pump beam intensity I_p by the relation:

$$n_p = I_p/\hbar\omega_p V_E \tag{19}$$

where the energy velocity V_E has been assumed to be equal to the group velocity $v_g = \dfrac{\delta\omega}{\delta k}$ for small intensities, neglecting renormalization effects.

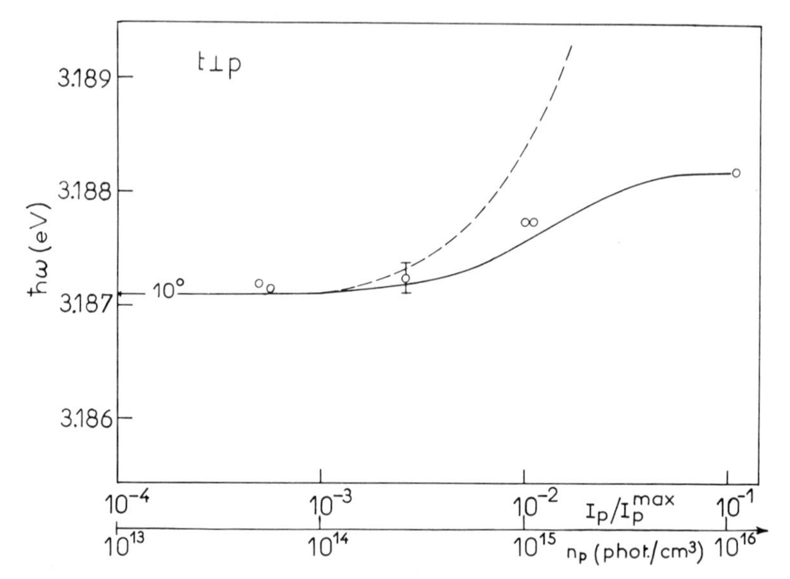

Figure 11. Spectral position of the maximum of the parametric emission as a function of the pump beam intensity for crossed polarizations of pump and test beams (R. Lévy *et al.*, 1986).

The dotted line of Fig. 11 gives the calculated spectral positions of the signal beam at which phase matching occurs, when using the dielectric function that we have discussed and the damping constants used to calculate Fig. 4.

Moreover, the nonradiative relaxation process that couples the two exciton states with crossed polarizations is assumed to be equal to $R = 10^{-4}$ eV. This is the minimum value necessary to obtain a positive shift of the maximum of the excitation spectra with increasing intensity as observed experimentally (Lévy *et al.*, 1986). We notice that the theoretical curve obtained with these parameters deviates from the experimental one, even at rather small intensities of the pump beam. The only way to correct this discrepancy is to assume that the different damping constants are in fact intensity-dependent.

We know from the calculation of the reflection coefficient near the exciton and biexciton resonances that the set of parameters given in Eq. 14 leads to a reflection anomaly at the energies corresponding to the transitions between exciton and biexciton states (Hönerlage *et al.*, 1986). This anomaly is very sensitive to the value of the population inversion lifetime τ and the dephasing constant Γ_{23}. All other constants are of minor influence.

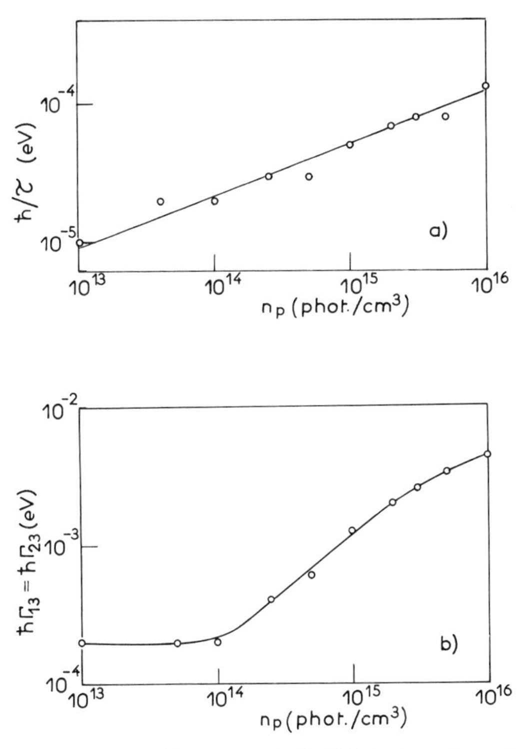

Figure 12. Damping constants (a) $\hbar\Gamma_{ii}(n_p)$ and (b) $\hbar\Gamma_{23}(n_p)$ as determined from Fig. 11 (R. Lévy *et al.*, 1986).

Since this anomaly has not been detected experimentally, we tried to fit reflection measurements and NDFWM simultaneously. We first adjusted $\hbar\Gamma_{ii}$ to a value that gives a calculated reflection anomaly smaller than the experimental error made in measuring the reflection coefficient, i.e. 2%. The intensity dependence of $\hbar/\tau$ is given in Fig. 12a. Using this result, we found that the calculated energy of the maximum of the excitation spectrum becomes smaller, but does not yet fit the experimental points. In order to improve the fit once more, we had to assume an intensity dependence of $\Gamma_{13} = \Gamma_{23} = \Gamma_{43}$ as given in Fig. 12b. This result has been cross-checked by decreasing the lifetimes τ even more. The same value of Γ_{23} was obtained, indicating that the result of Fig. 12a corresponds to a minimal value of $\hbar\Gamma_{ii}$. The choice of $\Gamma_{14} = \Gamma_{12}$ is not important.

Figs. 13 and 14 show the results obtained for parallel polarizations of the two laser beams. In this case, the parametric emitted beam and the test beam follow the renormalized polariton dispersion and two maxima are

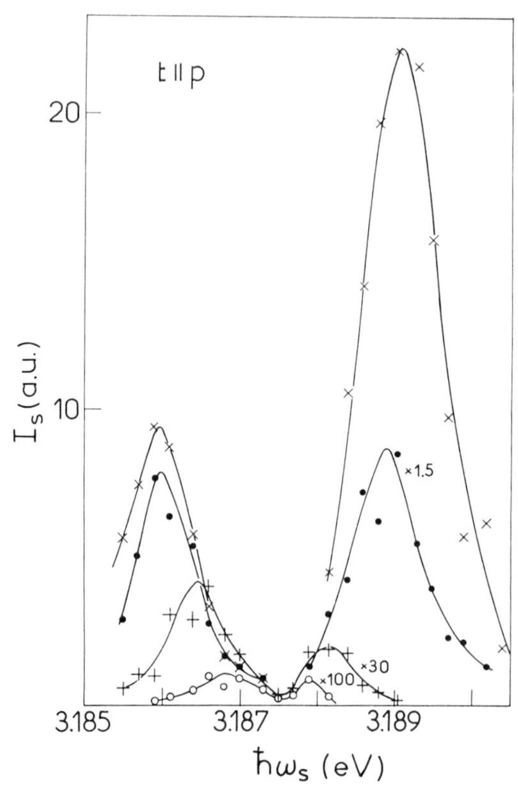

Figure 13. Excitation spectrum of the parametric emission for different intensities of the pump beam. Test and pump beams are polarized parallel to each other. The photon energy of the pump beam is fixed at 3.1845 eV (The same excitation conditions as in Fig. 10) (B. Hönerlage *et al*., 1983).

clearly observed in the excitation spectra. They correspond to the splitting of the HRS lines observed by Lévy *et al*. (1985a) and Grun *et al*. (1983). The relative height between the low- and high-energy maxima changes, indicating that the splitting is not symmetrical to a fixed center of gravity.

In Fig. 14, we give the spectral position of the maximum of the excitation spectrum of the signal emission as a function of the intensity of the pump beam I_p for parallel polarizations of test and pump beams. For the photon energy chosen, the phase-matching conditions are achieved at high-excitation intensities for two different photon energies. The full lines and dotted lines are the calculated positions for which phase matching is achieved using the same parameters as in Fig. 11. We notice that, at small intensities, using the damping constants just given, one single solution for phase

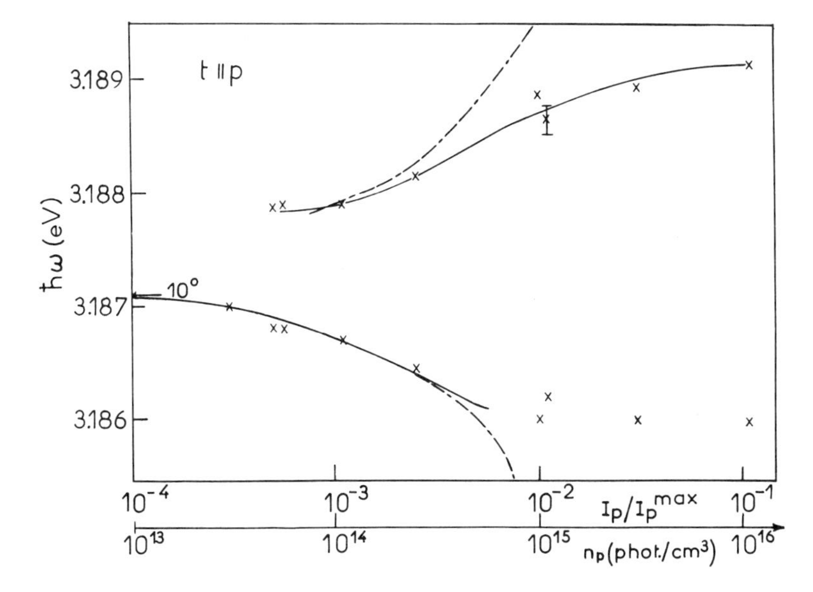

Figure 14. As Fig. 11, but for parallel polarizations of the pump and test beams (R. Lévy *et al.*, 1986).

matching exists as it is observed experimentally. For $n_p > 10^{15}$ photons/cm^3, no fit could be obtained for the lower induced branch by changing the damping constants. The spectral position of this branch is probably influenced by the two-photon absorption towards the biexciton state which is important around 3.186 eV. The signal emission is blocked to this value due to an additional resonance in χ. It is interesting to point out that the signal beam shows the same polarization behaviour for both branches of Fig. 14.

Similar results concerning the density dependence of transverse damping constants (Γ_{13}) have been obtained from two-photon absorption measurements by Chase *et al.* (1979), Peyghambarian *et al.* (1984) and Itoh *et al.* (1984). The lifetimes of the quasi particles ($1/\Gamma_{ii}$) do not show up, however, in these experiments since they are several orders of magnitude longer than the transverse relaxation times $1/\Gamma_{13}$. The observed increase of Γ_{13} and Γ_{ii} is attributed to mutual collisions between excitons and biexcitons, which give rise to the broadening of the resonance.

Apart from HR scattering and nondegenerate four-wave mixing, excitation-induced circular dichroism has also been used to study renormalization effects in CuCl (Haug, 1982; Kranz and Haug, 1986; Cho and Itoh, 1984; Kuwata, 1984; Kuwata *et al.*, 1981; Itoh and Kathono, 1982). In this case, a strong circularly polarized beam is sent on the crystal, which leads to an elliptical polarization of a linearly polarized test beam after its transmission through the sample. The nonlinear absorption of the biexcitons is at the origin of this effect.

III. THE DIELECTRIC FUNCTION IN THE QUASI-STATIONARY APPROXIMATION AND TRANSIENT EFFECTS

We have seen in chapter 8, Section II.B, that the damping constants and the exciton and biexciton lifetimes are important parameters which govern the dielectric function of the system. In addition, we have shown that they change when changing the intensity of excitation. Since we are working with pulsed excitation, it is important to study whether the dielectric properties of the system follow the pulsed excitation and, under which conditions, the finite lifetimes of the quasiparticles introduce a new dynamics. In order to do this, one has in principle to solve simultaneously the coupled Maxwell and Bloch equations (Bigot and Hönerlage, 1987) for the three-level system, since the dielectric function (and from this, the wavevector, the absorption and the reflection) is only defined for excitation with constant amplitude E_0. This is due to the fact that the convolution product (Eq. 10) involves a Fourier transformation when defining $\chi(\omega_t)$.

Since the procedure just outlined needs a lot of numerical calculations, we shall follow here a different approach and define a transient dielectric function $\epsilon(t, \omega)$, using a quasi-stationary approximation (Bigot *et al.*, 1985a; Hönerlage *et al.*, 1985b). It is worthwhile mentioning that, in this approximation, $\epsilon(t, \omega)$ does not describe intrinsic properties of the material, but depends on the time variation of the exciting field amplitude. We shall treat the problem as in chapter 8, Section II.A, but we shall restrict it to the interaction of the medium with a single electric field of the form:

$$\mathrm{E}(t) = \tilde{\mathrm{E}}(t) e^{-i\omega t} \tag{20}$$

The field amplitude $\tilde{\mathrm{E}}(t)$ varies slowly with respect to ω. Then, instead of using the Fourier series given in Eq. 11, we separate $\rho_{ij}(t)$ into slow- and fast-varying contributions in the form:

$$\rho_{ij}(t) = \sum_n \rho_{ij}^n(t) e^{in\omega t} \tag{21}$$

and, from Eq. 6, we now obtain a system of coupled linear differential equations for $\rho_{ij}^{n}(t)$. This system is truncated in the same way as was previously discussed, i.e. retaining resonant and antiresonant contributions with $|n| \leq 2$. This system of Bloch equations is solved numerically for a given amplitude variation $\tilde{\mathrm{E}}(t)$.

The convolution product of Eq. 10 is equivalent to the expression:

$$P(t) = \epsilon_0 \operatorname{Re} \int_{-\infty}^{t} \chi(t, t') \tilde{\mathrm{E}}(t') e^{i\omega t'}\, dt' \tag{22}$$

when neglecting the spatial dependence of the light field. Introducing $t' = t - \tau$, Eq. 22 can be approximated in the following way:

$$\begin{aligned} P(t) &= \epsilon_0 \operatorname{Re} \int_{0}^{\infty} \chi(t, \tau) \tilde{\mathrm{E}}(-\tau + t) e^{i\omega(t-\tau)}\, d\tau \\ &\simeq \epsilon_0 \operatorname{Re} \tilde{\mathrm{E}}(t) e^{i\omega t} \int_{0}^{\infty} \chi(t, \tau) e^{-i\omega\tau}\, d\tau \\ &= \epsilon_0 \operatorname{Re}\left[\tilde{\mathrm{E}}(t) \chi(t, \omega) e^{i\omega t}\right] \end{aligned} \tag{23}$$

Comparing now Eq. 23 to Eq. 9, we find as in the periodic case for the transient dielectric susceptibility $\chi(t, \omega)$:

$$\chi(t, \omega) = \frac{2N}{\epsilon_0 \tilde{\mathrm{E}}(t)} \left\{\left[\rho_{12}^{1}(t) + \rho_{21}^{1}(t)\right] \mu_{\mathrm{ex}} + \left[\rho_{23}^{1}(t) + \rho_{32}^{1}(t)\right] \mu_{\mathrm{Bi}}\right\} \tag{24}$$

In order to adopt the approximation of Eq. 23, $\tilde{\mathrm{E}}(t)$ has to vary "slowly" on a time scale which is determined by the variation of $\chi(t, \omega)$. Since it is related in Eq. 24 to the nondiagonal matrix elements of $\rho_{ij}^{1}(t)$, the important time constants are $1/\Gamma_{12}$ and $1/\Gamma_{23}$, which are in the picosecond range in CuCl (Vu Duy Phach *et al.*, 1977a; Vu Duy Phach *et al.*, 1977b). Therefore, the approach we have discussed is valid for nano- and sub-nanosecond pulses.

From Eqs. 2′ and 12, we obtain then the dielectric function $\epsilon(t, \omega)$, the wavevector $\mathbf{Q}(t, \omega)$ and the absorption coefficient $\alpha(t, \omega)$. Remembering that the complex refraction index $n = n' + in''$ is related to $\mathbf{Q}$ and α by:

$$n' = \frac{c\mathbf{Q}}{\omega} \quad \text{and} \quad n'' = \frac{c\alpha}{\omega} \tag{25}$$

we find for the time-dependent reflectivity R of the system:

$$R = \frac{(n' - 1)^2 + n''^2}{(n' + 1)^2 + n''^2} \tag{26}$$

Using $\hbar\Gamma_{\mathrm{ex}} = 10^{-6}$ eV, $\hbar\Gamma_{\mathrm{Bi}} = 5.10^{-7}$ eV and $\hbar\Gamma_{12} = \hbar\Gamma_{23} = 10^{-4}$ eV, we calculate the transient response of the system under pulsed excitation.

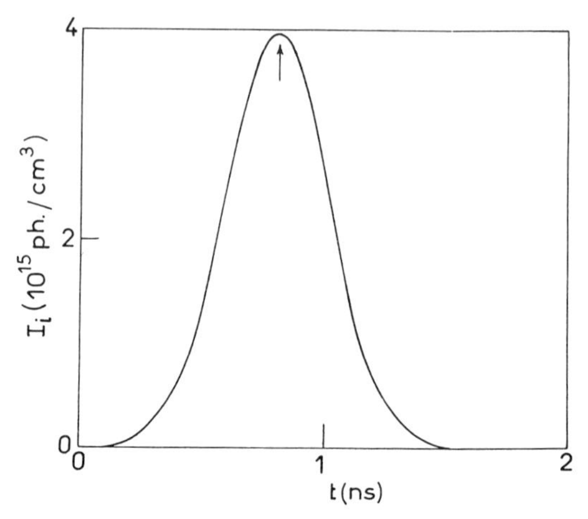

Figure 15. Temporal shape of the incident pulse (J. Y. Bigot *et al.*, 1985a).

As shown in Fig. 15, we assume a Gaussian pulse of the intensity $I_i(t)$, which reaches its maximum value at $t = 800$ ps and has a halfwidth of $\Delta = 600$ ps.

The photon energy of excitation, $\hbar\omega = 3.1865$ eV, is close to $E_{Bi}/2 = 3.186$ eV. The maximum incident intensity is $n_p = 4.10^{15}$ photons/cm^3. As indicated in Fig. 16a-b, **Q** and α show a memory effect that is more pronounced in the case of a broad biexciton resonance (solid lines for $\Gamma_{13} = 4.10^{-4}$ eV) than in the case of small dampings (dotted lines for $\Gamma_{13} = 2.5 \times 10^{-5}$ eV). It is also interesting to notice in Fig. 16b that α can even show optical gain at the end of the pulse. This effect vanishes if the exciton lifetime is longer than that of biexcitons ($\hbar\Gamma_{ex} \leq \hbar\Gamma_{Bi}$). As can be observed in Fig. 16c, the memory effect is due to the time-dependent inversion of population ($\rho_{22} - \rho_{33}$), which does not follow at all the incident pulse shape. As shown in Fig. 17, if a pulse is transmitted through a CuCl sample having the time-dependent variation of α shown in Fig. 16b, it will be deformed asymmetrically. Its shape changes depending on the damping constants and on the lifetimes used in this mean field calculation.

This time dependence of the nonlinear absorption and dispersion has important consequences for transmission measurements and for their interpretation in terms of optical bistability.

The time-resolved transmission measurements use the setup shown in Fig. 18. The excimer laser pumps a dye laser (BiBuQ in toluene) in a Hänsch configuration. The laser beam is further amplified in a second dye

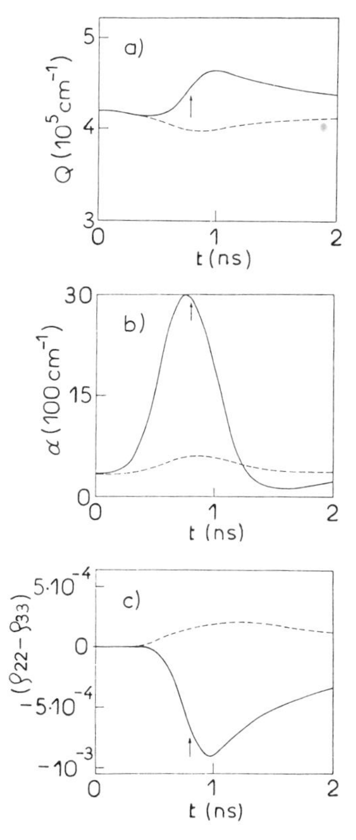

Figure 16. (a) Wavevector **Q**, (b) Absorption coefficient α and (c) population inversion $\rho_{22} - \rho_{33}$ as functions of time under pulsed excitation, using $\hbar\omega = 3.1865$ eV, $\hbar\Gamma_{13} = 4 \times 10^{-4}$ eV (full line) and $\hbar\Gamma_{13} = 2.5 \times 10^{-5}$ eV (dashed line). The maximum of the incident pulse is indicated by the arrow (J. Y. Bigot *et al.*, 1985a).

solution pumped by the same excimer laser. Great care is taken to keep the superradiant emission in the beam small ($\leq 1\%$) compared to the laser emission. The shape of the pulses is kept well defined in time with a width of about 6 ns. The laser emission has a spectral width of 0.4 meV (Bigot *et al.*, 1985b; Grun *et al.*, 1986).

After passing through a diaphragm, neutral density filters and a glan polarizer, the beam is focused onto monocrystals of CuCl (of 20 μm thickness) in a spot of 100 μm in diameter. The sample is cooled down to pumped liquid helium temperature. The power density can be varied up to 60 MW/cm^2. The transmission of the laser pulses through the sample is detected by a streak camera from Hadland, coupled to an OSA vidicon

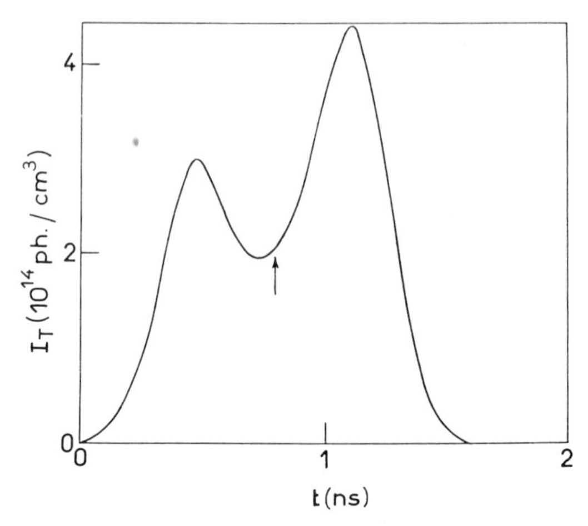

Figure 17. Transmitted pulse when calculated from Figs. 15 and 16. The maximum of the incident pulse is indicated by the arrow (J. Y. Bigot *et al.*, 1985a).

system. The overall time resolution of the camera can be varied from 8 to 130 ps. Reference pulses are, first, averaged over 20 shots in order to improve the signal-to-noise ratio. Then, for the same number of shots, the pulses transmitted through the crystal are recorded. Of course, this kind of measurement demands a high reproducibility of the laser pulse shape.

Fig. 19 shows the reference signal $I_i(t)$ as well as transmitted ones $I_T(t)$ for different laser energies. The transmitted signals are clearly deformed, the maximum of deformation being observed around half the biexciton energy. From these measurements, we determine the temporal dependence of the transmission $T(t) = I_T(t)/I_i(t)$ of the sample. It is given in Fig. 20 for different photon energies. Below the biexciton resonance, $\hbar\omega_1 < 3.186$ eV, nonlinear absorption is first observed. It turns into gain in the middle of the pulse and then changes back again into absorption. The system oscillates between absorption and gain in this spectral region (Fig. 20a). When the photon energy of the incident beam is increased, the gain vanishes and, above the biexciton resonance, pure nonlinear absorption is observed during the whole pulse (Fig. 20b).

With our dynamic model, we have tried to reproduce the observed transmitted pulse shape as well as the time variation of the transmission, using the mean field approximation and, therefore, neglecting propagation effects. Fig. 21a represents the simulated incident pulse shape for $\hbar\omega_1 =$ 3.1857 eV, i.e. slightly below the biexciton resonance. Fig. 21b gives the corresponding calculated pulse as transmitted through the sample. We

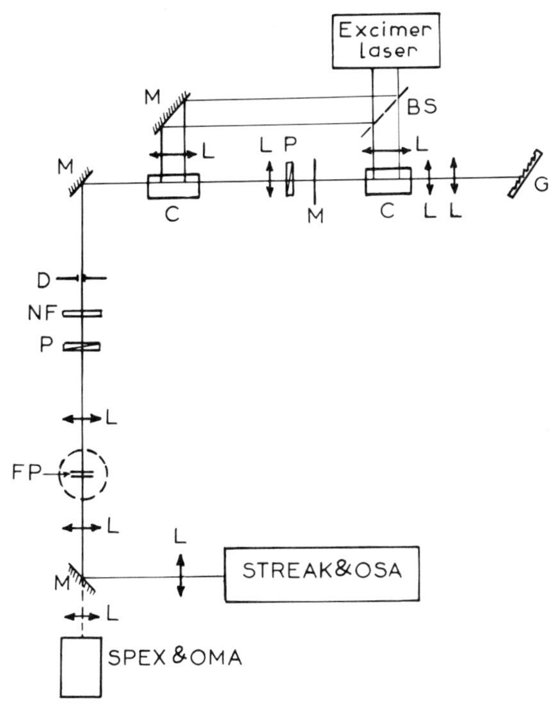

Figure 18. Experimental setup: BS: beam splitter; G: grating; L: lens; C: dye cell; M: mirror; P: polarizer; D: diaphragm; NF: neutral density filter; F.P.: Fabry–Perot (J. B. Grun *et al.*, 1986).

deduce from these curves the temporal variation of the transmission coefficient given in Fig. 22a. As in the experimental case represented in Fig. 20a, we obtain a strong nonlinear absorption at the beginning of the pulse, which turns over into gain in the middle of the pulse and then an oscillation between absorption and gain occurs. The lifetime of excitons and biexcitons are taken equal to 600 ps and 3 ns, corresponding to $\hbar\Gamma_{ex} = 10^{-6}$ eV and $\hbar\Gamma_{Bi} = 2 \times 10^{-7}$ eV, respectively. The dephasing times could not be kept at the values known from measurements at low intensities of excitation. The biexciton dephasing time Γ_{13} had to be considered as dependent on exciton and biexciton populations. We took the following expression:

$$\Gamma_{13} = \Gamma_{13}^{0} + a(\rho_{22} - \rho_{33}) \tag{27}$$

where $\rho_{22} - \rho_{33}$ represents the difference between exciton and biexciton populations. This type of damping function leads to a saturation of

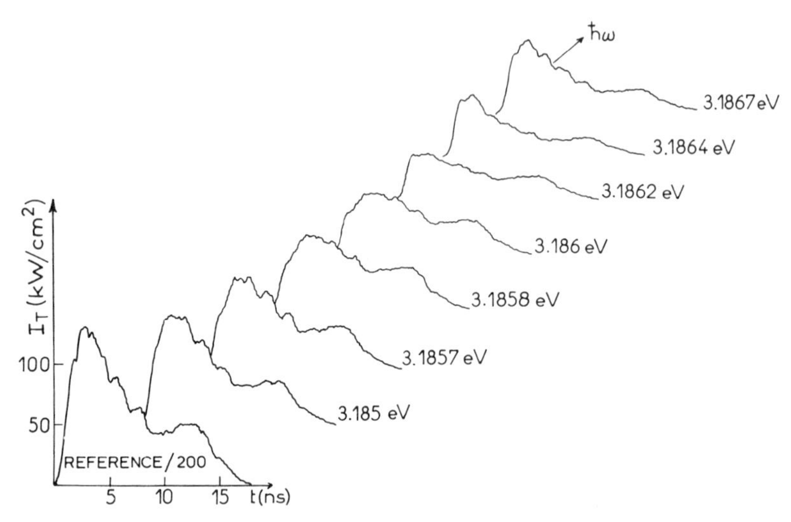

Figure 19. Intensities of the reference signal and of the signals transmitted through the sample for different laser energies (J. B. Grun *et al.*, 1986).

biexciton absorption which can simulate an induced biexciton emission. As a consequence, the imaginary part of the dielectric function oscillates, giving rise to gain and absorption during the pulse.

A similar approach has been used when exciting the crystal above half the biexciton energy (Fig. 22b). In this case, we are generating an important population of directly created biexcitons. By radiative recombination of these biexcitons, an appreciable population of excitons is also building up. Because of collisions between them, the lifetimes of biexcitons and excitons become equal and this will drastically affect the possibility of obtaining gain. As shown in Fig. 22b, the gain completely disappears if biexciton and exciton lifetimes become equal to 600 ps.

Finally, the experimental results of Fig. 20 have been interpreted by taking into account the population dynamics of excitons and biexcitons which modulate the nonlinear absorption of the system. Qualitatively, the same behaviour is observed when polycrystalline films are used instead of the monocrystalline samples previously studied. However, the memory effect is much more pronounced (Bigot *et al.*, 1986) in this case. The gain region, appearing when we excite the sample below half the biexciton energy, is delayed to the end of the pulse. It disappears when we excite above half the biexciton energy. These differences are simply explained in our model by the linear absorption of the polycrystalline samples, which is

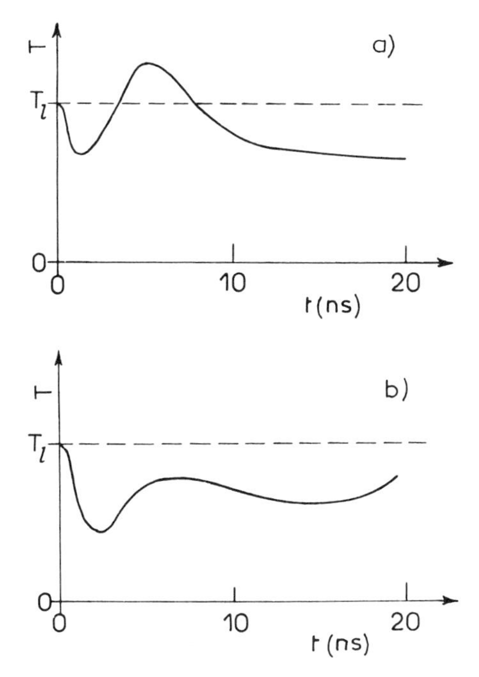

Figure 20. Time-dependent transmission of the sample for different excitation photon energies: (a) Below half the biexciton energy (3.1857 eV) (b) Above half the biexciton energy (3.187 eV) (J. B. Grun *et al.*, 1986).

about 10 times higher than in monocrystals. Therefore, when working with such polycrystalline films under nanosecond excitation, all the temporal variations we have observed are in fact transient effects, i.e. the response time of the system is usually longer than the pulse duration.

This is extremely important when looking experimentally for optical bistability, using a pulsed excitation. As will be discussed in more detail in Chapter 10 of this book, dispersive optical bistability has been observed in CuCl by placing monocrystalline samples in a Fabry–Perot resonator which supplies feedback (Lévy *et al.*, 1983). In these experiments, we used a pulse shape presenting secondary maxima which will be convenient to test the stability of the device.

In Fig. 23, we clearly observe dispersive optical bistability when we excite the device close to the biexciton resonance. We conclude that we have real bistability and not a simple memory effect from the fact that the oscillations of intensity in the incident pulse give rise to secondary loops in this $I_T(I_i)$ diagram which stay on the same branch of transmission. We find

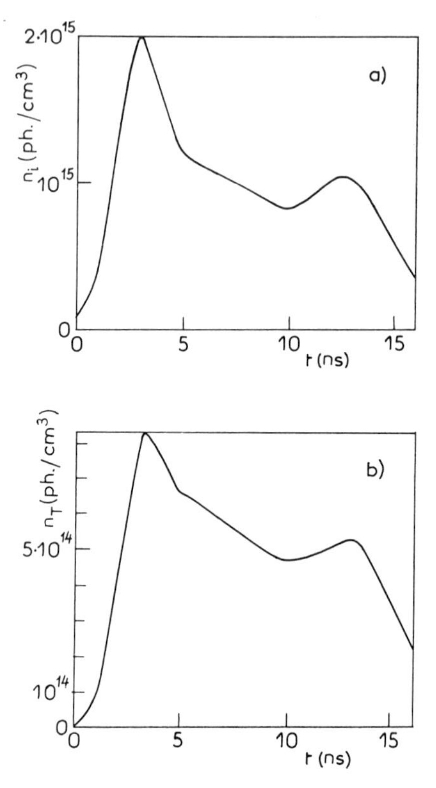

Figure 21. (a) Theoretical curve simulating the shape of the incident pulse ($\hbar\omega_l = 3.1857$ eV). (b) Corresponding calculated pulse transmitted through the sample, assuming exciton and biexciton lifetimes of the order of 600 ps and 3 ns, respectively (J. B. Grun *et al.*, 1986).

switching times of 260 and 450 ps, respectively (Bigot *et al.*, 1985b). Using another F.P. resonator, we observe the switching of the device at a lower intensity of excitation at a different photon energy. Fig. 24 shows the typical hysteresis loop obtained under these conditions. One can see that the switching times are much longer. The hysteresis turns counterclockwise when we excite below half the biexciton energy while the sense of revolution changes when we excite above the biexciton resonance (Fig. 24b).

These hysteresis curves can be fully explained using our model and introducing the damping constants deduced from the transmission measurements already discussed (Grun *et al.*, 1986). The different switching times between the bistable situations of Figs. 23 and 24a are mainly due to the

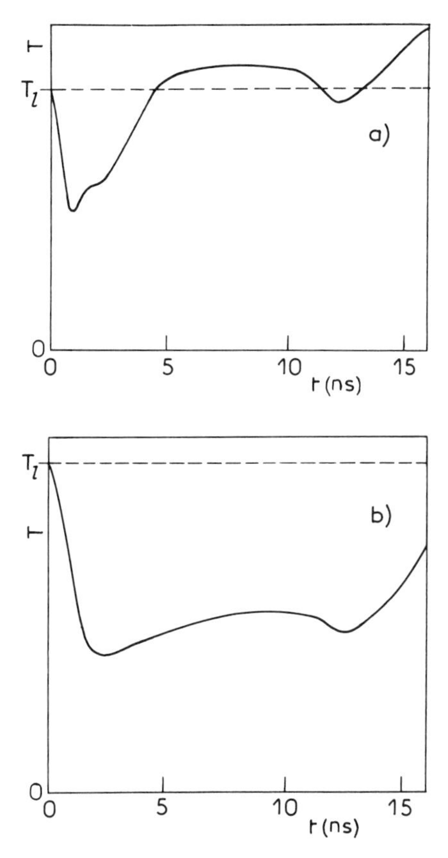

Figure 22. (a) Transmission coefficient calculated from the incident and transmitted intensities of Fig. 21. The calculation is made for $\hbar\omega_1 = 3.1857$ eV using $\hbar\Gamma_{13} = [2 \times 10^{-4} + 4(\rho_{22} - \rho_{33})]$ eV and assuming exciton and biexciton lifetimes of the order of 600 ps and 3 ns, respectively. (b) Transmission coefficient when exciting the crystal above half the biexciton energy. In this case, the biexciton lifetime is taken equal to the exciton lifetime (600 ps) (J. B. Grun *et al.*, 1986).

lifetimes of excitons and biexcitons which are the same in Fig. 23 (since $\hbar\omega_1 \sim E_{\text{Bi}}/2$) but different in Fig. 24a (since $\hbar\omega_1 < E_{\text{Bi}}/2$). In Fig. 24b, a purely transient behaviour of the absorption is observed since the change of the wavelength between Fig. 24a and 24b results in a change of the working point of the Fabry–Perot resonator. While it is situated in a minimum of transmission in Figs. 23 and 24a, it is on the maximum on Fig. 24b, and the intensity is not high enough to induce switching with the next Airy fringe.

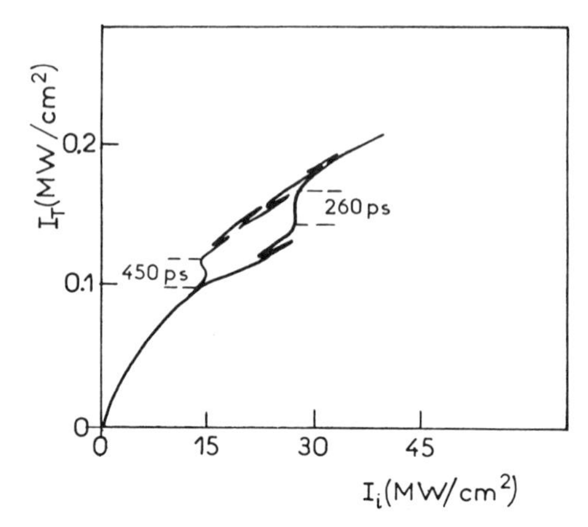

Figure 23. Experimental transmission cycle for a photon energy $\hbar\omega_1 = 3.1859$ eV (J. B. Grun *et al.*, 1986).

A more detailed analysis of the switching behaviour shows (Hönerlage *et al.*, 1986; Hönerlage and Grun, 1987) that the switching times cited above are not intrinsic quantities of the device, but depend also on the driving pulse shape and intensity as well as on the lifetimes of the quasi particles which change with the photon energy of the excitation.

If the same experiments are performed with polycrystalline samples, no bistable behaviour can be obtained under nanosecond pulsed excitation (Lévy *et al.*, to be published), but only transient effects show up.

The dynamics of the optical nonlinearities due to biexcitons and excitons influence the transmission of laser pulses through CuCl samples. The same observations have been made for other systems, where different quasi particles are involved (Hoang Xuan Nguyen and Zimmermann, 1984; Dagenais, 1984; Koch *et al.*, 1985; Toyozawa, 1979; Hanamura, 1981). These dynamic effects have also been observed in two-beam experiments: the time-dependence of the induced absorption due to exciton–biexciton transitions in CuCl has been studied by pump and probe experiments (Lévy *et al.*, 1979; Ojima *et al.*, 1978; Matsumoto and Shionoya, 1982; Leonelli *et al.*, 1986); the temporal variation of the index of refraction of CuCl has been investigated by measuring the shift with time of the interference fringes of a Fabry–Perot cavity containing CuCl (Lévy *et al.*, 1987); degenerate and nondegenerate four-wave mixing have also been shown to be influenced by population dynamics (Frindi *et al.*, 1986). We have suggested the possible use of these effects for pulse shaping and for inducing the switching of a test beam by a pump beam (Lévy *et al.*, 1985b).

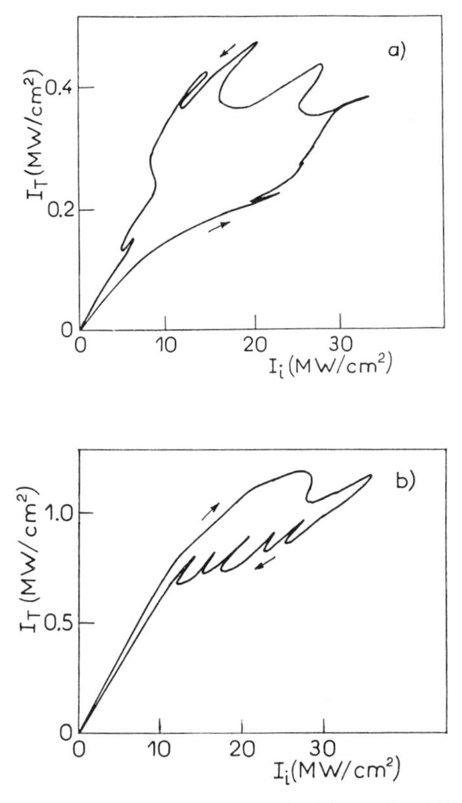

Figure 24. Experimental transmission cycle obtained with another F.P. cavity: (a) Photon energy: $\hbar\omega_l = 3.1857$ eV (b) Photon energy: $\hbar\omega_l = 3.1863$ eV (J. B. Grun *et al.*, 1986).

IV. CONCLUSION

Optical nonlinearities due to biexcitons play an important role in large, direct bandgap semiconductors, where their creation is enhanced by the presence of dipole-active excitons. These nonlinearities can be described theoretically by a three-level system with a density matrix formalism and they can be studied experimentally by different techniques: transmission measurements are appropriate for determining the nonlinear absorption, while hyper-Raman scattering or nondegenerate four-wave mixing are appropriate for the nonlinear dispersion of the material. Since one works close to the exciton and biexciton resonances in these experiments, these quasi particles are always really created to some extent. Their dynamics as well as the time- and intensity-dependence of their damping functions strongly influence the optical nonlinearities. They lead to an optical delay in the nonlinear response of a material with respect to a pulsed excitation envelope. This has important consequences on the switching times of optical bistables and may be used to change the shape of laser pulses.

ACKNOWLEDGMENTS

The authors are grateful to J. Y. Bigot and M. Frindi for many helpful discussions and for participating in some experiments and theoretical calculations presented in this chapter. They also thank C. Klingshirn, F. Fidorra and M. Wegener for their assistance and collaboration.

This work was supported by a contract with the Ministère des Postes, Téléphones, et Télécommunications of France, Centre National d'Etude des Télécommunications. Part of it has been carried out in the framework of an operation launched by the Commission of the European Communities under the experimental phase of the European Community Stimulation Action (1983–1985).

REFERENCES

Abram, I. (1983). *Phys. Rev. B* **28**, 4433.
Abram, I., and Maruani, A. (1982). *Phys. Rev. B* **26**, 4759.
Agrawal, G. P., and Carmichael, H. J. (1979). *Phys. Rev. A* **19**, 2074.
Bigot, J. Y., and Hönerlage, B. (1983). *Phys. Status Solidi B* **121**, 649.
Bigot, J. Y., and Hönerlage, B. (1987). *Phys. Rev. A* **36**, 715.
Bigot, J. Y., Miletic, J., and Hönerlage, B. (1985a). *Phys. Rev. B* **32**, 6478.
Bigot, J. Y., Fidorra, F., Klingshirn, C., and Grun, J. B. (1985b). *IEEE J. Quantum Electron.* **21**, 1480.
Bigot, J. Y., Frindi, M., Fidorra, F., Lévy, R., and Hönerlage, B. (1986). *Phys. Status Solidi B*, **138**, 622.
Bivas, A., Vu Duy Phach, Hönerlage, B., and Grun, J. B. (1977). *Phys. Status Solidi B* **84**, 235.
Blattner, G., Kurtze, G., Schmieder, G., and Klingshirn, C. (1982). *Phys. Rev. B* **25**, 7413.
Bowden, C. M., and Sung, C. C. (1979). *Phys. Rev. A* **19**, 2392.
Chase, L. L., Peyghambarian, N., Grynberg, G., and Mysyrowicz, A. (1979). *Opt. Commun.* **28**, 189.
Chase, L. L., Claude, M. L., Hulin, D., and Mysyrowicz, A. (1983). *Phys. Rev. A* **28**, 3696.
Chemla, D. S., and Maruani, A. (1982). *Progr. Quantum Electron.* **8**, 1.
Chemla, D. S., Maruani, A., and Batifol, E. (1979). *Phys. Rev. Lett.* **42**, 1075.

Cho, K., and Itoh, T. (1984). *Solid State Commun.* **52**, 287.
Dagenais, M. (1984). *Appl. Phys. Lett.* **45**, 1267.
Frindi, M., Hönerlage, B., and Lévy, R. (1986). *Phys. Status Solidi B*, **138**, 267.
Gogolin, A. A. (1974). *Sov. Phys. Solid State* **15**, 1824.
Gogolin, A. A., and Rashba, E. J. (1973). *Zh. Eksp. Teor. Fiz. Pisma Red.* **17**, 478.
Grun, J. B., Hönerlage, B., and Lévy, R. (1982). In *Excitons* (eds. E. J. Rashba and M. D. Sturge) North-Holland, Amsterdam, p. 459.
Grun, J. B., Hönerlage, B., and Lévy, R. (1983). *Solid State Commun.* **46**, 51.
Grun, J. B., Hönerlage, B., and Lévy, R. (1986). *Physica Scripta*, **13**, 184.
Hanamura, E. (1981). *Solid State Commun.* **38**, 939.
Haug, H. (1982), *Festkörperprobleme* **XXII** (*Advances in Solid State Physics* **22**), 149.
Haug, H. (1985), *J. Lumin.* **30**, 171.
Haug, H., and Schmitt-Rink, S. (1984). *Progr. in Quantum Electron.* **9**.
Haus, J. W., Bowden, C. M., and Sung, C. C. (1985). *Phys. Rev.* **A31**, 1936.
Henneberger, F., and May, V. (1982). *Phys. Status Solidi B* **109**, K 139.
Hoang Xuan Nguyen, and Zimmermann, R. (1984). *Phys. Status Solidi B*, **124**, 191.
Hönerlage, B., and Bigot, J. Y. (1984a). *Phys. Status Solidi B* **123**, 201.
Hönerlage, B., and Bigot, J. Y. (1984b). *Phys. Status Solidi B* **124**, 221.
Hönerlage, B., and Grun, J. B. (1987). *Europhysics Letters* **3**, 681.
Hönerlage, B., and Tomasini, F. (1986). *Solid State Commun.* **59**, 307.
Hönerlage, B., Bivas, A., and Vu Duy Phach (1978). *Phys. Rev. Lett.* **41**, 49.
Hönerlage, B., Rössler, U., Vu Duy Phach, Bivas, A., and Grun, J. B. (1980). *Phys. Rev. B* **22**, 797.
Hönerlage, B., Lévy, R., and Grun, J. B. (1982). *Opt. Commun.* **43**, 443.
Hönerlage, B., Bigot, J. Y., Lévy, R., Tomasini, F., and Grun, J. B. (1983). *Solid State Commun.* **48**, 803.
Hönerlage, B., Lévy, R., Grun, J. B., Klingshirn, C., and Bohnert, K. (1985a). *Phys. Rep.* **124**, 161.
Hönerlage, B., Lévy, R., Bigot, J. Y., and Grun, J. B. (1985b). *Proc. IVth Int. Symposium Ultrafast Phenomena in Spectroscopy*. Reinhardsbrunn GDR, Teubner Verlag, (1985), p. 160.
Hönerlage, B., Tomasini, F., Bigot, J. Y., Frindi, M., and Grun, J. B. (1986). *Phys. Status Solidi B* **135**, 271.
Hönerlage, B., Bigot, J. Y., Frindi, M., and Grun, J. B. *Proc. 18th Int. Conf. Phys. Semiconductors, Stockholm*, (1986), p. 1683.
Inoue, M. (1985). *Phys. Rev. B* **33**, 1317.
Itoh, T., and Kathono, T. (1982). *J. Phys. Soc. Japan* **51**, 707.
Itoh, T., and Suzuki, T. (1978). *J. Phys. Soc. Japan* **45**, 1939.
Itoh, T., Suzuki, T., and Ueta, M. (1978). *J. Phys. Soc. Japan* **44**, 345.
Itoh, T., Kathono, T., and Ueta, M. (1984). *J. Phys. Soc. Japan* **53**, 844.
Klingshirn, C., and Haug, H. (1981). *Phys. Rep.* **70**, 315, and references cited therein.
Koch, S. W., Schmitt, H. E., and Haug, H. (1985). *J. Lumin.* **30**, 232.
Kranz, H. H., and Haug, H. (1986). *J. Lumin.* **34**, 337.
Kurtze, G., Maier, W., Blattner, G., and Klingshirn, C. (1980). *Z. Phys. B* **39**, 95.
Kuwata, M. (1984). *J. Phys. Soc. Japan* **53**, 4456.

Kuwata, M., Mita, T., and Nagasawa, N. (1981). *Solid State Commun.* **40**, 911.
Leonelli, R., Mathae, J. C., Hvam, J. M., and Lévy, R. *Proc. 18th Int. Conf. Phys. Semiconductors, Stockholm*, (1986), p. 1437.
Lévy, R., Hönerlage, B. and Grun, J. B. (1979). *Phys. Rev. B* **19**, 2326.
Lévy, R., Bigot, J. Y., Hönerlage, B., Tomasini, F., and Grun, J. B. (1983). *Solid State Commun.* **48**, 705.
Lévy, R., Tomasini, F., and Grun, J. B. (1984). *J. Lumin.* **31 / 32**, 870.
Lévy, R., Hönerlage, B., and Grun, J. B. (1985a). *Helv. Phys. Acta* **58**, 252.
Lévy, R., Bigot, J. Y., Frindi, M., and Hönerlage, B. (1985b). *Phys. Status Solidi B* **132**, 495.
Lévy, R., Tomasini, F., Bigot, J. Y., and Grun, J. B. (1986). *J. Lumin.* **35**, 79.
Lévy, R., Bigot, J. Y., Frindi, M., and Wegener, M. To be published.
Lévy, R., Bigot, J. Y., and Hönerlage, B. (1987). *Solid State Commun.*, **61**, 331.
Maruani, A., and Chemla, D. S. (1981). *Phys. Rev. B* **32**, 841.
Maruani, A., Oudar, J. L., Batifol, E., and Chemla, D. S. (1978). *Phys. Rev. Lett.* **41**, 1372.
März, R., Schmitt-Rink, S., and Haug, H. (1980). *Z. Phys. B* **40**, 9.
Matsumoto, Y., and Shionoya, S. (1982). *Solid State Commun.* **41**, 147.
May, V., Henneberger, K., and Henneberger, F. (1979). *Phys. Status Solidi B* **94**, 611.
Nozue, Y., Itoh, T., and Cho, K. (1981). *J. Phys. Soc. Japan* **50**, 889.
Ojima, M., Kushida, T., Shionoya, S., Tanaka, T., and Oka, Y. (1978). *J. Phys. Soc. Japan* **45**, 884.
Peyghambarian, N., Sarid, D., Gibbs, H. M., Chase, L. L., and Mysyrowicz, A. (1984). *Optics Commun.* **49**, 125.
Sung, C. C., and Bowden, C. M. (1984). *J. Opt. Soc. Am. B* **1**, 395.
Tokihiro, T., and Hanamura, E. (1980). *Suppl. Prog. Theor. Phys.* **69**, 451.
Toyozawa, Y. (1979). *Solid State Commun.* **32**, 13.
Vu Duy Phach, Bivas, A., Hönerlage, B., and Grun, J. B. (1977a). *Phys. Status Solidi B* **84**, 731.
Vu Duy Phach, Bivas, A., Hönerlage, B., and Grun, J. B. (1977b). *Phys. Status Solidi B* **86**, 159.
Yariv, A. (1975). *Quantum Electronics*. 2nd ed. J. Wiley, New York.

9 OPTICAL PHASE CONJUGATION IN SEMICONDUCTORS

M. L. Claude, L. L. Chase, D. Hulin, and A. Mysyrowicz*

GROUPE DE PHYSIQUE DES SOLIDES DE L'E.N.S.—UNIVERSITÉ PARIS VII
TOUR 23, PLACE JUSSIEU, 75251 PARIS CEDEX 05
FRANCE

I. INTRODUCTION

One of the most spectacular recent developments of nonlinear optics is the demonstration of optical phase conjugation (OPC) in various media. This effect has stirred considerable interest in view of its wide field of potential applications: OPC has been considered or used for the transmission of information by optical means through disturbing media (such as turbulent

*Permanent address: Lawrence Livermore National Laboratory University of California, Livermore, CA 95550 (U.S.A.)

OPTICAL NONLINEARITIES
AND INSTABILITIES IN
SEMICONDUCTORS

ISBN 0-12-332915-9

atmospheric conditions), as an automatic tracking system of moving objects, in pattern recognition, in photolithography, in laser fusion, as a self-aligning mirror in a laser cavity, as an ultranarrow colour filter, in real-time holography (see, e.g., Fischer, 1982). It is clear that semiconductors, because of their compactness, rigidity and large optical nonlinearities, are going to play a major role in such applications.

On the other hand, OPC can also provide an invaluable tool for the investigation of the physical properties of various media. For example, in semiconductors, it can bring information concerning the rate of diffusion of excitation in the crystal, upon coherence memory, upon the origin of nonlinear responses of a medium, or as a sensitive spectroscopic method for identifying resonances of the system. In this chapter, a review of some of the properties of optical phase conjugation in intrinsic semiconductors is presented. The aim is not to give an exhaustive description of the present status of the field, but rather to stress some of the underlying physics, with a few selected examples.

A. What Is Optical Phase Conjugation?

Consider an optical wave propagating in the z-direction:

$$\mathbf{E}(\mathbf{r}, t) = \tfrac{1}{2}\left[\mathbf{e}_0(x, y) e^{i(\omega t - K_0 z + \phi_0)} + \text{c.c.}\right] \tag{1}$$

here $\mathbf{e}_0$ is the slowly varying complex amplitude of the field, $\mathbf{K}_0$ the optical wavevector, $\phi_0(x, y)$ the phase of the wave which is assumed to be a nearly plane wave.

The properties of the wave after reflection by an ordinary mirror are well known and can be found in textbooks of classical optics.

Consider now a so-called phase-conjugate mirror. By definition, such a mirror transforms the incident wave into a reflected wave $\mathbf{E}_r(r, t)$ having the same temporal variation but a reversed spatial phase.

$$\mathbf{E}_r(\mathbf{r}, t) = \tfrac{1}{2}\left[\mathbf{e}_0^*(x, y) e^{i(\omega t + K_0 z - \phi_0)} + \text{c.c.}\right] \tag{2}$$

Formally, this amounts to an operation of time inversion performed on the incident wave (leaving the spatial part unchanged), as can be immediately seen by writing explicitly the complete expression:

$$\begin{aligned}\mathbf{E}_r(\mathbf{r}, t) &= \tfrac{1}{2}\left[\mathbf{e}_0^*(x, y) e^{i(\omega t + K_0 z - \phi_0)} + \mathbf{e}_0(x, y) e_0^{-i(\omega t + K_0 z - \phi_0)}\right] \\ &= \mathbf{E}(\mathbf{r}, -t)\end{aligned} \tag{3}$$

Thus, the reflected wave $\mathbf{E}_r(\mathbf{r}, t)$ behaves as if it retraces the path of the original wave back in time, like in a movie played backwards. Most

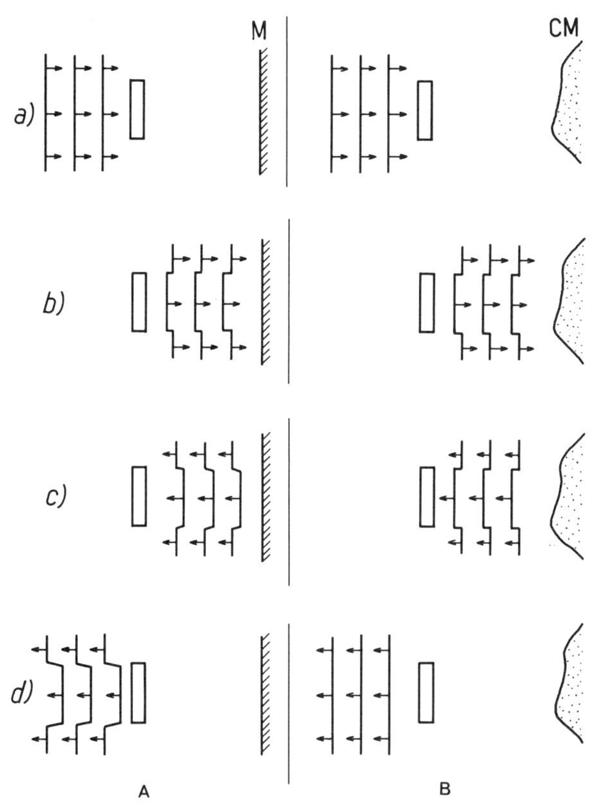

Figure 1. (A) Behaviour of a plane wave front. (a) Incident on a piece of glass (aberrator). (b) After transmission through the aberrator. (c) After reflection on an ordinary mirror (M). (d) On the way back after transmission through the aberrator. (B) Same as in (A) except that reflection occurs on a phase conjugate mirror (CM).

applications derive from this intriguing property, which leads to an automatic compensation of distortions of the phase front experienced by the primary wave on its way to the conjugate mirror. Fig. 1 illustrates this property with a simple example.

How is it possible to realize such a singular mirror? Early attempts have made use of adaptive optics, but only with moderate success, especially in the visible region of the spectrum. A breakthrough occurred when it was recognized by Zeldovich et al. (1972), that the back-scattered light due to stimulated Brillouin effect of an intense laser beam possessed these properties of conjugation. The generation of Brillouin scattering may be described as originating from the third-order nonlinear susceptibility of the medium.

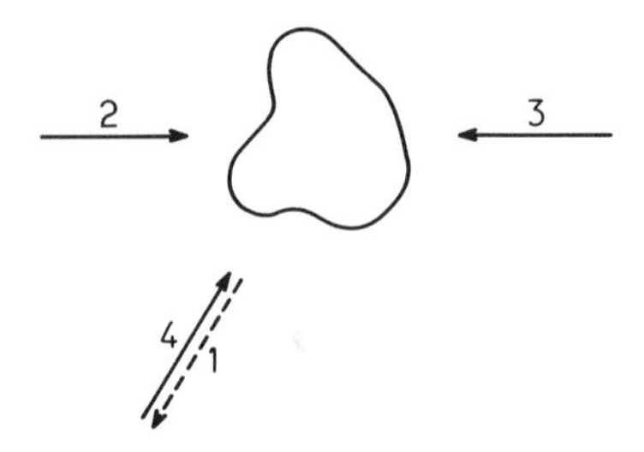

Figure 2. Geometric configuration for fully degenerate four-wave mixing. Probe beam 4 interacts in the nonlinear medium with pump beams 2 and 3 to generate a signal beam 1. All beams have a common frequency ω.

A more versatile method, also resulting from the third-order susceptibility $\chi^{(3)}$, has since been proposed by Hellwarth (1979) and is now the most commonly used in applications. It exploits a special configuration called fully degenerate four-wave mixing shown schematically in Fig. 2. Here, three incident beams with common frequency ω interact via the nonlinearity of the medium to generate a fourth coherent wave called the signal wave. Two of the incident waves (the pump waves labelled 2 and 3 here) are counterpropagating, while the third wave (the probe wave labelled 4) overlaps the pump waves in the nonlinear medium at an arbitrary angle.

Because of energy and momentum conservation, the generated signal wave 1 will have the same frequency ω as the incident beams and a propagation vector opposite to that of the probe beam. The third-order nonlinear polarization governing the generation of the signal wave in a cubic medium is then of the form (see, e.g., Bloembergen, 1965):

$$P_s^{(3)}(\omega) = \chi^{(3)}(\omega) : \mathbf{E}_2 \cdot \mathbf{E}_3 \cdot \mathbf{E}_4^* \quad (4)$$

As a consequence, the nonlinear medium activated by the two counterpropagating pump beams of frequency ω behaves like a conjugate mirror. The signal wave is generated by a polarization proportional to the complex conjugate of the probe field and possesses the properties of phase conjugation with respect to the incident probe wave.

It can be shown that in an isotropic medium, Eq. 4 can be cast in the form:

$$P_s^{(3)}(\omega) = A(\theta)(\mathbf{E}_2 \cdot \mathbf{E}_4^*)\mathbf{E}_3 + B(\theta)(\mathbf{E}_3 \cdot \mathbf{E}_4^*)\mathbf{E}_2 + C(\mathbf{E}_2 \cdot \mathbf{E}_3)\mathbf{E}_4^*, \quad (5)$$

i.e. it takes a vector form including the scalar products of the different incident fields.

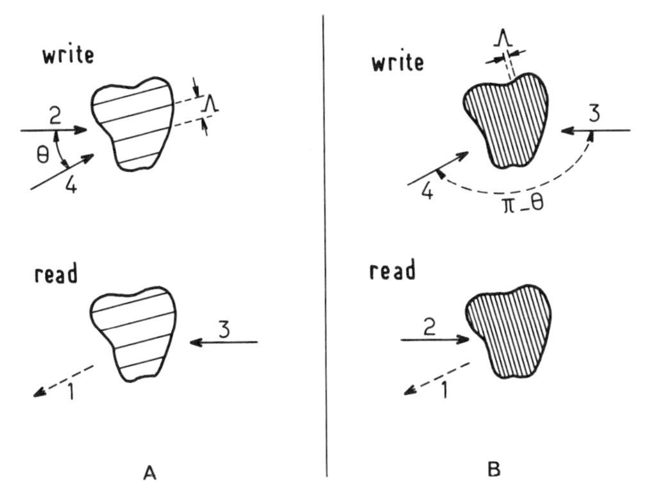

Figure 3. Holographic analogy for optical phase conjugation via 4 wave mixing. (a) Large-period grating. (b) Small-period grating.

Eq. 5 is appealing since it gives a direct interpretation of the physical phenomena taking place in the crystal. The first two terms can be described as real-time holographic processes: the probe field amplitude interferes with each of the pump fields to create a standing wave pattern. In turn this leads to a static volume grating of the refractive index (if the incident waves fall in a transparent region of the medium). The second beam is diffracted at the Bragg angle in the direction opposite to the incident probe beam. This interpretation in terms of real-time holographic writing and reading process explains how the reflectivity of the conjugate mirror $R_c = I_1/I_4$ can exceed 100% (where I_4 and I_1 are the probe and signal intensities). As illustrated in Fig. 3, two gratings are formed, corresponding to terms $A(\theta)$ and $B(\theta)$ in Eq. 5 with spacings $\Lambda_1 = \dfrac{\lambda}{2\sin\theta/2}$ and $\Lambda_2 = \dfrac{\lambda}{2\sin(\pi-\theta)/2}$ where θ is the angle between the probe and the first (second) pump beam and λ the wavelength inside the medium.

The third term has another physical origin, corresponding to a different ordering in the sequence of absorbed and emitted photons in each elementary process (see Fig. 4). It corresponds to a spatially uniform excitation of the medium, which can be viewed as a simultaneous absorption of two photons (one from each pump beam) giving a two-photon coherence at

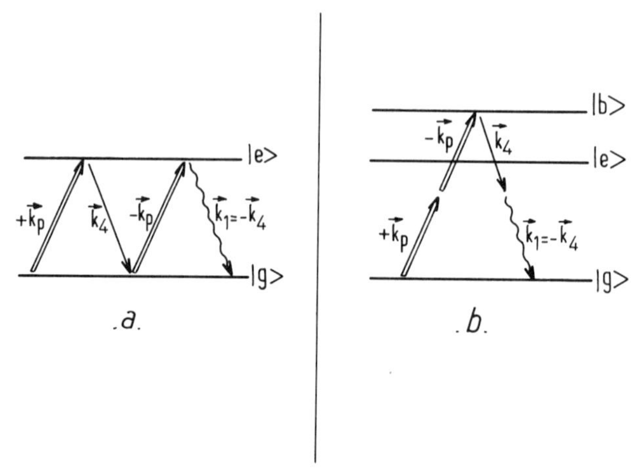

Figure 4. Schematic energy diagram for resonant processes leading to optical phase conjugation. (a) One-photon resonant process described in terms of simultaneous writing and reading of a population volume grating. (b) Two-photon resonant process corresponding to a spatially uniform coherent excitation of the system at frequency 2ω.

frequency 2ω, followed by the emission of two photons (one probe, one signal). This process should not be confused with that resulting from the standing wave pattern formed by the interference of the two incident pump beams, with period $\lambda/2$. Such a stationary index grating is formed indeed in the medium giving rise to a diffracted signal wave, but not in the direction $\mathbf{K}_1 = -\mathbf{K}_4$, and is therefore of no concern here.

The contribution from the third term in Eq. 5 is usually negligible compared to those originating from the grating and can only be observed under special conditions of strong enhancement due to a two-photon resonance. However, it has some unique properties. Since it results from a spatially uniform two-photon excitation of the medium, it leads to a phase-conjugate reflectivity that is independent of the incidence angle θ of the probe beam. Also it is free from effects associated with a motion of the excitation in the volume (see below). Further, it can be observed with a probe polarization vector orthogonal to both pump beam polarization vectors, in contrast with the conjugate signal originating from the grating terms, which requires a nonvanishing component of the probe vector along one of the pump polarization vectors, as can be readily seen from Eq. 5. In favourable cases, this property results in a perfect conjugator, for which the polarization vector of the reflected beam is always the complex conjugate of the incident probe polarization vector (irrespective of the angle of incidence and of the probe electric vector direction with respect to that of the pump beams). Such a property of polarization conjugation is not realized in conjugate mirrors relying upon a real-time holographic process, except for special circumstances such as small angle θ and particular polarization directions. Another interesting feature of the conjugate signal associated with a two-photon resonance is the possibility of obtaining information concerning the dephasing time of the two-photon resonance: by delaying the probe with respect to the pump beams and monitoring the signal intensity as a function of delay, one obtains direct access to its coherence (dephasing) time. Finally, it may be worth mentioning that the two-photon term C has recently attracted special interest since it could potentially lead to the production of so-called squeezed states of the radiation field (see Yuen, 1976) (for which the radiation noise is reduced below its shot noise level for one of the quadrature components of the field).

B. Resonant Enhancement

Usually, for incident light frequencies far from resonances, the nonlinear response of the system is relatively small, even in favourable situations. Typically, one expects phase-conjugate reflectivities R_c of the order 10^{-6} for an interaction length of 1 cm and pump intensities in the megawatt range. A well-known method of increasing the magnitude of the response is to tune the incident light frequency close to a resonance of the system. A corresponding enhancement of the nonlinear polarization occurs, as can be seen from an inspection of the microscopic expression of the nonlinear susceptibilities (see, e.g., Shen, 1984).

In a pure semiconductor at low temperature, the intrinsic electronic resonance of lowest energy is the exciton. As the incident laser frequency approaches the exciton, the response from the index grating increases. At resonance, a population grating of excitons is realized from the interference pattern of the probe and of the pump field amplitudes. In turn, this spatial modulation of the exciton population leads to a Bragg diffraction of the second pump beam. This diffraction occurs because of the change of absorption and corresponding dispersion associated with the population. The changes may be due to a saturation, a broadening or a shift of the exciton absorption line. A new aspect of the resonant situation is the fact that excitons are mobile particles, free to propagate throughout the sample volume. This will lead to a washing out of the grating, as the diffusion of particles takes place. During the finite lifetime of the excitons, the fringe contrast of the small-spaced grating will suffer a larger deterioration than that of the large-spaced grating, leading to unequal efficiencies for terms A and B in Eq. 5. One enters the large field of transient grating spectroscopy (see, e.g., Eichler, 1978, and Aoyagi et al., 1982) where information concerning diffusion constants of excitons can be obtained from diffraction efficiencies as a function of time delay between writing and reading steps or as a function of angle θ.

Another type of resonant situation leading to enhanced phase conjugation based on a two-photon coherence is associated with the presence of a biexciton level. Like in hydrogen atoms, excitons can form moleculelike entities consisting of two electron–hole pairs bound together. Such quasi particles can be created directly from the crystal ground state by a two-photon absorption process. As pointed out by Hanamura (1973), the corresponding transition cross section takes giant values, which makes it possible to obtain a predominant contribution from the two-photon coherence (third term in Eq. 5) to the conjugate signal. Note that the resonant two-photon enhancement occurs at photon energies smaller than the lowest free exciton energy since for a bound biexciton, the following relation holds: $E_B = 2\hbar\omega = 2E_x - B$ where E_B is the biexciton energy, and B is its dissociation energy into two free excitons of energy E_x.

Still another important physical process comes into play at higher excitations: if the density of generated excitonic particles is increased to the point that $Na_x^3 \simeq 0.1$ (N = exciton density, a_x = exciton Bohr radius), a Mott-like transition takes place in the medium, whereby the mutual screening among the bound particles leads to their dissociation into free electrons and holes. Large and abrupt changes in the optical properties of the crystal near the band edge accompany this phase transition. The intense exciton absorption lines disappear. Further, the presence of a large number of photogenerated electrons and holes leads to a reduction of the value of the band gap (band gap renormalization). The excited electrons and holes occupy the conduction and valence bands according to the Fermi statistics. A saturation of the optical transitions across the reduced gap near the band edge occurs at low electronic temperatures. The description of the optical properties of a crystal in the presence of a plasma is in general not simple and requires an involved theoretical treatment. Haug and Schmitt-Rink (1984) have developed a dielectric function formalism which reproduces well the experimental results in several semiconductors.

However, the calculations are not easily tractable and no general predictions can be made. An approximation, though it is an analytical expression, has been derived recently by Banyai and Koch (1986). A simplified description using only measurable material parameters has been given by Miller et al. (1981). Here, excitonic effects are ignored; a fraction of the absorbed photons are considered to be directly converted into free carriers. This assumption is realistic in small-gap materials where weakly bound excitons are easily ionized by thermal energy or internal fields. However, the band gap renormalization is ignored. This model yields a nonlinear refractive index varying like $1/\omega^3$ for $\hbar\omega \simeq E_g$ and predicts values of $\chi^{(3)}$ of the order of 1 esu for small-gap materials such as InSb.

There are other mechanisms that can be at the origin of phase conjugation. If a significant fraction of the absorbed energy is dissipated locally in a nonradiative process, a thermal grating is formed with the same periodicity as the initial interfering pattern of the input fields (see Martin and Hellwarth, 1979). This in turn leads to an index grating

$$\Delta n \sim \Delta T \frac{dn}{dT} \simeq \Delta T \frac{\partial n}{\partial E_g} \frac{\partial E_g}{\partial T} \tag{6}$$

Phase conjugation of thermal origin can be usually recognized by the long response time (typically of the order of ms).

Other important mechanisms are connected with the presence of impurities or defects in the crystal. In photorefractive materials, photoinduced charge carriers migrate to trap centers, where they recombine. This leads to the formation of a charge field with same periodicity, but possibly different phase, than the original interference pattern. Via the electrooptic effect, an index grating results. This mechanism can lead to very high efficiencies, requiring only mW in power, but the response time is very slow (see, e.g., Feinberg and Hellwarth, 1980, and Huignard and Micheron, 1976). Other impurity-related effects, such as bound excitons, can also be at the origin of very large nonlinearities (see Dagenais and Winful, 1984) and lead to efficient phase conjugate mirrors.

In what follows, we describe some typical examples of phase conjugation obtained in different intrinsic semiconductors.

II. OPTICAL PHASE CONJUGATION IN CuCl

Copper chloride provides an interesting system for the study of optical phase conjugation in a pure semiconductor, especially for the observation of conjugation due to a two-photon coherence. As emphasized previously, conjugate mirrors based on a two-photon coherent excitation of the nonlinear medium have several unique properties, but their experimental realization is not easily achieved, since this requires a strong two-photon resonance in the active medium. This condition is met in CuCl because of the presence of a biexcitonic state at an energy $E_M = 6.372$ eV at 4.2 K. The corresponding photon energy $\hbar\omega = E_M/2$ in a two-photon process falls in the transparency region of the crystal about 15 meV below the first strong absorption peak from the lowest Γ_5 exciton. The optical transition from the crystal ground state to the biexciton state has a giant two-photon transition cross section, allowing it to be discerned from the usual two-photon band-to-band absorption band. As was first pointed out by Hanamura (1973), this is due to the proximity of the exciton resonance acting as a nearly resonant intermediate state and also because of the particular type of electronic transition, which involves the simultaneous promotion of *two* electrons from the valence to the conduction band. There have been a great number of studies devoted to the biexciton in CuCl by two-photon absorption (Gale and Mysyrowicz, 1974), hyper-Raman scattering (Nagasawa et al., 1976, and Vu Duy et al., 1978) and active four-wave mixing (Maruani et al., 1978, and Chemla et al., 1979).

Recently, two groups (Chase et al., 1983 and 1984, and Claude et al., 1984, plus Mizutani and Nagasawa, 1983) have independently reported evidence for optical phase conjugation associated with the biexciton resonance. Results were obtained either in thin crystalline platelets typically 30 μm in thickness or in very thin (~ 1 μm) crystalline films (see Chase et al., 1983) deposited on a fused quartz substrate. Although phase conjugation was observed in both cases, the nonnegligible depletion of the pump, probe and signal beams due to tail one-photon absorption as well as two-photon self-absorption of the pump beams complicates the picture in thick crystals. By contrast, results obtained in thinner crystalline films are in better agreement with an analysis that assumes negligible losses of the input beams through the medium.

Fig. 5 shows the conjugate signal, as a function of input frequency, obtained in a 1.25 μm film held at 2 K for various polarization configurations of the input beams. All beams were derived from the same nitrogen laser pumped dye laser. Characteristics of the optical pulses are as follows: duration 5 ns, pump beam intensities ~ 1 MW/cm^2, probe beam intensity ~ 2 KW/cm^2, external input angle $\theta = 3°$.

Two distinct peaks are observed in the reflection efficiency plotted as a function of beam frequency ω (see Fig. 5). They have been analyzed under various combinations of the polarization vectors of the input beams, with the signal measured in each case in a polarization direction both parallel

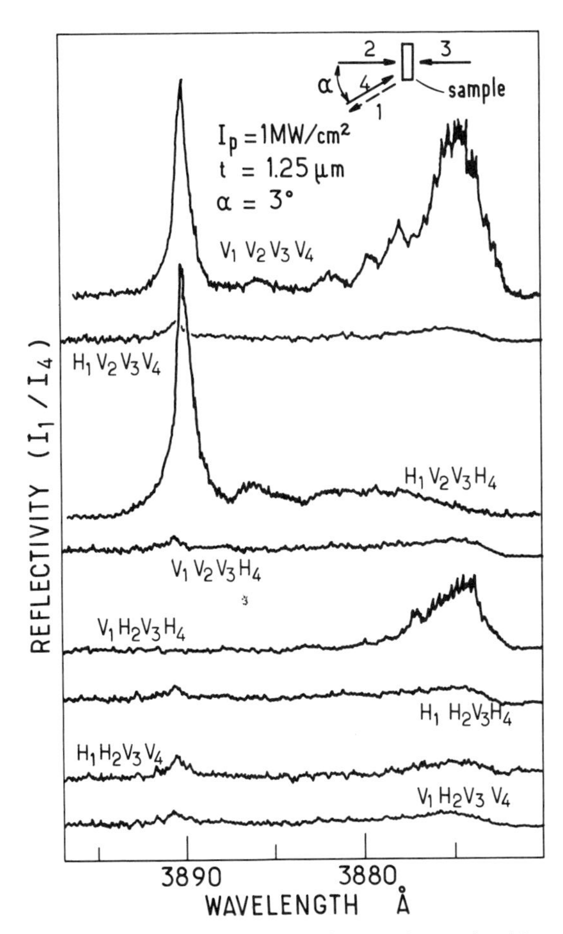

Figure 5. Laser wavelength dependence of optical phase conjugate signal intensity, obtained in a 1.25-μm–thick sample of CuCl at 2K. Intensities of pump beams 2 and 3 are $I_p \sim 10^6$ W/cm^2. $V_i(H_i)$ denotes a vertical (horizontal) polarization vector of beam i. 1 and 4 correspond to signal and probe beams, respectively. $I_4 \sim 5\% I_p$.

and perpendicular to that of the test beam. The conjugate nature of the signal beam has been verified by inserting a phase and polarization aberrator in the path of the probe beam. The signal beam at $\hbar\omega = 3.1862$ eV ($\lambda = 389$ nm) was well collimated after retraversing the aberrator, and its polarization property was nearly reconstructed to that of the original probe beam, even if a single pass through the aberrator caused an almost complete depolarization of the linearly polarized test beam.

One can draw several conclusions from the set of data of Fig. 5. If one first considers the peak located at higher energy $\hbar\omega = 3.198$ eV ($\lambda = 387.5$ nm), one notices that its position is slightly below the peak of the strong exciton absorption at $\hbar\omega = 3.200$ eV ($\lambda = 387.3$ nm). The maximum of this peak shifts to longer wavelengths for thicker samples, indicating that reabsorption of the signal and input beams by the strong excitonic resonance plays an important role in the spectral response. The signal can be attributed to the formation and reading of an exciton population grating. Indeed, referring to Eq. 5, one observes that the conjugate signals appear in the case corresponding to the large-spaced grating (fifth trace from the top), but not in the case of the small-spaced grating (seventh trace from the top). Knowing the sample thickness and index of refraction, one can evaluate the respective grating spacings to be $\Lambda = 37$ nm and $\Lambda = 1.4$ μm, respectively. On the other hand, from the value of the diffusion constant of excitons in CuCl (see Aoyagi, 1982) $\Delta = 330$ cm^2/sec and a typical exciton lifetime $\tau = 10^{-9}$ s, one obtains a diffusion length ~ 1 μm lying between both values. Thus, it is concluded that the resonance enhancement of the conjugate signal is connected with an exciton population.

Considering now the reflectivity peak at $\hbar\omega = 3.1862$ eV ($\lambda = 389$ nm), one can attribute its origin to a resonantly enhanced two-photon coherence. The signal peaks at the exact position of the biexciton state with giant two-photon transition cross section; its magnitude is independent of the direction of the probe beam vector with respect to that of the pumps (first and third traces from the top of Fig. 5) as expected from an ideal conjugator based on a two-photon resonance. Further, the signal at 3.1862 eV disappears if the signal wave is orthogonal to the probe wave, indicating that no appreciable contribution exists from a biexciton population grating. (Such a grating can be formed by interference of the test and each of the pump beams, giving rise to a spatially modulated population of biexcitons generated by two-photon absorption.)

Fig. 6 shows the conjugate reflectivity obtained on a 30-μm–thick crystal at 2 K. Here, the probe beam was incident at right angle from the pumps and the polarization vectors of probe and signal beams were parallel to each other but perpendicular to the pump beams' electric vectors. The upper part of the figure shows the transmission spectrum of a single pump beam through the same crystal, with similar pump intensity $I_2 = I_3 \simeq 10^6$ W/cm^2. The biexciton two-photon resonance is clearly apparent and the nonnegligible depletion of the pump intensity leads to the distortion of the conjugate signal in the bottom trace of the figure.

The absolute value of the conjugate reflectivity $R_c = I_1/I_4$ has been measured at $\hbar\omega = 3.1862$ eV by replacing the sample with a totally

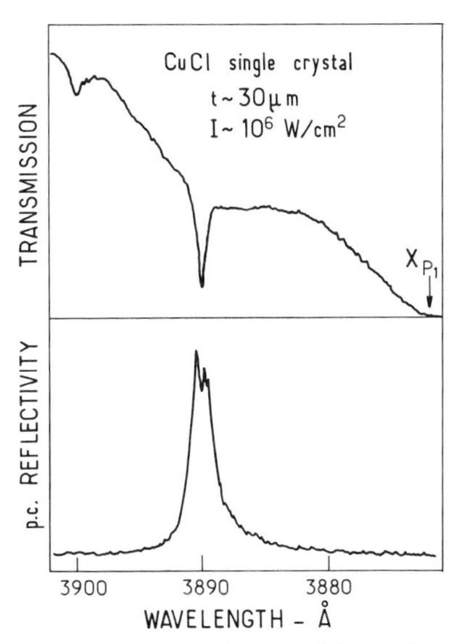

Figure 6. Conjugate reflectivity obtained on a 30-μm–thick sample crystal of CuCl at 2 K. The angle between pumps and probe is $\theta = 90°$. The polarization vectors are as follows: $\mathbf{E}_1 \parallel \mathbf{E}_4$; $\mathbf{E}_2 \parallel \mathbf{E}_3$; $\mathbf{E}_1 \perp \mathbf{E}_2$. Pump beam intensities are $I_p \sim 10^6$ W/cm^2. The upper trace shows the transmission spectrum of a single pump beam through the same crystal with similar intensity: $I \sim 10^6$ W/cm^2.

reflecting mirror. Values of several percent were obtained in one-micron–thick samples for $I_2 = I_3 \sim 10^6$ W/cm^2, and a maximum value $R_c = 20\%$ was achieved in a 30-μm–thick sample of poorer surface quality.

From reflectivity measurements it is possible to deduce the absolute value of $\chi^{(3)}$ at the two-photon resonance, using the following relation:

$$R_c = \tan^2\left(12\pi\omega E_2 E_3 l \chi^{(3)}_{xxxx}/n_c\right) \tag{7}$$

where l is the sample thickness and n_c the refractive index.

A value $\chi^{(3)} = 3.10^{-7}$ esu is obtained as long as the pump beams' intensities remain below 10^5 W/cm^2. A value of $\chi^{(3)}$ of 6.10^{-7} esu is deduced independently from the measured two-photon absorption coefficient. These values exceed by more than two orders of magnitude those deduced previously from four-wave mixing experiments at the biexciton resonance of CuCl (by Maruani et al., 1978, and Chemla et al., 1979). The measured values of $\chi^{(3)}$ suggest that conjugate signals close to 100% should be attainable in good-quality samples with thickness of 5 to 10 μm.

For pump intensities above 10^5 W/cm^2, a sublinear dependence of the signal (measured at its peak frequency value) versus pump intensity, with a slope ~ 0.5 is found, instead of the expected square law dependence observed for $I_p < 10^5$ W/cm^2. This saturationlike behaviour at the frequency peak of the signal is accompanied by a broadening of the resonance signal. A very similar behaviour is observed in the two-photon absorption lineshape of the biexciton in CuCl, and it has been explained by a collision broadening according to Chase et al. (1979). It is therefore likely that here, also, collisions between biexcitonic particles are at the origin of the saturationlike character of the signal.

Finally, in a series of measurements performed in a nearly degenerate regime, for which the probe beam is slightly detuned from the pump beam frequency set at the biexciton two-photon resonance of CuCl, Mizutani and Nagasawa (1983) have clearly observed the conjugate signal at the expected symmetric frequency with respect to the two-photon resonance (see Fig. 7).

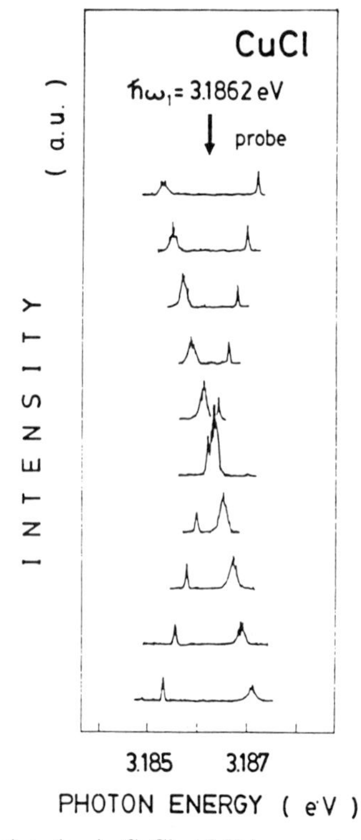

Figure 7. Optical phase conjugation in CuCl at 2 K in a nearly degenerate case. The energy of the excitation beams is set at 3.1862 eV (as indicated by the arrow). The phase-conjugated light ($2\hbar\omega_1 - \hbar\omega_2$) is seen as the broad peak and the scattering of the probe beam ($\hbar\omega_2$) as the narrow peak. (Reproduced from Mizutani and Nagasawa (1983) with permission of the authors.)

These results provide a good example of the large optical phase conjugation, based on a two-photon coherence, that can be obtained even in a nondegenerate case under favourable circumstances in a pure semiconductor.

III. OPTICAL PHASE CONJUGATION IN MICROCRYSTALS

Triggered by the work of Jain and Lind (1983), much attention has been given recently to cadmium sulfide-selenide glasses. These materials are well known to spectroscopists as low-pass, sharp-cutoff optical filters. They consist of microcrystalline particles of $CdSe_{1-x}S_x$ imbedded in an amorphous borosilicate glass matrix. They appear to be very attractive for applications since they are cheap, are readily available in good optical quality and have a high damage threshold. Further, the absorption edge of these filters can be tuned continuously between 500 nm and 1000 nm by changing the composition x. Finally, they should be free from the deleterious effects due to diffusion of excitation throughout the medium. It is therefore important to understand the physics of these systems.

Jain and Lind (1983) have reported room temperature conjugate reflectivities of up to 10% using pump intensities of the order of 1 MW/cm^2. An excellent aberration correction capability has been demonstrated. The optical phase conjugation has been studied for various polarization geometries. A striking feature is the fact that the signals generated by the small-spaced ($\Lambda \sim 1030$ Å) and large-spaced (3 μm) gratings are nearly equal, in sharp contrast with the case of pure CdS, where it is only observed for the large-spaced grating (Fig. 8). This behaviour is consistent with an excitation trapped in crystalline sites having typical dimensions between 10 and 100 nm.

Jain and Lind (1983) report relatively large values of the third-order susceptibility $\chi^3 \sim 10^{-9} - 10^{-8}$ esu. They have measured the response time of the conjugate signal by delaying one of the pump pulses with respect to the other input beams and monitoring the change of signal intensity with delay. A subnanosecond response time was found, which is corroborated by an observed narrowing of the signal pulse duration (from 8 ns to ~ 3.5 ns). Such a subnanosecond response time excludes a thermal effect as the origin of the induced grating.

The nonlinearity has been attributed to the formation of a plasma in the microcrystals through photon-induced electron–hole pair generation. From the amount of absorbed photon energy, an average carrier density in the range 10^{18} cm^{-3} is estimated. This value should exceed the critical density

POLARIZATION CONDITION	DFWM SIGNAL, RELATIVE UNITS	
	$CdS_{0.9}Se_{0.1}$ IN GLASS	CdS CRYSTAL
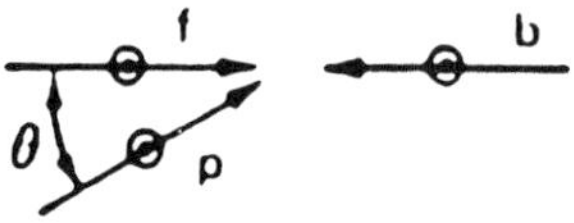 f, b, p, θ	1	1
 f, b, p A_{fp}, LARGE PERIOD (f–p) GRATING	0.35 (±0.07)	1
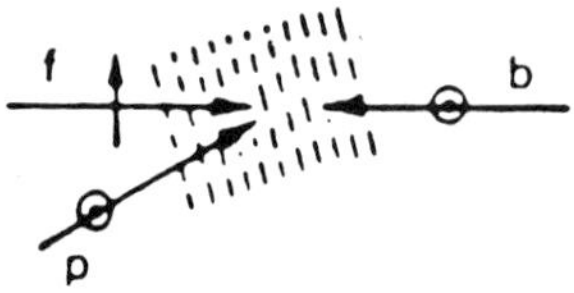 f, b, p A_{bp}, SMALL PERIOD (b–p) GRATING	0.45 (±0.08)	0
 f, b, p B, 2ω TEMPORAL COHERENCE	0	0

Figure 8. Relative intensities of OPC signal, as measured in a CdS_xSe_{1-x} glass and a CdS crystal for various polarization configurations of the input optical beams and $I_f = I_b \simeq 50$ MW/cm^2. (Reproduced from Jain and Lind (1983) with permission of the authors.)

for a Mott transition to a phase of unbound electron–hole pairs and make the interpretation very plausible. It should be noted that Jain and Lind (1983) have found a strongly sublinear dependence of the conjugate reflectivity as a function of pump intensity with a slope $m = 0.9$, instead of the expected square law dependence. They exclude a saturation of absorption as the mechanism responsible for this behaviour, since they failed to observe such a saturation in transmission experiments.

Roussignol et al. (1985) have repeated similar experiments using picosecond laser pulses. They find results that differ from those of Jain and Lind (1983) in several respects. These authors have explored the unsaturated regime, where the conjugate wave varied quadratically with the pump intensities. At higher pump values, they find the sublinear regime already reported by Jain and Lind (1983). However, they also observe a saturation in transmission appearing at the same pump intensities, in contrast to the results of Jain and Lind (1983). Roussignol et al. (1985) measured the response time of the conjugate wave, using a technique similar to that of Jain and Lind (1983). They find much longer response times, up to 40 ns. They attribute the origin of the nonlinearity to an impurity-related effect, whereby excited electrons get trapped with long decay time on impurity centers. This interpretation is derived from an observed long-lived luminescence ($\sim$ 400 ns) with nonexponential decay.

One could comment at that stage that both sets of data may be reconciled in the same model of an electron–hole plasma formed in the microcrystals. It is well known that the recombination rate of electron–hole pairs inside a plasma is density dependent, having terms proportional to n^2 and n^3 for the bimolecular and the Auger process, respectively. This can explain the different response times that depend upon initial plasma carrier density. Concerning the previously mentioned lack of saturation in transmission and conjugate reflectivity experiments, in the presence of a plasma, no general prediction is possible without a closer inspection of the system. The presence of excited carriers leads to a shrinkage of the forbidden gap (red shift of the absorption edge) the magnitude of which depends upon carrier density. On the other hand, the photoexcited carriers fill their respective bands, with a filling factor (saturation of the Bloch states near the extrema of the bands) dependent upon carrier density and temperature. Thus, depending upon the particular experimental situation (laser frequency, plasma density and temperature), either an increase or decrease of absorption can occur. The different laser frequencies and microcrystalline band gaps (due to different compositions) employed by Jain and Lind (1983), compared with those used by Roussignol et al. (1985), may thus account for the different behaviours observed by the two groups. In fact, a very recent experimental subpicosecond time-resolved pump-and-probe study by Peyghambarian et al. (1986) shows results that these authors interpret along the lines of the electron–hole plasma model. Large changes in the transmission of CdSe-microscrystallite, doped glasses are observed, which are either positive or negative, depending on the observation wavelength and delay between pump and probe pulses. Band filling and band-gap renormalization are explicitly considered in the discussion of the results.

In still another recent investigation, Cotter (1986) has reported OPC in CdSe doped glasses, using subpicosecond excitation. Cotter reports an order-of-magnitude enhancement of the fastest ($\sim$ 10 ps) phase-conjugate signal, with respect to the slower background contribution without evidence for saturation. Cotter excludes a thermal grating as the origin of the effect and attributes it to a pure electronic effect, from plasma trapping in the semiconductor microcrystals, in agreement with Jain and Lind (1983) and the conclusions of Peyghambarian et al. (1986).

In conclusion, colour glasses present a new physical system that seems very promising for obtaining large conjugate signals with fast response times, as well as other optical nonlinear effects, such as bistability. However, at this stage, more studies are required in order to have a better understanding of the physical phenomena at the origin of these nonlinearities.

IV. OPTICAL PHASE CONJUGATION ASSOCIATED WITH FREE CARRIERS

Some of the earliest phase conjugation experiments in semiconductors were based upon nonlinearities associated with free carriers, which may either be present in doped semiconductors or created through photoexcitation by the pump waves. Since most of this work has been extensively reviewed by Jain and Klein (1983), we shall simply summarize some of the significant physical mechanisms underlying the free-carrier–induced nonlinearities and the resulting properties of the phase conjugation processes. Patel et al. (1966) pioneered in the study of nonlinear optical effects associated with free carriers in highly doped materials (InSb, InAs and GaAs). In doped semiconductors, the nonparabolicity of the conduction and valence bands leads to a nonlinear response of the extrinsic carrier population through modification of the usual plasma contribution to the linear dielectric response. The theory of these effects was elaborated by Wolff and Pearson

(1966). The temporal response of phase conjugation in such systems is determined by the momentum relaxation time of the carriers, and very fast response can be obtained.

In pure semiconductors, nonlinearities can result from photoexcited carriers, in addition to the usual nonlinear susceptibility of the bound electrons. The magnitudes and behaviours of these two contributions are strongly dependent upon the relative magnitudes of the photon energy, $\hbar\omega$ and the optical bandgap, E_g. For $\hbar\omega \ll E_g$, free carriers must be produced by multiphoton absorption or avalanche ionization. In some experiments, this leads to phase conjugate reflectivities that vary with pump intensity, I_p, as a power law, often with a rather large exponent. For example, Watkins et al. (1980) observed a reflectivity of 10.6 μm light in Ge that varied as Ip (see Miller et al., 1981) at the highest intensities that they employed (see Fig. 9, Watkins et al., 1983). Although Watkins et al. (1981) attributed this dependence to multiphoton absorption, Jain and Klein (1983) suggest that

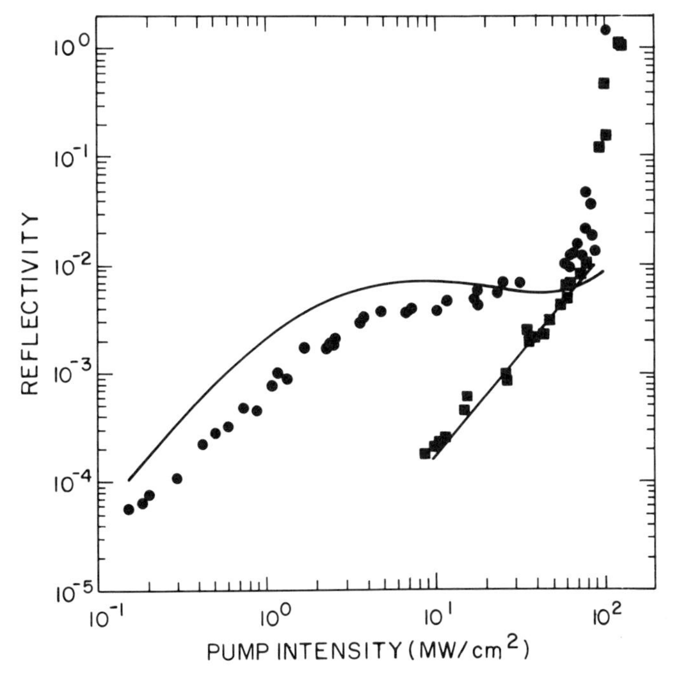

Figure 9. OPC observed in two 3-mm samples of Ge, one intrinsic (squares) and the other p-type (circles). Enhanced conjugate reflection due to saturable absorption sample is observed below 30 MW/cm^2 in the p-type. (Reproduced from Watkins et al. (1983) with permission of the authors.)

the precise mechanism is uncertain because of the limited intensity range over which this process could be observed without causing optical damage.

For $\hbar\omega > E_g$, other mechanisms of nonlinearity come into play, yielding some of the largest effective nonlinearities ever observed. The notion of an effective nonlinear susceptibility is often necessary in such cases because of the relatively long lifetime of the laser-created plasma, which can be many orders of magnitude larger than the laser pulse duration. For very short pulses, for example, the fluence of the pump waves is often more significant than their intensity.

Resonant excitation of electron–hole pairs may result in one or more of several physical processes that can lead to the production of absorptive or dispersive spatial gratings. For example, the generated plasma will contribute directly to the refractive index of the material. Another possibility arises from the modification of the optical response caused by band-filling effects, such as the Burstein–Moss shift of the optical bandgap. Absorption and refractive index gratings result from the associated "blocking" of interband transitions. The dynamic character of the phase conjugation signal resulting from such processes is quite variable since it is influenced by interband and intraband relaxation and scattering processes.

V. CONCLUSION

To summarize, we have discussed the origin of conjugate reflection in a few selected semiconductors. It has been shown that the nonlinearities responsible for optical phase conjugation may have different physical origins, depending on the particular semiconductor under investigation, on the wavelength of the input beams, on their polarization directions, and on their intensity. Optical phase conjugation can be used as a sensitive spectroscopic tool to study the resonances of the crystal. However, its main interest lies in the wide range of potential applications, arising from the property of time-reversal of the signal wave with respect to the incident probe wave.

REFERENCES

Aoyagi, Y., Segawa, Y., and Namba, S. (1982). *Phys. Rev. B* **25**, 1453.
Banyai, L., and Koch, S. W. (1986). *Z. Phys. B* **63**, 283.
Bloembergen, N. (1965). *Nonlinear Optics*. W. A. Benjamin, New York.
Chase, L. L., Peyghambarian, N., Grynberg, G., and Mysyrowicz, A. (1979). *Opt. Commun.* **28**, 189.

Chase, L. L., Claude, M. L., Hulin, D., and Mysyrowicz, A. (1983). *Phys. Rev. A* **28**, 3696.
——— (1984), *JOSA B*, **Vol. 1**, No. 3.
Chemla, D. S., Maruani, A., and Batifol, E. (1979). *Phys. Rev. Lett.* **42**, 1075.
Claude, M. L., Chase, L. L., Hulin, D., and Mysyrowicz, A. (1984). *Philos. Trans. R. Soc. London.* **A313**, 385–387.
Cotter, D. (1986). *IQEC 86*, San Francisco, to be published.
Dagenais, M., and Winful, H. G. (1984). *Appl. Phys. Lett.* **44**, 574.
Eichler, H. J. (1978). In *Festkörperprobleme XVII* (Advances in Solid State Physics 17), ed. J. Treusch. Vieweg Verlag, Braunschweig. pp. 241–263.
Feinberg, J., and Hellwarth, R. W. (1980). *JOSA* **70**, 599.
Fisher, R. A., ed. (1982). *Optical Phase Conjugation*. Academic Press, New York.
Gaskill, J. D., ed. (1982). Special issue of *Opt. Eng.* **21**, no. 2.
Gale, G. M., and Mysyrowicz, A. (1974a). *J. Physique* **35**, C 343.
——— (1974b). *Phys. Rev. Lett.* **32**, 727.
Hanamura, E. (1973). *Solid State Commun.*, **12**, 951.
Haug, H., and Schmitt-Rink, S. (1984). *Progr. Quantum Electron.* **9**, 3.
Hellwarth, R. W. (1977). *J. Opt. Soc. Am.*, **67**, 1.
Huignard, J. P., and Micheron, F. (1976). *Appl. Phys. Lett.* **29**, 591.
Jain, R. K., and Klein, M. B. (1983). *Optical Phase Conjugation* (ed. R. A. Fisher). Academic Press, New York.
Jain, R. K., and Lind, R. C. (1983). *JOSA* **73**, 647.
Martin, G., and Hellwarth, R. W. (1979). *Appl. Phys. Lett.* **34**, 371.
Maruani, A., Oudar, J. L., Batifol, E., and Chemla, D. S. (1978). *Phys. Rev. Lett.* **41**, 1372.
Miller, D. A. B., Seaton, C. T., Prise, M. E., and Smith, S. D. (1981). *Phys. Rev. Lett.* **47**, 197.
Mizutani, G., and Nagasawa, N. (1983). *J. Phys. Soc. Japan* **52**, 2251.
Nagasawa, N., Mita, T., and Ueta, M. (1976). *J. Phys. Soc. Japan* **41**, 929.
Patel, C. K. N., Slusher, R. E., and Fleury, P. A. (1966). *Phys. Rev. Lett.* **17**, 1011.
Peyghambarian, N., Olbright, G. R., and Fluegel, B. D. (1986). *IQEC 86*, San Francisco, to be published.
Roussignol, P., Ricard, D., Rustagi, K. C., and Flytzanis, C. (1985). *Opt. Commun.* **55**, 143.
Shen, Y. R. (1984). *The Principles of Nonlinear Optics*. J. Wiley, New York.
Vu Duy Phach, Bivas, A., Hönerlage, B., and Grun, J. B. (1978). *Phys. Status Solidi B* **86**, 159.
Watkins, D. E., Phipps, C. R., Jr., and Thomas, S. J. (1981). *Opt. Lett.* **5**, 248.
Watkins, D. E., Phipps, C. R., Jr., and Rigrod, W. W. (1983). *JOSA* **73**, 624.
Wolff, P. A., and Pearson, G. A. (1966). *Phys. Rev. Lett.* **17**, 1015.
Yuen, H. P. (1976). *Phys. Rev. A* **13**, 2226.
Zeldovich, Ya. B., Popovichev, V. I., Ragul'skii, V. V., and Faizullov, F. S. (1972). *JETP Lett.* **15**, 109.

10 NONLINEAR REFRACTION FOR CW OPTICAL BISTABILITY

B. S. Wherrett, A. C. Walker, and F. A. P. Tooley

DEPARTMENT OF PHYSICS
HERIOT-WATT UNIVERSITY
EDINBURGH EH14 4AS, U.K.

I. INTRODUCTION

A. Semiconductor Optical Bistability

The key to recent advances in the application of cw optical bistability has been the realisation that semiconductor materials exist in which the irradiance dependence of the refractive index is large enough to produce bistable

OPTICAL NONLINEARITIES
AND INSTABILITIES IN
SEMICONDUCTORS

ISBN 0-12-332915-9

characteristics at low power levels and in small samples (Miller et al., 1979; Gibbs et al., 1979a).

Two classes of bistability have received prominent attention, namely those associated with refractive index changes brought about (i) by carrier generation (optoelectronic) and (ii) by sample temperature changes (optothermal). In this chapter, some of the scaling rules for these mechanisms are discussed, rules that indicate which materials are likely to give the highest nonlinearities and at which laser frequencies one should operate.

The overall problem of semiconductor bistability is as follows. The material is described by a refractive index n, interband absorption α_i, and additional absorption (due for example to the created free carriers) α_p.

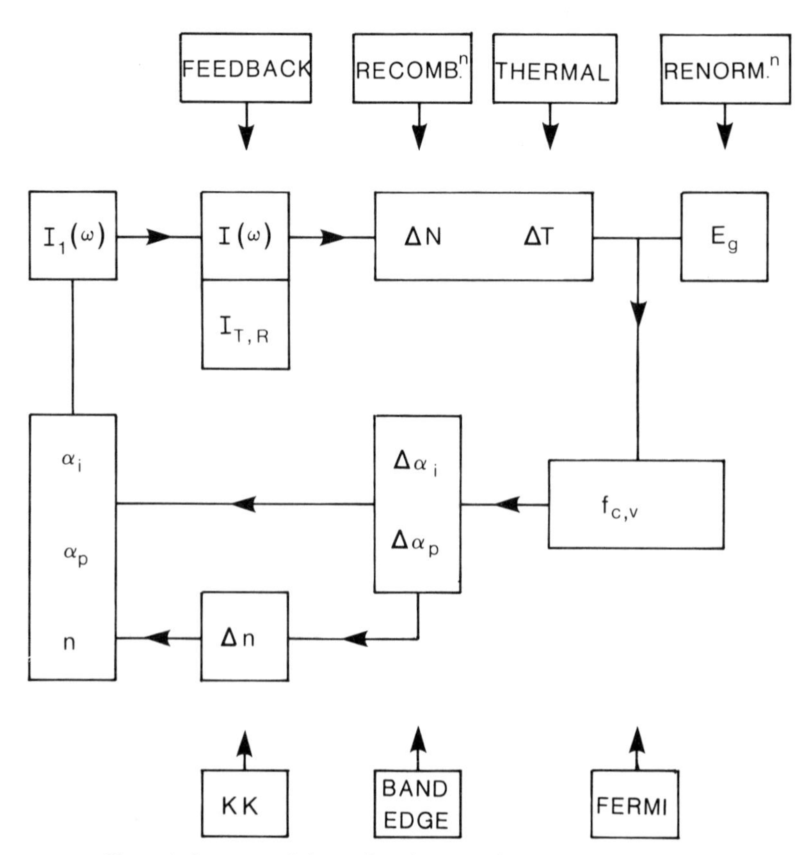

Figure 1. Summary of the semiconductor optical bistability problem.

Radiation I_1, at frequency ω, is incident on the material in a cavity formed either by the natural surface reflectivities or by external partial mirrors/coatings. The consequent internal irradiance $I(\omega)$ must then be determined and also the transmitted and reflected irradiances. This is the cavity feedback problem and it may involve steady-state or time dependence, plane-wave or nonuniform illumination.

The internal irradiance generates ΔN carriers and a temperature change ΔT. For steady state, one is therefore concerned with the carrier recombination mechanisms (trap, radiative, Auger, surface recombination) and/or the thermal losses. The band structure and the state populations alter as a consequence. It is normal to assume separate Fermi distributions of the photoexcited carriers within the valence and conduction bands f_v, f_c. Both thermal and renormalisation band-edge shifts occur as a result of the medium excitation. Given an understanding of all absorption processes, the changes $\Delta\alpha_i$ and $\Delta\alpha_p$ may be determined and, in turn, the refractive index change Δn calculated by a Kramers–Krönig transformation. A thermal expansion of the Fabry–Perot cavity may also be significant. One must cycle around this resulting computational loop self-consistently in order to produce the steady-state optical responses of the system, as shown in Fig. 1.

B. Optoelectronic and Optothermal Refractive Bistability

The cavity optical path length (nD) plays the dominant role in determining the Fabry–Perot response. For nonlinear systems, the path length change is given by

$$\frac{\Delta(nD)}{nD} = \frac{1}{n}\frac{\partial n}{\partial N}\Delta N + \left(\frac{1}{n}\frac{\partial n}{\partial T} + \frac{1}{D}\frac{\partial D}{\partial T}\right)\Delta T. \tag{1}$$

The first term, the electronic refractive contribution, is determined by the refractive index cross section (per generated carrier pair), $\partial n/\partial N = \sigma_n$. The second term is the thermo-optic contribution and is in general an order of magnitude larger than the final, thermal-expansion term. For bistability the change in optical path length, induced by the internal radiation, I, must be at least of order $\lambda_v/2\mathcal{F}$, where λ_v is the vacuum wavelength and $\mathcal{F}$ the cavity finesse (Wherrett et al., 1986; Miller, 1981). A low absorption is needed to attain a high finesse, but at the same time there must be sufficient absorption to give the required steady-state concentration or temperature changes.

$$\Delta N \simeq \frac{\alpha_i}{\hbar\omega} T_1 I, \qquad \Delta T \simeq (\alpha_i + \alpha_p)\frac{A}{\kappa} I. \tag{2}$$

Here we assume that trap recombination processes with a constant lifetime T_1 dominate the carrier equilibrium. A and κ are representative areas and thermal conductivities, respectively, which depend on the thermal geometry of the device (Wherrett et al., 1986). The total refractive index change is $\Delta n = (n_2 + n_2^T)I$, where

$$n_2 = \frac{\sigma_n}{\hbar\omega}\alpha_i T_i, \qquad n_2^T = \frac{\partial n}{\partial T}\frac{A(\alpha_i + \alpha_p)}{\kappa}. \tag{3}$$

Characteristic irradiance levels are

$$I_B = \frac{hc}{|\sigma_n| T_1}, \qquad I_B^T = \frac{\kappa\lambda_v}{|\partial n/\partial T|A}, \tag{4}$$

such that the critical switching irradiance, above which bistability is possible, is a product of I_B (or I_B^T) and a factor that depends only on the Fabry–Perot cavity parameters (αD and the reflectivities R_F, R_B). It is the scaling of $I_B^{(T)}$ that determines the relative merits of different materials.

II. OPTOELECTRONIC NONLINEARITIES

A. A Comparison of Nonlinearity Models

1. Historical Context

We have been able to introduce n_2 above because ΔN was assumed to be directly proportional to I. In practice, more complicated cases also exist (Miller et al., 1984; Walker et al., 1984) or it may not be possible to achieve large enough ΔN changes for bistability, due to saturation of the nonlinearity. In those cases for which the constant n_2 approximation does apply, it has been conventional to define an effective third-order nonlinear susceptibility, $\mathrm{X}_{\mathrm{eff}}^{(3)}$ (Wherrett and Higgins, 1982)

$$\mathrm{Re}\,\mathrm{X}_{\mathrm{eff}}^{(3)}(\omega, -\omega, \omega) = (n^2 c/4\pi^2)n_2. \tag{5}$$

This has enabled connections to be made between Δn and a number of third-order nonlinear processes observed for fifteen years before semiconductor optical bistability was demonstrated.

In a review of those nonlinear optical phenomena that had been considered prior to 1963, just three years after the first laser demonstration, Franken and Ward (1963) reported that "intensity-dependent" refraction should occur, but that it had not then been observed. In the previous year, however, an associated third-order nonlinearity due to third-harmonic generation in calcite was reported at a $\mathrm{X}^{(3)}$ value of 10^{-15} esu (Terhune et al., 1962). The first refractive susceptibilities, measured by birefringence studies in liquids, were of order 10^{-16} esu (H_2O) to 10^{-12} esu (CS_2) (Mayer and Gires, 1964; Maker et al., 1964).

One must turn to $(2\omega_1 - \omega_2)$ four-photon mixing experiments for the earliest relevant semiconductor studies. This mixing is close to being a nonlinear refraction process and would be so if ω_2 were adjusted to be equal to ω_1. In the large-gap material CdS, Maker and Terhune (1965) observed a correspondingly small $X^{(3)}$ of 5×10^{-12} esu. With the invention of the infrared CO_2 laser, however, Patel et al. (1966) were able to study small-gap semiconductors, achieving 8×10^{-10} esu in InSb with $\omega_1, \omega_2 \approx$ 1000, 1100 cm^{-1}, respectively (the InSb band edge corresponds to approximately 1900 cm^{-1}).

Between 1965 and 1975, interest in nonlinear refraction centered on self-focussing studies, primarily in liquids. Here, one is concerned with the spatial refractive index change brought about by an intense Gaussian beam and the resulting beam trapping and filamentation. The physical origin of self-focussing nonlinearities has been predominantly the molecular-orientational Kerr effect, inappropriate in solids, although as long ago as 1966, Javan and Kelley suggested that the saturation of anomalous dispersion in gases should lead to high Δn values.

Renewed interest in semiconductor nonlinear refraction came about as a culmination of several independent investigations. The first was the prediction of Szöke et al. (1969) that optical bistability should be possible for a saturating medium with feedback, and the demonstration by Gibbs et al. (1976) that optical bistability could indeed be observed experimentally (in sodium vapour), but that the origin was dispersive. The second was the growing interest, since the transient holography work of Woerdman and Bolger (1969), in the dynamic refractive index gratings established when noncolinear laser beams interfere across a semiconductor sample. The diffraction of a third beam allows one to study the dynamics of the excited carriers that produced the index grating or to achieve phase conjugation (Jarasiunas and Vaitkus, 1974; Bergmann et al., 1978). Third, the 1970s saw the development of the spin-flip Raman laser. In this instrument, one is pumping a semiconductor (albeit in a high magnetic field) at a frequency just below the band gap (Dennis et al., 1972). At that time $X^{(3)}$ refractive index effects associated with the Raman gain were known to be present, causing mode pulling, and the observed break-up of the transmitted pump-beam spatial profile was likely to have been due to nonlinear refraction.

It was not until the late 1970s, however, that the giant nonlinearities required for bistability were recognised. Thus, for InSb illuminated with continuous-wave (cw) CO-laser power levels, transmitted beam profile measurements and analysis (Miller et al., 1978; Weaire et al., 1979) indicated $X^{(3)}$ values as high as 10^{-2} esu for frequencies just below the band edge. Similar values have also been obtained from phase conjugation and degenerate four-wave mixing in cadmium mercury telluride (Jain, 1982). Finally, in 1979, reports were made of the first semiconductor optical bistability experiments (Gibbs et al., 1979a; Miller et al., 1979) with $X^{(3)}$ values of 10^{-5} to 10^{-4} esu in GaAs and 1 esu in InSb.

Since 1982, nonlinear refraction has been monitored in many materials, through beam profiling, four-wave mixing, transient grating and bistability experiments. Table 1 summarises many of these observations. In the

Table 1. Optoelectronic Nonlinear Refraction

Material	Temperature (K)	Wavelength (μm)	$-\sigma_n$ (cm^3)	$-\frac{\partial n}{\partial I}\left(\frac{\text{cm}^2}{\text{kW}}\right)$	Reference
CdHgTe	80	10.6	—	1	Miller et al., 1984
CdHgTe	300	10.6	—	5×10^{-4}	Craig and Miller, 1986
InSb	300	10.6	—	1.1×10^{-4}	Mathew et al., 1985
PbSnTe	50	10.6	—	$\sim 10^{-4}$	Kar et al, 1986
Te	300	10.6	$\sim 10^{-21}$	—	Staupendahl and Schindler, 1982
Ge	300	10.6	—	1.6×10^{-9}	Watkins et al., 1981
InSb	80	5.5	2×10^{-18}	1	Miller et al., 1981
InAs	80	3.1	—	2×10^{-2}	Poole and Garmire, 1984
InGaAsP	300	1.5	$\sim 10^{-20}$	—	Adams et al., 1985
InGaAsP	300	1.3	$\sim 10^{-20}$	—	Sharfin and Dagenais, 1985
Si	300	1.06	$\sim 10^{-21}$	—	Eichler, 1983
Si	300	1.06	—	1.2×10^{-7}	Jain and Klein, 1983
CdSe	300	1.06	—	$\sim 10^{-9}$	Jain and Klein, 1983
CdSe	300	1.06	—	$> 8 \times 10^{-10}$	Guha et al., 1985
CdTe	300	1.06	—	$\sim 10^{-9}$	Kremenitskii et al., 1979
GaAs (MQW)	300	0.86	2×10^{-19}	0.2	Miller et al., 1983
GaAs (MQW)	80	0.82	—	0.1	Ovadia et al., 1985
GaAs (MQW)	300	0.82	—	9×10^{-4}	Lee et al., 1986
GaAs	80	0.812	—	8.6×10^{-7}	Oudar et al., 1984
CdSSe	300	0.56	—	2×10^{-8}	Jain and Lind, 1983
CdS	300	0.53	—	3.4×10^{-9}	Jain and Klein, 1983
CdS	300	0.53	$\sim 10^{-19}$	—	Henneberger et al., 1987
GaSe	80	0.514	—	1.7×10^{-4}	Bakiev et al., 1983
CdS	2	0.487	—	0.1	Dagenais and Sharfin, 1986
CuCl	17	0.39	—	10^{-6}	Peyghambarian et al., 1983

experiments referred to in Table 1, the radiation wavelength may be close to the band-edge or to excitonic features, or two-photon interband absorption may be present.

A number of theoretical models have been used to interpret the various observations we have discussed. Early mixing work was explained in terms of (i) free-carrier nonparabolicity (Wolff and Pearson, 1966) and (ii) virtual interband transitions (Jha and Bloembergen, 1968; Wynne, 1969). Giant nonlinear refraction has been discussed (iii) using a direct saturation model (Miller et al., 1980), (iv) in terms of a dynamic Burstein–Moss effect (Wherrett and Higgins, 1982; Moss, 1980), (v) as a free carrier plasma effect (Jain and Klein, 1980), and (vi) as an enhanced nonparabolicity effect (Khan et al., 1981). In addition, a many-body treatment has been made by Haug and coworkers (Koch and Haug, 1981; Haug and Schmitt-Rink, 1984). This latter treatment is presented in detail elsewhere in this text.

2. Virtual-Transition Theory: Atoms

It is challenging from a fundamental physics point of view to try to understand why the above large range of experimental $X^{(3)}$ magnitudes exist for ostensibly similar phenomena. A quasi-dimensional analysis provides the simplest, first assessment of $X^{(3)}$ (Wherrett, 1984). Interest lies in near-resonance situations where specific optical transitions dominate the nonlinearity. Therefore, consider N discrete atoms per unit volume, characterised by two-level systems of energy difference E_{ba} and a transition dipole-moment er_{ba}. It is useful to remind ourselves of the contribution to the first-order susceptibility, $\Delta X^{(1)}$, and the linear refraction, Δn.

$$\Delta n = 2\pi \operatorname{Re}(\Delta X^{(1)})/n_0, \qquad \Delta X^{(1)} = \Delta P(\omega)/E(\omega). \tag{6}$$

Here $\Delta P(\omega)$ is the linear polarisation per unit volume associated with the transitions, which are considered to give a small contribution to the total refractive index n_0.

The radiation interaction Hamiltonian in the electric-dipole approximation (H') is to be included in the time-dependent Schrödinger equation:

$$H' = -erE(\omega)e^{-i\omega t} + \text{c.c.}, \qquad (H_0 + H')\Psi = i\hbar\partial\Psi/\partial t. \tag{7}$$

The expectation value of er, with respect to the perturbed state, gives us the polarisation ΔP.

$$\Psi(t) \simeq \Psi_a \exp(-iE_a t/\hbar) + c_b(t)\Psi_b \exp(-iE_b t/\hbar),$$

$$c_b(t) \simeq er_{ba}E(\omega)\frac{\exp[i(E_{ba} - \hbar\omega)t/\hbar]}{(E_{ba} - \hbar\omega)}.$$

Here, the rotating-wave approximation has been used to obtain just the dominant term in $c_b(t)$ and damping terms have been removed. The contribution to $\chi^{(1)}$ is thus

$$\mathrm{Re}\,\Delta\chi^{(1)} \simeq N\frac{|er_{ba}|^2}{(E_{ba} - \hbar\omega)}. \tag{8}$$

The form of this expression may be understood by considering the "event" to which it corresponds, namely, the temporary removal of a photon of energy $\hbar\omega$ from the incident field with the simultaneous electron excitation from level a to level b, followed by the re-emission of a photon accompanied by de-excitation back to the ground state (Fig. 2). The dipole moment $\langle er_{ba}\rangle$ represents the strength of each transition; the denominator is the energy mismatch in the intermediate state of the system in which a photon is removed and an electron is excited. The time interval for which this energy mismatch can be sustained is $\Delta t \simeq \hbar/(E_{ba} - \hbar\omega)$, from the uncertainty principle. It is not surprising that the effect of the event on the radiation, as expressed by $\chi^{(1)}$, is proportional to this "virtual-state" lifetime. The atomic density N is equally expected to be important.

Turning to the nonlinear coefficient: $\chi^{(3)} \simeq \Delta P(\omega)/E^3(\omega)$ in essence, third-order perturbation theory must be used to obtain the polarisation component of third-order in E. From a quasi-dimensional analysis, therefore, the susceptibility contribution from N transition systems must be of the form

$$\chi^{(3)} \simeq N|er_{ba}|^4\hbar^A E_{ba}^B F(\hbar\omega/E_{ba}), \tag{9}$$

with $A = 0$, $B = -3$. The dimensionless function F will contain various energy denominators associated with virtual-state lifetimes. If we set $F = 1$, then a characteristic base value for $\chi^{(3)}$ is obtained. In the atomic case with $N \simeq 10^{19}\,\mathrm{cm}^{-3}$, $r_{ba} \simeq 10^{-8}$ cm and $E_{ba} \simeq 10$ eV, a base-value of 10^{-18} esu is obtained.

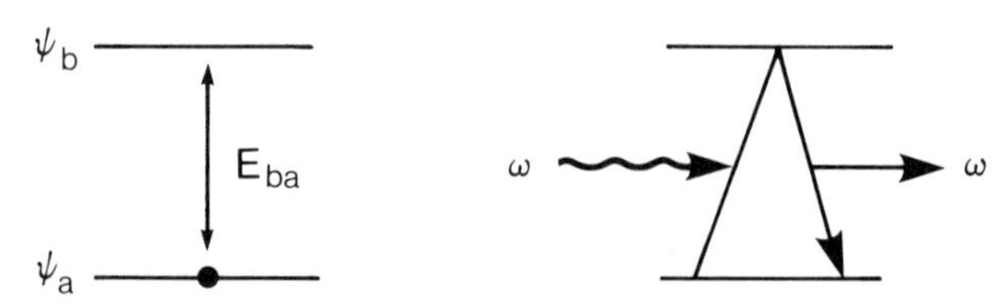

Figure 2. Schematic of the two-level system and resonant, absorptive contribution to the polarisation.

3. Virtual Transition Theory: Semiconductors

A similar analysis may be applied to the semiconductor case. First, however, the coordinate matrix elements should not be used for semiconductors, where the electronic Bloch states are extended, periodic functions. Instead, momentum matrix elements must be used. We shall concentrate on interband (v → c) transition contributions to $\mathrm{X}^{(3)}$

$$|er_{\mathrm{cv}}| \to |ep_{\mathrm{cv}}/m\omega_{\mathrm{cv}}| \simeq eP/E_{\mathrm{cv}}, \tag{10}$$

where P is known as the Kane momentum parameter and has dimensions of $[ML^3T^{-2}]$ and a value close to 10^{-19} esu in most materials. Second, the atomic density must be replaced by a sum over k-states per unit volume, and the two-level energy should be replaced, within the sum, by the k-dependent state energies. The latter energies are determined by the reduced effective mass for the interband transitions, which in turn is dominated by the band interaction between the conduction and valence bands, $H'_{\mathrm{band}} = \hbar \boldsymbol{k} \cdot \boldsymbol{p}/m$. The only parameters that appear in the summation are therefore matrix elements of p/m and the fundamental gap E_{g}, so that finally:

$$\mathrm{X}^{(3)} \propto e^4 \hbar^A E_{\mathrm{g}}^B P^C F(\hbar\omega/E_{\mathrm{g}}), \tag{11}$$

where F differs from that in Eq. 9, and one finds $A = 0$, $B = -4$, $C = 1$. We have in effect used Eq. 10 plus the substitution,

$$N \to E_{\mathrm{g}}^3/P^3, \tag{12}$$

this being the number of k-states per unit volume in an energy range of order E_{g}. Hence, for semiconductor band states:

$$\mathrm{X}^{(3)}_{\mathrm{semiconductor}} \propto e^4 P E_{\mathrm{g}}^{-4} F(\hbar\omega/E_{\mathrm{g}}). \tag{13}$$

For a small-gap semiconductor, $\mathrm{X}^{(3)} \approx F \times 10^{-9}$ esu. Over a range of semiconductors, P is essentially constant and the linear refractive index varies by a factor of less than two, whereas the bandgap can vary by a factor of ten or more. Thus the E_{g}^{-4} dependence is expected to dominate any scaling. The factor F must be considered if $\mathrm{X}^{(3)}$ magnitudes as large as those required for bistability are to be achieved. The likelihood of a large F is seen by considering Fig. 3. The mathematical contributions to $\mathrm{X}^{(3)}$ corresponding to the time-ordering diagrams will be proportional to the times for which each intermediate state can exist (Δt). In turn, each Δt is inversely proportional to the energy mismatch at the intermediate stage. But for the chosen schemes, the second intermediate state has precisely the same energy as the initial state. Thus, $\Delta E = 0$ and the individual contribu-

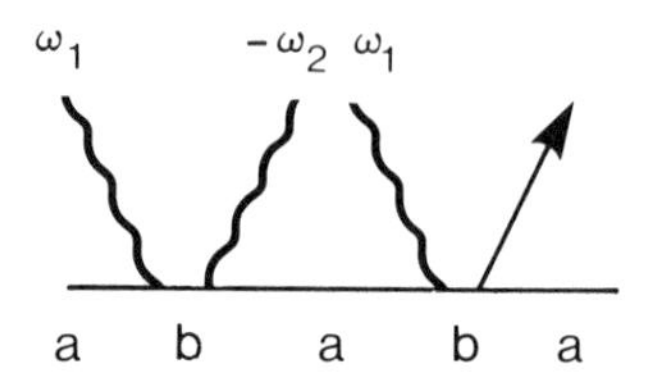

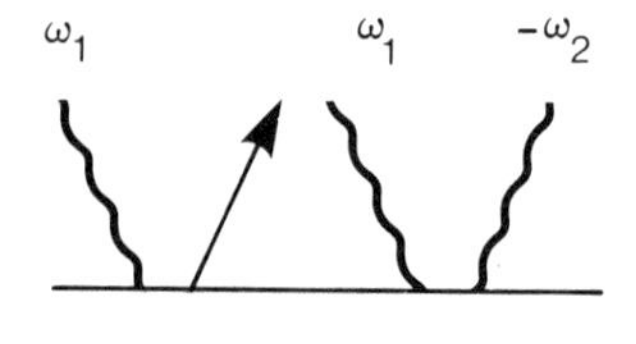

Figure 3. Resonant contributions to $(2\omega_1 - \omega_2)$-mixing and to nonlinear refraction.

tions to $\chi^{(3)}$ diverge to infinity. However, starting with four-stage virtual interband transitions for mixing at $(2\omega_1 - \omega_2)$, one obtains a nondiverging expression for $\chi^{(3)}$ dominated by the processes shown in Fig. 3. Summing over the two schemes, there is a partial cancellation of terms, following which ω_2 can be set equal to ω_1 without producing the above divergence (Wherrett, 1983). As a result, for N electrons at the bottom of the conduction band,

$$\chi_N^{(3)} \approx Ne^4 \frac{P^4}{E_g^7}\left(\frac{E_g}{E_g - \hbar\omega}\right)^3. \tag{14}$$

For a full valence band and empty conduction band, this virtual transition scheme model gives

$$\chi^{(3)} \approx -e^4 \frac{P}{E_g^4}\left(\frac{E_g}{E_g - \hbar\omega}\right)^{3/2}. \tag{15}$$

Note that this has precisely the scaling form predicted by dimensional analysis (Eq. 13) and that $\chi^{(3)}$ displays a minus three-halves resonance behaviour and is negative. The latter is significant experimentally because a negative nonlinear refraction leads to beam defocussing, which is stable by comparison with the catastrophic self-focussing.

The efficiency of resonance enhancement of virtual processes is demonstrated by $(2\omega_1 - \omega_2)$-mixing observations in InSb (MacKenzie et al., 1984). As the band edge is approached, $\chi^{(3)}$ values up to 10^{-6} esu are observed, some three orders of magnitude greater than are observed with midgap frequencies. These values are, however, still six orders of magnitude smaller than the refractive $\chi^{(3)}$ near resonance. To bridge this gap, one must set $\omega_2 = \omega_1$, thereby detecting the effect of real (rather than virtual) excitation of carriers.

4. Saturation Theory

By introducing two effective scattering times, we can make the link between the virtual transition and the real excitation pictures. By analogy with a two-level system, consider the optically coupled valence and conduction states that lie at the same $\boldsymbol{k}$. We can describe intraband scattering, due for example to electron–electron and electron–phonon interactions, by a damping or dephasing lifetime T_2 and describe interband recombination by a time T_1. Real (long-term) excitation becomes important for $(E_g/\hbar - \omega) \leq 1/T_2$ and, in optical mixing, for $(\omega_1 - \omega_2) \leq 1/T_1$. The same time-ordered diagrams dominate $X^{(3)}$ and can be calculated by using density matrix theory, with the essential result

$$X^{(3)} \approx X^{(3)}_{\text{virtual}} \frac{T_1}{T_2}. \tag{16}$$

The introduction of T_2 and T_1 is the basis of the direct saturation model of the nonlinearity. In those cases where the T_2 lifetime can be used, phenomenologically, to describe the band-tail absorption, one can re-express $X^{(3)}$ directly in terms of α, or ΔN in terms of the equilibrium excited-carrier concentration,

$$X^{(3)} \approx -\frac{e^2P^2}{E_g^4} nc\alpha T_1 \frac{E_g}{E_g - \hbar\omega}. \tag{17}$$

5. Dynamic Burstein–Moss Model

In the virtual transition picture, the excited carriers remain in the optically coupled states. More realistically, rapid intraband scattering occurs leading to thermal carrier distributions. This leads to the dynamic Burstein–Moss model for nonlinear refraction in which a thermal distribution of carriers, produced by interband absorption of unspecified origin, induces a consequent change in the linear optical properties. In Fig. 4, we show how the saturation of the band absorption leads to a reduction of the refractive index at frequencies below the bandgap, just as the saturation of a single, homogeneously broadened two-level system leads to saturation of the refractive index at all frequencies, with a reduction of the index just below the transition frequency. In this model a Kramers–Krönig analysis is employed to relate the changes in linear absorption and refraction.

A reduction by ΔN of the empty states (b) must give a negative contribution to Re $X^{(1)}$, because transitions into these states are removed.

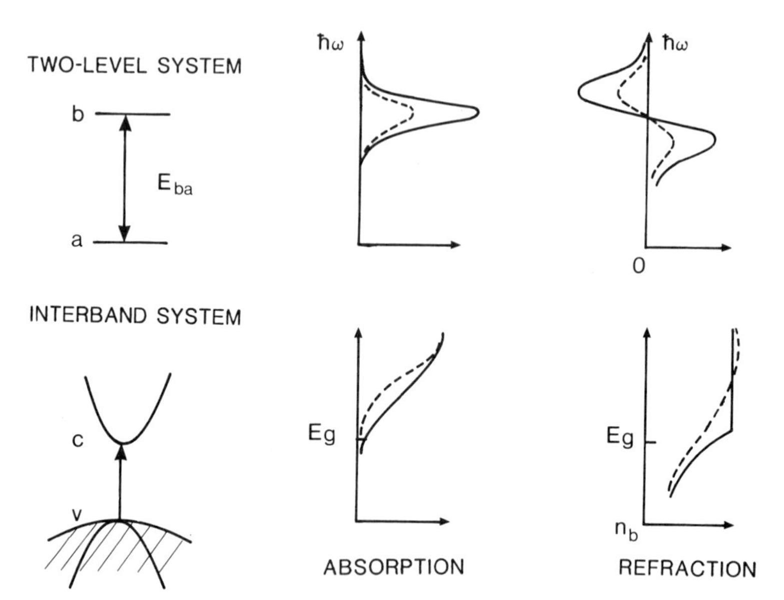

Figure 4. Saturation of absorption and refraction in (a) two-level, homogeneously broadened systems and (b) energy bands.

For two-level systems:

$$\operatorname{Re}\Delta \mathrm{X}_b^{(1)} = \frac{-\Delta N |er_{ba}|^2}{(E_{ba} - \hbar\omega)}. \tag{18}$$

Similarly, ΔN lower states (a) no longer contain electrons so that this number of states becomes available for radiative emission processes associated with b-to-a de-excitation; the corresponding susceptibility contribution is of the same magnitude and sign as $\Delta \mathrm{X}_b^{(1)}$. Thus, regardless of the recombination process, one obtains a refractive index change,

$$\Delta n = \frac{2\pi}{n_0}\Delta \mathrm{X}^{(1)} = \frac{2\pi}{n_0}(-2\Delta N)\frac{|er_{ba}|^2}{E_{ba}}\frac{E_{ba}}{E_{ba} - \hbar\omega}. \tag{19}$$

For a constant T_1-type recombination, the change in $\mathrm{X}^{(1)}$ or n is proportional to I and it can therefore be considered as an *effective* $\mathrm{X}^{(3)}$ or n_2 phenomenon, where

$$n_2 = \frac{\Delta n}{I} = \frac{-4\pi}{n_0}\frac{|er_{ba}|^2}{E_{ba}^2}\alpha T_1 \frac{E_{ba}^2}{\hbar\omega(E_{ba} - \hbar\omega)}. \tag{20}$$

The linear two-level absorption is,

$$\alpha = \frac{4\pi}{n_0} N \frac{\hbar\omega}{c} \frac{|er_{ba}|^2 T_2^{-1}}{(E_{ba} - \hbar\omega)^2 + \hbar^2 T_2^{-2}}. \tag{21}$$

Thus for $(E_{ba} - \hbar\omega) > \hbar T_2^{-1}$, the nonlinear index has essential form

$$n_2 \propto -\frac{N|er_{ba}|^4}{cn_0^2 E_{ba}^3} \frac{T_1}{T_2} \left(\frac{E_{ba}}{E_{ba} - \hbar\omega} \right)^3. \tag{22}$$

Transforming to the band case to give the index contribution due to the removal of heavy-hole–to–conduction transitions caused by the occupancy of ΔN conduction electrons in states at the bottom of the conduction band:

$$\Delta n = \frac{-2\pi}{n_0} \Delta N \frac{(eP)^2}{3E_g^3} \frac{E_g}{E_g - \hbar\omega}. \tag{23}$$

The factor 1/3 is included to account for averaging over all directions in k-space. Similar Δn contributions arise from the removal of the heavy-hole electrons and for light-hole–to–conduction transitions, etc. The n_2 coefficient can therefore now be expressed in terms of a sign, a material scaling factor, and a resonance enhancement.

$$n_2 \simeq -\frac{2\pi e^2}{3} \frac{\alpha_i T_1 P^2}{n_0 E_g^4} \frac{E_g^2}{\hbar\omega(E_g - \hbar\omega)}. \tag{24}$$

If the band-tail interband absorption coefficient, α_i, at frequencies just below $E_g/\hbar$, is treated as a result of T_2-broadening of the individual two-level interband systems, then the absorption coefficient is proportional to $1/[PT_2\sqrt{(E_g - \hbar\omega)}]$ and

$$n_2 \propto \frac{-e^4}{c} \frac{P}{n_0 E_g^4} \frac{T_1}{T_2} \left(\frac{E_g}{E_g - \hbar\omega} \right)^{3/2}. \tag{25}$$

This corresponds to Eqs. 15 and 16, obtained in the virtual transition analysis.

Eq. 25 applies to low carrier excitation and low temperatures (such that only those states at the band extrema are influenced). A very useful and simple expression may be obtained for the case in which relatively large numbers of carriers are excited, at 77 K or room temperature. In principle, Eq. 23 should be replaced by:

$$\Delta n = \frac{-2\pi}{n_0} \frac{(eP)^2}{3} \sum' \sum \frac{1}{E_g^2(k)[E_g(k) - \hbar\omega]}; \tag{26}$$

where Σ' sums over different bands, and Σ sums over band states that are occupied by photogenerated carriers. For example

$$\sum_{ck} \rightarrow \frac{1}{\pi^2} \int_0^{\infty} \frac{f_c(k) k^2 \, dk}{\left(E_g - \hbar\omega + \dfrac{\hbar^2 k^2}{2m_{ch}} \right)}, \tag{27}$$

$$\Delta N_c \rightarrow \frac{1}{\pi^2} \int_0^{\infty} f_c(k) k^2 \, dk. \tag{28}$$

These expressions will give the Δn contribution due to ΔN_c conduction carriers, influencing heavy-hole–to–conduction transitions (at reduced mass m_{ch}). $f_c(k)$ is the occupancy factor of the conduction states. In the Boltzmann limit,

$$f_c(k) = \exp\left[\frac{E_f - E_g - \dfrac{\hbar^2 k^2}{2m_c}}{k_b T} \right]. \tag{29}$$

The Fermi energy E_f is determined via the ΔN_c equation (Eq. 28). It drops out of Eqs. 26 and 27 to give

$$\Delta n_c = -\frac{2\pi}{n_0} \frac{(eP)^2}{3E_g^2} \frac{\Delta N_c}{k_b T} \frac{m_{ch}}{m_c} \frac{4}{\sqrt{\pi}} J\left(\frac{E_g - \hbar\omega}{k_b T} \frac{m_{ch}}{m_c} \right). \tag{30}$$

The function J is shown in Fig. 5

$$J(d) = \int_0^{\infty} \frac{x^2 e^{-x^2} \, dx}{(x^2 + d)}. \tag{31}$$

The form of the refractive index coefficient on this basis is

$$n_2 \propto -e^2 \frac{\alpha_i T_1 P^2}{n_0 E_g^3} \frac{1}{k_b T} \mathrm{f}\left(\frac{\hbar\omega}{E_g}, \frac{\hbar\omega}{k_b T} \right). \tag{32}$$

Again, narrow-gap materials will give favourably large coefficients.

Note that Eqs. 23, and 30 allow a determination of the refractive cross section per generated carrier, $\sigma_n = \partial n / \partial N$. This cross section is particularly useful in those cases where recombination cannot be described simply by a constant T_1-process.

Fig. 6 shows the observed n_2 bandgap enhancement in InSb obtained from beam-profiling experiments. The fit to the frequency dependence of

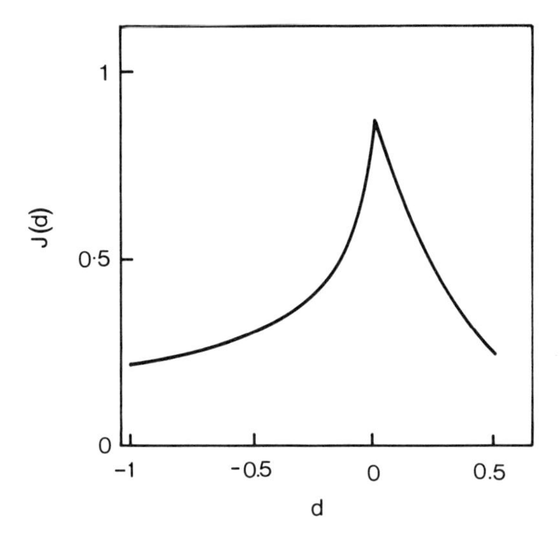

Figure 5. The function J, which dominates the frequency dependence of Δn_c

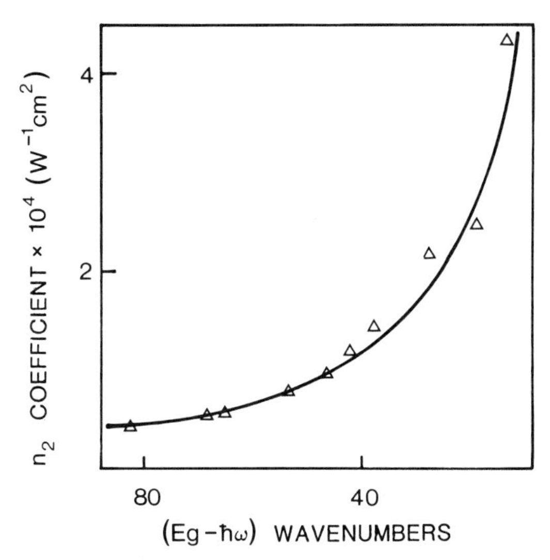

Figure 6. The nonlinear refraction coefficient, n_2, near the InSb band edge at 80 K (Miller et al., 1981). Full line corresponds to self-consistent calculations, accounting for the irradiance dependence of α_i (Higgins, 1983).

the experimental absorption, times that of J, is good. Equally, modelling the band tail to a T_2 broadening gives a reasonable fit for $T_2 = 10$ ps. The theory-to-experiment fits of the n_2 coefficient demand that $T_1 = 1\ \mu$s, manifesting the large enhancement that the T_1/T_2 factor gives.

Note that because the nonlinear coefficient is now described by the change in the linear index brought about by the relatively long-term saturation of transitions, we have in effect undertaken a Kramers–Krönig analysis of the linear absorption change (e.g. Levich, 1971)

$$\Delta n(\omega) = \frac{c}{\pi}\int_0^\infty \frac{\Delta\alpha(\omega')}{\omega'^2 - \omega^2}\,d\omega'. \tag{33}$$

6. Plasma Contribution

An additional but nonresonant contribution to $\Delta X^{(3)}$ comes from free-carrier transitions. Following classical Drude theory, applied to ΔN carriers of effective mass m_c in a medium of background refractive index n_0,

$$\epsilon = n_0^2\left(1 - \frac{\omega_p^2}{\omega^2}\right), \qquad \omega_p^2 = \frac{4\pi\Delta N e^2}{m_c n_0^2}. \tag{34}$$

The induced-carrier contribution to the total index may be written in terms similar to Eq. 30 by noting that $m_c \propto \hbar^2 E_g/P^2$ in semiconductors,

$$\Delta n_{fc} = -\frac{2\pi}{n_0}\frac{\Delta N_c e^2}{\omega^2 m_c} \simeq -\frac{2\pi}{n_0}\Delta N_c \frac{(eP)^2}{E_g^3}\left[\frac{E_g}{\hbar\omega}\right]^2. \tag{35}$$

7. Summary

In the preceding analysis, we have shown how much larger nonlinearities than those resulting from virtual transitions can be obtained by working with narrow-gap semiconductors, near-resonance conditions, and materials of long carrier relaxation time. The major reason for the huge value of the nonlinearities observed at near-bandgap frequencies by comparison to the small values for mixing and nonresonant nonlinear refraction is that the former case concerns active processes in which long-term excitation occurs rather than passive (virtual) processes in which the material plays only a catalytic role. Indium antimonide, with $E_g = 0.23$ eV at 80 K, $T_1 \simeq 500$ ns, and $\sigma_n \approx -10^{-18}$ cm^3, has proven to be particularly successful, with low irradiance (cw laser) bistability achievable. From Eqs. 4 and 30

$$I_B \simeq \frac{\hbar c}{e^2}\frac{n E_g^2 k_B T}{P^2 T_1}\frac{1}{J}. \tag{36}$$

It is noteworthy that the switching irradiance and power levels are inversely proportional to the carrier lifetime. The characteristic switching times, particularly for switch-OFF, are also dominated by the carrier lifetime so there is a trade-off between response speed and power requirements, although the energy absorbed in a given material during a switch process is not influenced by altering the recombination time (using ion-bombarded material for example).

B. Electronic Nonlinearities in Indium Antimonide

1. The Refractive Cross Section Near the Band Edge

The first observation of a giant nonlinearity was made in InSb by Miller et al. (1978). This bandgap resonant phenomenon was quantified in terms of a nonlinear refractive index n_2 coefficient from self-defocussing measurements by Weaire et al. (1979). Optical bistability was subsequently demonstrated at temperatures of both 5 K and 77 K.

The bandgap absorption edge of InSb, which lies near 7 μm wavelength at room temperature, moves to shorter wavelengths on cooling—passing through $\sim$ 5.5 μm at liquid nitrogen temperature (77 K). After further cooling, there is a reversal of this trend, such that near liquid helium temperatures ($\sim$ 5 K) it is again in the region of 5.5 μm. This has proven to be particularly convenient because of the existence of a high-power cw laser source in the 5.5 μm region: the CO laser, with $\sim$ 60 lines around 5 to 6 μm. It is therefore possible to choose a probe wavelength for either 5 K or 77 K operation that is sufficiently close to the band edge for strong resonance effects but not so close as to be fully absorbed.

Experimentally, the nonlinear coefficient increases rapidly as the band edge approaches as shown in Fig. 6.

On the basis of Boltzmann statistics being adequate for describing the equilibrium distribution of photoexcited carriers within the conduction and valence bands, the change in refractive index is given from Eqs. 30 and 35 of sections II.A.5 and 6 by:

$$\Delta n = \frac{-2\pi e^2}{n_0 \omega^2} \left\{ \frac{\Delta N_c}{m_c} \left[1 + Z\left(\frac{m_{ch}}{m} J(d_{hc}) + \frac{m_{cl}}{m} J(d_{lc}) \right) \right] + \frac{\Delta P_h}{m_h} \left[1 + Z \frac{m_{ch}}{m} J(d_{hh}) \right] + \frac{\Delta P_l}{m_l} \left[1 + Z \frac{m_{cl}}{m} J(d_{ll}) \right] \right\}, \tag{37}$$

where:

$$Z = \frac{4}{3\sqrt{\pi}} \frac{mP^2}{\hbar^2 k_b T} \left(\frac{\hbar\omega}{E_g} \right)^2,$$

$$J(d_{ij}) = \int_0^\infty \frac{x^2(e^{-x2})}{x^2 + d_{ij}} \, dx \qquad \text{and}$$

$$d_{ij} = \frac{E_g - \hbar\omega}{k_b T} \frac{m_{ci}}{m_j}.$$

Subscripts c, h, and l refer to the conduction, heavy-hole, and light-hole bands, respectively. ΔN and ΔP are the photogenerated electron and hole densities; m_{ch} and m_{cl} are reduced effective masses.

Eq. 37 shows the contributions to Δn of the three types of carriers that are photoexcited: electrons (ΔN_c), heavy holes (ΔP_h) and light holes (ΔP_l). The first term in each component represents the effect of the shift in the plasma frequency as a direct result of the increased free-carrier population. The remaining terms are the transition blocking, or saturation, contributions: two arising from free-electron blocking of transitions from the light and heavy valence bands to the conduction band and two more representing the consequence of hole generation within the two valence bands, i.e. the depletion of available electrons.

The relative importance of these terms can be deduced assuming the following parameter values:

$$m_c = 0.014m, \qquad m_h = 0.4m, \qquad m_l = 0.016m,$$

$$\frac{\Delta N_c}{\Delta P_h} \approx 1 + \left(\frac{m_l}{m_h} \right)^{3/2}, \qquad \frac{\Delta N_c}{\Delta P_l} \approx 1 + \left(\frac{m_h}{m_l} \right)^{3/2}.$$

For $T = 77$ K, $Z = 1200$, $E_g = 0.23$ eV (1870 cm^{-1}), $n_0 \approx 4$, $\omega = 3.43 \times 10^{14}$ rad s^{-1} (1820 cm^{-1}), Eq. 37 becomes:

$$\Delta n = \frac{-2\pi e^2 \Delta N_c}{n_0 \omega^2 m_c} \left\{ \frac{1}{0.014} [1 + (3.55 + 2.78)] + \frac{1}{0.4}(1 + 10.0) + \frac{1}{2.0}(1 + 2.9) \right\}.$$

It can be clearly seen that the dominant contribution to Δn comes from the first (free-electron) component. Thus, the change in refractive index is directly proportional to the density of free electrons generated in the conduction band. It is therefore convenient to define a nonlinear cross section:

$$\sigma_n = \frac{\Delta n}{\Delta N_c} = \frac{-2\pi e^2}{n_0 \omega^2 m_c}\left[1 + Z\left(\frac{m_{ch}}{m}J(d_{hc}) + \frac{m_{cl}}{m}J(d_{lc})\right)\right]. \quad (38)$$

Using the values detailed above for 77 K operation at 5.5 μm wavelength, $\sigma_n(th) = -1.76 \times 10^{-18}$ cm^3. Note that this value refers only to the low photogenerated carrier density limit. For higher levels of excitation, e.g., $\geq 10^{16}$ cm^{-3}, the complete Fermi distribution of carriers must be taken into account, along with the effects of band-gap renormalisation. (In practice, these two effects act in opposite senses. Neglecting them appears to be consistent with experiment for moderate carrier densities.)

It has been concluded that the magnitude of the induced nonlinear response is determined by the value of σ_n and the excited free-electron population. It follows that to model this response, one must determine the dynamic evolution of the photogenerated electrons (Walker et al., 1984).

A crucial parameter in this respect is the electron recombination rate. A review of published experimental data for excess carrier lifetimes in n-type InSb at 77 to 100 K shows a considerable spread of values, as summarised in Fig. 7. Nonetheless, there is a clear trend showing shortening electron lifetimes, T_1, at higher carrier densities (in excess of 5×10^{15} cm^{-3}) indicating an empirical relationship of the form:

$$T_1(N) = [r_1 + r_2(N_0 + \Delta N)]^{-1}, \quad (39)$$

where N_0 is the dark carrier density and ΔN the photoexcited density. This relation implies a monomolecular (e.g., trap) recombination at low carrier densities evolving to a bimolecular (e.g., radiative) recombination process at higher densities. The approximate fit to the data shown in Fig. 7 can be obtained using $r_1 = 1.5 \times 10^6$ s^{-1} and $r_2 = 1.5 \times 10^{-10}$ cm^3 s^{-1}, both rates being accurate to about $\pm 50\%$.

Using Eq. 39 for T_1, the equilibrium excess carrier density can be calculated for any internal irradiance, from $\Delta N = \alpha_i T_1(N) I / \hbar\omega$. Thus, $\Delta N(I)$ and, hence, the refractive index change can be obtained from

$$\Delta N(I) = \frac{\left[(r_1 + r_2 N_0)^2 + \dfrac{4 r_2 \alpha_i I}{\hbar\omega}\right]^{1/2} - (r_1 + r_2 N_0)}{2 r_2}. \quad (40)$$

The value of T_1 at low ΔN, of the order of 670 ns, is consistent with recent dynamic measurements on optical bistability (OB). For example, OB switch rise and fall times, which should be of order T_1 or more (assuming small switch increments) are observed to be in the range 500 ns–2 μs. In

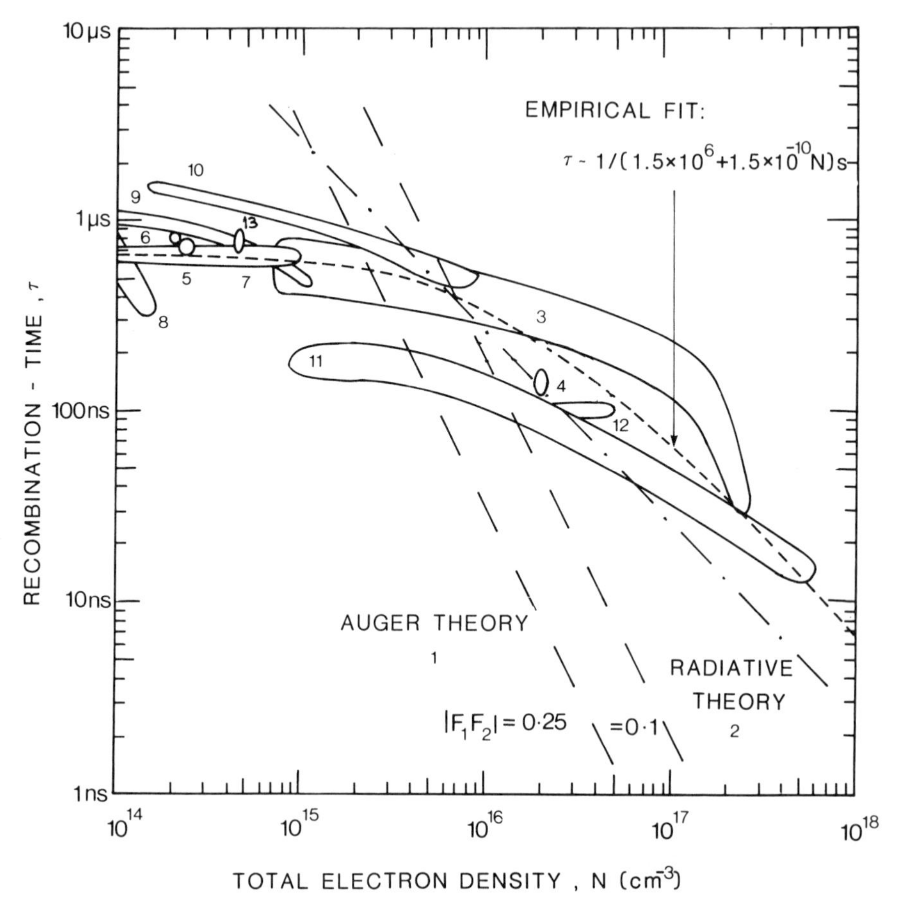

Figure 7. Electron lifetime measurements for *n*-InSb, at temperatures 77–100 K. 1. Beattie and Landsburg (1960); 2. Fossum and Ancker-Johnson (1975); 3. Schneider et al. (1976); 4. Johnston et al. (1980); 5. Wertheim (1956); 6. Laff and Fan (1961); 7. Berkeliev et al. (1964); 8. Blaut-Blachev et al. (1975); 9. Malyutenko et al. (1980); 10. Slusher et al. (1969); 11. Dick and Ancker-Johnson (1972); 12. Danishevskii et al. (1969); 13. Hollis et al. (1967).

addition, measurements of gain bandwidths of an InSb transphasor (Al-Attar et al., 1986b) indicate that a lifetime of 230 ± 25 ns is obtained at $\Delta N \sim 1.2 \times 10^{16}$ cm^{-3}, compared to 300 ± 150 ns predicted by Eq. 39. Finally, Eq. 40 has been used a number of times in modelling OB devices (Tooley et al., 1985; Al-Attar et al., 1986a; MacKenzie et al., 1984) and the results shown to be consistent with experimental observations.

The maximum value for n_2 occurs at low irradiance levels when $T_1 = 670 \pm 350$ ns. This value can be used to deduce an experimental estimate for σ_n from the earliest data of Miller et al. (1981). Using $\sigma_n = n_2\hbar\omega/\alpha_i T_1$ and the measured $n_2 \approx -2.5 \times 10^{-4}$ cm^2 W^{-1} and $\alpha_i \approx 8$ cm^{-1} (at $\bar{\upsilon} = 1820$ cm^{-1}), then it follows that σ_n(expt) $= -2.3 \times 10^{-18}$ ($\pm 50\%$) cm^3 —in excellent agreement with the value calculated above.

There have been four InSb studies from which experimental σ_n values can be deduced. MacKenzie et al. (1984) investigated phase conjugation effects in InSb based on degenerate four-wave mixing and deduced that $X^{(3)} \sim 1.1 \pm 0.3$ esu at 1823 cm^{-1} and ~ 80 K. Using their measured absorption coefficient, $\alpha_i = 15$ cm^{-1}, to determine the operating point on the band edge, this translates into a value σ_n(expt) $= -3.8 \times 10^{-18}$ ($\pm 30\%$) cm^3 at 1820 cm^{-1} and 77 K. Walker et al. (1984) and Tooley et al. (1985) made comparisons of measured optical bistable responses with a model based on the carrier dynamics we have described. Satisfactory fits between theory and experiment were obtained assuming σ_n(expt) $= -1.9 \times 10^{-18}$ cm^3 at 1820 cm^{-1}. Finally, Al-Attar et al. (1986a) measured the dependence of OB switching powers on illumination wavelength and obtained a good fit using the calculated dependence of σ_n shown in Fig. 8 (based on similar theory to that just described but assuming an 1880 cm^{-1} bandgap and neglecting the plasma contribution). This indicates a value of $\sigma_n = -1.2 \times 10^{-18}$ cm^3 at 1820 cm^{-1}. (In modelling the experimental results, this latter calculation did use full Fermi distributions and included certain bandgap renormalisation effects and hence σ_n became dependent upon ΔN). It can be concluded from these various studies that the nonlinear cross section has a value around $\sigma_n = -2 \times 10^{-18}$ ($\pm 50\%$) cm^3, near 1820 cm^{-1}.

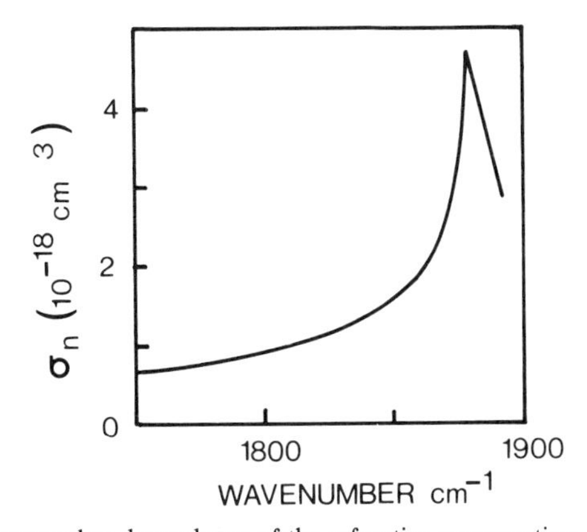

Figure 8. Wavenumber dependence of the refractive cross section, σ_n, for InSb.

One final material parameter of importance is the magnitude of free-carrier absorption. At 77 K around 5.5 μm, the hole absorption cross section, s_p, in InSb is $\approx$ 200 times greater than the electron cross section, s_n, and therefore dominates. The value $s_p \sim 1.5 \pm 0.5 \times 10^{-15}$ cm^2 (Kurnick and Powell, 1959) is sufficiently large to give significant absorption at the carrier densities being generated in OB experiments—consequently reducing the cavity finesse at higher irradiance levels. Walker et al. (1984) have shown that this increasing intraband absorption dominates the interband absorption saturation that is occurring simultaneously. This effect has also been included in modelling the experimental OB results.

2. The Two-Photon Nonlinearity in InSb (300 K)

The bandgap absorption edge in InSb at 300 K lies at a wavelength of 6.9 μm. The nearest powerful source at a longer wavelength is the CO_2 laser, operating around 9 to 11 μm. These wavelengths are too far from the absorption edge to generate free-carrier pairs by single-photon absorption. However, two-photon absorption can be sufficiently strong to generate significant free carrier populations, particularly when using high-peak power, pulsed irradiation (10–100 kW cm^{-2}). As is the case in 77 K, 5.5 μm operation, these carriers directly modify the refractive index and consequently an optically bistable response can be obtained in a cavity (Kar et al., 1983; Wei Ji et al., 1986b).

Once again, the free-hole contributions to the nonlinear response can be neglected and σ_n can be calculated from Eq. 37. Using $T = 300$ K and $\lambda = 10.6$ μm ($Z = 136$), this gives:

$$\sigma_n = \frac{-2\pi e^2}{n_0 \omega^2 m_c}[1 + (0.27 + 0.22)];$$

i.e.

$$\sigma_n(th) = -1.3 \times 10^{-18}\,\text{cm}^{-3}.$$

It can be seen in this case that the transition blocking contributions are reduced to become approximately half of the plasma term, mainly as a result of the weaker resonance and higher temperature (lower Z). Nonetheless, because of the ω^{-2} dependence of the plasma contribution, the final σ_n value is roughly similar to the 77 K, 5.5 μm case.

Again, a knowledge of free-electron lifetimes is required in order to compare the predicted σ_n value with experimental results. It is accepted that recombination in InSb at room temperatures is dominated by an Auger process and is therefore strongly dependent on carrier densities. For intrin-

Table 2. Measured Intrinsic Electron Lifetimes, *n*-InSb Room Temperature

Lifetime, τ_i (ns)	Temperature	References
~ 100	290 K (extrapolated)	Wertheim, 1956
60	300 K	Goodwin, 1960
28	290 K	Hollis et al., 1967
53–75	294 K	Fujisada, 1974
120	295 K	Kravchenko et al., 1975
110	room temperature	Gibson et al., 1976
58	room temperature	Cristoloveanu and Viktorovich, 1977
57	291 K	Bruhns and Kruse, 1980
47	room temperature	Johnston et al., 1980

sic material (electron density N_i) the Auger rate is given by:

$$r = (2 + \Delta N/N_i)(1 + \Delta N/N_i)/2\tau_i, \tag{41}$$

where τ_i is the low-excitation carrier lifetime for intrinsic material. Table 2 summarises the values for τ_i deduced from published data; the average value is 70 ± 30 ns. With a value for r it is possible to calculate the irradiance required to achieve a particular excess carrier density:

$$I = \left(\frac{2\hbar\omega r \Delta N}{\beta} \right)^{1/2}, \tag{42}$$

where β is the two-photon absorption coefficient. The theoretical value for β is around 5 cm MW^{-1}, while experimental measurements cover the range 0.2 to 5 cm MW^{-1} (Gibson et al., 1976; Johnson et al., 1980).

The results of OB studies have been modelled by Wei Ji et al. (1986a) using Eqs. 40 and 41 and including the effects of free-carrier absorption, on the basis of the cross sections $s_p + s_n = 6 \times 10^{-16}$ cm^2. A good fit was obtained assuming $\beta\tau_i = 1.4 \times 10^{-14}$ s cm W^{-1} and $\sigma_n = -2 \times 10^{-18}$ cm^3. This can be compared with the self-defocussing measurements of Mathew et al. (1985), who found $dn/dI = -0.11 \times 10^{-6}$ $\mathrm{cm}^2\ \mathrm{W}^{-1}$ when $I = 2.1 \times 10^5$ W cm^{-2}. Again assuming that $\beta\tau_i = 1.4 \times 10^{-14}$ s cm W^{-1}, then this is equivalent to $\sigma_n = -1.3 \times 10^{-18}$ cm^3 in approximate agreement. If $\beta = 5$ cm MW^{-1} is assumed the most probable value, then these results imply $\tau_i = 3$ ns—a value inconsistent with both Table 2 and the observed OB switching times of 20 to 100 ns (Wei Ji et al., 1986a and b). Instead, it would appear that either the much lower value, $\beta \sim 0.2$ cm MW^{-1}, is appropriate or that σ_n must be very much smaller than predicted. This discrepancy remains to be resolved.

III. OPTOTHERMAL NONLINEARITIES

A. Scaling Rules

We shall concentrate on direct-gap semiconductors, under near-resonant operation. The bandgap dependence of the irradiance required for optothermal bistability, I_B^T, is contained in the factor $\lambda_v/|\partial n/\partial T|$ in Equation (4) (Wherrett et al., 1986).

The size of the bandgap is expected to play only a small role in optothermal refractive bistability. Indeed, the thermal mechanism has been observed at similar irradiance levels in a number of materials, including InSb ($E_g \simeq .23$ eV) and ZnSe ($E_g \simeq 2.5$ eV). Once again, the irradiance for bistable switching is inversely proportional to the characteristic switch-OFF (device-relaxation) time.

Near the band edge, the nonlinear index is taken to be the sum of a contribution due to the temperature dependence of the edge itself plus a background thermo-optic coefficient:

$$\frac{\partial n}{\partial T} = \frac{\partial n}{\partial E_g}\frac{\partial E_g}{\partial T} + \left(\frac{\partial n}{\partial T}\right)_b. \tag{43}$$

Fig. 9 shows the energy-gap temperature shifts of crystalline semiconductors. Note that they differ by less than a factor of two over the vast majority of materials; $\partial E_g/\partial T = -3 \times 10^{-4}$ to -6×10^{-4} eV K^{-1}. To a rough approximation, empirically $\partial E_g/\partial T \propto (E_g)^{1/2}$.

The coefficient $\partial n/\partial E_g$ is obtainable from a Kramers–Krönig transformation of the edge absorption; for direct-gap materials (McLean, 1960)

$$\alpha(\hbar\omega, E_g) = \frac{2e^2\hbar^{1/2}}{nm^2c\omega}\left[\frac{2m_r}{\hbar^2}\right]^{3/2}|\boldsymbol{\epsilon}.\boldsymbol{p}_{cv}|^2(\omega - \omega_g)^{1/2}. \tag{44}$$

More generally it is useful to rewrite α in the scaling form:

$$\alpha = C_1\frac{e^2}{\hbar c}\frac{E_g}{n_0P}\frac{\left(\dfrac{\hbar\omega}{E_g} - 1\right)^{1/2}}{\dfrac{\hbar\omega}{E_g}}. \tag{45}$$

C_1 is a dimensionless constant that depends on the detail of the band structure. The material is characterised by its bandgap, background refractive index n_0 and the Kane interband momentum matrix element P. The

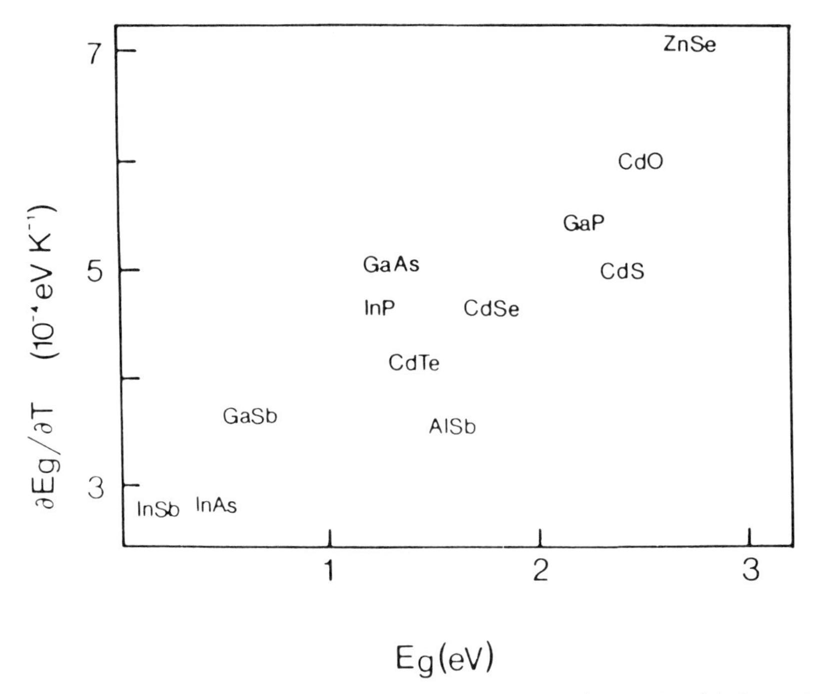

Figure 9. Thermal bandgap coefficients for a range of semiconductors (Landolt-Börnstein, 1982).

refraction change associated with a ΔE_{g} edge shift is

$$\Delta n = \frac{c}{\pi} \int_0^\infty \frac{\alpha(E_{\mathrm{g}} + \Delta E_{\mathrm{g}}) - \alpha(E_{\mathrm{g}})}{(\omega')^2 - \omega^2} \, d\omega'. \tag{46}$$

In the scaling form that picks out the material and frequency dependences

$$\frac{\partial n}{\partial E_{\mathrm{g}}} = C_1 \frac{e^2}{n_0 P E_{\mathrm{g}}} \, \mathrm{g}\left(\frac{\hbar\omega}{E_{\mathrm{g}}}\right). \tag{47}$$

The function g is plotted in Fig. 10:

$$g(x) = \mathrm{Re}\left[\frac{4(2 - \sqrt{1+x} - \sqrt{1-x}) + \left(\frac{1}{\sqrt{1+x}} - \frac{1}{\sqrt{1-x}}\right)x}{2x^2}\right]$$

Just beneath the band edge, the dominant frequency dependence of $\partial n / \partial E_{\mathrm{g}}$

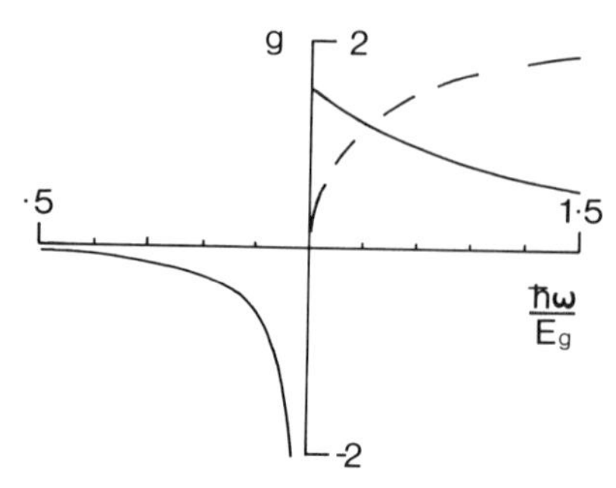

Figure 10. The function of $(\hbar\omega/E_g)$ that determines the model frequency dependence of $\partial n/\partial E_g$ (see text).

is $(E_g - \hbar\omega)^{-1/2}$. Hence, the approximate scaling relevant to near–band-edge $\partial n/\partial T$ values is

$$\partial n/\partial T \approx E_g^{-1/2}\left(\frac{E_g}{E_g - \hbar\omega}\right)^{1/2}.$$

That is, one does not expect dramatically different coefficients for small and large gap materials. The $E_g^{-1/2}$-dependence is consistent with the empirical results summarised in Table 3.

Table 3. Optothermal Nonlinear Refraction

Material	Wavelength (μm)	n	$\frac{\partial n}{\partial T}$ (K^{-1})	Reference
PbSnSe	10.6	4	-8×10^{-3}	Kar et al., 1986
CdHgTe	10.6	4	-1×10^{-3}	Craig et al., 1985
Ge	10.6	4	4×10^{-4}	Golik et al., 1983
InSb	10.6	4	4×10^{-4}	Walker and Hunter, 1985
InSb	5.5	4	4×10^{-3}	Tooley et al., 1985
Si	1.06	3.5	2.4×10^{-4}	Eichler, 1983
GaAs	0.85	3.5	7×10^{-4}	Miller et al., 1986
CdSSe	0.647	~ 2.5	3×10^{-6}	Gibbs et al., 1979b
$GeSe_2$	0.633	2.5	2.5×10^{-4}	Hajto and Janossy, 1983
GaSe	0.514	3	2.5×10^{-4}	Antonioli et al., 1979
ZnSe	0.514	2.5	2.5×10^{-4}	Kar and Wherrett, 1986
ZnS	0.514	2.4	1.4×10^{-4}	Harris et al., 1977
HgS	0.514	~ 3	8×10^{-5}	Ayrault et al., 1973
ZnSe	0.476	2.7	5×10^{-4}	Kar and Wherrett, 1986

B. Thermal Effects in Indium Antimonide

In a material with an optoelectronic nonlinearity, such as InSb, the heating effect of either the absorbed optical power or any externally applied heat input can result in a number of significant changes to the parameters relevant to a near-bandgap probe beam. In InSb, the principal consequence of a temperature rise (ΔT) is a decrease in the bandgap energy ($d\bar{v}/dT = -2\ \text{cm}^{-1}\ \text{K}^{-1}$) which gives rise to the following three changes:

i) An increase in absorption α_i to a value given approximately by: $\alpha_i \exp(\Delta T/17.1)$ (assuming $\Delta T \leq 10$ K and the operating wavenumber ($\bar{v}$) to be within the range $1802\ \text{cm}^{-1} \leq \bar{v} \leq 1840\ \text{cm}^{-1}$).

ii) An increase in the linear refractive index n_0 (described already), which we have measured as $dn_0/dT = 1.2 \times 10^{-3}\ \text{K}^{-1}$ in this wavelength region.

iii) An increase in the magnitude of σ_n, given by $d\sigma_n/dT = 2.6 \times 10^{-20}\ \text{cm}^3\ \text{K}^{-1}$, assuming equivalence between a thermal shift of the bandgap and a change in operating wavelength.

All these effects have been taken into account when modelling the response of a bistable InSb etalon to pulsed incoherent illumination (Tooley et al., 1985). In addition, the absorption change has been exploited in demonstrating induced absorption bistability (Wherrett et al., 1984).

Finally, the dependence of n_0 upon temperature has permitted a demonstration of optothermal refractive bistability in InSb at room temperature using cw 10.6 μm wavelength radiation (Walker and Hunter, 1985).

C. Thermal Effects in Zinc Selenide

A number of devices have been developed using the II–VI semiconductor ZnSe. These have been optothermal devices, mostly relying on the thermo-optic coefficient in this material. ZnSe has an unusually high refractive index for a material transmitting over most of the visible spectrum: $n_0 \approx 2.6$ rising to ≈ 2.7 in the green, i.e., near its (blue) absorption edge. Small percentage variations in n_0 can give significant changes in optical thickness and, thus, sensitive interference devices can be constructed. Two forms of ZnSe have been exploited: bulk polycrystalline material and evaporated thin films.

1. Bulk Polycrystalline ZnSe

CVD-grown polycrystalline ZnSe has a steep bandgap absorption edge near 470 nm. The temperature dependence of the bandgap ($dE_g/dT \sim -6.5 \times 10^{-4}\ \text{eV K}^{-1}$) gives rise to both refractive index and absorption

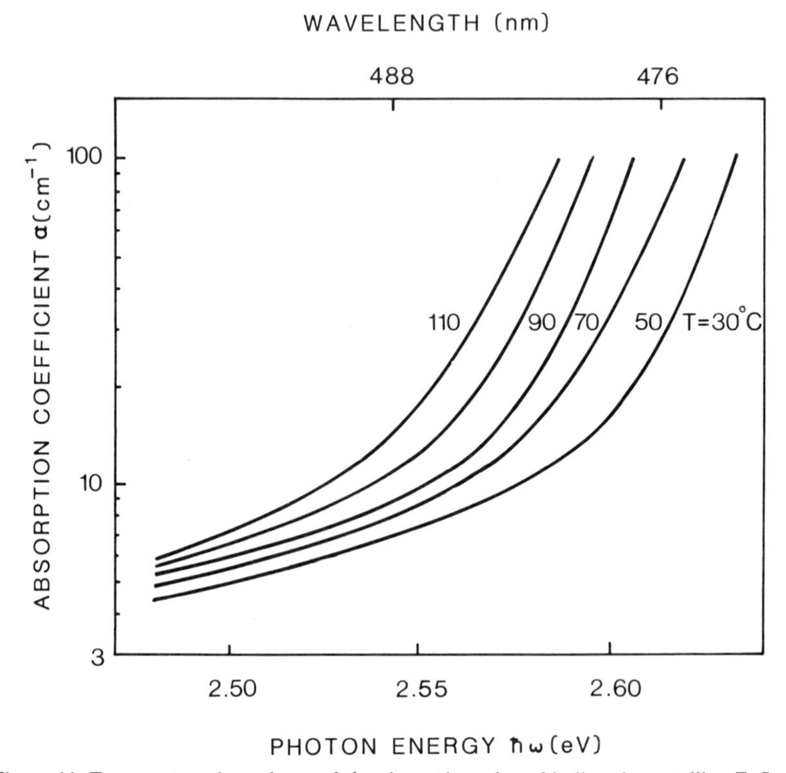

Figure 11. Temperature dependence of the absorption edge of bulk polycrystalline ZnSe.

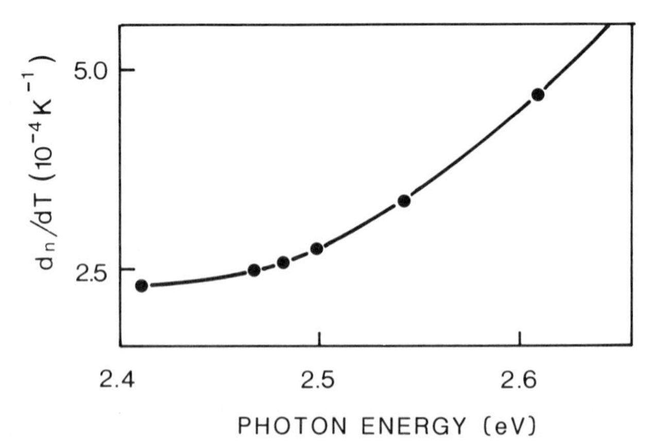

Figure 12. Temperature dependence of the ZnSe thermo-optic coefficient in the band-edge region.

nonlinearities. Each has been exploited to demonstrate refractive and induced-absorption bistability (Kar and Wherrett, 1986; Taghizadeh et al., 1985).

A valuable radiation source for studying bandgap resonant effects in ZnSe is the argon-ion laser which has 8 lines in the range 501.7 nm to 454.5 nm, the strongest outputs in this range being at 488 nm and 476.5 nm. Figs. 11 and 12 show the variation with temperature, in this wavelength region, of both the absorption coefficient and the refractive index (Kar and Wherrett, 1986).

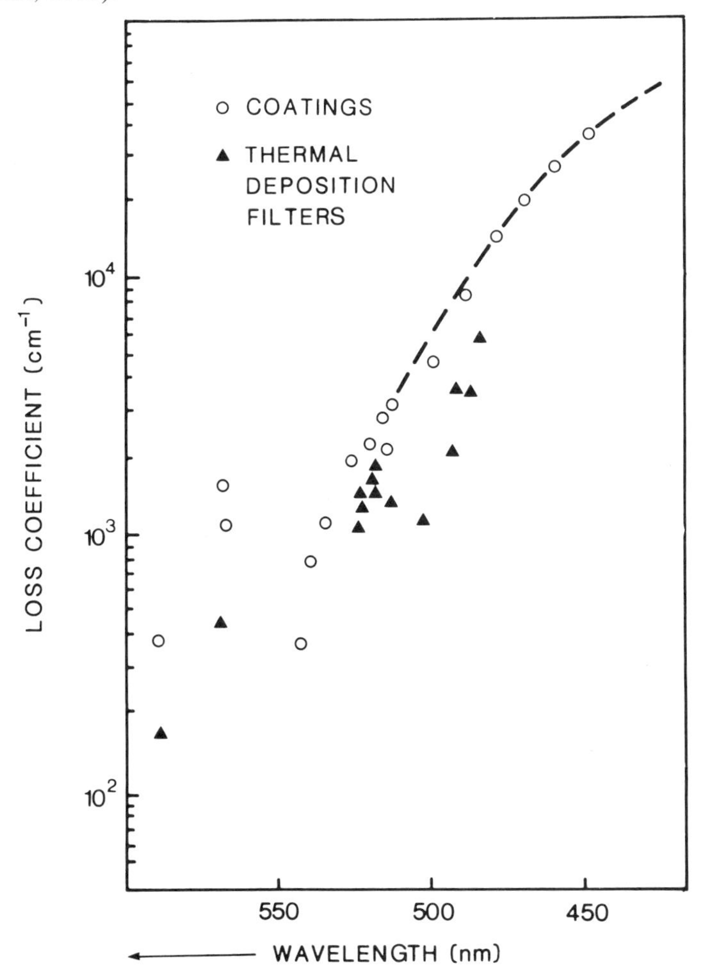

Figure 13. Loss coefficients of thin-film ZnSe determined from transmission–reflection analysis of band-pass filters and of individual coatings.

Table 4. Absorption and Thermo-Optic Coefficients Measured for ZnSe Thin Films

Wavelength (nm)	α (cm^{-1})	n_0	$\partial n/\partial T$ (K^{-1})
514.5	~ 2000	2.72	1.5×10^{-4}
632.8	~ 200	2.47	1.0×10^{-4}

The thermo-optic coefficient, $\mathrm{d}n/\mathrm{d}T$, is enhanced at the band edge, reaching $4.5 \times 10^{-4}\ K^{-1}$ around 476 nm.

2. Evaporated Thin-Film ZnSe

ZnSe films, 0.5 to 1.5 μm thick, have been incorporated into multilayer interference OB devices. To study the properties of these (thermally evaporated) films in more detail, single ZnSe layers have been grown on glass substrates for spectroscopic investigation. This material has been found to have a broad absorption edge (Wherrett and Smith, 1987), of a form lying between the sharp edge of polycrystalline material and the extended tail of amorphous ZnSe (see Fig. 13). The absorption coefficient remains above 100 cm^{-1} over most of the visible spectrum.

Because of the extended absorption, longer wavelengths have been used to illuminate the thin-film devices—commonly 514.5 nm and also 632.8 nm. The broad absorption tail also leads to weak wavelength dependence of the parameters relevant to OB—as shown in the data of Table 4, deduced from the available experimental results (Chow et al., 1986; Janossy et al., 1985; Mathew, 1987).

SUMMARY

We have shown that by using real excitation of an optical medium, large changes can be induced in the optical constants of the material with relatively low irradiance levels. For example ~ 1% changes in a refractive index can be induced with ~ 0.1 to 10 kW cm^{-2} irradiances and absorption coefficients raised by several orders of magnitude under the appropriate conditions. These results contrast strongly with the purely dielectric, virtual transition, nonlinear phenomena (e.g., second harmonic generation, optical Kerr effect) which typically require over 1 MW cm^{-2} irradiances. As a result of the larger-magnitude nonlinearities, it is possible to operate optically nonlinear devices and study various phenomena (e.g., optical bistability) with low-level (e.g., milliwatt) continuous illumination.

Two underlying mechanisms have been identified and studied: optoelectronic and optothermal nonlinearities. The optoelectronic nonlinearities are the result of interband electronic transitions in a semiconductor inducing significant excess free-carrier populations which, in turn, modify both optical constants (e.g., by partially blocking further transitions or generating a significant plasma contribution to the dielectric constant). With such a nonlinearity, both its magnitude and dynamics are determined by the recombination rates of the photoexcited carriers. We have not addressed the role of Coulombic effects, in particular of excitonic features. These are described elsewhere in this volume. Very high optoelectronic nonlinearities can be generated close to exciton features (Miller et al., 1983; Ovadia et al., 1985). It is not yet clear, however, whether bistability can be achieved at sufficiently low power levels to take full advantage of these nonlinearities, because of their saturation at low-excitation concentrations. Optothermal nonlinearities arise from the inevitable relaxation of energy absorbed by a semiconductor (or any other material) to the lattice in the form of heat. The temperature rise induced can be sufficient again to change significantly the optical constants. In this case, the dynamics (and sensitivity) are normally determined by the environment of the medium, such as the degree of heat-sinking. This provides considerable flexibility for controlling the magnitude of the nonlinear response and engineering a variety of optical devices.

REFERENCES

Adams, M. J., Westlake, H. J., O'Mahony, M. J., and Hendry, I. D. (1985). *IEEE J. Quantum Electron.* **21**, 1498.

Al-Attar, H. A., Hutchings, D., Russell, D., Walker, A. C., MacKenzie, H. A., and Wherrett, B. S. (1986a). *Opt. Commun.* **58**, 433.

Al-Attar, H. A., MacKenzie, H. A., Tooley, F. A. P., and Walker, A. C. (1986b). *IEEE J. Quantum Electron.* **QE-22**, 663.

Antonioli, G., Bianchi, D., and Franzosi, P. (1979). *Appl. Opt.* **18**, 3847.

Ayrault, B., Langlois, H., Lecocq-Mayer, M. C., and Lefin, F. (1973). *Phys. Status Solidi A* **17**, 665.

Bakiev, A. M., Dneprovskii, V. S., Kovalyuk, Z. D., and Stadnik, V. A. (1983). *Sov. Phys. Dokl.* **28**, 579.

Ball, D., and Preier, H. M. (1986). *Springer Proc. Phys.* **8**, 144.

Beattie, A. R., and Landsberg, P. T. (1960). *Proc. R. Soc. London A* **258**, 486.

Bergmann, E. E., Bigio, I. J., Feldman, B. J., and Fisher, R. A. (1978). *Opt. Lett.* **3**, 82.

Berkeliev, A. D., Galavanov, V. V., and Naledov, D. M. (1964). *Sov. Phys. Solid State* **6**, 2264.

Blaut-Blachev, A. N., Ivleva, V. S., Korotin, V. G., Krivnagov, S. N., Selyanina, V. I., and Śmetannikovam, Yu. S. (1975). *Sov. Phys. Semicond.* **9**, 247.
Bruhns, H., and Kruse, H. (1980). *Phys. Status Solidi B* **97**, 125.
Chow, Y. T., Wherrett, B. S., Van Stryland, E. W., McGuckin, B. T., Hutchings, D., Mathew, J. G. H., Miller, A., and Lewis, K. (1986). *J. Opt. Soc. Am.* **B3**, 1535.
Craig, D., and Miller, A. (1986). *Opt. Acta* **33**, 397.
Craig, D., Dyball, M. R., and Miller, A. (1985). *Opt. Commun.* **54**, 383.
Cristoloveanu, S., and Viktorovich, P. (1977). *Phys. Status Solidi A* **39**, 109.
Dagenais, M., and Sharfin, W. F. (1986). *Springer Proc. Phys.* **8**, 122.
Danishevskii, A. M., Patrin, A. A., Ryvkin, S. M., and Yaroshetskii, I. D. (1969). *Sov. Phys. JETP* **29**, 781.
Dennis, R. B., Pidgeon, C. R., Smith, S. D., Wherrett, B. S., and Wood, R. A. (1972). *Proc. R. Soc. London A* **331**, 203.
Dick, C. L., and Ancker-Johnson, B. (1972). *Phys. Rev. B* **5**, 526.
Eichler, H. J. (1983). *Opt. Commun.* **45**, 62.
Fossum, H. J., and Ancker-Johnson, B. (1975). *Phys. Rev. B* **8**, 2850.
Franken, P. A., and Ward, J. F. (1963). *Rev. Mod. Phys.* **35**, 23.
Fujisada, H. (1974). *J. Appl. Phys.* **45**, 3530.
Gibbs, H. M. (1985). *Optical Bistability: Controlling Light with Light*. Academic Press, New York.
Gibbs, H. M., McCall, S. L., and Venkatesan, T. N. C. (1976). *Phys. Rev. Lett.* **36**, 1135.
Gibbs, H. M., McCall, S. L., Venkatesan, T. N. C., Gossard, A. C., Passner, A., and Wiegmann, W. (1979a). *Appl. Phys. Lett.* **35**, 6.
Gibbs, H. M., Venkatesan, T. N. C., McCall, S. L., Passner, A., Gossard, A. C., and Wiegmann, W. (1979b). *Appl. Phys. Lett.* **34**, 511.
Gibson, A. F., Hatch, C. B., Maggs, P. N. D., Tilley, D. R., and Walker, A. C. (1976). *J. Phys. C* **9**, 3259.
Golik, L. L., Grigor'yants, A. V., Elinson, M. I., and Balkarei, Yu. I. (1983). *Opt. Commun.* **46**, 51.
Golubev, G. P., Dnepravskii, V. S., Kovalyuk, Z. D., and Stadnik, V. A. (1985). *Sov. Phys. Solid State* **27**, 265.
Goodwin, D. W. (1960). *Proc. Int. Conf.: Solid State Phys. in Elec. and Telecomm., Brussels 1958*, vol. 2. Academic Press, London, p. 759.
Guha, S., Van Stryland, E. W., and Soileau, M. J., (1985). Opt. Lett. **10**, 285.
Hajto, J., and Janossy, I. (1983). *Philos. Mag. B* **47**, 347.
Harris, R. J., Johnson, G. T., Kepple, G. A., Krok, P. C., and Mukai, H. (1977). *Appl. Opt.* **16**, 436.
Haug, H. and Schmitt-Rink, S. (1984). *Philos. Trans. R. Soc. London A* **313**, 221.
Henneberger, F., Puls, J., Rossmann, H., Spiegelberg, Ch., Kretzschmar, M., and Haddard, I. (1987). *Physica Scripta*, **T13**, 195.
Higgins, N. A. (1983). Ph.D. thesis, Heriot-Watt University, unpublished.
Hollis, J. E. L., Choo, S. C., and Heasell, E. L. (1967). *J. Appl. Phys.* **38**, 1626.
Jain, R. K. (1982). *Opt. Eng.* **21**, 199.
Jain, R. K., and Klein, M. B. (1980). *Appl. Phys. Lett.* **37**, 1.
Jain, R. K., and Klein, M. B. (1983). *Optical Phase Conjugation* (ed. R. A. Fisher), ch. 10, 369. Academic Press, New York.

Jain, R. K., and Lind, R. C. (1983). *JOSA* **72**, 647.
Janossy, I., Taghizadeh, M. R., Mathew, J. G. H., and Smith, S. D. (1985). *IEEE J. Quantum Electron.* **QE-21**, 1447.
Jarasiunas, K., and Vaitkus, J. (1974). *Phys. Status Solidi A* **23**, K19.
Javan, A., and Kelley, P. L. (1966). *IEEE J. Quantum Electron.* **QE-2**, 470.
Jha, S. S., and Bloembergen, N. (1968). *Phys. Rev.* **171**, 891.
Johnston, A. M., Pidgeon, C. R., and Dempsey, J. (1980). *Phys. Rev.* **22**, 825.
Kar, A. K., and Wherrett, B. S. (1986). *J. Opt. Soc. Am. B* **3**, 345.
Kar, A. K., MacKenzie, H. A., Wei Ji, Reid, J. J. E., Grisar, R., Ball, D., and Preier, H. M. (1986). Springer, *Proc. Phys.* Vol. 8, 144.
Kar, A. K., Mathew, J. G. H., Smith, S. D., Davis, B., and Prettl, W. (1983). *Appl. Phys. Lett.* **42**, 334.
Khan, M. A., Bogart, T. J., Kruse, P. W., and Ready, J. F. (1981). *Opt. Lett.* **5**, 469.
Koch, S. W., and Haug, H. (1981). *Phys. Rev. Lett.* **46**, 450.
Kravchenko, A. F., Morozov, B. V., and Skok, E. M. (1975). *Sov. Phys. Semicond.* **8**, 1324.
Kremenitskii, V., Odoulov, S., and Soskin, M. (1979). *Phys. Status Solidi A* **51**, K63.
Kurnick, E. H., and Powell, J. M. (1959). *Phys. Rev.* **116**, 597.
Laff, R. A., and Fan, H. Y. (1961). *Phys. Rev.* **121**, 53.
Landolt-Börnstein, (1982). *Numerical Data and Functional Relationships in Science and Technology*, Group III, vols. 17a, b. Springer-Verlag, Berlin.
Lee, Y. H., Chavely-Pirson, A., Rhee, B., Gibbs, H. M., Gossard, A. C., and Wiegmann, W. (1986). *Appl. Phys. Lett.* **49**, 1505.
Levich, B. G. (1971). In *Theoretical Physics*, **2** North-Holland, Amsterdam.
MacKenzie, H. A., Al-Attar, H. A., and Wherrett, B. S. (1984). *J. Phys. B* **17**, 2141.
McLean, T. P. (1960). *Prog. Semicond.* **5**, 55.
Maker, P. D., and Terhune, R. W. (1965). *Phys. Rev. A* **137**, 801.
Maker, P. D., Terhune, R. W., and Savage, C. M. (1964). *Phys. Rev. Lett.* **12**, 507.
Malyutenko, V. K., Bolgov, S. S., Pipa, V. I., and Chaikin, V. I. (1980). *Sov. Phys. Semicond.* **14**, 4. Mathew, J. C. H. (1987). Private communication.
Mathew, J. G. H., Kar, A. K., Heckenberg, N. R., and Galbraith, I. (1985). *IEEE J. Quantum Electron.* **QE-21**, 94.
Mayer, G., and Gires, F. (1964). *C. R. Hebd. Seances Acad. Sci.* **258**, 2039.
Miller, A., Parry, G., and Daley, R. (1984). *IEEE J. Quantum Electron.* **QE-20**, 710.
Miller, A., Steward, G., Blood, P., and Woodridge, K. (1986). *Opt. Acta* **32**, 387.
Miller, D. A. B. (1981). *IEEE J. Quantum Electron.* **QE-17**, 306.
Miller, D. A. B., Mozolowski, M. H., Miller, A., and Smith, S. D. (1978). *Opt. Commun.* **27**, 133.
Miller, D. A. B., Smith, S. D., and Johnston, A. M. (1979). *Appl. Phys. Lett.* **35**, 658.
Miller, D. A. B., Smith, S. D., and Wherrett, B. S. (1980). *Opt. Commun.* **35**, 221.
Miller, D. A. B., Seaton, C. T., Prise, M. E., and Smith, S. D. (1981). *Phys. Rev. Lett.* **47**, 197.
Miller, D. A. B., Chemla, D. S., Eilenberger, D. J., Smith, P. W., Gossard, A. C., and Wiegmann, W. (1983). *Appl. Phys. Lett.* **42**, 925.
Moss, T. S. (1980). *Phys. Status Solidi B* **101**, 555.
Oudar, J. L., Abram, I., and Minot, C. (1984). *Appl. Phys. Lett.* **44**, 689.

Ovadia, S., Gibbs, H. M., Jewell, J. L., Sarid, D., and Peyghambarian, N. (1985). *Opt. Eng.* **24**, 565.
Patel, C. K. N., Slusher, R. W., and Fleury, A. (1966). *Phys. Rev. Lett.* **17**, 1010.
Peyghambarian, N., Gibbs, H. M., Rushford, M. C., and Weinberger, D. A. (1983). *Phys. Rev. Lett.* **51**, 1692.
Poole, C. D., and Garmire, E. (1984). *Opt. Lett.* **9**, 356.
Schneider, W., Groh, H., and Hubner, K. (1976). *Z. Phys. B* **25**, 29.
Sharfin, W. F., and Dagenais, M. (1985). *Appl. Phys. Lett.* **46**, 819.
Slusher, R. E., Giriat, W., and Brueck, S. R. J. (1969). *Phys. Rev.*, **183**, 758.
Staupendahl, G., and Schindler, K. (1982). *Opt. Quantum Electron.* **14**, 157.
Szöke, A., Daneu, V., Goldhar, J., and Kurnit, N. A. (1969). *Appl. Phys. Lett.* **15**, 376.
Taghizadeh, M. R., Janossy, I., and Smith, S. D. (1985). *Appl. Phys. Lett.* **46**, 331.
Terhune, R. W., Maker, P. D., and Savage, C. M. (1962). *Phys. Rev. Lett.* **8**, 401.
Tooley, F. A. P., Walker, A. C., and Smith, S. D. (1985). *IEEE J. Quantum Electron.* **QE-21**, 1340.
Walker, A. C., and Hunter, J. J. (1985). Heriot-Watt University, internal report (unpublished).
Walker, A. C., Tooley, F. A. P., Prise, M. E., Mathew, J. G. H., Kar, A. K., Taghizadeh, M. R., and Smith, S. D. (1984). *Philos. Trans. R. Soc. London A* **313**, 357.
Watkins, D. E., Phipps, C. R., and Thomas, S. J. (1981). *Opt. Lett.* **6**, 76.
Weaire, D., Wherrett, B. S., Miller, D. A. B., and Smith, S. D. (1979). *Opt. Lett.* **4**, 831.
Wei Ji, Kar, A. K., Keller, U., Mathew, J. G. H., and Walker, A. C. (1986a). *Springer Proc. Phys.* **8**, 94.
Wei Ji, Kar, A. K., Mathew, J. G. H., and Walker, A. C. (1986b). *IEEE J. Quantum Electron.* **QE-22**, 369.
Wertheim, C. K. (1956). *Phys. Rev.* **104**, 662.
Wherrett, B. S. (1983). *Proc. R. Soc. London A* **390**, 373.
Wherrett, B. S. (1984). *Philos. Trans. R. Soc. London A* **313**, 213.
Wherrett, B. S., and Higgins, N. A. (1982). *Proc. R. Soc. London A* **379**, 67.
Wherrett, B. S., and Smith, S. D. (1987). *Physica Scripta*, **T13**, 189.
Wherrett, B. S., Hutchings, D., and Russell, D. (1986). *J. Opt. Soc. Am. B* **3**, 551.
Wherrett, B. S., Tooley, F. A. P., and Smith, S. D. (1984). *Opt. Commun.* **52**, 301.
Wiegmann, W. (1986). *Appl. Phys.* **49**, 1505.
Woerdman, J. P., and Bolger, B. (1969). *Phys. Lett. A* **30**, 164.
Wolff, P. A., and Pearson, G. A. (1966). *Phys. Rev. Lett.* **17**, 1015.
Wynne, J. J. (1969). *Phys. Rev.* **138**, 1296.

11 OPTICAL INSTABILITIES IN SEMICONDUCTORS: THEORY

S. W. Koch

OPTICAL SCIENCES CENTER AND PHYSICS DEPARTMENT
UNIVERSITY OF ARIZONA
TUCSON, ARIZONA

I. INTRODUCTION

Most semiconductor materials show strongly enhanced optical nonlinearities in the spectral vicinity of their fundamental absorption edge (Klingshirn and Haug, 1981; Gibbs, 1985). Microscopically, these effects are caused by interaction processes in the many-body system of the laser-excited electron–hole pairs or by thermal effects. A consistent theoretical description of the optical semiconductor properties therefore requires the quantum-mechanical computation of the response function of the material, i.e., of the susceptibility χ or, equivalently, of the dielectric function ϵ using many-body

ISBN 0-12-332915-9

theory (Haug and Schmitt-Rink, 1984). In general, the resulting optical susceptibility depends on the distribution function of the excited electron–hole pairs. Under quasi-equilibrium conditions, the carriers assume a Fermi-distribution within their bands, which is characterized by a quasi-chemical potential and by an effective temperature. The chemical potential then is determined by the total number of particles, leading to a density dependence of the optical susceptibility and therefore to density-dependent absorption and refractive index spectra.

The microscopic expression for the susceptibility is the material equation that describes the semiconductor nonlinearities. The process of carrier generation through light absorption couples the electron–hole pair density to the electromagnetic field. The electromagnetic field in turn is described by the macroscopic Maxwell equations, in which the polarization field depends on the value of the electron–hole pair density through the equation for the susceptibility. This set of equations (i.e., the microscopic equation for the susceptibility together with the macroscopic equations for the carrier density and for the light field) constitutes the combined microscopic and macroscopic approach to consistently describe the semiconductor nonlinearities.

The present chapter deals almost exclusively with the macroscopic theory. For the microscopic calculation of the nonlinearities, we refer to Haug's Chapter 3 of this book. The discussion of semiconductor instabilities is subdivided into four parts. In Section II the coupled equations for the light field and for the electron–hole pair density are derived without specifying the boundary conditions. The effect of increasing absorption optical bistability is discussed in Part III. Here only a single pass of the light through the medium takes place, allowing the formation of spatial excitation kink structures in the light-propagation direction. Dispersive and decreasing absorption optical bistability in semiconductor etalons is analyzed in Section IV, and new results are presented for the example of electron–hole plasma nonlinearities in room-temperature GaAs IV.A. In Section IV.B, we briefly discuss the optical bistability through biexciton formation in CuCl. The recent results on temporal instabilities of an induced absorber in a ring resonator are summarized in Section V.

II. THE COUPLED TRANSPORT EQUATIONS

The starting point for a macroscopic description of the optical semiconductor instabilities is Maxwell's equation for wave propagation in a dielectric

$$\frac{\partial^2 E_0}{\partial z^2} - \frac{1}{c^2}\frac{\partial^2 E_0}{\partial t^2} = \frac{4\pi}{c^2}\frac{\partial^2 P_0}{\partial t^2}, \tag{1}$$

where propagation in z-direction has been assumed and where P_0 is the polarization field of the medium. For the analysis of most experimental situations, one is only interested in the evolution of the slowly varying amplitude E of the field

$$E_0 = Ee^{-i(\omega t - kz)} + \text{c.c.} \tag{2}$$

Inserting Eq. 2 and the equivalent expression for P_0 into Eq. 1 and dropping all higher-order terms (slowly varying amplitude approximation), one obtains

$$\left(\frac{c}{n_0} \frac{\partial}{\partial z} + \frac{\partial}{\partial t} + \frac{c\alpha_0}{2n_0} \right) E = i \frac{2\pi\omega}{n_0^2} P_{\text{nl}}. \tag{3}$$

Here n_0 is the linear refractive index, α_0 is the linear absorption coefficient, and P_{nl} is the nonlinear part of the polarization, which can be written as

$$P_{\text{nl}} = \chi_{\text{nl}}(N) E \tag{4}$$

where

$$\begin{aligned} \chi_{\text{nl}}(N) &= \chi(N) - \chi(N = 0) \\ &= \frac{\epsilon_{\text{nl}}}{4\pi} \\ &= \chi'_{\text{nl}} + i\chi''_{\text{nl}} \end{aligned} \tag{5}$$

is the nonlinear part of the optical semiconductor susceptibility. As mentioned in the introduction to this chapter, the susceptibility depends on the density N of excited electron–hole pairs, on the effective plasma temperature, etc. Here and in the following, we restrict our discussion to electronic optical nonlinearities. In the case of additional thermal effects, one simply has to replace N by N, T as the argument of χ and one has to deal with the dynamic temperature equation additionally to the density rate equation to be given shortly. Note, that in any case, the imaginary part χ''_{nl} of the susceptibility is directly related to the change of the absorption coefficient,

$$\begin{aligned} \Delta\alpha &= \alpha_0 - \alpha(N) \\ &= \alpha(N = 0) - \alpha(N) \\ &= \frac{4\pi\omega\chi''_{\text{nl}}}{cn_0}. \end{aligned} \tag{6}$$

The real part of the susceptibility determines the changes of the refractive index

$$\Delta n \cong \frac{2\pi\chi'_{\text{nl}}}{n_0}. \tag{7}$$

These quantities are available from the microscopic calculations.

Through $\chi_{nl}(N)$, the field amplitude E is nonlinearly coupled to the electron–hole pair density N. This density in turn couples to the light intensity

$$I = \frac{|E|^2 c n_0}{8\pi} \tag{8}$$

through a rate equation like

$$\frac{\partial N}{\partial t} = -\frac{N}{\tau} + \frac{\alpha(\omega, N)}{\hbar\omega} I + \nabla D \nabla N, \tag{9}$$

where τ is the carrier relaxation time and D is the electron–hole pair diffusion coefficient.

Eqs. 3, 4 and 9, together with an expression for $\chi(N)$, determine the nonlinear optical response of semiconductors. Note that this approach corresponds to an adiabatic elimination of the polarization dynamics, an approximation that is well justified for the present purposes because of the extremely fast phase destroying processes in semiconductors, which typically occur on a subpicosecond time scale.

III. INTRINSIC OPTICAL BISTABILITY WITH INCREASING ABSORPTION

Optical bistability is the effect that one may have two (meta-) stable values of the light intensity transmitted through a nonlinear material for one value of the input intensity. The actually realized transmission depends on the excitation history, i.e., one reaches a different state if one either decreases the incident intensity I_0 from a sufficiently high original level or if one increases I_0 from zero. This scenario resembles in many respects that of a first-order phase transition, in which metastability and hysteresis also occur (see, e.g., Graham, 1973; Haken, 1977). In fact, optical bistability has been discussed as an example of a first-order nonequilibrium phase transition in a system far from thermodynamic equilibrium (see, e.g., Graham and Schenzle, 1981; Koch, 1984; Lugiato, 1985).

From a conceptual point of view, the simplest example of optical bistability is obtained if one considers a medium whose absorption increases with increasing excitation density. Bistability in such a system may occur without any additional feedback mechanism. There are numerous mechanisms that may cause induced absorption in semiconductors and other systems (for a review see, e.g., Gibbs, 1985). In the present chapter, we shall concentrate on the induced absorption observed in semiconductors such as CdS at low temperatures as a consequence of the bandgap reduction. However, most of the macroscopic features to be discussed are quite general and may very well also occur in other systems.

Taking the microscopic mechanism of bandgap reduction and assuming that the semiconductor is excited at a frequency below the exciton, i.e., in the absorption tail, one has only weak absorption for low intensities. Nevertheless, if the exciting laser is sufficiently strong, even this weak absorption generates a density of electron–hole pairs causing a reduction of the semiconductor bandgap. Eventually, the band edge shifts below the frequency of the exciting laser giving rise to a substantially increased one-photon absorption coefficient. Hence, one has an absorption which increases with increasing carrier density (Schmidt et al., 1984; Koch, 1984; Koch et al., 1985). An example for this induced absorption is reproduced in Fig. 1.

The coupling between carrier density N and light intensity I is described by Eq. 9, which in the stationary, spatially homogeneous case leads to

$$\alpha(N) = \frac{N}{I}\frac{\hbar\omega}{\tau}. \tag{10}$$

This relation can be bistable, as indicated by the straight line in Fig. 1, which is the rhs of Eq. 10. The slope of this straight line is inversely proportional to I and the intersection points with the curve $\alpha(N)$ are the

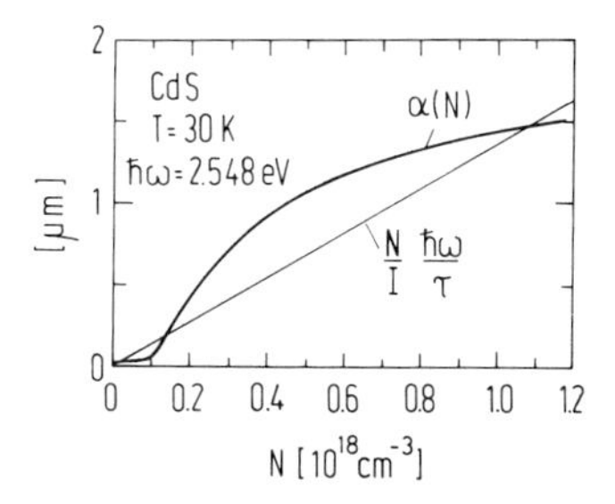

Figure 1. Density dependence of the absorption $\alpha(N)$ computed for CdS at a plasma temperature $T = 30$ K for an excitation energy $\hbar\omega = 2.548$ eV. The straight line is the rhs of Eq. 10. The intersections of this line with $\alpha(N)$ are the graphic solutions of Eq. 10. For different values of I, one obtains one, two or three intersections, indicating monostability, the boundaries of the bistable regime or optical bistability, respectively. In the case of three intersections, the intermediate solution can be shown to be unstable (from Haug et al., 1986).

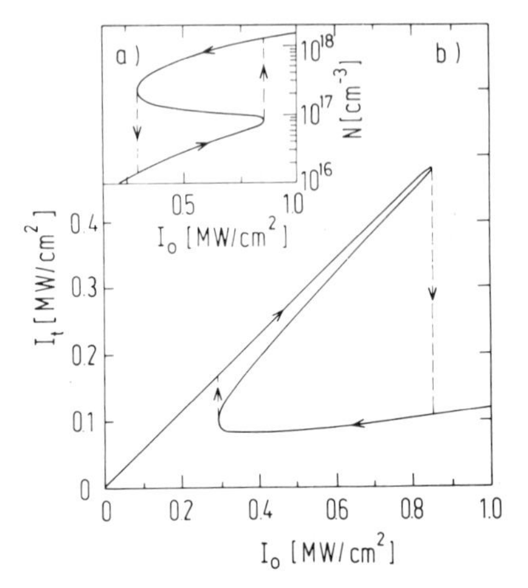

Figure 2. Stationary solutions of Eqs. 9 and 12 for the electron–hole density N (a) and for the transmitted intensity I_t (b) versus the incident intensity using the absorption coefficient shown in Fig. 1 (from Schmidt et al., 1984).

graphical solutions of Eq. 10, which clearly show the occurrence of intrinsic optical bistability without resonator feedback (see Fig. 2a).

It is worthwhile to stress at this point that for such an induced absorption bistability, only a single pass of the light beam through the medium is required. Hence, one has no superposition of forward and backward travelling waves as, e.g., in a Fabry–Perot resonator (see Section IV for details) and one may therefore neglect nonlinear dispersive effects if one is only interested in the characteristic variation of the transmitted intensity and not in transverse beam-profile variations or in diffraction effects. In this case, one can rewrite Eq. 3 using Eqs. 6 and 8 as

$$\left[\frac{\partial}{\partial t} + \frac{c}{n_0}\frac{\partial}{\partial z}\right] I = -\frac{c}{n_0}\alpha(N) I \tag{11}$$

where $z = 0,\ L$ are the sample front and end faces, respectively. If N is constant, the steady-state solution of Eq. 11 is Beer's law

$$I(z) = I(z = 0)e^{-\alpha(N)z}$$

showing that the transmitted intensity is high if α is low and vice versa. The resulting intensity loop (Schmidt et al., 1984) is plotted in Fig. 2b.

The described situation is experimentally relevant if $D\tau/L^2 \gg 1$, i.e., in the diffusion-dominated case or for very thin semiconductor samples. Induced absorption bistability in CdS for these conditions has indeed been observed experimentally (Rossmann et al., 1983; Bohnert et al., 1983), and the experimental results are properly described by the outlined theory (Schmidt et al., 1984).

The analysis of spatial inhomogeneities, which become important in the weak-diffusion limit, $D\tau/L^2 \ll 1$, requires a solution of the coupled transport Eqs. 9 and 11. These equations may be simplified somewhat considering that the temporal response of semiconductor nonlinearities, in which real excitation of carriers is involved, is essentially determined by the relaxation time τ of the electron–hole pairs. The light field I follows quasi-instantaneously all temporal changes of N. Therefore, one can set $\partial I/\partial t \equiv 0$ to obtain

$$\frac{\partial I(z,t)}{\partial z} = -\alpha[N(z,t)]\,I(z,t). \tag{12}$$

Eqs. 9 and 12 have been solved numerically by Koch et al. (1984) for a temporally varying incident intensity. The incident intensity as well as the resulting transmitted intensity are shown in Fig. 3a for vanishing electron–hole pair diffusion, $D = 0$. We see that the transmitted intensity increases proportionally to the incident intensity in the short-time regime after the onset of the excitation. The local electron–hole density (Fig. 3b) increases continuously, until immediately behind the front face of the crystal a value is reached which corresponds to the end of the lower branch of the bistable hysteresis loop (Fig. 2a). The still-increasing incident light intensity causes a jump of the local electron–hole density in the region behind the front face of the crystal to the upper bistable branch. The spatial distribution of $N(z,t)$ assumes the shape of a kink (inset to Fig. 3b). Now, one has a high density and, thus, a high absorption in a part of the crystal. The total absorption has increased in comparison to a situation without a density kink. This gives rise to a drop of the transmitted intensity causing the first peak in the temporal dependence of the transmitted light pulse (Fig. 3a). Since the incident intensity is still increasing, the electron–hole density rises too, however, in a spatially quite inhomogeneous fashion, because the local values of $N(z,t)$ follow the two different branches of the density hysteresis loop. Immediately behind the kink, the density increases according to the lower branch, until it reaches the instability point of that branch and it jumps to the upper branch. Hereby, another fraction of the medium is switched to the state of high absorption. By this mechanism, the density kink moves stepwise deeper into the sample, causing stepwise increasing total absorption, which in turn leads to the successive drops of the transmitted intensity. This partial sample switching explains the structures of $I_t(t)$ shown in Fig. 3a. The transmitted intensity follows again the

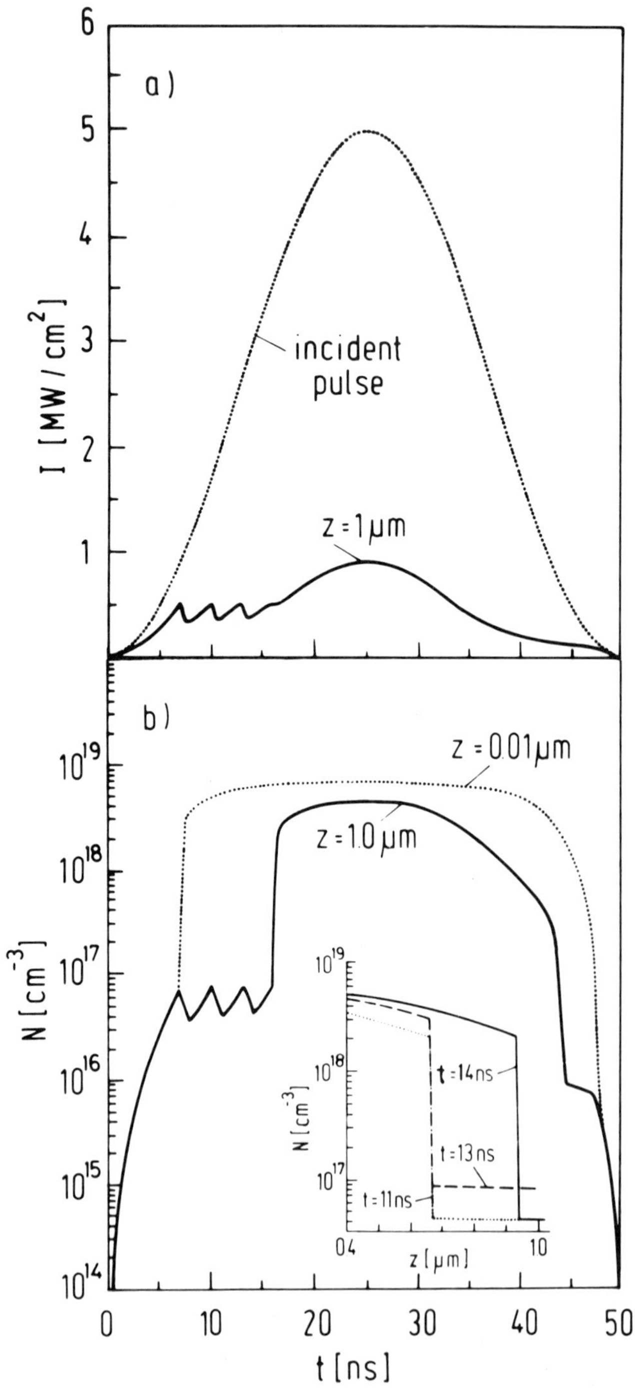

Figure 3. Numerical results for a CdS crystal platelet of 1 μm thickness assuming the absorption shown in Fig. 1. As functions of time are plotted a) the incident pulse and the pulse that is transmitted through the crystal, b) the local electron–hole density immediately behind the front face of the crystal ($z = 0.01$ μm) and at $z = 1$ μm. The inset shows the spatial density distribution for different times (from Koch et al., 1985).

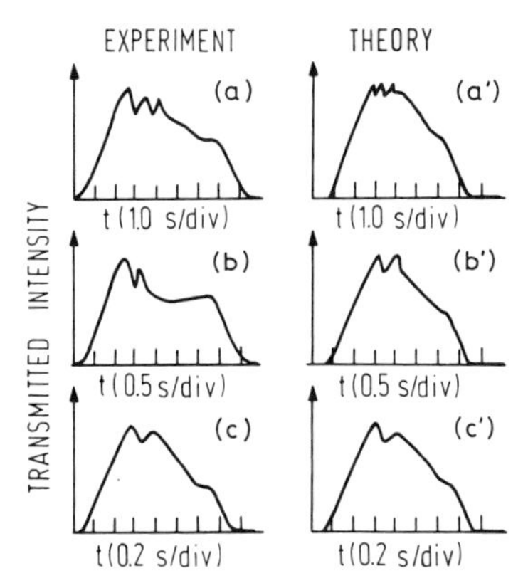

Figure 4. The light intensity transmitted through a semiconductor-microcrystallite-doped glass is plotted versus time for a time-dependent excitation in the form of a triangular pulse. (a) shows experimental results and (b) shows the theoretical predictions. The sawtooth structure is a consequence of the stepwise kink movement through the medium (see text) (from Gibbs et al., 1985).

incident intensity, after the kink has completely moved through the crystal, which is now totally in the state of high absorption.

The theoretically predicted sawtooth structures in the transmitted light intensity have been observed in semiconductor-microcrystallite-doped glasses (Gibbs et al., 1985). The results are reproduced in Fig. 4, where experimental findings are compared to the theoretical predictions for the condition of a thermal bandgap reduction.

A detailed, combined analytical and numerical investigation of the structure, formation and motion of the density kinks in increasing absorption optical bistability has been reported by Lindberg et al. (1986a). These authors stress that the kink formation and the almost discontinuous propagation cannot be discussed by steady-state properties. The sawtooth structures in the transmitted intensity are genuinely dynamic features. For a simple ramp model of the absorption coefficient and for a linearly increasing incident intensity

$$I_0(t) = t\bar{I},$$

where $\bar{I}$ is a constant, one finds analytic estimates for the kink position z_c and for the switching intervals, e.g.,

$$z_c \cong \frac{a}{t_0}.$$

In this relation, a is the proportionality factor, and t_0 is the time at which

I_0 reaches the critical value for switch-up of the density to the upper bistable branch. These estimates have been confirmed by numerical studies. It has also been found (Lindberg et al., 1986), that the position z_c of the kink is stabilized long before the complete density profile has been established.

It is worthwhile to mention at the end of this chapter that one may regard the discussed density kinks as a realization of spatial two-phase coexistence of regions with high and low electron–hole pair densities. Such phase-coexistence phenomena are quite characteristic for first-order phase transitions in equilibrium as well as in nonequilibrium systems (see, e.g., Koch, 1984; Haug and Koch, 1985).

IV. OPTICAL BISTABILITY IN SEMICONDUCTOR ETALONS

Dispersive optical bistability or bistability through decreasing absorption may be obtained if a semiconductor is brought into an optical resonator, which introduces additional feedback for the light field. In the following, we consider a Fabry–Perot resonator consisting of two lossless mirrors of reflectivity R and transmissivity $T = 1 - R$. The nonlinear semiconductor material fills the space between the mirrors. In many practical applications, the mirrors are actually the endfaces of the semiconductor crystal itself, and R is just the natural reflectivity, or R is increased through additional high reflectivity coatings evaporated onto these surfaces.

To treat the feedback introduced by the mirrors, it is useful to decompose the complex field amplitude E into the forward and backward propagating parts (see, e.g., Bonifacio and Lugiato, 1976 and 1978; Marburger and Felber, 1978; Agrawal and Charmichael, 1979; Miller, 1981; and Gibbs, 1985, also for further references)

$$\begin{aligned} E &= E_{\mathrm{F}} + E_{\mathrm{B}} \\ &= \xi_{\mathrm{F}} e^{-i\phi_{\mathrm{F}}} + \xi_{\mathrm{B}} e^{i\phi_{\mathrm{B}}}, \end{aligned} \tag{13}$$

where the amplitudes ξ and the phases ϕ are real quantities. The boundary conditions of the Fabry–Perot resonator can be written as

$$\begin{aligned} E_{\mathrm{F}}(z=0) &= \sqrt{T}\,E_0 + \sqrt{R}\,E_{\mathrm{B}}(z=0) \\ E_{\mathrm{B}}(z=0) &= \sqrt{R}\,E_{\mathrm{F}}(z=L)\,e^{i\beta/2-\alpha_{\mathrm{tot}}/2} \\ E_{\mathrm{F}}(z=L) &= \frac{E_t}{\sqrt{T}} = E_{\mathrm{F}}(z=0)\,e^{i\beta/2-\alpha_{\mathrm{tot}}/2}, \end{aligned}$$

yielding

$$\left|\frac{E_t}{E_0}\right|^2 = \frac{T^2}{(e^{\alpha_{\rm tot}/2} - Re^{-\alpha_{\rm tot}/2})^2 + 4R\sin^2(\beta/2)}. \tag{14}$$

The optical thickness $\alpha_{\rm tot}$ as well as the phase shift β have to be computed using Eqs. 3, 4, 7 and 13. The total phase shift of the light after passing through the resonator can be written as

$$\beta = \phi_{\rm F}(z = L) - \phi_{\rm B}(z = L) - 2\delta, \tag{15}$$

where 2δ contains all carrier-density-independent phase shifts of the linear medium and of the mirrors. Inserting Eq. 13 into Eq. 3 and using Eqs. 6 and 7, one obtains

$$\left[\frac{\partial}{\partial z} + \left(\frac{n_0}{c}\frac{\partial}{\partial t}\right) + \frac{\alpha(\omega, N)}{2}\right]\xi_{\rm F/B} = 0 \tag{16a}$$

and

$$\left[\frac{\partial}{\partial z} + \frac{n_0}{c}\frac{\partial}{\partial t}\right]\phi_{\rm F/B} = {}^{-}/_{+}\frac{\omega\Delta n(\omega, N)}{c}, \tag{16b}$$

where the minus sign is for $\phi_{\rm F}$.

Most semiconductor bistability experiments are done for resonator lengths $L \cong 0.5$–$10\ \mu$m. Under these conditions, the round-trip time for the light in the resonator is substantially shorter than the carrier relaxation time τ and one is justified to adiabatically eliminate the dynamics of the light field (bad cavity limit),

$$\begin{aligned}\frac{\partial \xi}{\partial t} &\cong 0,\\ \frac{\partial \phi}{\partial t} &\cong 0.\end{aligned} \tag{17}$$

Solving Eq. 16 for these conditions yields

$$\xi_{\rm F/B}(z) = \xi_{\rm F/B}(0)\exp\left\{-\frac{1}{2}\int_0^z dz'\,\alpha[N(z')]\right\} \tag{18a}$$

and

$$\phi_{\rm F/B}(z) = {}^{-}/_{+}\frac{\omega}{c}\int_0^z dz'\,\Delta n[N(z')], \tag{18b}$$

showing, that $\alpha_{\rm tot}$ and β in the resonator Eq. 14 are given by

$$\alpha_{\rm tot} = \int_0^L dz\,\alpha[N(z)] \tag{19}$$

and

$$\frac{\beta}{2} = -\left\{\delta + \frac{\omega}{c}\int_0^L dz\, \Delta n[N(z)]\right\}. \tag{20}$$

The spatial carrier distribution $N(z)$ has to be computed from Eq. 9.

For practical applications to semiconductor etalons, it is often a good approximation to neglect the spatial density variations (diffusion dominated case). Then one can trivially evaluate the integrals in Eqs. 19 and 20, and the resonator formula 14 becomes

$$I_t = I_0 T^2 \frac{1}{[e^{\alpha(\omega, N)L/2} - Re^{-\alpha(\omega, N)L/2}]^2 + 4R\sin^2[\delta + \omega\,\Delta n(\omega, N)L/c]}. \tag{21}$$

This equation is coupled to the spatially averaged rate Eq. 9

$$\frac{dN}{dt} = -\frac{N}{\tau} + \frac{\alpha(\omega, N)}{\hbar\omega} I, \tag{22}$$

where I has to be taken as the intensity inside the resonator

$$I = I_t \frac{1 + R}{T}. \tag{23}$$

Note, that the incident intensity I_0 in Eq. 21 may be slowly time-dependent on the time scale of the resonator round-trip time.

A. Electron–Hole Plasma Bistability

To explicitly solve Eqs. 21–23 for an example of practical interest, we use the band-edge nonlinearities of room-temperature GaAs computed in the framework of the plasma theory of Banyai and Koch (1986). Evaluating the generalized Elliott formula for the parameters of room-temperature GaAs (Lee et al., 1986), one obtains the absorption spectra shown in Fig. 5c. The corresponding dispersive changes Δn are plotted in Fig. 5d. Figs. 5a and b show the experimental spectra for room-temperature GaAs, demonstrating quantitative agreement between theory and experiment. Inserting the computed $\alpha(N)$ and $\Delta n(N)$ into Eqs. 21 through 23, one can directly simulate nonlinear optical device performance. We assume a time-dependent triangular incident laser pulse $I_0(t)$ and obtain bistable hysteresis loops by plotting transmitted intensity versus input intensity. For a resonator filled with GaAs at room temperature, we find the results shown in Fig. 6a–c for different temporal pulse widths t_0. The curves 1–4 in each figure are obtained for slightly different resonator lengths L, i.e., for different detun-

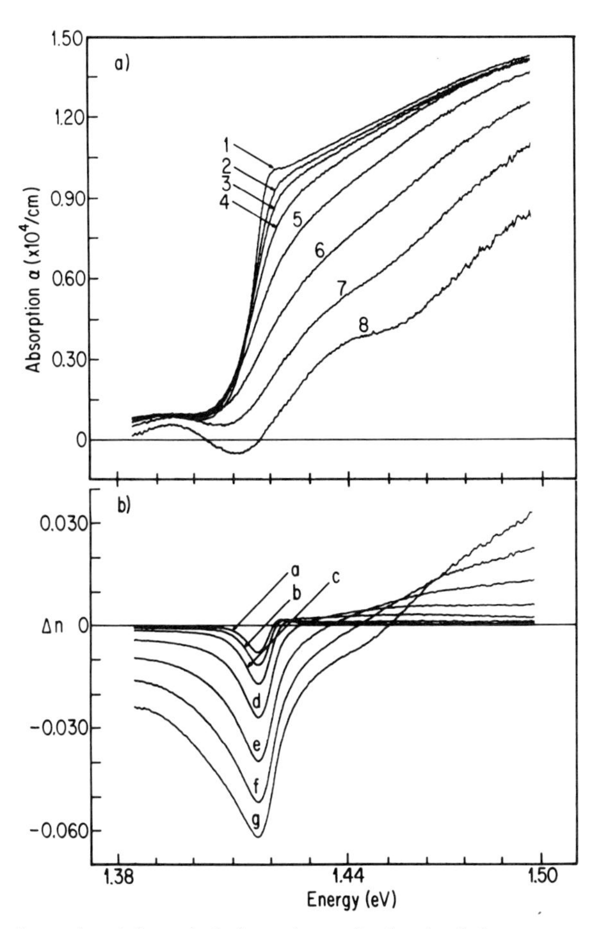

Figure 5. Experimental and theoretical absorption and refractive index spectra are compared for room temperature GaAs. (a) The experimental absorption spectra are obtained for different excitation intensities I (mW): (1) 0; (2) 0.2; (3) 0.5; (4) 1.3; (5) 3.2; (6) 8; (7) 20; (8) 50 using quasi-cw excitation directly into the band and a 15 μm excitation spot size. The oscillations in curve (8) are a consequence of imperfect antireflection coating. (b) The dispersive changes Δn are obtained through a Kramers–Kronig transformation of the absorptive changes $\alpha(I = 0) - \alpha(I)$. The agreement with direct measurements of dispersive changes has been tested for the same conditions using a 299-Angstrom multiple-quantum-well sample. (c) Calculated absorption spectra for different electron–hole pair densities N (cm^{-3}): (1) 10^{15} (linear spectrum); (2) 8 10^{16}; (3) 2 10^{17}; (4) 5 10^{17}; (5) 8 10^{17}; (6) 10^{18}; (7) 1.5 10^{18}. The unrenormalized gap is $E_g^0 = 1.420$ eV and the exciton Rydberg is $E_R = 4.2$ meV. (d) Kramers–Kronig transformation of the calculated absorption spectra (from Lee et al., 1986).

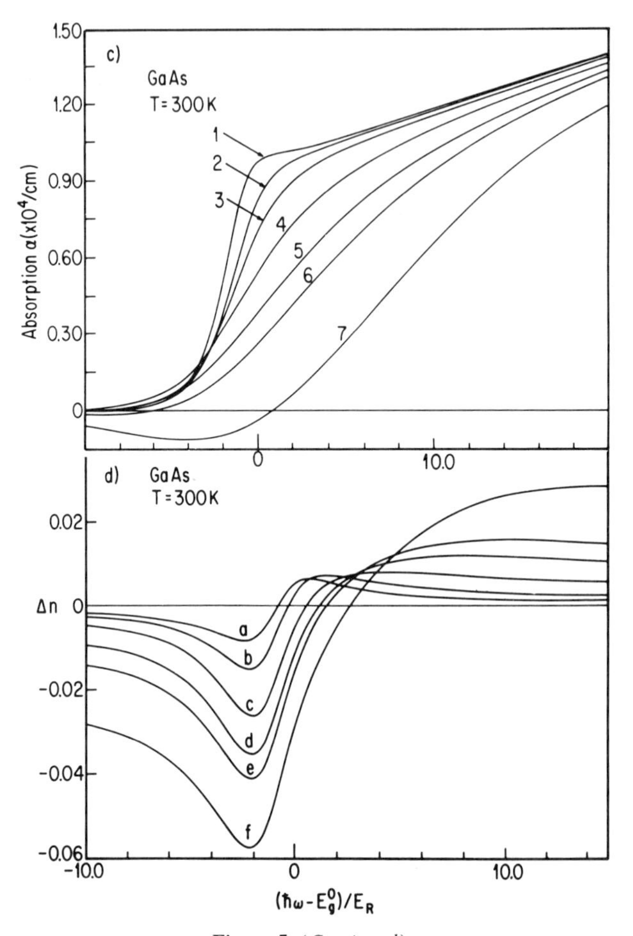

Figure 5. (*Continued*)

ings of the excitation frequency ω with respect to the nearest resonator eigenfrequency ω_R, $\omega_R < \omega$. In Figs. 6a through c, curves 2 and 3 show well-developed bistable loops similar to those observed in experiments. These loops get wider for shorter pulses. This is the well-known effect of dynamic hysteresis. This effect can even produce seemingly bistable behavior, as in curves 1 and 4, which vanishes for longer pulses. A comparison with the corresponding steady-state result shows that Fig. 6c essentially represents the truly bistable behavior. For more details, see Warren et al. (1987).

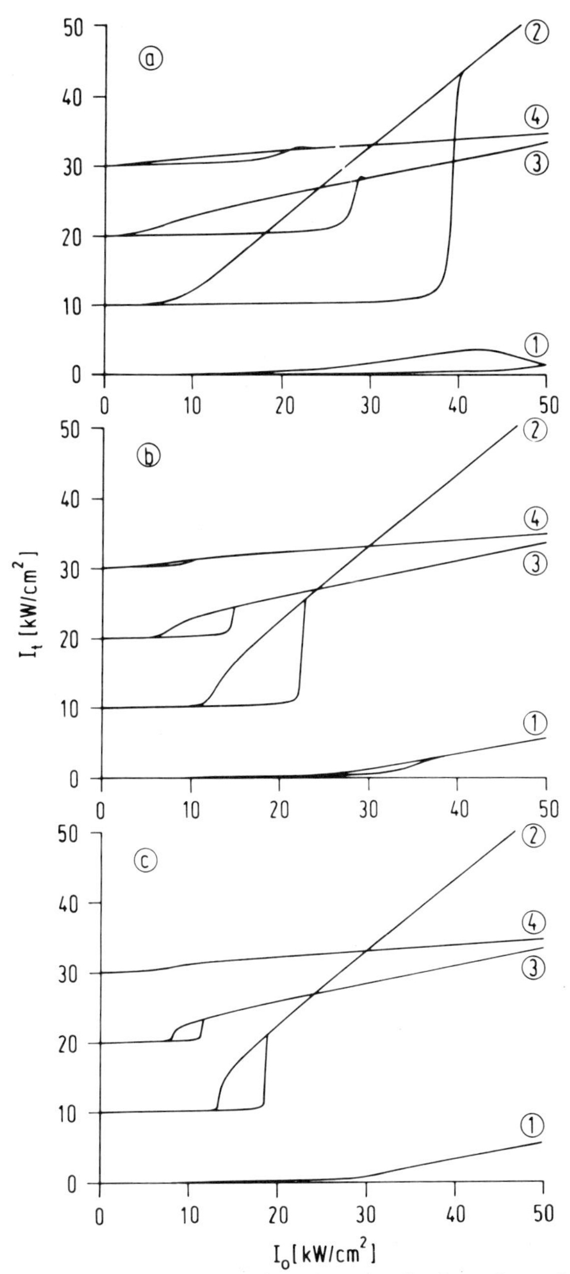

Figure 6. Transmitted intensity versus input intensity using the absorption and refractive index spectra shown in Fig. 5. Eqs. 21–23 have been solved for a triangular incident laser pulse $I_0(t)$ of full width $t_0 = 0.1\ \mu s$ (a), $1\ \mu s$ (b) and $10\ \mu s$ (c). The excitation energy $\hbar\omega = 1.4032$ eV. The resonator length $L = 2.046\ \mu m$ (1), $2.042\ \mu m$ (2), $2.038\ \mu m$ (3), and $2.034\ \mu m$ (4) which give rise to the detunings of -0.0170 eV (1), -0.0142 eV (2), -0.0115 eV (3), and -0.0080 eV (4). The mirror reflectivity $R = 0.9$ and the carrier relaxation time $\tau = 10$ ns. The baseline for the transmitted intensity in curves 2, 3 and 4 has been shifted by 10, 20 and 30 kW/cm², respectively (from Waren et al., 1987).

B. Biexcitonic Bistability

Originally, we applied the outlined combined microscopic and macroscopic approach (Koch and Haug, 1981; Haug et al., 1981) to study optical bistability due to the biexcitonic nonlinearity in wide-gap semiconductors (see also Hanamura, 1981). This nonlinearity is caused by the (virtual) generation of excitonic molecules through absorption of two photons well below the band edge of, e.g., CuCl. Details of this process are discussed in Chapters 8 and 9 of this book as well as in the recent review article by Hönerlage et al. (1985), and they will not be repeated here.

An approximate analytic expression for the dielectric function $\epsilon(\omega)$ for the molecule formation has been derived by März et al. (1980). This result has been used by Koch and Haug (1981) to evaluate the resonator formula (21) for this nonlinearity and to predict optical bistability in CuCl. Even though more recent calculations showed that the dielectric function of März et al. (1980) is incomplete (see, e.g., Haug and Schmitt-Rink, 1984), the prediction of biexcitonic bistability nevertheless has stimulated considerable theoretical and experimental interest, as is documented, e.g., in the review article by Hönerlage et al. (1985). The most spectacular result of these investigations has been the experimental observation of the predicted bistability almost simultaneously by two independent groups (Levy et al., 1983 and Peyghambarian et al., 1983).

The effects of random fluctuations on the two-photon bistability in CuCl have been analyzed by Haug et al. (1982) and by Schmidt et al. (1983). The results have been summarized by Koch (1984).

V. INSTABILITIES OF AN INDUCED ABSORBER IN A RING CAVITY

The discussion in section III shows that a semiconductor whose absorption increases with increasing carrier density (induced absorber) shows intrinsic optical bistability without a resonator geometry. Additional instabilities can be observed in such a system if additional feedback is provided (Lindberg et al., 1986b). Because absorption decreases the light in the propagation direction, the analysis is most transparent for the case of an unidirectional ring resonator as feedback device, and assuming the diffusion-dominated situation, $D\tau/L^2 \gg 1$, for the medium.

The ring resonator is schematically plotted in Fig. 7. The boundary conditions lead to the following equation for the light field at the front face

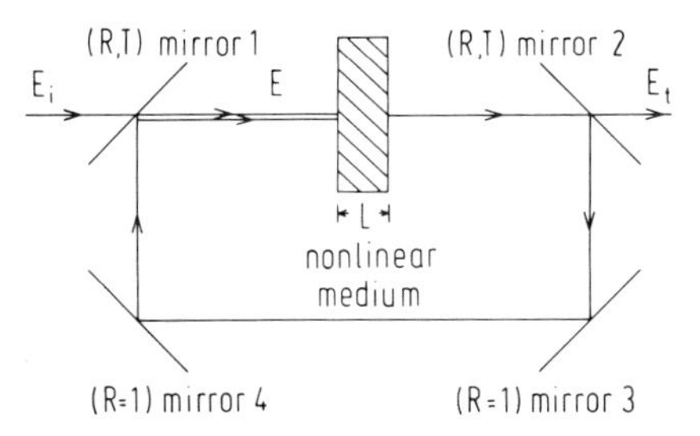

Figure 7. Schematic drawing of a ring resonator.

of the sample (see, e.g., Gibbs, 1985).

$$E(t + \tau_R) = \sqrt{T}\,E_i + RE(t)\exp\left[-i\left(\delta - \Delta n(N)\frac{\omega L}{c}\right) - \frac{\alpha(N)L}{2}\right], \tag{24}$$

where δ is the linear round-trip phase shift.

The stationary solutions of Eqs. 22 and 24 allow a variety of topologically different output-input characteristics for an induced absorber in a resonator (see Fig. 8). From the dynamic equations, one also finds higher, oscillatory instabilities in which a temporally constant (cw) incident field E_i is converted into a time-dependent output field. As long as E_i is too low to switch the medium in the resonator to high absorption, the round-trip feedback is high, and the field in the cavity as well as the transmitted field increase with each round-trip. This increase continues until the internal intensity exceeds the switch-up value causing a strong rise of the absorption and, hence, a cutoff of the feedback. From then on, the field decreases until it reaches the switch-down value at which the medium returns to low absorption and the cycle begins to repeat itself. A numerical solution of Eqs. 22 and 24 has been reported by Lindberg et al. (1986b), and an example of an oscillatory output for cw input is reproduced in Fig. 9 for different ratios of the round-trip time τ_R and the medium response time τ.

The dynamic behavior of the induced absorber in a ring cavity for the limit $\tau \ll \tau_R$ has been studied by Lindberg et al. (1986b) and by Haug et al. (1986). In this case, the density dynamics can be eliminated adiabatically and the system can be described by a discrete map. For a certain regime of input-field values, this map has no fixed points, i.e., no steady-state solutions exist and one always obtains a temporally oscillating output field. The pattern of these oscillatory solutions shows strong locking behavior. Denoting by p the number of maxima that occur in q round-trips, after which the

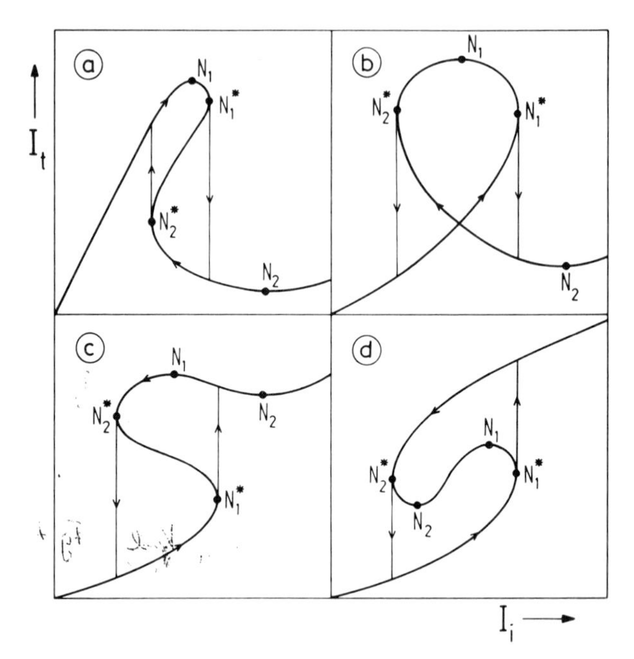

Figure 8. Topologically different output-input characteristics for a semiconductor with induced absorption in a ring resonator (schematically). Here, N_1 and N_2 are the points at which the derivative of the transmitted intensity $I_t = Te^{-\alpha(N)}N/\alpha(N)$ with respect to N vanishes and N_1^* and N_2^* are the points at which the derivative of $I_i = \frac{N}{\alpha(N)}\left[(1 - Re^{-\alpha/2})^2 + 4Re^{-\alpha/2}\sin^2\left(\frac{\theta - \Delta n}{2}\right)\right]$ with respect to N vanishes. The various curves of transmitted versus incident intensity are obtained for different sequences of N_1, N_2, N_1^* and N_2^*, respectively (from Lindberg et al., 1986b).

pattern repeats itself, one defines the effective frequency of the oscillations as the ratio p/q. Each class of solutions with a given effective frequency is stable for a certain range of input field values, where, however, the extension of the respective stability regime decreases with increasing p. In a large range of parameter values, it has been shown (Haug et al., 1986) that a complete Farey-tree structure exists for the locked regions, i.e., between two modes p/q and p'/q', with $pq' - p'q = \pm 1$, one always has the mode $(p + p')/(q + q')$. In another regime of parameter values, mode coexistence has also been found (Lindberg et al., 1986b).

These theoretical predictions have been verified recently by Klingshirn et al. (1986) in a hybrid experiment using a CdS platelet and a delayed electro-optical feedback. The observed self-pulsing shows indeed the predicted frequency locking according to a Farey-tree construction. If the

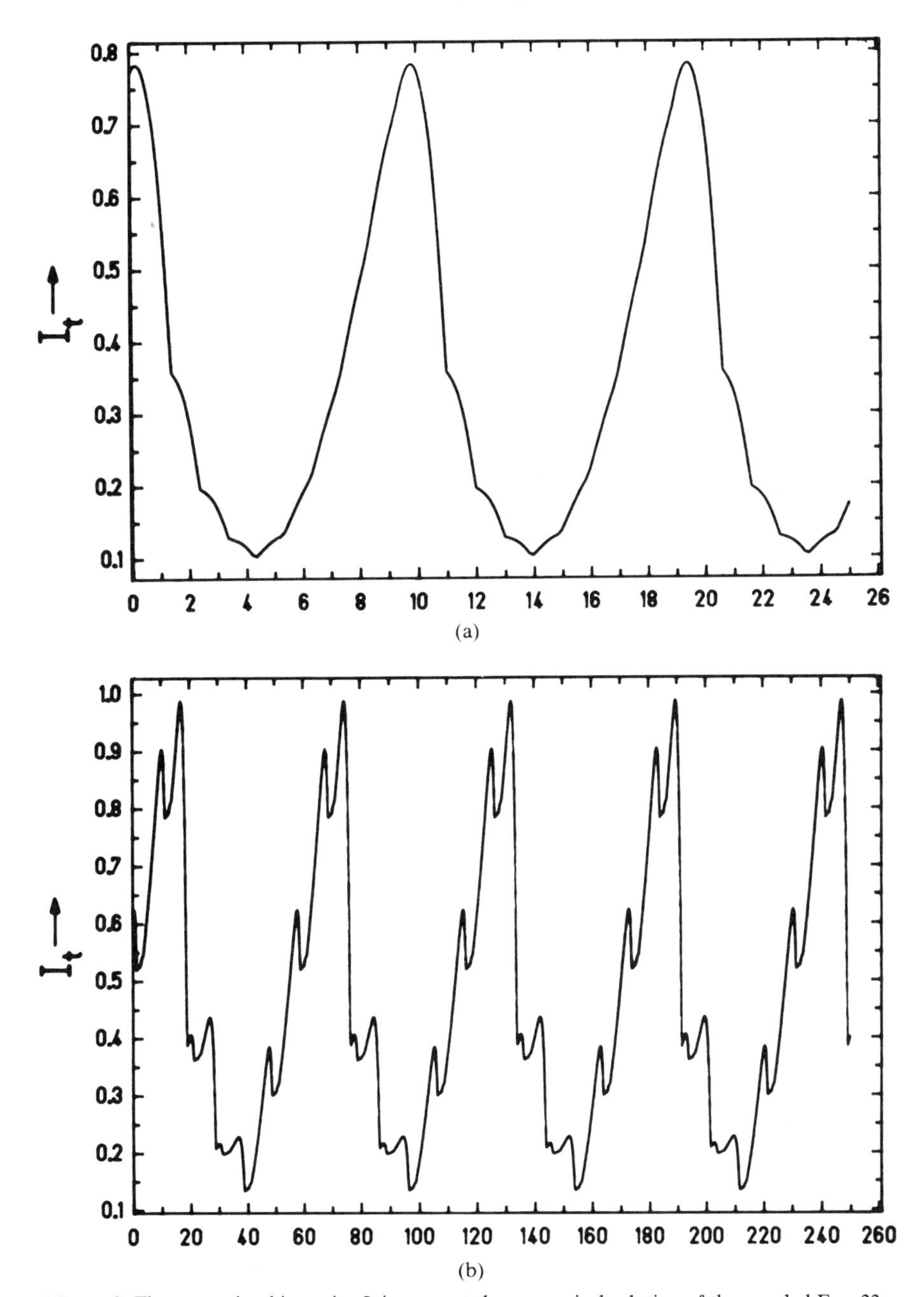

Figure 9. The transmitted intensity I_t is computed as numerical solution of the coupled Eqs. 22 and 24. The resulting oscillations are plotted versus time for a constant value of the incident field E_i in the unstable regime and for different ratios of the round-trip time τ_R/τ (a) 1; (b) 10; (c) 100 (from Lindberg et al., 1986b).

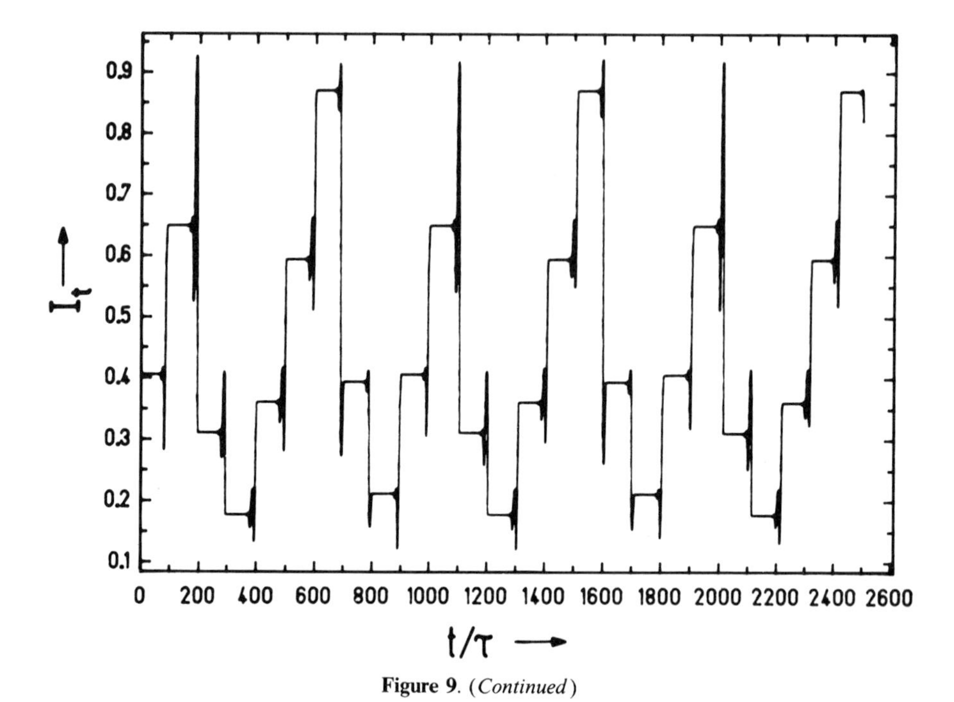

Figure 9. (*Continued*)

induced absorber is in a monostable mode, one can get also chaotic output (Galbraith et al., 1986) as has been seen experimentally by Wegener (1986) using the above-mentioned electro-optical system.

ACKNOWLEDGMENTS

The author thanks L. Banyai, H. M. Gibbs, H. Haug, C. Klingshirn, Y. H. Lee, M. Lindberg, G. R. Olbright, N. Peyghambarian, and M. Warren for stimulating collaboration. He acknowledges financial support from the Optical Circuitry Cooperative at the Optical Sciences Center, University of Arizona, Tucson, Arizona.

REFERENCES

Agrawal, G. P., and Charmichael, H. J. (1979). *Phys. Rev. A* **19**, 2074.
Banyai, L., and Koch, S. W. (1986). *Z. Phys. B* **63**, 283.
Bohnert, K., Kalt, H., and Klingshirn, C. (1983). *Appl. Phys. Lett.* **43**, 1088.
Bonifacio, R., and Lugiato, L. A. (1976). *Opt. Commun.* **19**, 172.
Bonifacio, R., and Lugiato, L. A. (1978). *Phys. Rev. A* **18**, 1129.
Galbraith, I., Lindberg, M., Kranz, H., and Haug, H. (1986). Preprint.

Gibbs, H. M. (1985). *Optical Bistability: Controlling Light with Light*. Academic Press, New York.
Gibbs, H. M., Olbright, G. R., Peyghambarian, N., Schmidt, H. E., Koch, S. W., and Haug, H. (1985). *Phys. Rev. A* **32**, 692.
Graham, R. (1973). In *Springer Tracts in Modern Physics*, vol. 66, Springer-Verlag, Berlin.
Graham, R., and Schenzle, A. (1981). *Phys. Rev. A* **23**, 1302.
Haken, H. (1977). *Synergetics: An Introduction*. Springer-Verlag, Berlin.
Hanamura, E. (1981). *Solid State Commun.* **38**, 939.
Haug, H., and Koch, S. W. (1985). *IEEE J. Quantum Electron.* **QE-21**, 1385.
Haug, H., and Schmitt-Rink, S. (1984). *Progr. Quantum Electron.* **9**, 3.
Haug, H., Koch, S. W., März, R., and Schmitt-Rink, R. (1981). *J. Lumin.* **24 / 25**, 621.
Haug, H., Koch, S. W., Neumann, R., and Schmidt, H. E. (1982). *Z. Phys. B* **49**, 79.
Haug, H., Koch, S. W., and Lindberg, M. (1986). *Physica Scripta*, **T13**, 178.
Hönerlage, B., Levy, R., Grun, J. B., Klingshirn, C., and Bohnert, B. (1985). *Phys. Rep.* **124**, 161.
Klingshirn, C., and Haug, H. (1981). *Phys. Rep.* **70**, 315.
Klingshirn, C., and Wegener, M. (1986). Preprint.
Koch, S. W. (1984). *Dynamics of First-Order Phase Transitions in Equilibrium and Nonequilibrium Systems*, Lecture Notes in Physics, vol. 207. Springer-Verlag, Berlin.
Koch, S. W., and Haug, H. (1981). *Phys. Rev. Lett.* **46**, 450.
Koch, S. W., Schmidt, H. E., and Haug, H. (1984). *Appl. Phys. Lett.* **45**, 932.
Koch, S. W., Schmidt, H. E., and Haug, H. (1985). *J. Lumin.* **30**, 232.
Lee, Y. H., Chavez-Pirson, A., Koch, S. W., Gibbs, H. M., Park, S. H., Morhange, J., Jeffrey, A., Peyghambarian, N., Banyai, L., Gossard, A. C., and Wiegmann, W. (1986) *Phys. Rev. Lett.* **57**, 2446.
Lévy, R., Bigot, J. Y., Hönerlage, B., Tomasani, F., and Grun, J. B. (1983). *Solid State Commun.* **48**, 1511.
Lindberg, M., Koch, S. W., and Haug, H. (1986a). *Phys. Rev. A* **33**, 407.
Lindberg, M., Koch, S. W., and Haug, H. (1986b). *J. Opt. Soc. Am. B* **3**, 751.
Lugiato, L. A. (1985). In *Progress in Optics*, vol. XXI (E. Wolf, ed.) North-Holland, Amsterdam.
Marburger, J. H., and Felber, F. S. (1978). *Phys. Rev. A* **17**, 335.
März, R., Schmitt-Rink, S., and Haug, H. (1980). *Z. Phys. B* **40**, 9.
Miller, D. A. B. (1981). *IEEE J. Quantum. Electron.* **QE-17**, 306.
Peyghambarian, N., Gibbs, H. M., Rushford, M. C., and Weinberger, D. A. (1983). *Phys. Rev. Lett.* **51**, 1692.
Rossmann, H., Henneberger, F., and Voigt, I. (1983). *Phys. Status Solidi B* **115**, K63.
Schmidt, H. E., Koch, S. W., and Haug, H. (1983). *Z. Phys. B* **51**, 85.
Schmidt, H. E., Haug, H., and Koch, S. W. (1984). *Appl. Phys. Lett.* **44**, 787.
Warren, M., Koch, S. W., and Gibbs, H. M. (1987), invited paper for special issue of IEEE Computer Society publication "Computer", eds. Bachman, T. E., and Parrish, E. A. (to be published)
Wegener, M. (1986). Private communication to H. Haug.

12 SEMICONDUCTOR OPTICAL NONLINEARITIES AND APPLICATIONS TO OPTICAL DEVICES AND BISTABILITY

N. Peyghambarian and H. M. Gibbs

Optical Sciences Center
University of Arizona
Tucson, Arizona

I. INTRODUCTION

Optical bistable devices and logic gates are of current interest for their potential applications in future optical computers and signal processing

OPTICAL NONLINEARITIES
AND INSTABILITIES IN
SEMICONDUCTORS

ISBN 0-12-332915-9

systems (Gibbs, 1985; Gibbs *et al.*, 1986). These devices take advantage of the nonlinearity of optical materials and combine it with a feedback mechanism for operation. The most common device consists of a nonlinear medium between two partially reflecting mirrors forming a nonlinear etalon. Depending on whether the feedback is provided electronically or optically, the bistability is hybrid (Smith, 1980; Smith and Tomlinson, 1981) or all-optical (Peyghambarian and Gibbs, 1986; Peyghambarian and Gibbs, 1985), respectively. In usual dispersive or absorptive optical bistability, the feedback is external (e.g., external Fabry–Perot mirrors). Induced absorption or increasing absorption bistability occurs when the feedback is intrinsic (or internal) (Toyozawa, 1978; Epshtein, 1978; Bohnert *et al.*, 1983; Rossmann *et al.*, 1983; Nguyen and Zimmermann, 1984; Miller, 1984; Schmidt *et al.*, 1984; Goldstone and Garmire, 1984; Haus *et al.*, 1985; Gibbs *et al.*, 1985).

Optical gating has also attracted much attention. Optical bistable devices may perform gating operations under different operating conditions. More than one input beam is employed for optical gating. For example, to demonstrate a NOR gate, the transmission of a probe pulse is monitored in the presence or absence of two input pulses. The probe transmission, which is the gate output, is adjusted (in the absence of any inputs) to the "high" state by matching its peak wavelength to one of the Fabry–Perot peaks. The input pulse changes the index of refraction of the material, shifts the Fabry–Perot peak away from the probe wavelength and thereby forces the gate output to the "low" state. The second input pulse shifts the Fabry–Perot peak even further and the probe transmission stays in the "low" state. Various gating operations such as AND, OR and NOR have been demonstrated in bulk GaAs and GaAs–AlGaAs MQWs (Peyghambarian and Gibbs, 1985).

Semiconductors are attractive materials for this application because of their large nonlinearity, rapid response time, room-temperature operation, and small size. Various nonlinear mechanisms in semiconductors have been employed. Free-exciton saturation in GaAs–AlGaAs multiple-quantum-wells (Morhange *et al.*, 1986); screening of the continuum states and bandfilling in bulk GaAs (Koch *et al.*, 1986); biexciton generation in CuCl (Peyghambarian *et al.*, 1983; Levy *et al.*, 1983); bound-exciton saturation in CdS (Dagenais and Winful, 1984); and bandfilling in InSb (Miller *et al.*, 1979; Kar *et al.*, 1983; Seaton *et al.*, 1983; Sarid *et al.*, 1984), InAs (Poole and Garmire, 1984), and CdHgTe (Miller *et al.*, 1984); and thermal nonlinearity in ZnS and ZnSe interference filters (Karpushko and Sinitsyn, 1978; Karpushko and Sinitsyn, 1982; Olbright *et al.*, 1984; Smith *et al.*,

Table 1

Material	Temp. (K)	λ (μm)	Switching Intensity	Pulse Energy (Gating)	Speed	Nonlin.
GaAs bulk	300	0.88	5 mW 5 kW/cm^2	3 pJ	30 ps	Bandfilling and Screening
GaAs MQW	300	0.82–0.88	5 mW 5 kW/cm^2	3 pJ	—	Free Exciton Saturation
InSb	77	5	8 mW 40 W/cm^2	—	500 ns	Bandfilling
InAs	77	3	3 mW 75 W/cm^2	—	200 ns	Bandfilling
CdHgTe	300	10.6	$500 \frac{\text{kW}}{\text{cm}^2}$	—	400 ns	Bandfilling
CdS	2	0.48	0.75 mW 330 W/cm^2	—	2 ns	Bound Exciton
CuCl	77	0.4	$10 \frac{\text{MW}}{\text{cm}^2}$	—	100 ps	Biexciton
InGaAsP/ InP	300	1.3	—	20 pJ(elect.) 1 fJ(opt.)	1 ns	Hybrid
SEED	300	0.84–0.86	0.67 μW to 3.7 mW	1 nJ 18 fJ/μm^2	1.5 ms–400 ns	Hybrid
ZnS, ZnSe	300	0.5–0.7	5 mW	—	1 ms–10 μs	Thermal

1984; Jin *et al.*, 1986; Taghizadeh *et al.*, 1986) have been successfully used to demonstrate bistability. The operating parameters and the origins of the nonlinearities in various semiconductors are summarized in Table 1. Such tables should be used with care because the reported parameters are usually not the optimized values and sometimes the switching speed may not correspond to the switching power indicated. Tables such as this should be used as a guide for selection of the proper material for the desired wavelength and temperature and to provide an overall range of switching powers and speeds.

II. OPTICAL BISTABILITY IN BULK GaAs AND GaAs MQWs

Optical bistability has been observed in bulk GaAs and GaAs MQWs at room temperature, as shown in Figs. 1 and 2 (Gibbs *et al.*, 1982). Comparison of bulk GaAs and GaAs MQWs indicates no significant difference in switching power and performance at room temperature. However, MQWs can operate at various wavelengths, depending on the well size. The minimum switching power obtained to date at room temperature is about 3 mW, which corresponds to about 3 kW/cm^2 intensity.

The similarity in performance of room-temperature optical bistability in bulk and MQWs is striking because the exciton resonance is more pronounced in MQWs (Miller *et al.*, 1982), which should imply a stronger refractive nonlinearity than for bulk GaAs on the sides of the resonance. However, detailed measurements of intensity-dependent absorption spectra show that the magnitude of nonlinear index in bulk and MQWs is similar despite the fact that the exciton resonance is much more pronounced in MQWs. The origin of the nonlinearity is different in bulk and MQW structures. These results will be described in the next section.

The large room-temperature nonlinearity in a MQW and the shift of the exciton resonance toward shorter wavelengths with reduced well thicknesses suggest the use of a diode laser instead of a dye laser as a light source for bistability. The operating wavelength of a 53-Å MQW is compatible with that of available diode lasers. A Hitachi HLP 1400 single-mode laser diode operating at 830 nm with approximately 6-mW power was used to observe bistability in a room-temperature MQW consisting of 300 periods of 53-Å

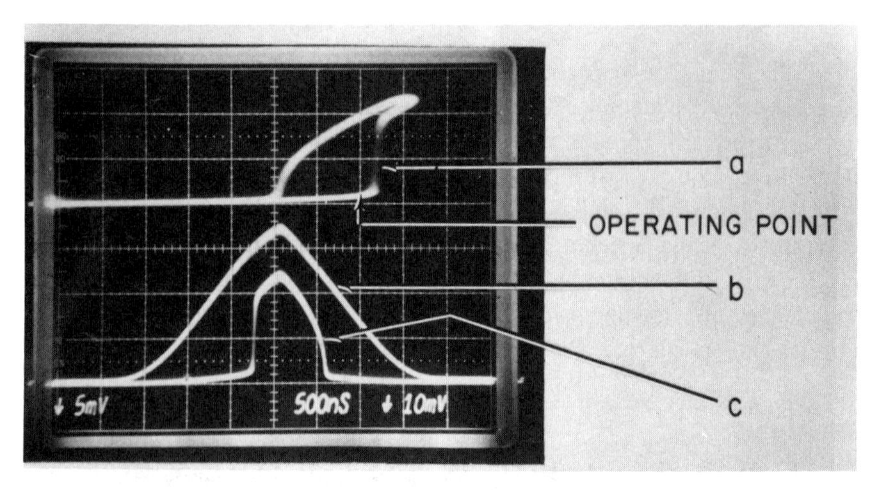

Figure 1. Optical bistability in bulk GaAs. Trace (a) is the output intensity versus input intensity. Trace (b) is the input intensity versus time. Trace (c) is the output transmission versus time.

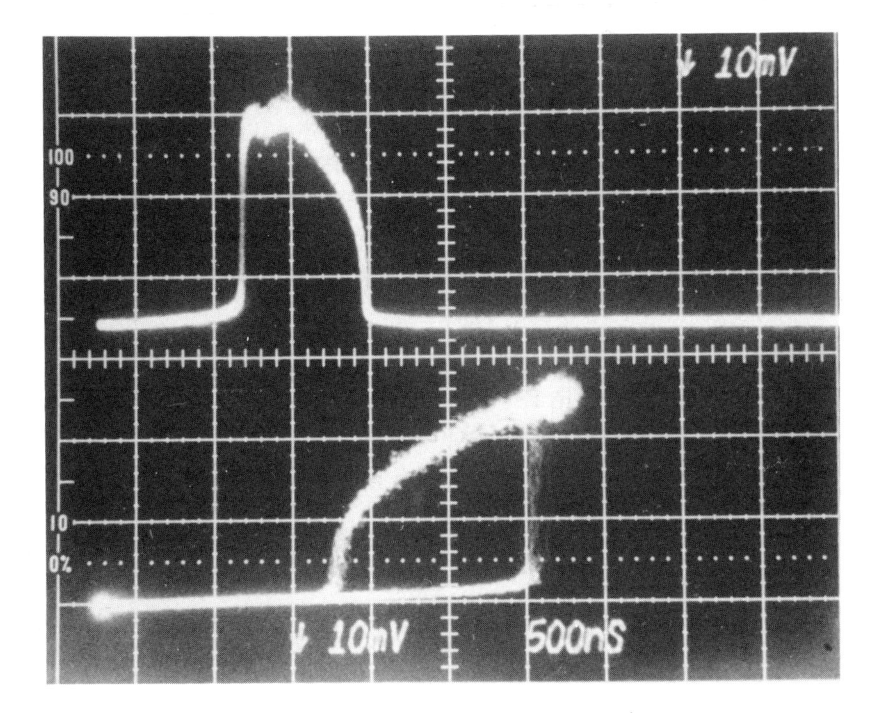

Figure 2. Room-temperature optical bistability in a GaAs–AlGaAs MQW etalon. The top trace is the transmission of the device (output) versus time. The bottom trace is the output versus input, showing hysteresis. Input is a triangular pulse.

$Al_{0.3}Ga_{0.7}As$ (Tarng *et al.*, 1984; Gibbs *et al.*, 1983). GaAs–AlGaAs MQWs have also been the subject of investigation for increasing absorption optical bistability (Miller *et al.*, 1984).

III. ORIGIN OF OPTICAL NONLINEARITIES OF BULK GaAs AND GaAs MQWs

Nonlinear absorption changes in the vicinity of the exciton frequency are measured and the results are compared to a plasma theory to better understand the physics of the nonlinearity and the magnitude of the nonlinear index, Δn, in bulk GaAs and MQWs (Morhange *et al.*, 1986; Koch *et al.*, 1986). The experimental configuration is a pump-and-probe scheme where a single-frequency pump beam is tuned above the bandgap of the material and a broad-band probe beam monitors the pump-induced changes in the transmission of the sample. The pump beam generates a high-density electron–hole plasma at room temperature. A Kramers–Kronig transformation of measured changes in the absorption coefficient, $\Delta\alpha$, gives dispersive changes Δn. These experimental values are then compared with a plasma theory that takes into account many-body effects associated with high-density carriers. (For more information concerning the complete theoretical model, see the chapter by S. W. Koch in this book and Banyai and Koch, 1986.) Fig. 3 shows the experimental and theoretical results for bulk GaAs. The top curve (1) in Fig. 3(a) is an unpumped absorption spectrum of bulk GaAs and shows a small exciton feature. At small pump intensities, most of the negative absorption change comes from the exciton saturation, but the net effect is not very strong because it is partly cancelled by the positive contribution from the bandgap reduction. As the pump intensity increases, broad absorption changes in the band begin to take place and become dominant. The maximum refractive index change was approximately -0.06 at about 3 to 4 meV below the exciton energy. The nonlinear dispersive changes far below the band edge show nonresonant characteristics rather than resonant exciton behavior. The rising slopes at the high-energy end of Fig. 3(b) are artifacts arising from the finite integration limits (1.38–1.50 eV) set in the Kramers–Kronig transformation.

Comparison of the experimental results with theory allows one to analyze the relative contribution of the different nonlinear effects. Bandfilling and screening of the continuum-state Coulomb enhancement are almost equally efficient in generating the observed dispersive changes. The bandfilling is mainly responsible for the broad low-frequency tails of Δn, whereas the

Figure 3. Room-temperature bulk GaAs optical nonlinearities: experiment and theory. (a) Experimental absorption spectra for different excitation intensities I (mW): 1) 0; 2) 0.2; 3) 0.5; 4) 1.3; 5) 3.2; 6) 8; 7) 20; 8) 50 on a 15-μm diameter spot. (b) Nonlinear refractive index changes corresponding to the measured absorption spectra. The curves (a)–(g) in Fig. 3(b) are obtained by the Kramers–Kronig transformation of the corresponding experimental data (2–8) in Fig. 3(a). For example, curve a is obtained by transformation of the difference between spectra 2 and 1; similarly curve b is obtained by transformation of the difference spectra between 3 and 1, etc. (c) Calculated absorption spectra for different electron–hole pair densities N (cm^{-3}): 1) 10^{15}; 2) 8×10^{16}; 3) 2×10^{17}; 4) 5×10^{17}; 5) 8×10^{17}; 6) 10^{18}; 7) 1.5×10^{18}. $E_g^0 = 1.433$ eV and $E_R = 4.2$ meV. (d) Calculated nonlinear refractive index changes. The curves (a)–(f) in Fig. 3(d) are obtained from the curves (2)–(7) in Fig. 3(c), respectively.

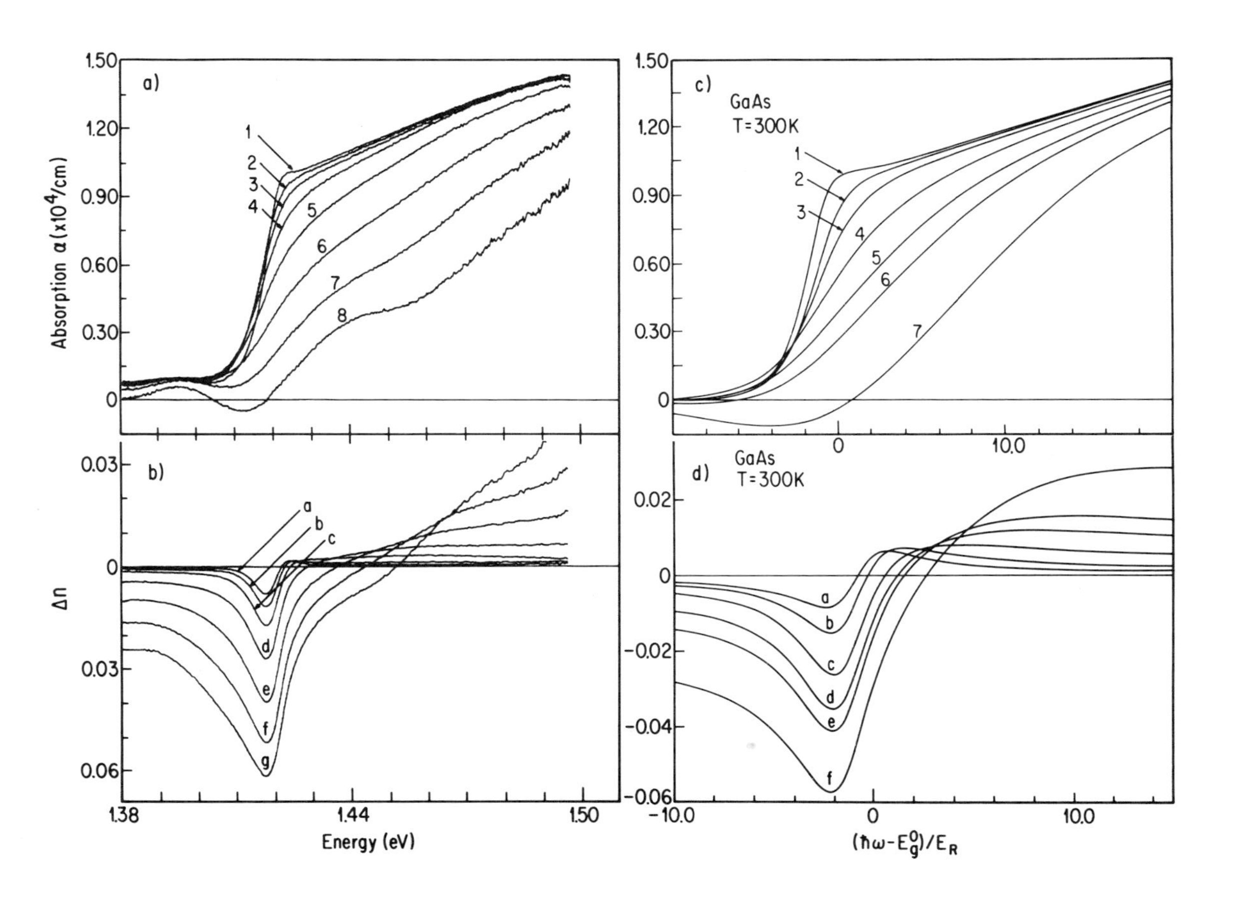
a)
b)
c)
d)
GaAs
T=300K
GaAs
T=300K
Absorption α(x10⁴/cm)
Δn
Energy (eV)
(ħω−E⁰g)/E_R
1.50
1.20
0.90
0.60
0.30
0
0.03
0.06
1.38
1.44
0.02
-0.02
-0.04
-0.06
-10.0
10.0

screening causes the sharp structure in the vicinity of the bandgap. The absorptive changes obtained through the bleaching of bound states (exciton ionization) are largely compensated through the red shift of the continuum states caused by the bandgap reduction. In other words, the negative Δn from exciton saturation is largely offset by a positive Δn from bandgap reduction at low intensities. The nonlinearities leading to a Δn sufficient for bistability are roughly equal contributions from the bandfilling and screening of the Coulomb enhancement of the continuum states.

The nonlinearity in MQWs arises from bleaching of exciton resonance with small contribution from bandfilling and continuum states screening. It is concluded that for intensities required for optical bistability, the nonlinearity is mainly excitonic for MQWs with well size W of 76 Å and 152 Å, which are either smaller or slightly larger than the exciton radius a. For a 299-Å MQW, ($W \cong 2a$) where the MQW behaves more like bulk but still has distinct excitonic features, the nonlinearity is caused by the combination of exciton bleaching, bandfilling, and screening of the continuum states. The measured nonlinear indices are rather similar, within a factor of three, for all these samples. The magnitudes of the $\Delta n/N$ at low carrier densities for several MQWs were measured. It varied from the maximum of $3.8 \times 10^{-19}\,\text{cm}^3$ for the 76 Å MQW to the minimum of $1.2 \times 10^{-19}\,\text{cm}^3$ for bulk GaAs, for a carrier density of $N = 3 \times 10^{16}\,\text{cm}^{-3}$. The value of $\Delta n/N$ varied from $2.4 \times 10^{-20}\,\text{cm}^3$ for the 76 Å MQW to $1.3 \times 10^{-20}\,\text{cm}^3$ for the bulk, for a carrier density of $N = 3 \times 10^{18}\,\text{cm}^{-3}$. N is the density of carriers generated by pump and $\Delta n/N$ is the change in refractive index per each electron–hole pair.

The dynamics of high-density excitons and carriers have also been extensively studied in both bulk GaAs and GaAs MQWs (Shank *et al.*, 1979; Shank *et al.*, 1982; Oudar *et al.*, 1984; Peyghambarian *et al.*, 1984; Hulin *et al.*, 1986; Knox *et al.*, 1985; Oudar *et al.*, 1985).

IV. OPTICAL LOGIC GATES OF BULK GaAs AND GaAs MQWs

Nonlinear etalons can perform various optical logic operations such as AND, OR and NOR. The principle of such operations is the shifting of the Fabry–Perot transmission peak in response to input pulses. The nonlinear medium must be such that the absorption of one input pulse changes the refractive index at the probe wavelength enough to shift the transmission peak of the etalon by about one instrument width (Jewell *et al.*, 1986). For example, if the etalon is initially tuned to the probe wavelength, a NOR-gate operation results if the control pulse is able to detune the etalon from the probe wavelength. The speed of the gate can be very rapid, with the potential for picosecond decisionmaking at a gigahertz repetition rate.

Optical gating in GaAs has also progressed considerably. One-ps switch-on time of a GaAs MQW gate has been demonstrated (Migus *et al.*, 1985). This time indicates how fast the etalon peak shifts in response to an input pulse with an above-bandgap frequency. Recovery of the gate requires removal of the carriers produced by the input pulse. This recovery takes more than 10 ns in the usual GaAs and MQW etalons because of the long lifetime of carriers. The recovery time has been shortened to 200 ps (Lee *et al.*, 1986) (detector limited) using surface recombination in an etalon with no AlGaAs outside layers (the AlGaAs windows are normally used to stop etching of the GaAs substrate). A 30-ps recovery and 70-ps cycling time have also been demonstrated in similar etalons (Jewell *et al.*, 1986).

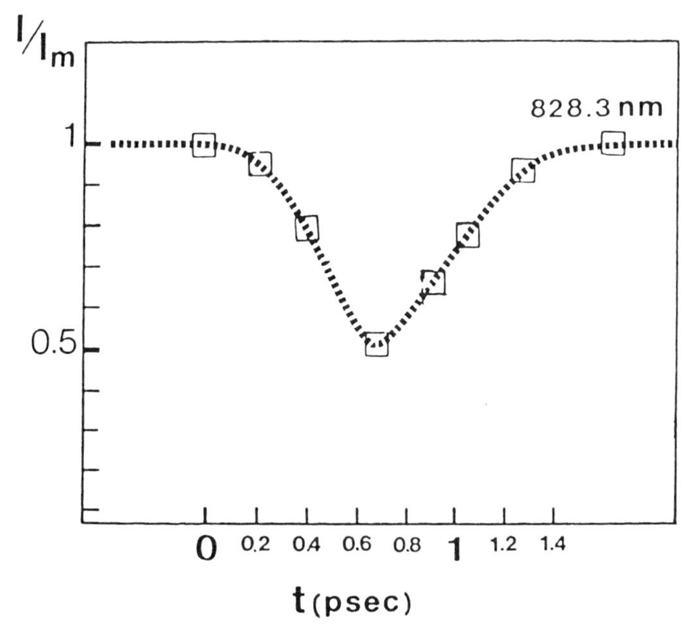

Figure 4. Picosecond switching of a GaAs optical logic gate. Optical Stark effect has been employed to obtain these switching speeds.

If the input pulse frequency is tuned below the exciton, in the transparency region of the material, then the electric field of the laser may be used to switch the etalon. The advantage of this optical-Stark-effect switching is that the carrier lifetime is no longer a limitation as carriers are not excited in this scheme. A 1-ps recovery using this effect has been demonstrated as shown in Fig. 4 (Migus *et al.*, 1986), which shows NOR-gate switch-on and-off times of 1 ps.

V. OPTICAL NONLINEARITIES OF CdS_xSe_{1-x}-DOPED GLASSES

Glasses doped with microcrystallites of semiconductors CdS_xSe_{1-x} are emerging as a new class of nonlinear materials for application in nonlinear signal-processing and optical devices. The crystallite sizes range from 50 to 1000 Å. The rapid response time (a few tens of picoseconds), room-temperature operation and inexpensiveness of materials make them exceptionally attractive for nonlinear device applications.

These materials have been available commercially for many years as sharp-cut color filters, manufactured by Corning and Schott, for use as linear blocking elements. The importance of semiconductor-doped glasses for nonlinear optical applications was realized first by researchers at Hughes Laboratories in 1983 who reported degenerate four-wave mixing in these glasses (Jain and Lind, 1983). During 1986, the origin of the nonlinearity (Olbright *et al.*, 1986; Flytzanis *et al.*, 1986), the magnitude of the nonlinear index at various wavelengths in the vicinity of the bandgap (Olbright and Peyghambarian, 1986) and the femtosecond dynamics of such glasses have been investigated (Peyghambarian *et al.*, 1986; Nuss *et al.*, 1986). Phase conjugation, four-wave mixing (Flytzanis *et al.*, 1986; Stegeman *et al.*, 1986; Cotter, 1986; Etchepare *et al.*, 1986) and luminescence (Stegeman *et al.*, 1986) studies were reported. The use of semiconductor-doped glasses as nonlinear waveguides has also been proposed (Stegeman *et al.*, 1986; Sarid *et al.*, 1985) and has become an attractive challenge for researchers in this field. In this chapter, the focus is on the work that has been performed in Arizona.

The magnitude of the nonlinear index at various wavelengths in the vicinity of the bandgap of the semiconductor microcrystallites has been measured using an interferometric technique (Olbright and Peyghambarian, 1986). Fig. 5 shows the result of this single-beam measurement in which the change in the index of refraction, Δn, is plotted as a function of incident

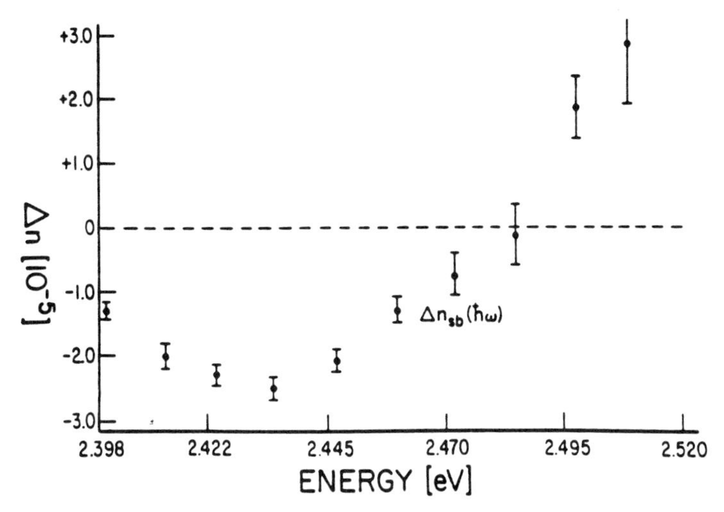

Figure 5. Interferometrically measured changes in the index of refraction of a $CdS_{0.9}Se_{0.1}$-doped glass.

photon energy for an input intensity of $I = 3\ MW/cm^2$. The error bars increase for photon energies above the bandgap because the larger absorption produces poor fringe contrast and fringe quality.

The physical origin of the optical nonlinearities has been studied using a two-beam pump-probe experiment (Olbright *et al.*, 1986). The transmission of a weak probe beam is monitored as a function of wavelength for various pump intensities. The pump-induced changes in the probe transmission spectra at room temperature are plotted in Fig. 6a for various pump intensities. The spectrum labeled iii) corresponds to the highest pump intensity, while the spectrum labeled i) is the linear absorption spectrum. With increasing incident intensity, a large shift of the absorption spectrum to higher energies is observed. This blue shift cannot arise from pump-induced thermal effects, which would instead produce a red shift. The observed blue shift of the spectra is attributed to the bandfilling effects where the states on the bottom of the conduction band are filled with electrons and states on the top of the valance band are filled by holes. The change in absorption coefficient $\Delta\alpha(\omega)$ obtained by subtracting spectrum i) from iii), i.e., $\Delta\alpha(\omega) = \alpha(\omega, I = 3\ MW/cm^2) - \alpha(\omega, I = 200\ kW/cm^2)$, is displayed in Fig. 7a by the solid curve. The corresponding dispersion of the change in the refractive index is obtained by a Kramers–Kronig transformation.

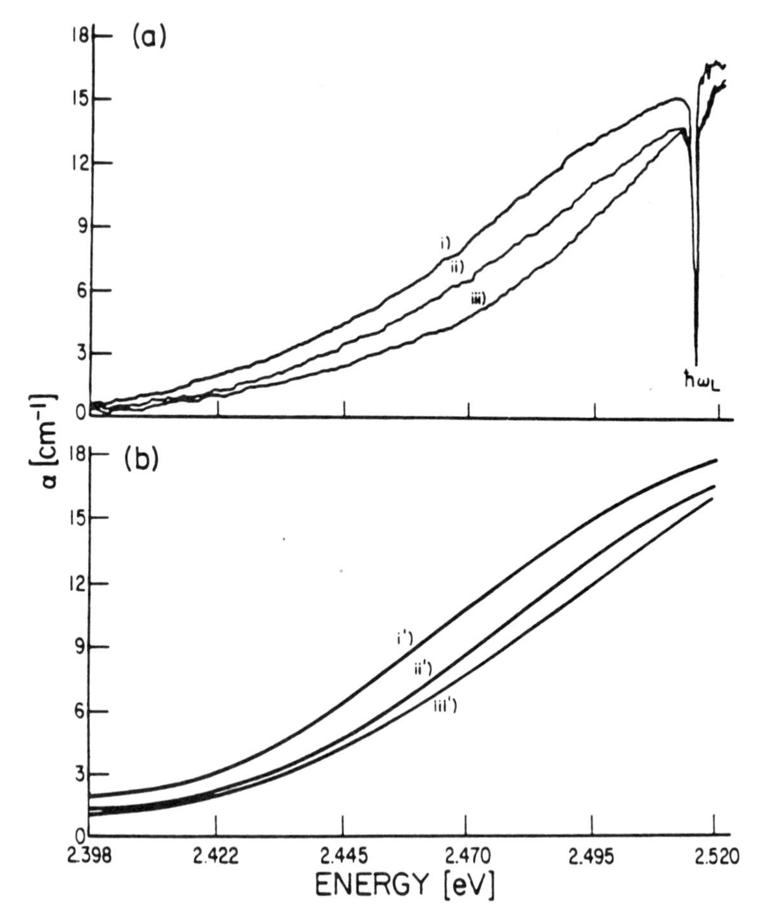

Figure 6. Room-temperature absorption spectra for $CdS_{0.9}Se_{0.1}$-doped glass. (a) Experimental results: i) 200 kW/cm^2 (linear spectrum), ii) 1.2 MW/cm^2, iii) 3 MW/cm^2. (b) Theoretical results: i′) $N = 0$, ii′) $N = 4 \times 10^{17}\ \mathrm{cm}^{-3}$, iii′) $N = 1 \times 10^{18}\ \mathrm{cm}^{-3}$.

The experimental results have been compared with a theoretical model developed for the optical nonlinearities in highly excited semiconductors. The Maxwell–Garnet equation (Genzel and Martin, 1973) is used to express the average dielectric function ϵ^{av} of the composite material (glass plus embedded semiconductor crystallites),

$$\epsilon^{\mathrm{av}}(\omega) = \epsilon_{\mathrm{g}} \frac{\epsilon(\omega)(1 + 2p) + 2\epsilon_{\mathrm{g}}(1 - p)}{\epsilon(\omega)(1 - p) + \epsilon_{\mathrm{g}}(2 + p)}, \tag{1}$$

where $\epsilon_{\mathrm{g}} = 2.25$ is the dielectric constant of the host glass, $p = 1.25 \times 10^{-3}$

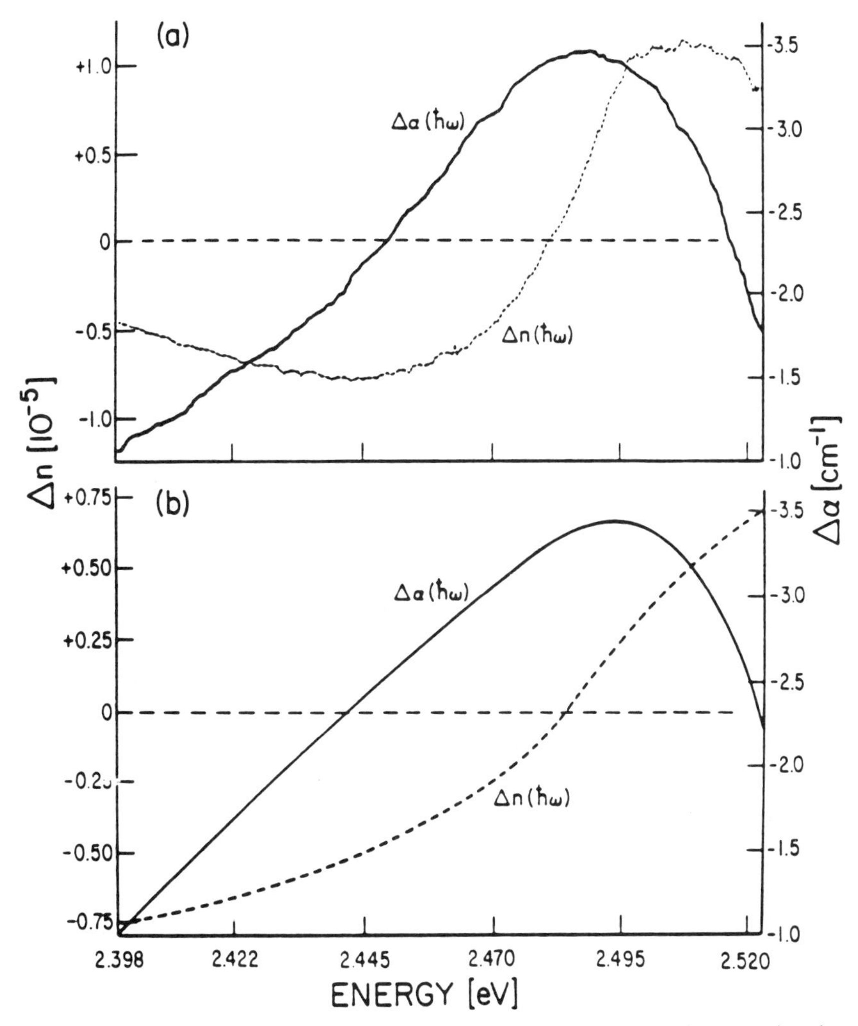

Figure 7. Absorptive changes and corresponding dispersive changes, of $CdS_{0.9}Se_{0.1}$-doped glass. (a) Experimental results: $\Delta\alpha = \alpha(I = 3\ MW/cm^2) - \alpha(I = 200\ kW/cm^2)$. (b) Theoretical results: $\Delta\alpha = \alpha(N = 10^{18}\ cm^{-3}) - \alpha(N = 0)$.

is the fraction of the total volume occupied by semiconductor crystallites, and $\epsilon(\omega)$ is the complex dielectric function of the semiconductor crystallites. For calculating $\epsilon(\omega)$, again the plasma theory is used (Banyai and Koch, 1986). The detailed comparison between theory and experiment is shown in Fig. 7, where we compare the theoretical and experimental change in the absorption and the corresponding refractive indices $\Delta\alpha(\omega)$ and $\Delta n(\omega)$, respectively.

VI. FEMTOSECOND DYNAMICS OF CdS_xSe_{1-x}-DOPED GLASSES

The dynamics of the optical nonlinearities in these materials have also been investigated using femtosecond light pulses (Peyghambarian *et al.*, 1986; Nuss *et al.*, 1986). The experiment uses a high-repetition–rate femtosecond pump-probe technique. The output of a balanced CPM-ring dye laser is amplified to 2 μJ with a copper vapor laser at 7.3 kHz repetition rate. The amplified pulse, with a duration of 200 fs, is used to generate a white-light continuum in an ethylene glycol jet for the probe pulse while a fraction of the amplified light at 620 nm is used as the pump pulse. Fig. 8 displays transmission of the probe as a function of wavelength for various time delays between pump and probe. Spectrum labeled -450 fs is taken when the probe precedes the pump and therefore is representative of the unexcited semiconductor, showing the band-edge absorption of the microcrystallites. Spectra labeled -100 fs and 0 ps clearly show a well-resolved shift of the absorption edge to lower energies, indicating the onset of carrier screening and bandgap renormalization and tail-state broadening. At 200 fs, this red shift is almost completely dominated by a blue shift of the band-edge. This suggests that carriers initially created with a few LO phonon energies above the bandgap have relaxed to the bottom of the band and initiated the bandfilling mechanism. After 400 fs, the carrier relaxation is complete and appears as a large blue shift of the absorption edge. Spectra taken at 400 fs and up to 1 ps are practically the same with similar magnitude of the blue shift. After a few picoseconds, the blue shift starts to recover, indicating the onset of electron–hole recombination. The blue shift almost completely recovers in $\cong$ 50 ps. However, a small portion of the shift does not fully recover and persists for times in excess of 500 ps. This may be attributed to carriers confined to traps near the band-edge.

These experimental results are compared with the above-mentioned plasma theory. For this comparison, a knowledge of the injected carrier density, which is obtained by solving the time-dependent carrier-concentration rate equation for 200-fs duration excitation and 30-ps recombination lifetime, is required. The solution yields a carrier density that initially follows the excitation pulse and then becomes nearly constant for an interval of the first few picoseconds of excitation. The peak carrier density is 0.006 N_0, where $N_0 = (\alpha I_0 \tau_r)/(\hbar\omega)$ is the steady-state carrier density. Here, α is the absorption coefficient at photon energy $\hbar\omega$, I_0 is the peak pulse intensity, and τ_r is the recombination lifetime. The absorption spectra is then computed for various plasma temperatures T_p, using the carrier

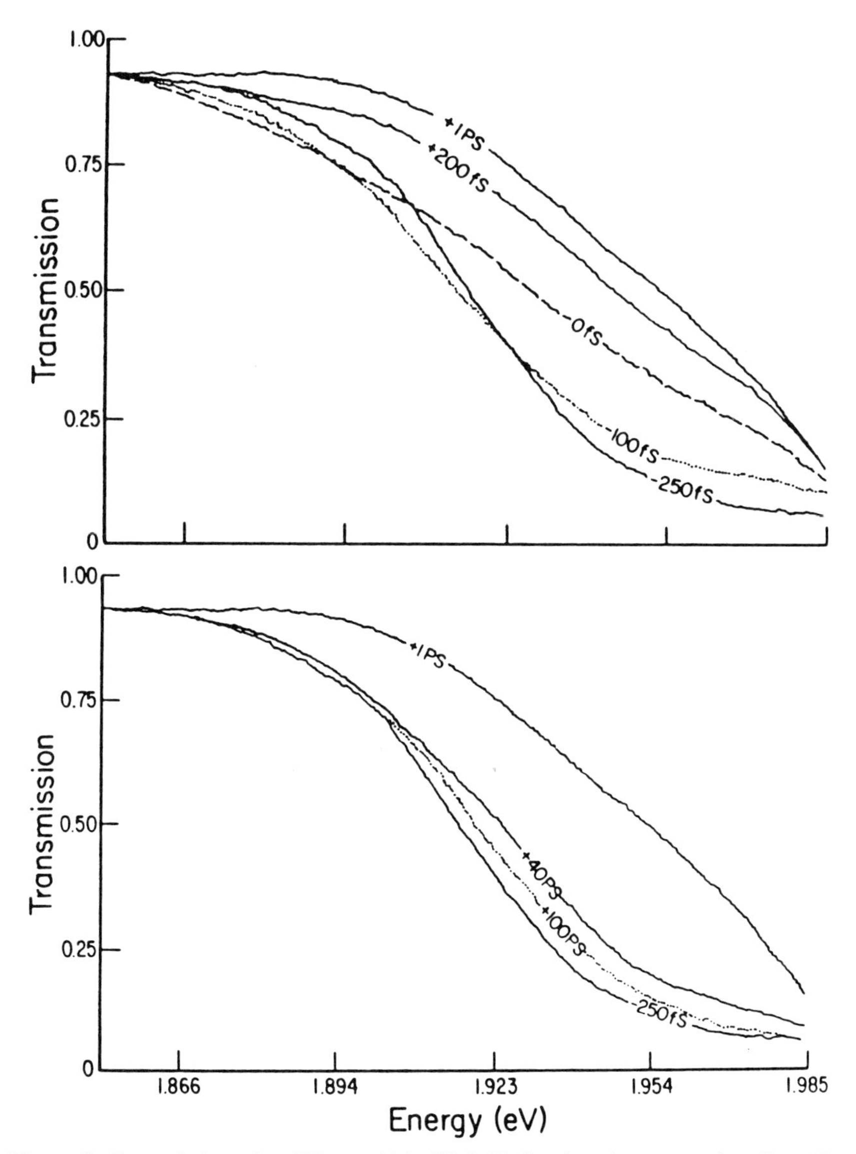

Figure 8. Transmission of a 750-μm–thick CS 2-64 Corning sharp-cut color filter. The pump-probe delay is varied. The pump intensity is 0.5 GW/cm^2.

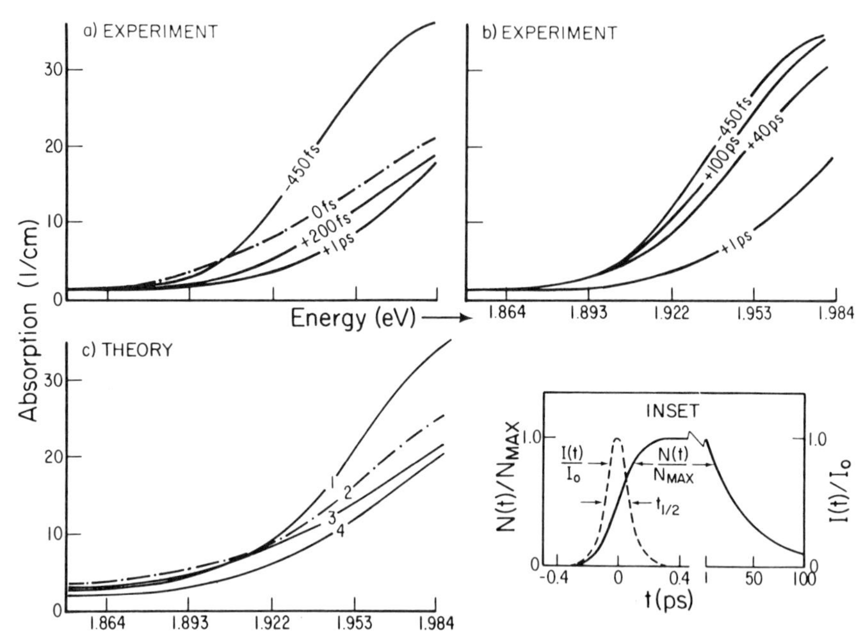

Figure 9. (a, b) Room-temperature absorption spectra of a 750-μm–thick glass doped with $CdS_{0.2}Se_{0.8}$ microcrystallites for various pump-probe delays. (c) Plasma theory calculation: 1) $T = 300$ K, $N = 10^{15}$ cm^{-3}; 2) $T = 550$ K, $N = 5 \times 10^{17}$ cm^{-3}; 3) $T = 450$ K, $N = 9.5 \times 10^{17}$ cm^{-3}; 4) $T = 350$ K, $N = 10^{18}$ cm^{-3}. Inset: Relative temporal evolution of the excitation pulse, $I(t)$, and the carrier concentration, $N(t)$, $N_{max} = 0.006\ N_0$.

density just calculated. In Fig. 9c, the results of this theoretical calculation are presented and are compared with the experimental results of Fig. 8, which are replotted as absorption (instead of transmission) in Figs. 9a and 9b. Curve 1 is the computed low-excitation $N = 10^{15}$ cm^{-3} absorption (linear spectrum) which is scaled to the -450 fs curve of Fig. 9a. Curve 2 is computed for $T_p = 550$ K and $N = 5 \times 10^{17}$ cm^{-3}. This curve should be compared with the spectrum labeled 0 fs in Fig. 9a (it should be noticed that the carrier density reaches 0.5 $N_{MAX} = 5 \times 10^{17}$ cm^{-3} for $\Delta t = 0$ fs). However, a calculated spectrum with a nonthermal carrier distribution also gives a spectrum similar to Curve 2. This suggests that spectrum labeled 0 fs either originates from a distribution of carriers at quasiequilibrium with a high plasma temperature or a nonthermal distribution. Curves 3 and 4 are computed for different plasma temperatures $T_p = 450$ and 350 K, respectively, with $N = 9.5 \times 10^{17}$ and 10^{18} cm^{-3}, respectively, and should be compared with spectra labeled 200 fs and 1 ps in Fig. 9a (in 200 fs, the carrier density reaches 0.95 $N_{max} = 9.5 \times 10^{17}$ cm^{-3}). These results are in

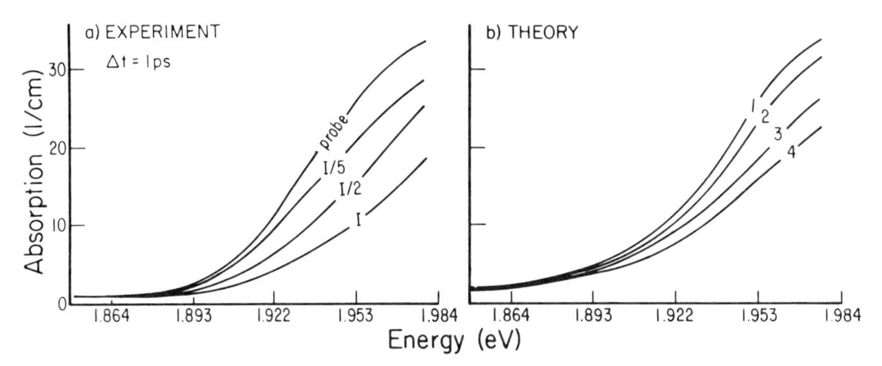

Figure 10. (a) Experimental absorption spectra of a $CdS_{0.2}Se_{0.8}$ glass at a fixed pump-probe delay of 1 ps for various pump intensities, $I = 500$ MW cm^2. (b) Plasma theory calculation for $T_p = 350$ K. Curve 1) $N \cong 10^{15}$ cm^{-3}; 2) $N = 2 \times 10^{17}$ cm^{-3}; 3) $N = 5 \times 10^{17}$ cm^{-3}; 4) $N = 10^{18}$ cm^{-3}.

good qualitative agreement with the experimental results, indicating that plasma cooling by LO phonon scattering and the corresponding thermalization of the carrier temperature to near lattice temperature take place in 400-fs (the 400 fs spectrum is similar to the 1-ps spectrum). This also implies that bandfilling is mainly responsible for the steady-state optical nonlinearity of these materials.

In a complementary series of experiments, the time delay between the pump and probe pulses was kept constant at 1 ps and the pump intensity (and thereby the carrier density) was varied. The delay of 1 ps was chosen because carrier temperature reaches the lattice temperature in times shorter than 1 ps and therefore a steady state can be assumed. The results of these measurements are shown in Fig. 10a, where a blue shift of the absorption spectrum was observed as the pump intensity was increased. A good agreement between the experiment and theory was obtained. The calculated spectra are shown in Fig. 10b. These results reconfirm that bandfilling nonlinearity is the dominant mechanism in steady state, consistent with the results of our nanosecond experiments in a $CdS_{0.9}Se_{0.1}$-doped glass (Figs. 6 and 7).

VII. INCREASING ABSORPTION OPTICAL BISTABILITY AND "KINKS" IN CdS_xSe_{1-x}-DOPED GLASSES

Koch, Schmidt and Haug (Koch *et al.*, 1984) have solved the transport equations for the light intensity and the excitation density for a system that exhibits increasing absorption optical bistability. If the input light intensity is increased linearly at a rate appropriate for the excitation lifetime, a discontinuity in the excitation density occurs between the front of the sample, which has switched "off," and the back, which is still "on." This kink jumps discontinuously along the beam propagation direction. Consequently, the transmitted intensity has a sawtooth temporal dependence: the transmission drops each time the kink jumps deeper into the absorber, dramatically demonstrating the local nature of increasing absorption bistability which needs no cavity. The occurrence of kinks in increasing absorption optical bistability has been demonstrated (Gibbs *et al.*, 1985) in a semiconductor-doped glass. Thermal nonlinearity of the doped glass was employed. Fig. 11 shows the experimental observation of kinks and comparison with the theoretical model. A single sudden drop in I_{T} may result from either entire-sample or partial-sample switchdown. But two or more sudden drops assure that a kink has jumped deeper into the sample. The sequence (a) to (c) in Fig. 11 shows fewer kinks the faster the input rise time. Pronounced changes (most noticeably blooming) in the far-field profile occur, but they are suppressed here by collecting all of the transmitted light. Good agreement between theory and experiment is obtained.

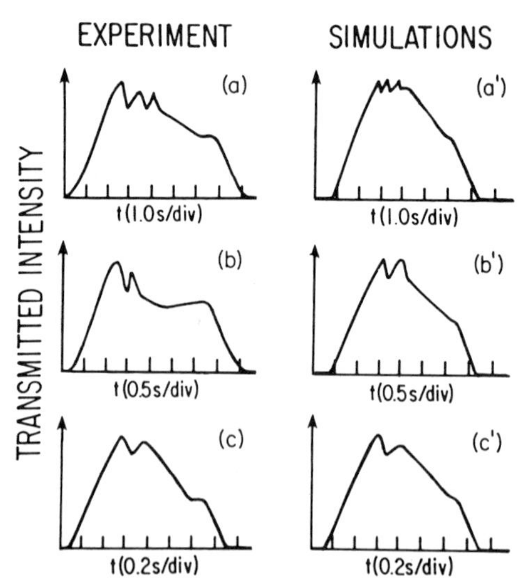

Figure 11. (a), (b), (c) Experimental kinks in a Corning CS 3-70/No. 3384 sharp-cut color filter of thickness $L \cong 2$ mm and beam radius $r_0 \cong 14$ μm for the same peak input intensity. The rise time of the triangular input pulse is decreased from (a) to (c). (a')–(c'): Numerically calculated normalized $I_I(t)$ and $I_T(t)$ for three different values of $\Delta t/\tau$: (a') 350, (b') 175 and (c') 70.

VIII. CuCl BIEXCITON OPTICAL NONLINEARITIES AND BISTABILITY

Bistability in CuCl is of particular interest because of predictions of picosecond switch-off times using the biexciton two-photon resonance (Hanamura, 1981; Koch and Haug, 1981; Sung and Bowden, 1984). This fast response of the system is the result of the virtual formation of biexcitons with two-photon excitation detuned from resonance and is interesting in terms of both the physics of the process and its potential application for picosecond optical logic and signal processing. We have observed optical limiting and bistability (Peyghambarian *et al.*, 1983) in CuCl using a 12-μm–thick CuCl film sandwiched between 90% dielectrically coated mirrors; we observed limiting at 15 K with an input intensity of about 14 MW/cm^2. Our results are also consistent with the results obtained by the Strasbourg group (Levy *et al.*, 1983; Bigot *et al.*, 1985). (See also the chapter by Hönerlage *et al.* in this volume.)

Recently, we have been able to epitaxially grow single-crystal thin films of CuCl (Olbright and Peyghambarian, 1986). Single-crystal structure is achieved by thermal deposition of CuCl onto a sodium chloride (NaCl) substrate whose space lattice and lattice constant match that of CuCl. X-ray diffraction analysis reveals single-crystal structure of these CuCl thin films. Luminescence spectra arising from the decay of excitonic molecules to exciton states are substantially narrower in the single-crystal thin films compared with polycrystalline thin films grown on amorphous substrates. Under resonant excitation at 7 K, the two-photon absorption spectra for single-crystal thin films of CuCl exhibit a reduction in linewidth and an increase in the peak absorption as compared with those for CuCl polycrystalline thin films (Peyghambarian *et al.*, 1986) (Fig. 12). The CuCl single-crystal thin films are promising for optical bistability and gating at higher temperatures.

IX. OPTICAL COMPUTING AND PATTERN RECOGNITION IN ZnS AND ZnSe INTERFERENCE FILTERS

There is a growing interest in the study of ZnS and ZnSe thin-film interference filters (IFs) for their potential use in optical signal processing and optical computing. Considerable research has been devoted to their basic properties such as the nonlinear mechanism, switching powers, switching times, filter design and crosstalk since the first observation of optical

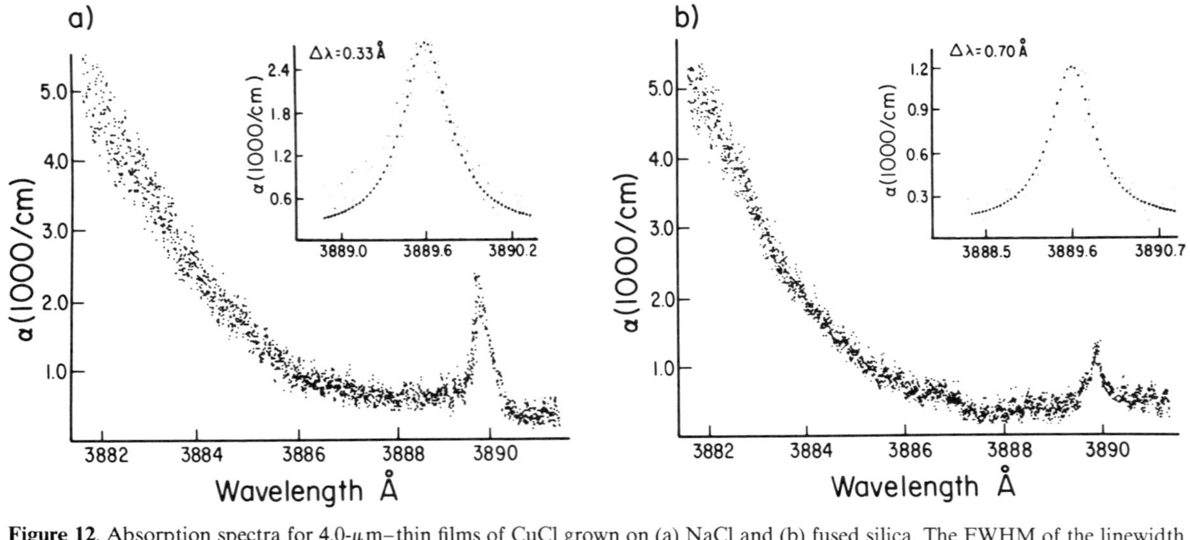

Figure 12. Absorption spectra for 4.0-μm–thin films of CuCl grown on (a) NaCl and (b) fused silica. The FWHM of the linewidth in (a) is 0.33 Å and in (b) is 0.70 Å. Inset: Lorentzian fit to absorption peak, $\alpha(\omega) = \alpha_0(\Delta\omega)^2/[(\omega_{xx} - \omega)^2 + (\Delta\omega)^2]$, $\omega = 2\pi c/\lambda$.

bistability in these filters by Karpushko and Sinitsyn (Olbright *et al.*, 1984; Smith *et al.*, 1984; Jin *et al.*, 1986; Karpushko and Sinitsyn, 1982). Cascading, gain and a triple-bistable-element loop have been demonstrated (Smith, Janossy *et al.*, 1985a; Smith, Walker *et al.*, 1985b). The attempt to achieve parallel operation of multiple "pixels" (logic gates) has also begun. We previously reported the observation of simultaneous bistable switching of multiple pixels on ZnS and ZnSe IFs and gave a brief report on the first demonstration of the recognition of a simple pattern (Jin *et al.*, 1986). That experiment has been extended to simple symbolic substitution requiring symbol scription and fan out using ZnS IFs (Gibbs and Peyghambarian, 1986; Tsao *et al.*, 1987).

One important aspect of a bistable device is fan-out capability. Defining fan out as the ratio of the change in output to the change in input, we have obtained a fan out of 4 using a ZnS interference filter (Tsao *et al.*, 1987). A 40-mW beam held the filter's operating point just below bistable switch-up; a 2-mW input then yielded an 8-mW change in output power. In the symbolic-substitution experiment, two ZnS IFs are operated in cascade. A portion of the output beam from the first stage and a holding beam are used to switch the next bistable filter. Both stages are driven by the CW output from a Coherent Innova 90-4 argon laser operating at 514 nm; the beam is modulated by a rotating half-wave plate with a total noise level of about 3%.

Contrast can be defined as the ratio of the upper-branch transmission just after switch-on to the lower-branch transmission just before switch-on. The largest contrast we have observed in ZnS IFs is about 4, while for ZnSe it is about 3. This may be because of the higher transmission of ZnS filters, which gives them an additional advantage of a larger absolute output. It may be advantageous to use ZnSe IFs for low-power applications (e.g., in the last stage of a system). ZnS IFs could be used to switch a number of pixels on a ZnSe IF when a large fan out is needed or when the noise level requires a high absolute switching signal.

Speed and parallelism are two major advantages of optics for signal processing and computing. The ideal all-optical computing device should consist of very fast optical logic gates operating in a powerful architecture. This is a great challenge to both the architects and device designers.

Symbolic substitution logic (SSL), proposed by A. Huang (Huang, 1983), is one of the parallel algorithms intended to take advantage of the intrinsic parallelism of optics. The two main stages of SSL are: (1) pattern recognition and (2) symbol scription. To carry out SSL, one needs first to encode the input information as some particular binary pattern on an input mask.

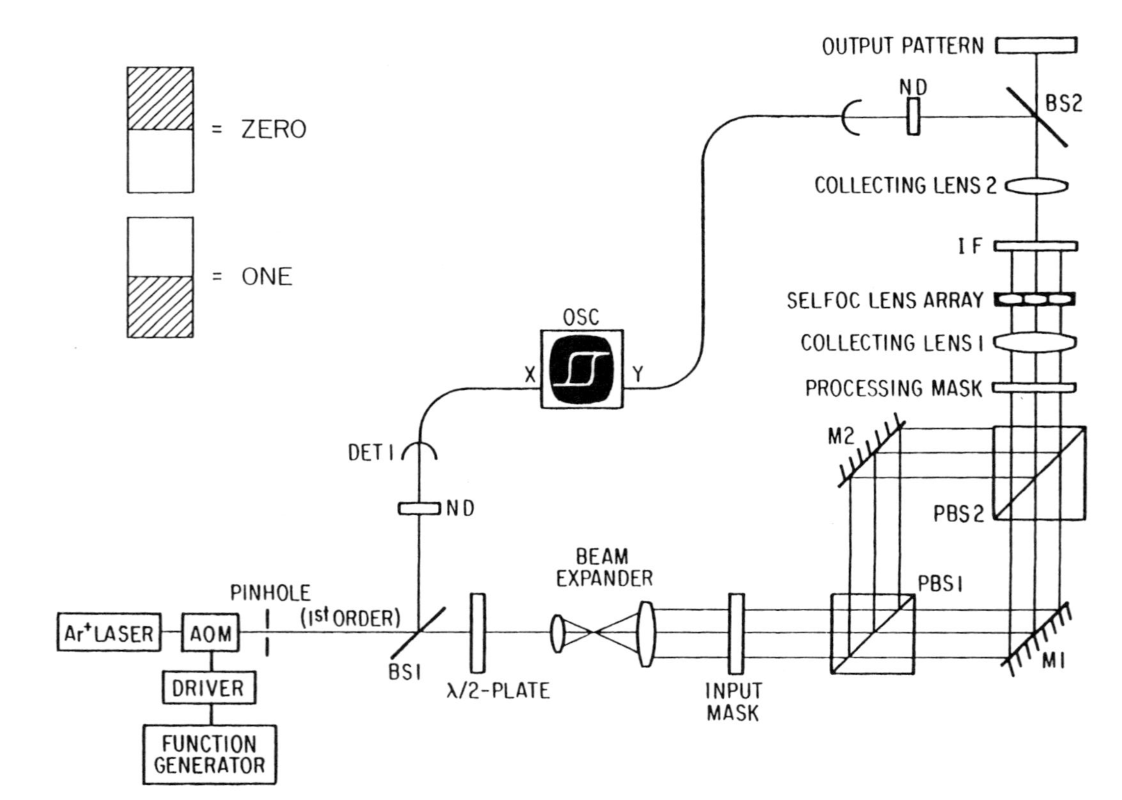

Figure 13. Pattern recognition experiment using two-dimensional optical bistable "AND"-gate array. (Inset: Binary encoding in the space domain.)

An example is given in the inset of Fig. 13 where the binary 1 (0) is represented by a bright spot on the top (bottom) and a dark spot on the bottom (top). Copies of this input are then made; one can then make relative shifts of these copies to identify the occurrence of one or more particular patterns. The identification decisions are made by an array of optical logic gates on which the input and shifted copies are superimposed. The output answers are used to generate a new pattern according to predefined substitution rules. A fan-out capability of 2 is required to implement this algorithm.

Pattern recognition, the first step in SSL, has been executed using existing nonlinear bistable devices plus linear optical components. The setup is shown in Fig. 13, where the input is encoded on a mask (35-mm film consisting of transparent and opaque spots). The goal is to identify the number and locations of the pattern formed by two bright spots located next to each other horizontally. This is done as just described, where the shift is one pixel horizontally; the two shifted patterns are superimposed with a beam combiner. A single spot in the combined pattern can have one of three light levels: no light, one unit of light or two units of light. The logic decision is made by focusing the combined pattern onto a 2×3 AND-gate array, using a Nippon Sheet Glass fly's eye fiber-optic lens array, resulting in 1-mm pixel spacing on a ZnSe IF. (The switching power per pixel is about 2 mW.) The output is a bright spot in the right-hand pixel wherever the input has bright spots adjacent horizontally. Note that all the locations of this pattern are identified by a single logic operation, independent of the array size; this is a clear utilization of the parallel nature of optics.

The Mach–Zehnder configuration used in the experiment results in equal optical path lengths for both inputs to the AND-gate array, which is important if one uses very fast gates driven by fast pulses. The orthogonal polarization of the two inputs prevents interference between the two input beams. After receipt of the input, the time required to complete the pattern recognition is the switching time of the gates and the propagation delay from the input mask to the gate array. Thus, a large array of fast optical gates should give a large throughput.

A very simple but complete symbolic substitution has also been achieved (see Fig. 14). The desired pattern is the simultaneous occurrence of bright spots in the lower left-hand and upper right-hand corners of an arbitrary 2×2 array. When that pattern occurs, the symbol-scription part generates an output pattern consisting of a bright top row and a dark bottom row. This is accomplished by an AND-gate operation of the output of the

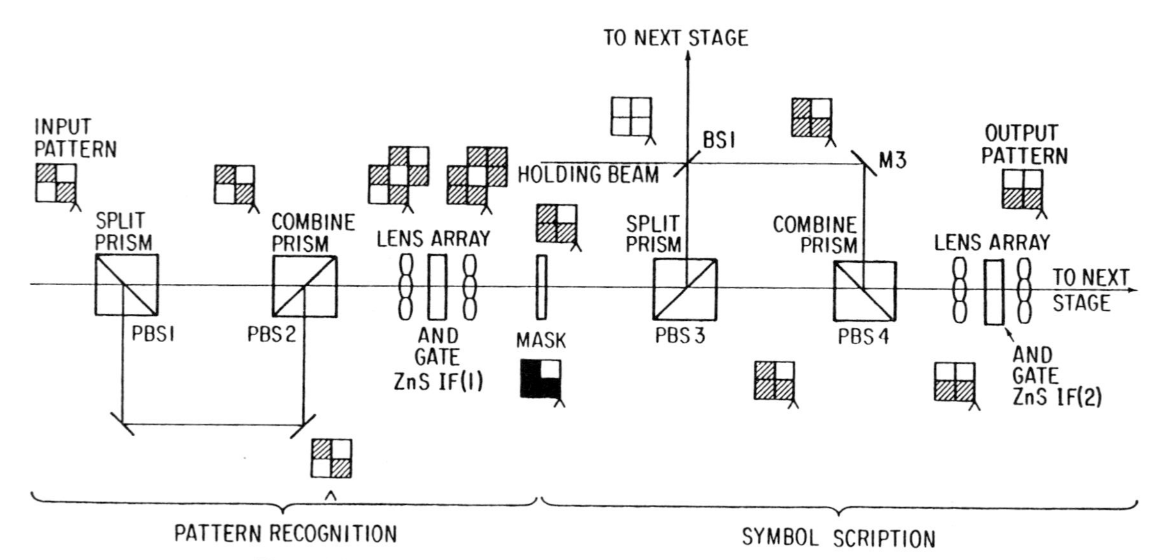

Figure 14. Symbolic substitution experiment using ZnS bistable devices.

recognition stage with the strong holding beams. If the desired pattern is not present, the output is completely dark.

Although the computations accomplished by these experiments are admittedly simple, they do illustrate pattern recognition, cascading and symbol scription. They also required considerable expansion of single-beam techniques: beam division, multiple-beam focusing and reimaging, and nonlinear etalon uniformity and stability. Techniques that are easily extendable to massive parallelism (100×100 or greater) were used wherever possible.

X. CONCLUSION

For optical signal processing and computing applications, many switches or gates should operate simultaneously. GaAs NOR-gate arrays are the most promising for this parallel processing implementation. Assume 10^4 pixels on 1 cm^2 requiring 10 pJ per bit operation and operating once every nanosecond. This requires 100 W of laser power and results in 100 W/cm^2 heat load. It yields 10^{13} bit operations per second, more than would be performed by a CRAY. Much work is required to convert the array's capability for many operations in parallel into a programmed or programmable system able to make useful computations, to recognize patterns or to learn. However, the first few steps have been already taken as displayed in Fig. 15, which shows an array that has exhibited NOR-gate and bistable operations (Venkatesan *et al.*, 1986). This was a 100×100 array of pixels, each 9 μm $\times$ 9 μm in size, formed by reactive ion etching.

For implementation of nonlinear gates and bistable switches in commercial systems, several problems have to be solved: uniformity of gate characteristics anywhere on the array, cascadability, fan-out capability and addressability to independent pixels on an array, to mention just a few. However, as our knowledge about the physics of the nonlinearity and operation of these devices grows, we should be able to model accurately the operation of a single device, thereby optimizing the operating parameters. This should lead to realization of low-power operation, cascading and gain. Materials with large nonlinearities should be sought and implemented to reduce the power requirement. Fly's-eye lens arrays, holograms and spatial light modulators should allow beam divisions and addressing. New algorithms and architectures that take advantage of nonlinear etalons should be designed.

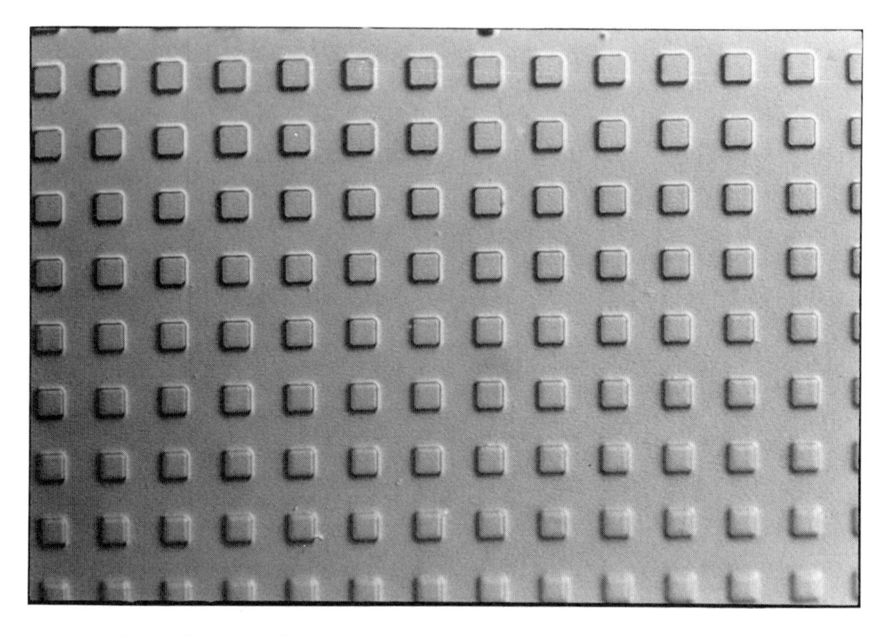

Figure 15. A two-dimensional array of optical logic gates, etched in GaAs.

ACKNOWLEDGMENTS

This research has been performed in collaboration with D. Hulin, A. Mysyrowicz, A. Migus and A. Antonetti (femtosecond studies of GaAs optical gates); S. W. Koch (theoretical work on GaAs and color filter nonlinearities and kinks); H. Haug (color filter kinks); M. Warren, Y. Lee and A. Chavez (bulk GaAs nonlinearities); J. Morhange, S. H. Park, A. Jeffery and M. Derstine (GaAs MQW nonlinearities); G. Olbright, B. Fluegel and F. Jarka (femtosecond dynamics of doped glasses); R. Jin, M. T. Tsao, L. Wang, R. Sprague, G. Gigioli, H. Kulcke, Y. Li and H. Chou (optical computing in interference filters); A. C. Gossard, W. Wiegmann, J. English, H. Morkoc, P. K. Bhattacharya and D. Weinberger (GaAs and MQW growth); and D. Weinberger, A. Yariv and V. Venkatesan (GaAs etching). The research was supported by NSF grants ECE 8610170 and ECE 8317410, the U. S. Air Force Office of Scientific Research, the U. S. Army Research Office and the Optical Circuitry Cooperative.

REFERENCES

Apanasevich, S. P., Karpushko, F. V., and Sinitsyn, G. V. (1984). *Sov. J. Quantum Electron.* **14**, 873.

Banyai, L., and Koch, S. W. (1986). *Z. Phys. B* **63**, 283.

Bigot, J. Y., Fidorra, F., Klingshirn, C., and Grun, J. B. (1985). *IEEE J. Quantum Electron.* **QE-21**, 1480.

Bohnert, K., Kalt, H., and Klingshirn, C. (1983). *Appl. Phys. Lett.* **43**, 1088.

Cotter, D. (1986). Post deadline paper PD19, IQEC '86, San Francisco, California, June 9–13, 1986.

Dagenais, M., and Winful, H. G. (1984). *Appl. Phys. Lett.* **44**, 574.

Epshtein, E. M. (1978). *Sov. Phys. Tech. Phys.* **23**, 983.

Etchepare, J., Grillon, G., Thomazeau, I., Hamoniaux, G., and Orszag, A. (1986). In *Ultrafast Phenomena* (eds. G. R. Fleming and A. E. Siegman). Springer-Verlag, Berlin.

Flytzanis, C., Hache, F., Ricard, D., and Roussiginol, P. (1986). Paper TuNN1, IQEC '86, San Francisco, California, June 9–13, 1986.

Genzel, L., and Martin, T. P. (1973). *Surf. Sci.* **34**, 33.

Gibbs, H. M. (1985). *Optical Bistability: Controlling Light with Light.* Academic Press, New York.

Gibbs, H. M., and Peyghambarian, N. (1986). International Optical Computing Conference, Jerusalem, July 1986.

Gibbs, H. M., McCall, S. L., Passner, A., Gossard, A. C., Wiegmann, W., and Venkatesan, T. N. C. (1979). *IEEE J. Quantum Electron.* **QE-15**, 108D.

Gibbs, H. M., Tarng, S. S., Jewell, J. L., Weinberger, D. A., Tai, K., Gossard, A. C., McCall, S. L., Passner, A., and Wiegmann, W. (1982). *Appl. Phys. Lett.* **41**, 221.

Gibbs, H. M., Jewell, J. L., Moloney, J. V., Tai, K., Tarng, S. S., Weinberger, D. A., Gossard, A. C., McCall, S. L., Passner, A., and Wiegmann, W. (1983). Presented at the International Workshop on Optical Phase Conjugation and Instabilities, September 20–25, 1982, Corsica, *J. Physique* (Paris) **44**, C2–C195.

Gibbs, H. M., Olbright, G. R., Peyghambarian, N., Schmidt, H. E., Koch, S. W., and Haug, H. (1985). *Phys. Rev. A* **32**, 692.

Gibbs, H. M., Mandel, P., Peyghambarian, N., and Smith, S. D. eds. Optical Bistability 3 Springer-Verlag, Berlin, 1986.

Goldstone, J. A., and Garmire, E. (1984). *Phys. Rev. Lett.* **53**, 910.

Hanamura, E. (1981). *Solid State Commun.* **38**, 939.

Haus, J. W., Sung, C. C., Bowden, C. M., and Cook, J. M. (1985). Optical Materials Conference, Gaithersburg, Maryland, May 7–9, 1985, NBS Publication No. 697, ed. A. Feldman.

Huang, A. (1983). In *Proc. IEEE 10th International Optical Computing Conference*, p. 13.

Hulin, D., Mysyrowicz, A., Antonetti, A., Migus, A., Masselink, W. T., Morkoc, H., Gibbs, H. M. and Peyghambarian, N. (1986). *Phys. Rev. B* **33**, 4389.

Jain, R. K., and Lind, R. C. (1983). *J. Opt. Soc. Am.* **73**, 647.

Jewell, J. L., Lee, Y. H., Duffy, J. F., Gossard, A. C., Wiegmann, W. and English,

J. H. (1986). In *Optical Bistability* 3 (ed. H. M. Gibbs, P. Mandel, N. Peyghambarian and S. D. Smith). Springer-Verlag, Berlin.
Jewell, J. L., Lee, Y. H., Warren, M., Gibbs, H. M., Peyghambarian, N., Gossard, A. C., and Wiegmann, W. (1985). *Appl. Phys. Lett.* **46**, 918.
Jin, R., Wang, L., Sprague, R. W., Gibbs, H. M., Gigioli, G., Kulcke, H., Macleod, H. A., Peyghambarian, N., Olbright, G. R., and Warren, M. (1986). In *Optical Bistability* 3, (ed. H. M. Gibbs, P. Mandel, N. Peyghambarian and S. D. Smith). Springer-Verlag, Berlin.
Kar, A. K., Mathew, J. G. H., Smith, S. D., Davis, B., and Prettl, W. (1983). *Appl. Phys. Lett.* **42**, 334.
Karpushko, F. V., and Sinitsyn, G. V. (1978). *J. Appl. Spectrosc.* (USSR) **29**, 1323.
Karpushko, F. V., and Sinitsyn, G. V. (1982). *Appl. Phys. B* **28**, 137.
Knox, W. H., Fork, R. L., Downer, M. C., Miller, D. A. B., Chemla, D. S., Shank, C. V., Gossard, A. C., and Wiegmann, W. (1985). *Phys. Rev. Lett.* **54**, 1306.
Knox, W. H., Hirlmann, C., Miller, D. A. B., Shah, J., Chemla, D. S., and Shank, C. V. (1986). *Phys. Rev. Lett.* **56**, 1191.
Koch, S. W., and Haug, H. (1981). *Phys. Rev. Lett.* **46**, 450.
Koch, S. W., Schmidt, H. E., and Haug, H. (1984). *Appl. Phys. Lett.* **45**, 932.
Koch, S. W., Lee, Y. H., Chavez-Pirson, A., Gibbs, H. M., Morhange, J., Park, S. H., Carty, T., Jeffrey, A., Peyghambarian, N., Batdrof, B., Banyai, L., Gossard, A. C., and Wiegmann, W. (1986). Postdeadline paper PD14, IQEC '86, San Francisco, California, June 9–13, 1986.
Lee, Y. H., Warren, M., Olbright, G., Gibbs, H. M., Peyghambarian, N., Venkatesan, T., Smith, J. S., and Yariv, A. (1986). *Appl. Phys. Lett.* **48**, 754.
Lévy, R., Bigot, J. Y., Hönerlage, B., Tomasini, F., and Grun, J. B. (1983). *Solid State Commun.* **48**, 705.
Migus, A., Antonetti, A., Hulin, D., Mysyrowicz, A., Gibbs, H. M., Peyghambarian, N., and Jewell, J. L. (1985). *Appl. Phys. Lett.* **46**, 70.
Migus, A., Hulin, D., Antonetti, A., Mysyrowicz, A., Gibbs, H. M., Peyghambarian, N., Masselink, M., and Morkoc, H. (1986). Postdeadline paper ThU9, CLEO 1986, San Francisco, California, June 9–13.
Miller, A., Parry, G., and Daley, R. (1984). *IEEE J. Quantum Electron.* **QE-20**, 710.
Miller, D. A. B. (1984). *J. Opt. Soc. Am. B* **1**, 857.
Miller, D. A. B., Smith, S. D., and Johnston, A. (1979). *Appl. Phys. Lett.* **35**, 658.
Miller, D. A. B., Chemla, D. S., Eilenberger, D. J., Smith, P. W., Gossard, A. C., and Tsang, W. T. (1982). *Appl. Phys. Lett.* **41**, 679.
Miller, D. A. B., Gossard, A. C., and Wiegmann, W. (1984). *Opt. Lett.* **9**, 162.
Morhange, J. F., Park, S. H., Peyghambarian, N., Jeffrey, A., Gibbs, H. M., Lee, Y. H., Chavez-Pirson, A., Koch, S. W., Gossard, A. C., English, J. H., Masselink, M., and Morkoc, H. (1986). Annual Meeting of the Optical Society of America, Seattle, Washington, Oct. 19–24, 1986.
Nguyen, H. X., and Zimmermann, R. (1984). *Phys. Status Solidi B* **124**, 191.
Nuss, M. C., Zinth, W., and Kaiser, W. (1986). Paper MBB2, IQEC '86, San Francisco, California, June 9–13, 1986.
Olbright, G. R., and Peyghambarian, N. (1986). *Solid State Commun.* **58**, 332.
Olbright, G., Peyghambarian, N., Gibbs, H. M., Macleod, A., and Van Milligen, F. (1984). *Appl. Phys. Lett.* **45**, 1031.

Olbright, G. R., Peyghambarian, N., Koch, S. W., and Banyai, L. (1986). Paper TuNN2, IQEC '86, San Francisco, California, June 9–13, 1986.

Oudar, J. L., Hulin, D., Migus, A., Antonetti, A., and Alexandre, F. (1985). *Phys. Rev. Lett.* **55**, 2074.

Oudar, J. L., Migus, A., Hulin, D., Grillon, G., Etchepare, J., and Antonetti, A. (1984). *Phys. Rev. Lett.* **53**, 384.

Peyghambarian, N., and Gibbs, H. M. (1985a). *Opt. Eng.* **24**, 68.

Peyghambarian, N., and Gibbs, H. M. (1985b). *Optics News* II, 7.

Peyghambarian, N., and Gibbs, H. M. (1986). *Physics Today* **39**, S-57.

Peyghambarian, N., Gibbs, H. M., Rushford, M. C., and Weinberger, D. (1983). *Phys. Rev. Lett.* **51**, 1692.

Peyghambarian, N., Gibbs, H. M., Jewell, J. L., Antonetti, A., Migus, A., Hulin, D., and Mysyrowicz, A. (1984). *Phys. Rev. Lett.* **53**, 2433.

Peyghambarian, N., Olbright, G. R., Fluegel, B., and Koch, S. W. (1986a). Post-deadline paper PD20, IQEC '86, San Francisco, California, June 9–13, 1986.

Peyghambarian, N., Olbright, G. R., Weinberger, D. A., Gibbs, H. M., and Fluegel, B. D. (1986b). *J. Lumin.* **35**, 241.

Poole, C. D., and Garmire, E. (1984). *Appl. Phys. Lett.* **44**, 363.

Rossmann, H., Henneberger, F., and Voigt, J. (1983). *Phys. Status Solidi B* **115**, K63.

Sarid, D., Gibbons, W., and Warren, M. (1985). *Proceedings of the Optical Bistability* 3 *Conference*, Tucson, Arizona, Dec. 2–4, 1985.

Sarid, D., Jameson, R. S., and Hichernell, R. K. (1984). *Opt. Lett.* **9**, 159.

Schmidt, H. E., Haug, H., and Koch, S. W. (1984). *Appl. Phys. Lett.* **44**, 787.

Seaton, C. T., Smith, S. D., Tooley, F. A. P., Prise, M. E., and Taghizadeh, M. R. (1983). *Appl. Phys. Lett.* **42**, 131.

Shank, C. V., Fork, R. L., Leheny, R. F., and Shah, J. (1979). *Phys. Rev. Lett.* **42**, 112.

Shank, C. V., Fork, R. L., Greene, B. J., Weisbuch, C., and Gossard, A. C. (1982). *Surf. Sci.* **113**, 108.

Smith, P. W. (1980). *Opt. Eng.* **19**, 456.

Smith, P. W., and Tomlinson, W. J. (1981). *IEEE Trans. Spectrosc.* **18**, 16.

Smith, S. D., Mathew, J. G. H., Taghizadeh, M. R., Walker, A. C., Wherrett, B. S., and Hendry, A. (1984). *Opt. Commun.* **51**, 357.

Smith, S. D., Janossy, I., MacKenzie, H. A., Mathew, J. G. H., Reid, J. J. E., Taghizadeh, M. R., Tooley, F. A. P., and Walker, A. C. (1985a). *Opt. Eng.* **24**, 569.

Smith, S. D., Walker, A. C., Tooley, F. A. P., Mathew, J. H., and Taghizadeh, M. R. (1985b). *Optical Bistability* 3, (ed. H. M. Gibbs, P. Mandel, N. Peyghambarian and S. D. Smith). Springer-Verlag, Berlin.

Stegeman, G. I., Seaton, C. T., Hetherington, W. M., Boardman, A. D., and Egan, P. (1986). In *Nonlinear Optics*: *Materials and Devices* (ed. C. Flytzanis and J. L. Oudar). Springer-Verlag, Berlin.

Sung, C. C., and Bowden, C. M. (1984). *Phys. Rev. A* **29**, 1957.

Taghizadeh, M. R., Tooley, F. A. P., and Mathew, J. G. H. (1986). In *Optical Bistability* 3, (eds. H. M. Gibbs, P. Mandel, N. Peyghambarian and S. D. Smith). Springer-Verlag, Berlin.

Tarng, S. S., Gibbs, H. M., Jewell, J. L., Peyghambarian, N., Gossard, A. C., Venkatesan, T., and Wiegmann, W. (1984). *Appl. Phys. Lett.* **44**, 360.

Toyozawa, Y. (1978). *Solid State Commun.* **28**, 533.

Tsao, M. T., Wang, L., Jin, R., Sprague, R. W., Gigioli, G., Kulcke, H. M., Li, Y. D., Chou, H. M., Gibbs, H. M., and Peyghambarian, N. (1987). *Opt. Eng.*, Jan.

Venkatesan, T., Wilkins, B., Lee, Y. H., Warren, M., Olbright, G., Gibbs, H. M., Peyghambarian, N., Smith, J. S., and Yariv, A. (1986). *Appl. Phys. Lett.* **48**, 145.

Wherrett, B. S., Hutchings, D., and Russell, D. (1986). *J. Opt. Soc. Am. B* **3**, 351.

13 ELECTRIC FIELD DEPENDENCE OF OPTICAL PROPERTIES OF SEMICONDUCTOR QUANTUM WELLS: PHYSICS AND APPLICATIONS

D. A. B. Miller and D. S. Chemla

AT & T BELL LABORATORIES
HOLMDEL, NEW JERSEY

S. Schmitt-Rink

AT & T BELL LABORATORIES
MURRAY HILL, NEW JERSEY

OPTICAL NONLINEARITIES
AND INSTABILITIES IN
SEMICONDUCTORS

ISBN 0-12-332915-9

I. INTRODUCTION

The ability to grow sophisticated layered semiconductor structures offers many new opportunities in semiconductor physics and devices. A recent and unexpected benefit is that this technology is applicable to novel optical effects and devices. One type of layered structure that has received much attention in this context is the quantum well (QW), and in this chapter we shall concentrate on the physics and applications of the electric field dependence of QW optical properties. This chapter is complementary to that by Chemla et al., which discusses the nonlinear optical properties of QWs.

The effects of quantum confinement of electrons and holes in QWs are dramatically apparent in their linear optical properties. The relatively smooth absorption spectrum of bulk semiconductors is broken into a rich structure that directly reflects the quantum confinement. This structure is further enhanced by exceptionally strong excitonic effects. Application of electric fields can further enrich the structure, with behavior such as the quantum-confined Stark effect that is quite unlike that seen in bulk materials. Such remarkable effects are not only of interest from the point of view of physics, but are also relatively practical. They have stimulated a number of speculative new devices, such as optical modulators and self–electro-optic effect devices, that can operate at room temperature, with exceptionally low operating energy densities, and are compatible with laser diodes and semiconductor electronics.

In this chapter, we shall attempt to outline the necessary physics to understand the electroabsorption and related effects in QWs, and we shall summarize the principles and prospects of the various devices proposed so far.

II. LINEAR OPTICAL PROPERTIES OF QUANTUM WELLS

The most obvious consequence of the quantum confinement of carriers in quantum wells is that the single-particle (i.e., electron and hole) states are discrete in the direction perpendicular to the layers. This results in the basic steplike density of states for the single particles at low energies. Understanding these single-particle states is essential to understanding the optical properties. After discussing these states, we shall discuss the linear optical absorption. Interband optical absorption always involves the creation of an electron–hole pair, and these always interact via their Coulomb attraction.

This interaction means that the optically created electron and hole cannot be fully described in terms of the simple single-particle states, and we must introduce the concept of the exciton. Excitonic effects are already well known in bulk materials, but they are particularly strong in quantum wells. We shall very briefly discuss the excitonic states (which are treated in greater detail in Chapter 4 by Chemla et al.). Then we shall set up the formalism under which we shall discuss the theory of electroabsorption for electric fields perpendicular to the quantum well layers.

A. Single-Particle States in Quantum Wells

Single-particle states in semiconductor heterostructures are usually described in terms of the "envelope function approximation" (Bastard, 1981). This approximation assumes that the change of composition only produces small perturbations of the atomic potential and that the carrier dynamics are still determined by the band parameters of the bulk semiconductors involved in the heterostructure. The effects of the sharp interfaces are accounted for by "macroscopic" boundary conditions satisfied by the carrier envelope wave function, i.e. the continuity of the envelope wave function and the conservation of the probability current at the interfaces. Although this could appear as a crude approximation in systems where large changes of composition occur over one interatomic distance, so far it has been very successful in explaining a number of transport and optical phenomena.

Let us consider first the ideal case of a single layer of a direct, low band-gap semiconductor, 1, whose thickness is L_z and which is embedded in a direct, larger band-gap semiconductor, 2, i.e. $E_{1g} < E_{2g}$. We define E_{jg} and $m_{je,h}$ as the band gaps and the electron and hole masses in the two materials, $j = 1, 2$. We assume that the two materials have simple spherical parabolic bands

$$E_{jc}(k) = +\frac{E_{jg}}{2} + \frac{\hbar^2}{2m_{je}}k^2 \tag{1a}$$

$$E_{jv}(k) = -\frac{E_{jg}}{2} - \frac{\hbar^2}{2m_{jh}}k^2 \tag{1b}$$

The two semiconductors are usually called, respectively, the quantum well and the barrier materials. The heterostructure is completely defined if the discontinuities at the conduction and valence band extrema $V_{c,v}$ are known. Presently no first principles theory can accurately predict how the band-gap discontinuity $E_{2g} - E_{1g}$ is shared between the two bands, so that the $V_{c,v}$ have to be experimentally measured for each system (see, e.g., R. C. Miller et al., 1984).

The carriers are free to move in the quantum-well material along the plane of the layers. Their motion in the z-direction perpendicular to the layers is determined for each type of carrier by a one-dimensional Schroedinger equation

$$-\frac{\hbar^2}{2m_{e,h}}\frac{\partial^2\phi_{ne,h}}{\partial z_{e,h}^2} + V_{e,h}(z_{e,h})\phi_{ne,h} = E_{ne,h}\phi_{ne,h} \tag{2}$$

where e and h refer to the electron and hole equations, respectively, E_n is the energy of the nth level, and $V(z)$ is the structural potential (i.e. the square "quantum-well" potential due to the band discontinuities $V_{c,v}$). Note that $m_{e,h}$ also changes as we move from one material to the other. The solutions of this equation are (Schiff, 1968)

$$\sqrt{E_n}\tan\left[\frac{\pi}{2}\sqrt{E_n}\right] = \left[\frac{m_1}{m_2}(V - E_n)\right]^{1/2} \tag{3a}$$

for odd n, and

$$\sqrt{E_n}\,\mathrm{cotan}\left[\frac{\pi}{2}\sqrt{E_n}\right] = \left[\frac{m_1}{m_2}(V - E_n)\right]^{1/2} \tag{3b}$$

for even n. Here we have chosen for the unit of energy the energy of the first eigenstate of the QW for an infinitely high barrier, $E_\infty = \dfrac{\hbar^2}{2m_1}\left[\dfrac{\pi}{L_z}\right]^2$. Written in terms of these normalized parameters, Eq. 3 is universal and can be solved once and for all. It is found that there is always at least one bound state and a continuum of unbound states for a QW. For reasonable values of $V_{c,v}$ there are several bound states with a sinusoidal dependence in the quantum well and exponential decay in the barriers. In fact, it can be shown from Eq. 3 that for $1 < V < 4$ (with V measured in units of E_∞), there are two bound states, for $4 < V < 9$, there are three bound states, and so on. The low-lying bound states are well confined with little penetration into the barrier material. Typical wavefunctions and potential structures are shown in Fig. 1.

These states are two-dimensional, with a steplike density of states (including spin) given by

$$g(E) = \frac{m_{e,h}}{\pi\hbar^2}\theta(E - E_n) \tag{4}$$

where θ is the Heaviside (step) function. The continuum of states corre-

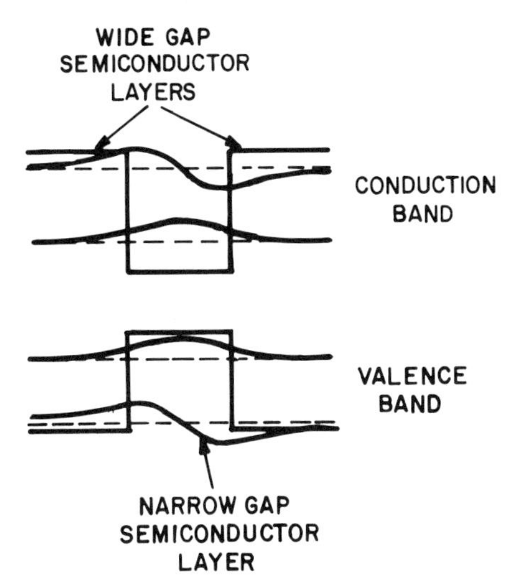

Figure 1. Energy structure and single-particle wavefunctions for a simple quantum well, showing two confined states in the conduction and valence bands (not to scale).

sponds to energies above the potential discontinuities. It is essentially three-dimensional, but it can show resonances analogous to those encountered in optics for Fabry–Perot etalons. These resonances occur when the thickness of the quantum-well layer equals an integer number of wavelengths of the running wave (Bastard, 1984). Such resonances have been observed in real systems (Zucker et al., 1984).

The ideal situation we have described applies quite well to the simple conduction band of III–V compounds. The case of the valence band is much more complicated. In bulk III–V semiconductors, the valence band consists of an upper $J = 3/2$ multiplet, fourfold degenerate at the zone center, and a $J = 1/2$ spin-orbit split-off band. For the usual growth direction, the normal to the layer is the (001) crystalographic axis, which becomes the natural axis for quantization of the angular momentum. The hole masses to be used in Eq. 3 are the masses for this direction; they are $m_{\text{lh,hh}} = m_0/(\gamma_1 \pm 2\gamma_2)$ respectively for the light (lh) ($J_z = 1/2$) and heavy (hh) hole ($J_z = 3/2$) in terms of the Luttinger parameters (Luttinger, 1956). The degeneracy at the zone center is lifted by the quantum confinement and at $k = 0$ the squares of the magnitudes of the optical transition matrix elements from the hh and lh band to the conduction band are in the ratio 3/4 and 1/4 for light polarized parallel to the layer and 0 and 1 for light polarized normal to the layer.

The situation in the plane of the layers requires more care to analyze; the simplest description uses the 4×4 Luttinger Hamiltonian with incorporation of the confinement potential in the z-direction (Chang and Schulman, 1983; Ekenberg and Altarelli, 1984; Broido and Sham, 1985). The solution leads to a set of spin-degenerate subbands with highly nonparabolic in-plane dispersion. It is easy to see why the hole subbands must be nonparabolic. The hole masses in the plane of the layers at the zone center in a situation with this symmetry are found to be $m_{\|\mathrm{hh,lh}} = m_0/(\gamma_1 \pm \gamma_2)$, respectively, for the hh and lh. If the bands are to be parabolic, this implies that the in-plane curvature of the hh subbands is larger than that of the lh ones and, thus, that band crossing would occur even for the lowest subbands. This is of course prevented by the confinement potential, which, when included in the Hamiltonian, forces the curves to repel each other at finite k, resulting in the nonparabolic dispersion we have mentioned. It is interesting to note that the character of the subbands is also mixed away from $k = 0$. If one is interested in deeper hole states, more than four subbands have to be considered and the problem quickly becomes very difficult to solve. The dispersion of the hole subbands in the plane of the layers is of course important for a number of transport properties, but it is also important for optical processes because it enters in the relative electron–hole dynamics in excitons, as we shall discuss.

In many practical cases, the sample consists of a regular stack of layers rather than a single QW surrounded by two very thick barriers. The structure of the energy spectrum of such a "superlattice" can be described in the envelope function model using an approach similar to that used in the Kronig–Penney model of the energy bands of a lattice (Bastard, 1981). Now it is found that the eigenvalue spectrum consists of a set of narrow bands for the low-lying levels and a continuum for energies above the potential discontinuities. The low-lying subbands will exhibit a small dispersion in the z-direction. However, if the thickness of the barrier and the height of the potential step are large enough, the wave functions of the first few levels do not overlap substantially from one well to the other. The dispersion along the normal becomes negligible, and the wave functions approach those of a single quantum well; this will be the case for all conditions and levels that we shall consider in this chapter. Then one can safely consider that the properties of the sample are the same as those of a set of independent quasi–two-dimensional quantum wells. For the high-energy levels, the properties are really three-dimensional, although strongly anisotropic.

B. Excitons and Optical Absorption

Before proceeding to the discussion of optical absorption, we shall consider the two-particle, electron–hole states as these are the states involved in interband optical transitions. In the most general sense of the term, all electron–hole pair states can be referred to as excitons, although it is more common usage to reserve the term *exciton* for the bound electron–hole states, and we shall generally adopt the latter convention. (The term *bound exciton* refers to an exciton bound to something else, such as an impurity.)

In the cases that we consider here, even the bound electron–hole pair states have envelope functions much larger than the unit cell. Then, the behavior of the electron and hole envelope functions is like that of the electron and proton wavefunctions of a hydrogen atom. The bound electron–hole states ("excitons") correspond to bound hydrogenic states, and the unbound states are the ionized or "scattering" states of the hydrogen atom that correspond to classical hyperbolic orbits of electron and proton. In the QW, in contrast to bulk materials, the excitons are confined in the direction perpendicular to the layers. This confinement actually makes excitonic effects stronger in the QW.

We can write the Hamiltonian of an electron and a hole in a QW, including the effects of external electric field **F**, as (D. A. B. Miller et al., 1984b, 1985a)

$$H = H_{\mathrm{e}z} + H_{\mathrm{h}z} + H_{\mathbf{r}xy} + H_{\mathbf{F}} + H_C + H_{\mathbf{R}xy} + E_{\mathrm{g}} \tag{5}$$

$H_{\mathrm{e}z}$ and $H_{\mathrm{h}z}$ are the single-particle Hamiltonians as in Eq. 2 for the electron and hole, respectively, for the direction perpendicular to the layers including the structural potentials of the QW but without field (as solved in Eq. 3). The motion in the x–y plane has been transformed into relative ($\mathbf{r}_{xy}$) and center of mass ($\mathbf{R}_{xy}$) coordinates in the usual manner, and $H_{\mathbf{r}xy} = -(\hbar^2/2\mu)\,\partial^2/\partial\mathbf{r}_{xy}^2$ and $H_{\mathbf{R}xy} = -(\hbar^2/2M)\,\partial^2/\partial\mathbf{R}_{xy}^2$ are the electron and hole relative motion and center of mass motion kinetic energy operators, respectively, in the plane of the layers. The fact that the hole masses are different in the plane of the layers from their values perpendicular to the layers is implicit in the definitions of the reduced mass $\mu = m_{\mathrm{e}}m_{\|\mathrm{h}}/(m_{\mathrm{e}} + m_{\|\mathrm{h}})$ and the total mass $M = m_{\mathrm{e}} + m_{\|\mathrm{h}}$. $H_{\mathbf{F}} = -e\mathbf{F}\cdot(\mathbf{r}_{\mathrm{e}} - \mathbf{r}_{\mathrm{h}})$ is the net potential energy from the applied field, where $\mathbf{r}_{\mathrm{e}}$ and $\mathbf{r}_{\mathrm{h}}$ are the electron and hole positions. $H_{\mathrm{C}} = -e^2/\epsilon|\mathbf{r}_{\mathrm{e}} - \mathbf{r}_{\mathrm{h}}|$ is the Coulomb energy of the electron and hole from their mutual attraction. Here, e is the electronic charge and ϵ is the appropriate dielectric constant. Although we have chosen to group the terms in the Hamiltonian H in a manner convenient for our discussion, H contains all kinetic and potential energies for the electron and hole envelope functions. This is the Hamiltonian of a confined hydrogenic system with applied field.

In the theory that we shall present explicitly in this chapter for the effects of fields perpendicular to the layers, we shall take the approach of solving for electron–hole eigenstates of H. In reality, there are no true eigenstates in the presence of a uniform field, because the particles may always tunnel to some lower energy state, but for all the perpendicular field cases that we shall consider here, the states are sufficiently long-lived that this quasi-eigenstate approach is adequate. Even without field, the states only have finite lifetime anyway because of other effects not included in H, such as scattering by phonons.

The photon energy for the creation of an electron–hole pair in a given state is the appropriate eigen energy of H. In this chapter, we presume that the photon field is sufficiently weak that the electron–hole pair states are not modified by virtual or real absorption of photons; such effects are discussed in Chapter 4 by Chemla et al. With these simplifications, the calculation of the linear absorption, with or without field, reduces to calculation of all the electron–hole pair states of H, followed by evaluation of their optical absorption strengths. The optical absorption is simply deduced from elementary time-dependent perturbation theory. In practice, the states are also broadened by mechanisms not included in H, such as phonon scattering, but we shall make no attempt to include these in the theory; where necessary, we shall include them semiempirically.

The first simplification we may make is to drop the term $H_{\mathbf{R}xy}$; for electron–hole pairs created by direct optical absorption, there can be negligible center of mass momentum because the photon momentum is so small, so we may neglect this kinetic energy term. If we neglect the Coulomb attraction term H_{C}, we lose the excitonic effects. Interband absorption then effectively creates free and noninteracting electron–hole pairs. The resulting optical absorption for the bound subbands of the QW, as described by the imaginary part, ϵ_2, of the dielectric constant, ϵ, is (D. A. B. Miller et al., 1986a, b)

$$\epsilon_2(\hbar\omega) \propto g_{2\mathrm{D}} \sum_{n,m} \left| \int_{-\infty}^{\infty} dz\, \phi_{\mathrm{e}n}(z) \phi_{\mathrm{h}m}(z) \right|^2 \theta(\hbar\omega - E_{\mathrm{g}} - E_{\mathrm{e}n} - E_{\mathrm{h}m}) \quad (6)$$

where $\phi_{\mathrm{e}n}(\phi_{\mathrm{h}m})$ is the z-wavefunction of the nth (mth) electron (hole) subband with associated subband energy $E_{en}(E_{hm})$, $\hbar\omega$ is the photon energy, and $g_{2D} = \mu/2\pi\hbar^2$ is the joint two-dimensional density of states.

Eq. 6 corresponds to a set of steps with edges at the energy separation of the corresponding electron and hole subbands. In the absence of field, for an infinitely deep well or for symmetric valence and conduction band parameters, the electron and hole wave functions are orthogonal and the selection rule $n_c = n_v$ is strictly satisfied. In general, this selection rule does not strictly apply; however, the overlap integral is usually small for $n \neq m$. The corresponding transitions appear as weak spectral features in the absorption profile; they are often called "forbidden transitions" in the literature. This picture is also simplified in that it presumes that the character of the hole states does not change with momentum in the plane, and the set of steps from one type of hole can simply be added to that from the other type of hole. Although this is certainly not correct, the theory of electroabsorption has not so far included these effects, and for much of the basic understanding of electroabsorption, this simplified approach will suffice.

If we include H_C, then, just as in bulk materials, we observe excitonic effects. The excitons are generally confined in the quantum wells. This confinement makes the exciton smaller in all directions, giving it a larger binding energy. It can therefore "complete" several classical orbits before being destroyed by optical phonon ionization even at room temperature, thus giving a relatively sharp room-temperature absorption resonance. Additionally, it is also generally true that the optical absorption strength of a given electron–hole pair state is proportional to the probability of finding the electron and hole in the same unit cell in the state in question. The smaller QW exciton thus has larger absorption strength than excitons in comparable bulk materials (Chemla and Miller, 1985). The lowest (1S) exciton is by far the strongest, and it is the only one that we shall consider here explicitly. The free electron–hole pair states near to the band-gap energy also have enhanced optical absorption strength, just as in bulk materials, because the probability of finding electron and hole in the same place is enhanced by the "hyperbolic" orbit.

One useful limit is that in which the separations between the confined levels of the electron and of the hole are large compared to the Coulomb energy of the electron–hole pair. In this "strong confinement" case, the Coulomb interaction cannot significantly perturb the motion in the z-direction (although it still profoundly influences it in the plane of the layers), and we may separate the electron–hole wavefunction in the form

$\Psi(z_e, z_h, \mathbf{r}_{xy}) = \phi_e(z_e)\phi_h(z_h)U(\mathbf{r}_{xy})$, where $U(\mathbf{r}_{xy})$ is the relative motion wavefunction of the electron and hole in the plane.

It is now straightforward to solve for the 1S exciton wavefunctions even in the presence of field by variational techniques (D. A. B. Miller et al., 1985a). In this limit, which is actually obeyed by many situations of practical interest, the optical absorption can be written

$$\epsilon_2(\hbar\omega) \propto \sum_{n,m,q} \left| \int_{-\infty}^{\infty} dz\, \phi_{en}(z)\phi_{hm}(z) \right|^2 |U_{n,m,q}(\mathbf{r}_{xy} = 0)|^2 \times \delta(\hbar\omega - E_g - E_{en} - E_{hm} - E_{n,m,q}) \tag{7}$$

where q refers to all the possible two-particle relative motion states in the x–y-plane for given states n and m perpendicular to the plane, with associated energy $E_{n,m,q}$. The two Eqs. 6 and 7 will form the basis for the coming theoretical discussion of perpendicular-field electroabsorption. It is easily shown that Eq. 7 reduces to Eq. 6 when the relative motion wavefunctions are plane waves with the associated free-particle density of states.

III. PARALLEL-FIELD ELECTROABSORPTION IN QUANTUM WELLS

For electric fields in the plane of the QW layers, the electroabsorption is qualitatively similar to that in bulk semiconductors, both theoretically (Lederman and Dow, 1976; D. A. B. Miller et al., 1985a) and experimentally (D. A. B. Miller et al., 1985a). We shall start therefore by summarizing bulk electroabsorption very briefly.

Bulk electroabsorption is often handled within a time-dependent formalism different from that used by us here to treat the perpendicular field electroabsorption, although the two are ultimately equivalent. We shall not review these methods here, as they have been covered elsewhere (Dow and Redfield, 1970). The simplest bulk electroabsorption is the Franz–Keldysh (FK) effect (Franz, 1958; Keldysh, 1958). In the FK effect, absorption appears below the optical absorption edge when an electric field is applied to the semiconductor, and ripples are observed in the absorption above the band-gap energy (Tharmalingam, 1963). This is often explained as a photon-assisted tunneling between the bands, hence the time-dependent approach. These simple models neglect the electron–hole Coulomb attraction. (In the same approximation, we shall present a model for perpendicular-field electroabsorption in QWs (the quantum-confined Franz–Keldysh effect, QCFK, which shows how that electroabsorption is related to the FK effect.)

The inclusion of excitonic effects enhances the FK effect even well below the absorption edge (Dow and Redfield, 1970; Merkulov and Perel, 1973). The exciton resonances also now broaden and shift to lower energies with field. The shift is a hydrogenic Stark shift. The broadening is a lifetime broadening because the exciton can now be destroyed in short times by field ionization. In practice, the shift is comparatively small (~ 10% of the binding energy) before the resonance is effectively destroyed by the field at fields of the order of the classical ionization field. Hence, the broadening often dominates the electroabsorption spectra.

In the ideal theoretical case of a two-dimensional exciton, the electroabsorption has been analyzed with results that are qualitatively similar to those for the three-dimensional case (Lederman and Dow, 1976). The ionization time has also been compared for the two- and three-dimensional cases (D. A. B. Miller et al., 1985a). The two-dimensional exciton is somewhat harder to ionize because of its larger binding energy and also because of some geometric effects, but the general behavior is similar. There has not so far been any analysis of the actual QW situation, where the exciton has finite thickness, but there is no reason to expect that the behavior will be qualitatively different. The only experimental work (D. A. B. Miller et al., 1985a; Knox et al., 1986) shows broadening of the

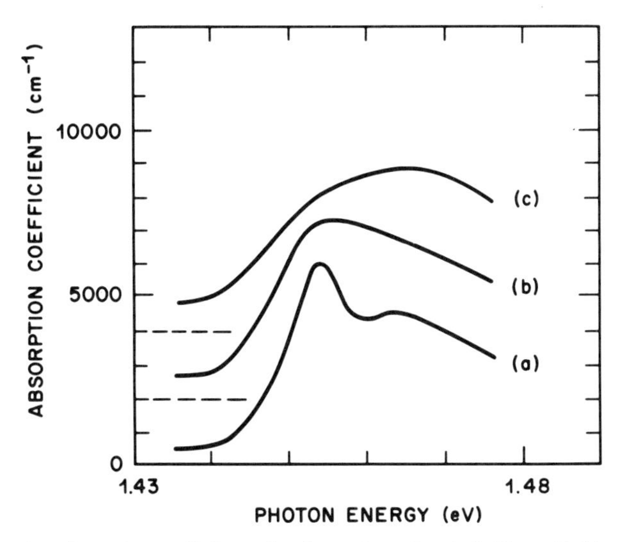

Figure 2. Spectra of quantum-well absorption for various electric fields applied in the plane of GaAs/AlGaAs quantum well layers (D. A. B. Miller et al., 1985a). (a) 0 V/cm; (b) 1.6×10^4 V/cm; (c) 4.8×10^4 V/cm.

excitons with field as expected (see Fig. 2) for GaAs wells with AlGaAs barriers.

It has been possible to demonstrate that this parallel field electroabsorption is very fast (Knox et al., 1986). In these experiments, an electric field transient was generated by shorting the electrodes with carriers created by absorption of a short light pulse, and the changes in absorption were measured to occur with a time response $\lesssim 300$ fs.

IV. PERPENDICULAR-FIELD ELECTROABSORPTION IN QUANTUM WELLS

The effects of fields perpendicular to the layers are qualitatively different from electroabsorption in bulk materials. The absorption edge shows a clear shift to lower energies while still remaining abrupt, with clearly resolved excitonic peaks at very high fields. We shall approach the explanation of this behavior in two stages. First, we shall consider the electroabsorption neglecting the Coulomb interaction, just as was done for the FK effect. In fact, we shall show that in this case, the resulting QW electroabsorption can be described as a quantum-confined Franz–Keldysh effect (QCFK) even although the behavior for thin wells is quite unlike the FK effect (D. A. B. Miller et al., 1986a). This approach is useful in understanding the QW absorption and its relation to the bulk behavior, but it cannot explain the other remarkable phenomenon in perpendicular field QW electroabsorption, namely that the exciton peaks remain resolved at very high fields (e.g., 100 times the classical ionization field). For this, we must include the Coulomb interaction in a model called the quantum-confined Stark effect (QCSE) (D. A. B. Miller et al., 1984b, 1985a).

A. Quantum-Confined Franz–Keldysh Effect

The importance of the QCFK is largely conceptual, so we shall consider the extreme case of an infinitely deep QW. Then all the states are bound, the wavefunctions all are identically zero at the walls of the well, and the problem can be solved exactly with Airy functions as the solutions. Fig. 3 shows the resulting wavefunctions for a particular illustrative case.

We may evaluate the optical absorption using Eq. 6. Fig. 4 shows the resulting optical absorption for two different QW thicknesses and bulk material, calculated for a hypothetical material such as GaAs but with only one (heavy) hole band for simplicity. At zero field, the bulk material shows the usual parabolic absorption edge, whereas the QWs show steplike

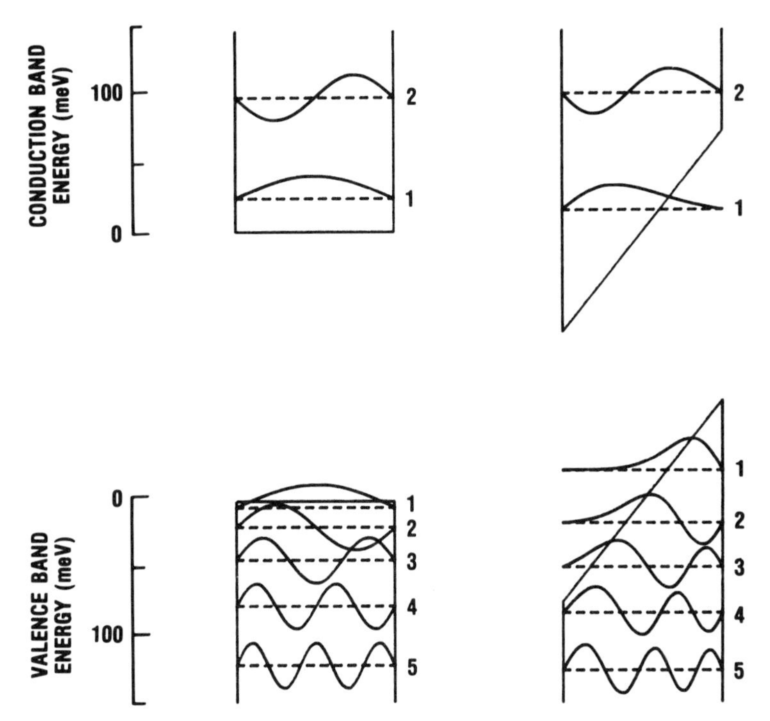

Figure 3. Valence and conduction energy levels and normalized wave functions in a 150 Å GaAs-like quantum well at 0 and 10^5 V/cm, plotted together with the net confining potential, including the effect of field (D. A. B. Miller et al., 1986a).

absorption. With applied field, the thin (100 Å) QW shows a movement to lower photon energy of the first absorption step with some loss in height, and "forbidden" transitions appear at higher photon energies, corresponding in this case to transitions from the second and third hole levels to the first electron level. The reasons for this behavior in the spectra are easy to see from the nature of the wave functions in Fig. 3. With field, the first electron and hole levels are losing overlap, so the resulting transition becomes weaker. The higher hole levels are also now no longer orthogonal to the first electron level, and so they start to acquire finite overlap. Note that the sum of the heights of all the transitions between a given electron level and all the hole levels is conserved as the field is applied. This sum rule can be proved quite generally (D. A. B. Miller et al., 1986b).

The behavior of the thick (300 Å) QW in Fig. 4 is calculated in exactly the same manner, but its behavior is much more complex and bears a very close resemblance to the bulk spectrum with field. In this case, the wavefunctions are very strongly perturbed by the field, to such an extent that the

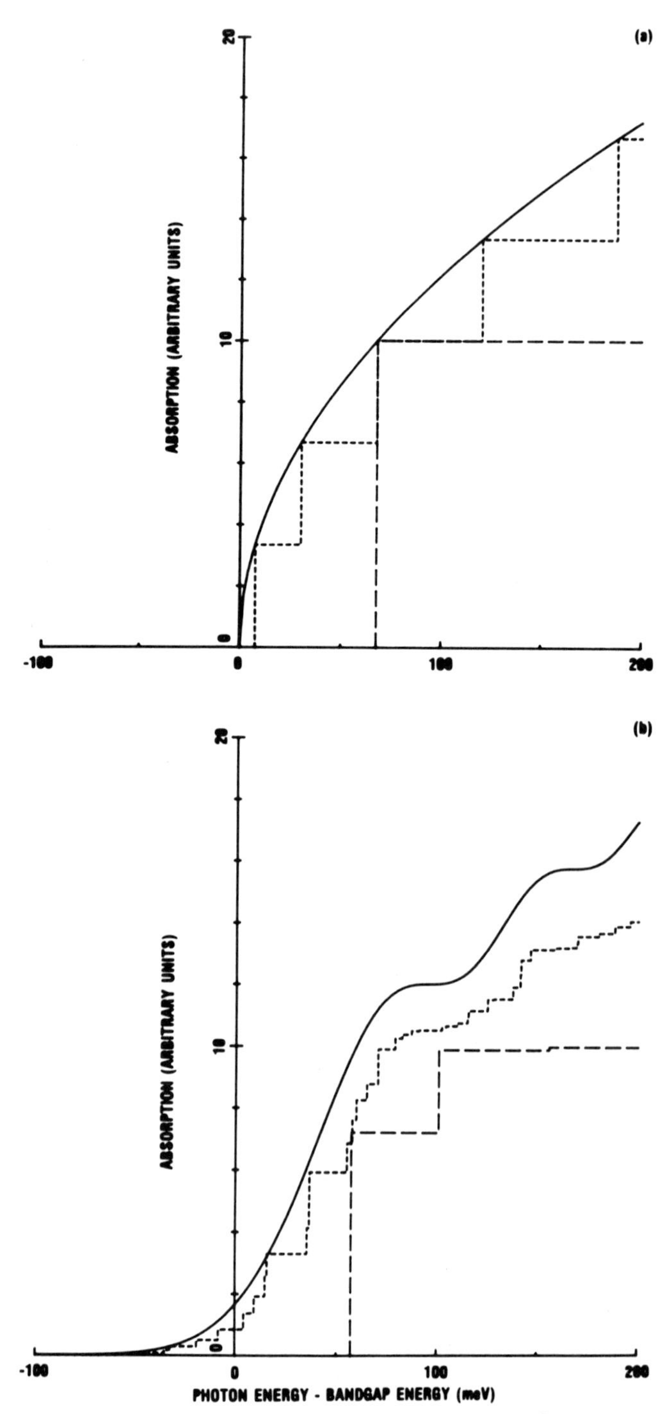

Figure 4. Theoretical absorption of GaAs-like material in a 100 Å quantum well (long dashed lines), a 300 Å quantum well (short dashed lines), and bulk material (solid lines). (a) At zero field. (b) At 10^5 V/cm (D. A. B. Miller et al., 1986a).

"forbidden" transitions are actually stronger than the "allowed" transitions. The Franz–Keldysh oscillations above the band edge are quite clearly apparent, and the weak FK absorption tail below the edge is reproduced over many orders of magnitude in a similar steplike fashion. In fact, it can be proved rigorously that the QW electroabsorption in this approximation tends to the FK effect as the QW becomes thick.

The change from strongly quantized behavior, as in the thin well, to quasi-FK behavior, as in the thick well, actually takes place rather rapidly with increasing well thickness because of the scaling of the dimensionless problem. The natural dimensionless units for solving the problem of an infinite well are the confinement energy E_∞ of the first level for the energy unit and, for the field unit, a potential drop of one confinement energy over the well width L_z. Since $E_\infty \propto 1/L_z^2$, a given actual field corresponds to a dimensionless field that scales as L_z^3.

Apart from showing the connection between the QW and bulk electroabsorptions, the QCFK also gives an alternative (and equivalent) picture of the FK effect without recourse to a time-dependent model. Now the FK effect can be understood as the transitions between the eigenstates of a thick semiconductor slab in the presence of the field. This also leads to a simple interpretation of the FK oscillations seen above the band edge (D. A. B. Miller et al., 1986a).

B. Quantum-Confined Stark Effect

As already mentioned, when excitonic effects are included in bulk electroabsorption, the excitonic absorption resonances broaden rapidly with field, with comparatively little Stark shift. In Fig. 5, we show spectra of quantum-well electroabsorption (Weiner et al., 1985; D. A. B. Miller et al., 1986b) for electric field perpendicular to the layers. Not only do we see large shifts as expected from the discussion of the QCFK above, but also the exciton peaks remain well resolved up to very high fields, in strong contrast, for example, to the spectra in Fig. 2. The spectra in Fig. 5 were taken in a waveguide sample so that the two distinct polarizations of the optical electric vector in this system, namely parallel and perpendicular to the layers, could both be used. As we have discussed, in the perpendicular polarization, the heavy-hole–to–conduction transitions are forbidden; this has nothing to do with the applied static electric field, although it is apparently not destroyed by it. The loss of the heavy-hole transitions is useful because then the spectra are particularly simple and oscillator strengths can be extracted more reliably.

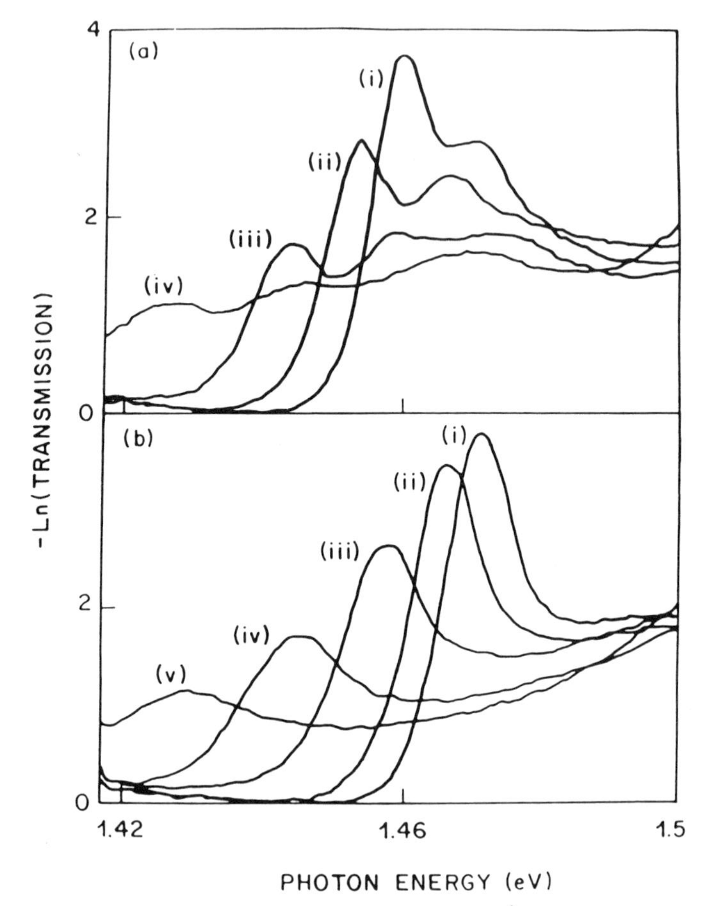

Figure 5. Absorption spectra of a waveguide containing 94 Å GaAs quantum wells, as a function of electric field applied perpendicular to the layers (Weiner et al., 1985; D. A. B. Miller et al., 1986b). (a) Incident optical polarization parallel to the plane of the layers for fields of (i) 1.6×10^4 V/cm, (ii) 10^5 V/cm, (iii) 1.3×10^5 V/cm, and (iv) 1.8×10^5 V/cm. (b) Incident optical polarization perpendicular to the plane of the layers for fields of (i) 1.6×10^4 V/cm, (ii) 10^5 V/cm, (iii) 1.4×10^5 V/cm, (iv) 1.8×10^5 V/cm, and (v) 2.2×10^5 V/cm.

To understand the exceptional behavior of the QW absorption with field perpendicular to the layers, we shall start by solving the complete problem of the confined hydrogenic system in an electric field, including the Coulomb term H_C. The final resulting model (D. A. B. Miller et al., 1984b, 1985a; Brum and Bastard, 1985) can be called the quantum-confined Stark effect (QCSE).

Using the strong confinement approximation we have discussed, the electron–hole wavefunction separates. Consequently, the problem separates, and we can proceed by (i) solving the separate electron and hole problems for particles in quantum wells skewed by the field to give the wavefunctions ϕ_e and ϕ_h with associated energies E_e and E_h, on the assumption that the Coulomb attraction between the electron and the hole is not strong enough to perturb the z motion significantly, and (ii) solving for the in-plane motion variationally using these ϕ_e and ϕ_h. We shall do this calculation only for the lowest exciton state as this is the most extreme and is the state that gives rise to the resonance that we can observe experimentally. Consequently, we use a 1S-like orbital $U_\lambda(\mathbf{r}_{xy}) = \sqrt{2/\pi}\,\lambda^{-1}\exp(-r/\lambda)$ with adjustable parameter λ. We minimize the energy E_B, which we refer to as the exciton binding energy (D. A. B. Miller et al., 1985a), where

$$E_B = \langle\Psi|H_{\mathbf{r}xy} + H_C|\Psi\rangle \tag{8}$$

Now the energy of the electron–hole pair in the presence of the field becomes the sum of the three energies calculated separately and the band-gap energy,

$$E = \langle\Psi|H|\Psi\rangle \simeq E_e + E_h + E_B + E_g \tag{9}$$

and we have a complete solution of $H|\Psi\rangle = E|\Psi\rangle$ within our approximations.

There are many methods of solving the single-particle problem. If we approximate the states as being true time-independent eigenstates, it can be solved variationally (Bastard et al., 1983) or by using the exact Airy function solution for the infinite well (with adjusted thickness to account for the penetration into the barriers) (D. A. B. Miller et al., 1984b, 1985a). It is also important that the strictly unbound nature of the states can be treated. One method that enables the quasi-bound energy levels and wavefunctions to be calculated for this problem is the tunneling resonance technique (see, e.g., D. A. B. Miller et al., 1984b, 1985a, 1986b); this also yields the width of the state, i.e. the lifetime of the quasi-bound state, which is crucial for a full understanding of the QCSE as we shall discuss later. Other methods have been proposed (see, e.g., Austin and Jaros, 1985; Klipstein et al., 1986; Hiroshima and Lang, 1986; Borondo and Sanchez-Dehesa, 1986; Singh and Hong, 1986; Ahn and Chuang, 1986). Because the states of interest are usually reasonably long-lived, all of these methods give similar results for the energy levels and wavefunctions.

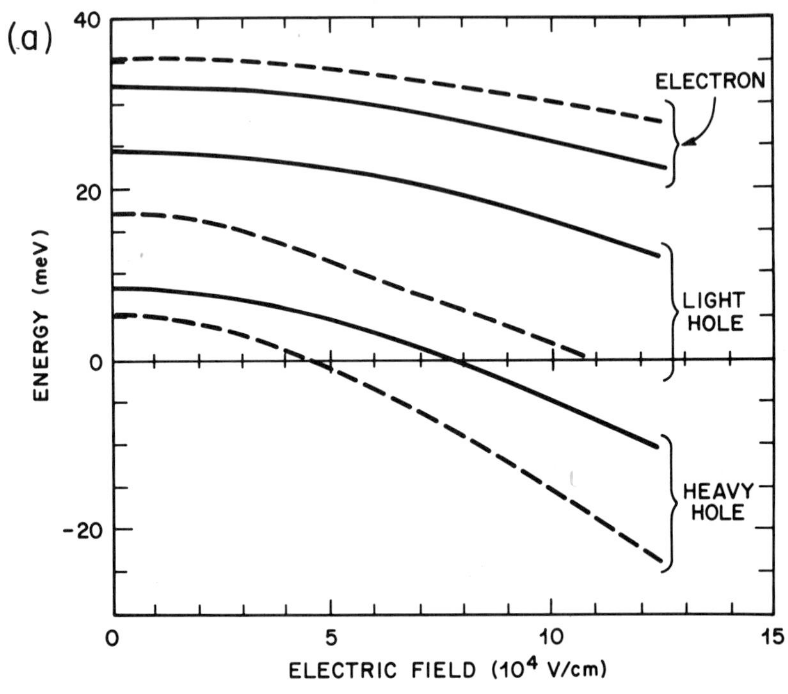

Figure 6. Calculated energies for a 95 Å GaAs quantum well with $Al_{0.32}Ga_{0.68}As$ barriers as a function of electric field perpendicular to the layers (D. A. B. Miller et al., 1985a). (a) Energies of the first electron, heavy-hole and light-hole states relative to the potential at the center of the quantum well. (b) Exciton binding energy, E_B. The solid lines assume a 57 : 43 ratio of conduction to valence band discontinuities, and the dashed lines are for a 85 : 15 ratio.

Calculations using the tunneling resonance technique for the single-particle problems are shown in Fig. 6a. The heavy hole shifts more than the light hole or electron because of its heavier mass; the heavy mass states are of lower confinement energy and are more easily perturbed. The calculations are performed for two ratios of conduction to valence band discontinuities, of which the 57 : 43 ratio is nearer to currently accepted values. In fact, the actual shift of the transitions (which varies with the sum of the electron and hole energies) is relatively insensitive to the discontinuity ratio. For smaller valence band discontinuity, the hole wavefunctions penetrate more into the barriers and also have smaller confinement energies associated with these larger wavefunctions; consequently, they are more easily perturbed by the field. The electron with larger conduction band discontinuity behaves in exactly the opposite fashion, giving a compensating effect. The appearance of "forbidden" transitions with increasing field does

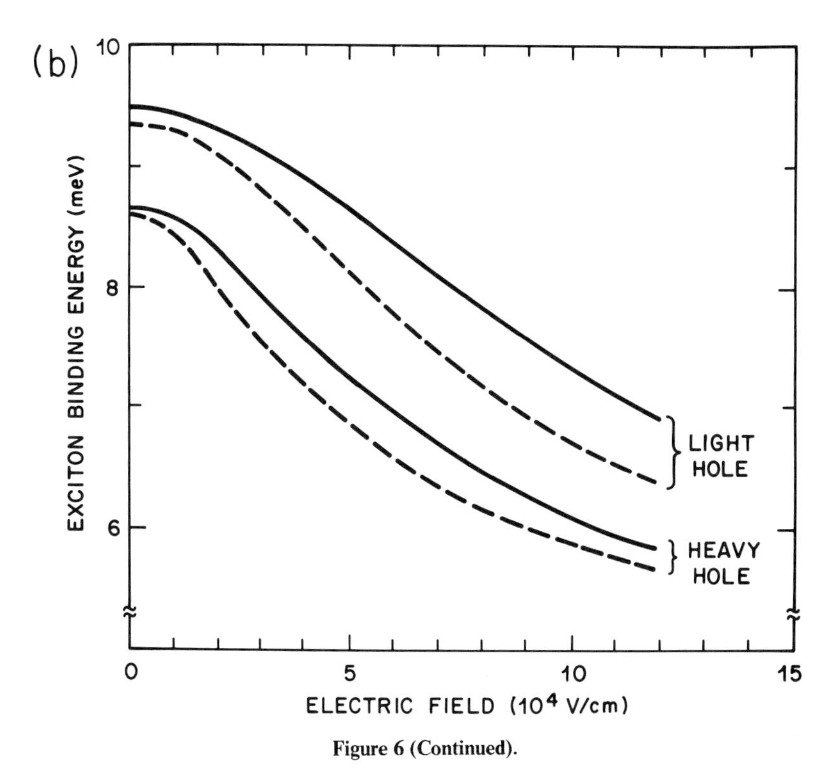

Figure 6 (Continued).

however enable the band-gap discontinuities to be measured (Yamanaka et al., 1986). Using the forbidden transition energies, it is possible to measure the energy difference between two subband energies in the same band, an energy that is sensitive to the discontinuity.

Variational calculations for the exciton binding energy, E_B, are shown in Fig. 6b. The absolute shift in this energy is not very large, although associated with it is a predicted significant increase of the exciton radius (D. A. B. Miller et al., 1985a). This energy decreases because the electron and hole are being separated by the field; hence, their attraction is weaker; hence, the larger orbit. Note, however, that even at these large fields, the exciton binding energy still exceeds the bulk exciton value (~ -4 meV). We can expect the corrections to the energy of the relative motion to be less for all other in-plane motion states and, hence, since the energy correction is relatively small for the lowest exciton, the rest of the absorption will shift almost bodily to lower energies, just as in the QCFK.

Good comparisons between experimental and theoretical shifts have been reported in both GaAs–AlGaAs QWs (D. A. B. Miller et al., 1984b, 1985a;

Alibert et al., 1985) and InGaAs–InP QWs (Bar-Joseph et al., 1987). Clear QCSE has also been observed in GaSb–AlGaSb QWs (Miyazawa et al., 1986) and InGaAs–GaAs strained layer QWs (Van Eck et al., 1986). Shifts accompanied by significant broadenings have also been seen in InGaAs–InAlAs QWs (Wakita et al., 1986a).

Thus far, we have modeled the shift of the exciton peak with field, but we have not yet explained the most unusual aspect of this QW electroabsorption, namely, why the exciton peaks should persist to such high fields. The reason is simply that the field ionization of the exciton is inhibited by the walls of the QW. The electrons and holes are pulled to the opposite walls of the QW by the field, but they do not tunnel rapidly through these walls. The tunneling time of the individual electrons and holes can be deduced directly from the tunneling resonances calculations from the width of the tunneling resonances (D. A. B. Miller et al., 1984b, 1985a). For the conditions considered here, this time is usually $\gg 1$ ps and, hence, has little effect on the exciton, which is destroyed in ~ 300 fs at room temperature from phonon ionization, as discussed in Chapter 4. (The Coulomb attraction of the electron and the hole is itself too weak to prevent the tunneling and need not be considered here.) Hence, the electron and hole can "complete" several classical orbits before dissociation, and the resonance is retained. The reason for the large shifts of the peaks is that the particle is not destroyed by the field and, hence, very large fields can be used. It is important therefore that the QW be thin compared to the bulk exciton diameter; otherwise, the exciton could be effectively field ionized by pulling the electron and hole to opposite sides of the well. The resulting shifts are simple Stark shifts of a strongly confined hydrogenic system; hence, the title QCSE. To see similar effects with an actual hydrogen atom would require confinement within less than 1 Å and fields of the order of 10^{10} V/cm.

It is interesting to note that for GaAs, the thickness of ~ 300 Å at which we could expect to lose the suppression of field ionization is approximately that at which the FK effect is recovered in the QCFK for the fields that we use here. Hence, the transition from QW to bulklike electroabsorption both with and without excitons should occur under approximately similar conditions at least for this material.

The theory also enables us to calculate the change in absorption strength with field, at least in principle. In practice, absorption sum rules help in making comparison to actual data (D. A. B. Miller et al., 1986b). It can be shown that the total area under the absorption spectrum is conserved under electroabsorption; the spectra in Figs. 2 and 5 both obey this rule em-

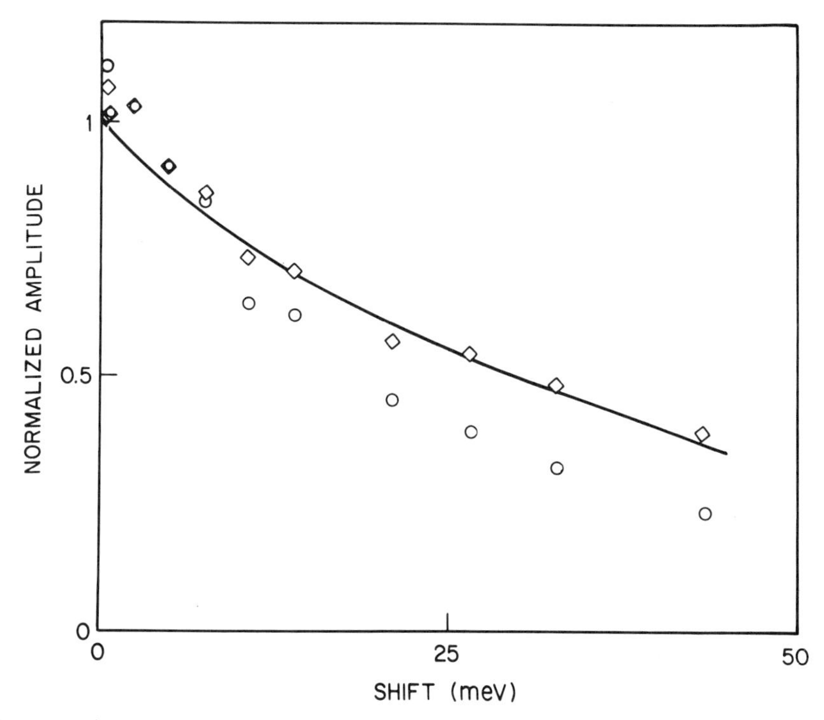

Figure 7. Comparison of the (light-hole) exciton absorption peak height (circles) and area (diamonds) with the calculated squared electron–hole overlap integral (D. A. B. Miller et al., 1986b).

pirically. Furthermore, under the strong confinement approximation, the area under the excitonic peak is approximately proportional to the overlap $I = |\int dz\, \phi_e(z)\phi_h(z)|^2$, even although the exciton becomes larger, because the oscillator strength lost from the lowest exciton state will be recovered in other excitonic states approximately in the same spectral region. Fig. 7 shows a comparison of I with the exciton area (i.e., height $\times$ width) and the exciton height for the light-hole exciton peak in Fig. 5. The exciton area follows I well, within experimental error, although the peak height does not. The small increase in the exciton width may be partly due to these other excitonic transitions, but may also be due to the fact that layer thickness fluctuations will cause larger broadening as the electron and hole are pushed against the walls of the well (Hong and Singh, 1986). The increase in width cannot be explained as a lifetime broadening resulting from tunneling of electrons or holes through the barriers (D. A. B. Miller et al., 1986b; Hong and Singh, 1986).

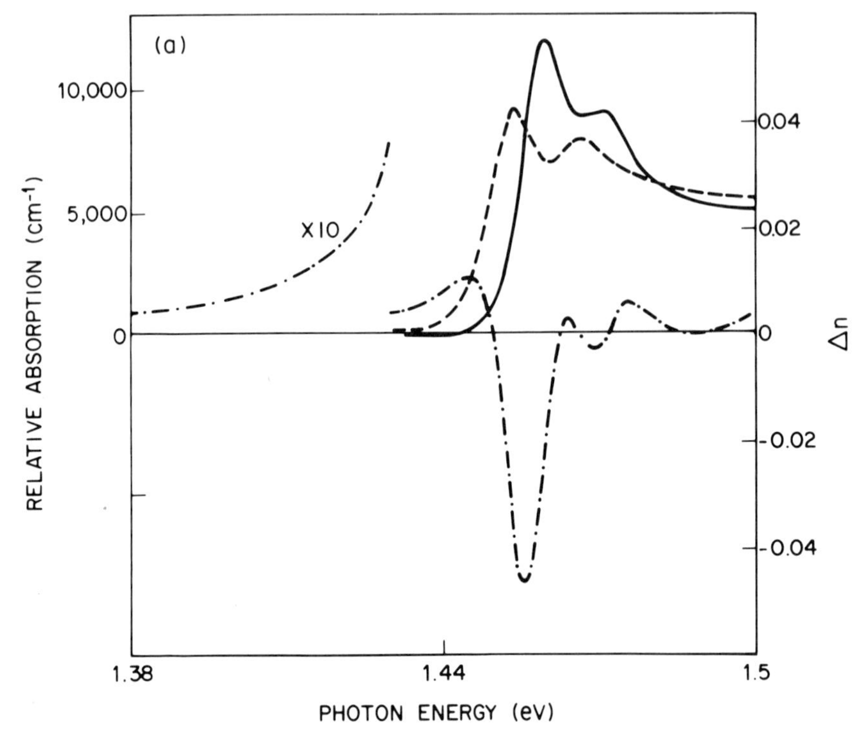

Figure 8. Field-induced changes in refractive index (Weiner et al., 1987b) for optical polarization in the plane of the layers. The solid and broken curve are the absorption spectra taken respectively at zero field and at 6.5×10^4 V/cm perpendicular to the layers. The dot–dashed curve is the calculated change in refractive index.

The changes in refractive index associated with the changes in absorption can be calculated using the Kramers–Kronig relations directly from the experimental absorption spectra (Weiner et al., 1987b) subject to the constraints imposed by the sum rules (see Fig. 8). It is also possible to calculate these from first principles (Yamamoto et al., 1985; Nagai et al., 1986), but it is difficult to include all the effects of excitons since in principle all the excitonic states would have to be known at all fields. Changes in refractive index associated with the QCSE have been measured (Glick et al., 1986; Nagai et al., 1986).

The QCSE can also be seen in structures other than "rectangular" quantum wells. Recently, Islam et al. (1987) have investigated perpendicular field electroabsorption in pairs of coupled quantum wells, with good agreement between experiment and theory for both the shifts and the changes in optical absorption strength, although some of the higher confined levels show evidence of valence band mixing. It has also recently been proposed that the piezoelectric field generated in certain strained-layer superlattices will cause a built-in QCSE that will modify the states (Mailhiot and Smith, 1986).

Related effects can be seen in luminescence, and these effects simulated some of the early interest in electric field dependence of QW optical properties (Mendez et al., 1982; Yamanishi and Suemune, 1983; Bastard et al., 1983; Miller, R. C., and Gossard, 1983). In experiments in narrow QWs, the luminescence is observed to quench with field, with an associated *decrease* in luminescence lifetime (Kash et al., 1985). In this case, the decreasing lifetime is attributed to tunneling of the particles out of the wells before radiative recombination occurs. In thicker QWs, the luminescence also becomes weaker with field, but the luminescence lifetime *increases* with field, and strong QCSE shifts are seen in the luminescence peaks (Polland et al., 1985; Kan et al., 1986). In this case, the particles do not tunnel rapidly out of the wells, and the increased lifetime and decreased luminescence intensity is due to the decreased luminescence strength that is the complement of the decreased optical absorption strength of the lowest transition with field. In the simplest model, this decrease results from the decreased electron–hole overlap integral as the particles are pulled to the opposite sides of the well. Excitonic effects could also be included in this theory, although these would not be relevant for high particle densities.

V. APPLICATIONS

QW structures have several general features that make them promising candidates for novel optical devices. Good electronic devices and laser diodes can be made out of the same materials, making QWs attractive for a variety of integrated optoelectronics. The wavelengths near the band-gap energy of the QW, where many of the interesting optical effects are observed, are compatible with laser diode light sources. Perhaps most general of all is the fact that QWs and other related layered structures offer a new degree of freedom in device design because we have a new way of engineering physical properties. When we combine these general features with the specific electro-optical effects discussed in this chapter, several devices of serious practical interest result.

A. Modulators

The changes of absorption with electric fields either parallel or perpendicular to the QW layers at room temperature are sufficient to produce significant changes in optical transmission in samples only microns thick. For the case of the parallel field, the contrast in absorption coefficients is limited, but relatively low fields (10^4 V/cm) are required. In this case, such d.c. fields can be applied with pairs of electrodes formed into the surface of material with a low carrier concentration. This effect may be of use in very-high–speed electro-optic sampling (Knox et al., 1986), where the absolute sensitivity and the low capacitance structure are advantageous.

In the case of the QCSE, the shift of the absorption without large broadening allows a significant absorption (e.g., 5000 cm^{-1}) to appear for photon energies just below the band-gap energy where the material was previously substantially transparent. The resulting large contrast in absorption coefficients is attractive for modulators where large contrast ratios are

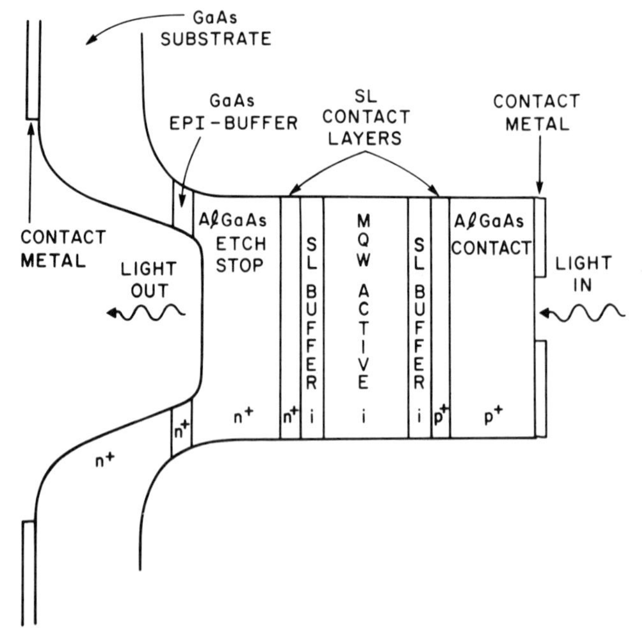

Figure 9. Typical diode structure for a QCSE modulator (not to scale). The superlattice (SL) buffer layers are optional. The multiple quantum well (MQW) layers are contained within the depletion region of the diode. The substrate is removed with a selective etch underneath the modulator (Wood et al., 1984).

desired without large insertion loss, such as in communications systems. The QCSE devices require fields $\sim 5 \times 10^4$ to 2×10^5 V/cm. The practical way of achieving these fields is to grow a diode such that the QWs are in the depletion region of the diode. The required fields can then be applied by reverse biasing the diode. A typical structure is shown in Fig. 9. Here the light propagates perpendicular to the layers, and changes in transmission of a factor of > 2 are possible for about 1 μm of QWs. Such modulators have been demonstrated in GaAs–GaAlAs (Wood et al., 1984), InGaAs–InGaAlAs (Wakita et al., 1985a), InGaAs–InAlAs (Wakita et al., 1985b), InGaAs–GaAs (a strained layer system) (Van Eck et al., 1986), and InGaAs–InP (Bar-Joseph et al., 1987). GaSb–AlGaSb also seems a promising system for QCSE modulators (Miyazawa et al., 1986). In such GaAs–GaAlAs modulators made on GaAs substrates, the GaAs substrate must usually be removed by a chemical etch because it is opaque at the operating wavelength. GaSb–AlGaSb on GaSb substrates have the same problem. It has however proved possible to make robust and complex structures in this way be epoxying the material to a sapphire substrate (see, e.g., D. A. B. Miller et al., 1986c). An alternative solution is to grow an integral mirror. Boyd et al. (1987) have successfully fabricated a QW modulator grown on a dielectric mirror composed of alternating layers of GaAs–GaAlAs superlattice and AlAs. The device operates as a modulator in reflection, the light to be modulated passing through the QW, off the mirror, and back through the modulator again. This double pass also improves the contrast ratio of the modulator.

Modulators in which the light propagates perpendicular to the layers are obviously attractive for two-dimensional arrays of modulators. This property has been exploited in the integrated SEED array to be discussed (D. A. B. Miller et al., 1986c). An electrically addressed spatial light modulator has also been proposed (Goodhue et al., 1986), in which charge-coupled devices would be formed over a multiple QW modulator structure, hence making an array of electrically controlled modulators.

An alternative geometry is to propagate the light in the plane of the QWs by confining the light in a waveguide containing a few QWs (Tarucha et al., 1985; Weiner et al., 1985; Wood et al., 1985b; Tarucha and Okamoto, 1986; Arakawa et al., 1986; Wakita et al., 1986b). Such modulators can also operate with low-voltage drive (see, e.g., Weiner et al., 1987a), and with coupled QW structures (Islam et al., 1987). In some of these waveguide structures (Weiner et al., 1985; Islam et al., 1987), modulation with low loss and high contrast can also be observed at the zero-field exciton absorption peak wavelength. This is because the overlap between electron and hole can be made very small for the $n = 1$ "allowed" transition by applying a large field, hence making the material relatively transparent. One reason behind the design of the coupled quantum-well modulator was to make a structure that would modulate in this manner. The waveguide structure is also suitable for integrating with a laser, because a QW laser can be fabricated from the same layer structure run in forward rather than reverse bias. Such integrated modulator–laser structures have been demonstrated (Tarucha and Okamoto, 1986; Arakawa et al., 1986).

Both perpendicular and waveguide modulators have been demonstrated to speeds of ~ 100 ps (Wood et al., 1985a, 1985b). The ultimate speed limitation of the QCSE is probably subpicosecond, but the practical limit is the time taken to charge up the capacitance of the device in the demonstrations so far.

QCSE modulators are in general attractive for a number of reasons. Physically, they can be quite small, with current dimensions limited by fabrication techniques so far employed. Small size helps keep the drive energy low. The energy required to drive the modulator is fundamentally the energy required to charge the volume of the modulator to the operating field, which in turn is the same thing as $(1/2)\ CV^2$. For a field of $\sim 5 \times 10^4$ V/cm, this energy is ~ 1.5 fJ for a cubic micron, i.e. 1.5 fJ/μm^2 for a 1-μm–thick modulator, making this a very-low–energy mechanism. The small size also means that there are no velocity mismatch problems at high speeds. (In long waveguide modulators using electro-optic materials, the electrical and optical waves do not propagate at the same speed, giving a mismatch that limits the bandwidth of the modulator.) It also appears that QCSE modulators may have relatively low chirp, i.e. the refractive index changes associated with the absorption changes may be relatively small in the spectral regions attractive for absorption modulation so that little excess bandwidth will be introduced by the undesired phase modulation (Weiner et al., 1987b).

B. Self–Electro-Optic Effect Devices

The energy densities of a number of fJ/μm^3 for QCSE modulation are extremely low as we have discussed, and it would be desirable to take advantage of them in nonlinear optical devices where high energy requirements are a significant problem. Even absorption saturation in semiconductors, a relatively low-energy mechanism by the standards of nonlinear optics, requires ~ 100 fJ/μm^3 for comparable changes in optical absorption at room temperature at these wavelengths. One way of utilizing the

QCSE for devices that can operate with optical inputs and outputs is to use a photodetector to drive a QCSE modulator, and this is the general principle of the self–electro-optic effect device (SEED) (D. A. B. Miller, 1985; D. A. B. Miller et al., 1984a, 1985b, 1986b, 1986c; Weiner et al., 1985, 1987a). Functionally, it behaves as a nonlinear optical device as far as the optical beams are concerned, although it will generally also have an electrical power supply and the option of electrical control as well.

Typical problems of systems that combine such optical and electronic functions are that they are clumsy, slow, and heavy energy consumers. The clumsiness can come from the fact that the various components (e.g., photodetectors, amplifiers, and modulators or light sources) rely on different technologies so that they cannot be integrated. The inability to integrate results in parasitics, particularly capacitance, associated with the packaging. These parasitics slow down the system and increase the energy requirements. Existing modulators and light sources also tend to require relatively large energies to drive them; the necessity of providing this energy limits the speed of operation in many cases. With the possible exception of the existing light sources and modulators, none of the components are themselves fundamentally slow or energy-expensive. Both photodetectors and transistors can operate on timescales of 10's of picoseconds, and the operating energies of these components can be low, although driving the interconnections between the components can consume significant energies and can slow down operation. The SEED is an attempt to solve the problems of such optoelectronic systems, firstly, by using the QCSE to give low-energy optical outputs and, secondly, to eliminate the problems of parasitics and electrical interconnections by making very intimately integrated devices in which the only electrical communication is within the integrated device itself. The technology that enables all the components to be integrated within one structure is layered semiconductor growth. The reasonable prospect with SEEDs is for devices that can operate at speeds and energies comparable with good electronic devices, but with optical inputs and outputs that can exploit the ability of optics for communicating information.

The simplest form of SEED is one in which the QW diode is itself the only photodetector. The diodes used for QCSE modulators make good photodetectors (D. A. B. Miller et al., 1984a; Wood et al., 1985b; Larsson et al., 1985, 1986), with internal quantum efficiencies near unity. A circuit that can show optically bistable behavior is shown in Fig. 10. To show optical bistability, the diode should be illuminated at a wavelength near the zero field exciton peak (e.g., near 1.46 eV in Fig. 4a). When there is no light

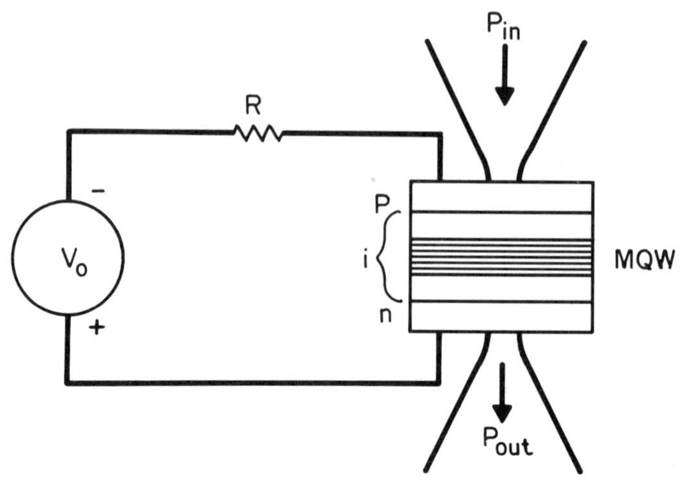

Figure 10. Simple circuit for an optically bistable SEED (D. A. B. Miller et al., 1984a).

shining on the diode, all the bias voltage is across the QWs, and the absorption is relatively low. As the light power is increased, photocurrent is generated, and, consequently, voltage is dropped across the resistor, reducing the voltage across the QWs and, hence, *increasing* the absorption. This increase in absorption results in yet more photocurrent, giving more voltage across the resistor, less voltage on the QWs, more absorption, and so on. This positive optoelectronic feedback mechanism can become strong enough to cause switching into a high absorption state. The resulting input–output characteristic is bistable, as shown in Fig. 11. This bistability is an example of the class of bistability from increasing absorption (D. A. B. Miller, 1984).

The existence of optical bistability has some interesting corollaries. If the optical power is held constant and the supply voltage is varied, the voltage across the diode will also exhibit electrical bistability (D. A. B. Miller et al., 1985b). This bistability can be viewed as resulting from the fact that the diode shows negative differential resistance; as the voltage increases across the diode, the (photo)current *decreases* because the absorption decreases. Another consequence of the negative differential resistance is that the system can be made to oscillate at constant input voltage and optical power by replacing the load resistor with an inductor (D. A. B. Miller et al., 1985b). The resulting oscillations are seen on both the voltage across the modulator and in the transmitted optical beam. Oscillation frequencies > 1 MHz have been observed.

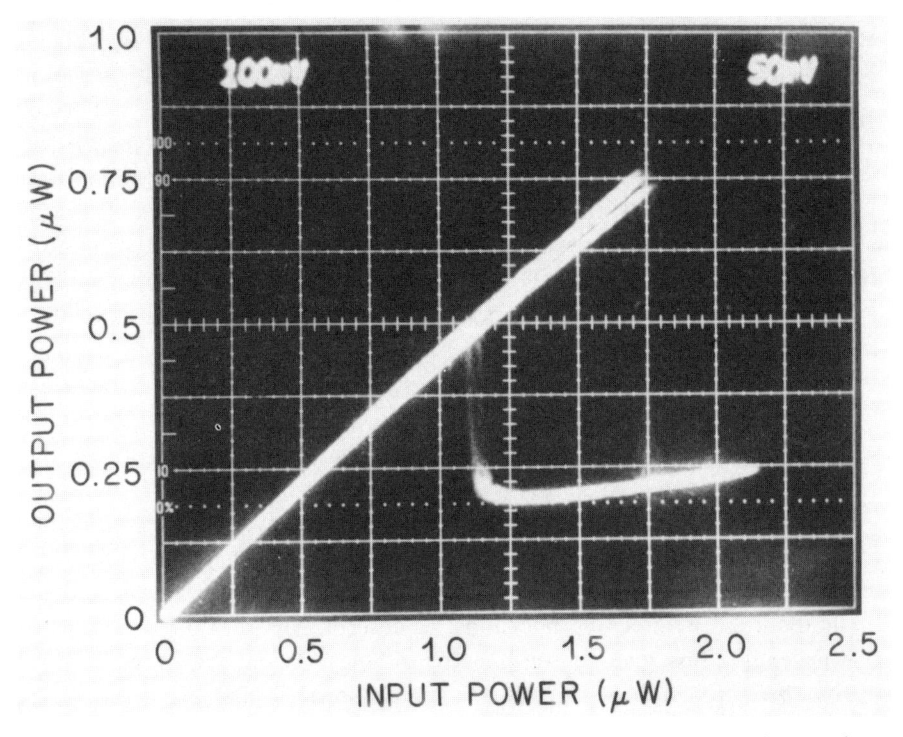

Figure 11. Superimposed optical input–output characteristics of four devices in an integrated SEED array (D. A. B. Miller et al., 1986c).

The power at which switching takes place is determined by the resistor, with larger resistors giving lower power. The switching time is approximately the resistance-capacitance time constant of the load resistor and the device capacitance. The power-speed product is constant over many orders of magnitude. For example, a switching time of 30 ns has been observed with an optical power of 1.6 mW in a 100-μm device with a 47-kΩ load resistor and an estimated device capacitance of 0.6 pF (D. A. B. Miller et al., 1985b). The photodetection is known to be capable of operating on a timescale of hundreds of picoseconds (Larsson et al., 1986), as is the modulation; consequently, faster operation is to be expected in smaller devices with improved packaging.

To allow SEEDs to be fabricated in arrays and to permit scaling to much smaller devices without incurring stray capacitances, an integrated structure has been developed (D. A. B. Miller et al., 1986c). In this structure, the function of the resistor is performed by another photodiode, transparent to the light used to illuminate the QWs, that is grown on top of the QW diode. The effective value of this load "resistor" is set by the amount of short-wavelength light shining on this top load diode and, hence, is not fixed during manufacture. Arrays of devices with built-in load "resistors" can be fabricated by etching mesas into this structure and by taking common top and bottom contacts off all the mesas. There are thus only two electrical connections to the whole array. Uniform 2×2 arrays of 200×200-μm devices have been reported (D. A. B. Miller et al., 1986c). Fig. 11 shows the characteristics of the four devices in the array superimposed. Recently also fully functional 6×6 arrays of 60×60-μm devices have been demonstrated (Livescu et al., 1987), with performance that scales with area as expected.

The SEEDs discussed so far have all employed positive feedback. By choosing to operate at longer wavelengths where the absorption increases with increasing voltage, the feedback becomes negative. As a result, if the diode is driven by a current source, the voltage across the diode will adjust itself so as to set the photocurrent equal to the source current. Since the photocurrent is directly proportional to absorbed optical power at a given wavelength, this gives a self-linearized optical modulator in which the absorbed optical power is linearly proportional to the driving current (D. A. B. Miller et al., 1984c, 1985b). If the current source is itself another reverse-biased photodiode, the power transmitted by the QW diode will decrease linearly as the power on the other diode increases, giving an inverting light-by-light modulation. It has been proposed that this could be exploited with the integrated SEED arrays to make optically addressed spatial light modulators (D. A. B. Miller, 1985, Livescu et al., 1987).

Another application of the negative feedback is to make an optical level shifter (Miller et al., 1984c, 1985b). Here, if the QW diode is driven with a constant current, it will attempt to absorb a constant amount of optical power, even as the input power is varied. This therefore tends to subtract a constant power from the transmitted beam, shifting the level of the optical signal (see Fig. 12).

The various examples of SEEDs demonstrated so far are only a few of the many possible devices. Many different electronic components can be integrated with the QW modulator using the layered growth technology, including, for example, phototransistors (D. A. B. Miller et al., 1985). Such structures would introduce gain into the system without the need for positive feedback. This could be useful for optical logic systems as it would remove the need for biasing near some critical threshold in order to obtain the necessary gain. Various other applications of such transistor structures are possible.

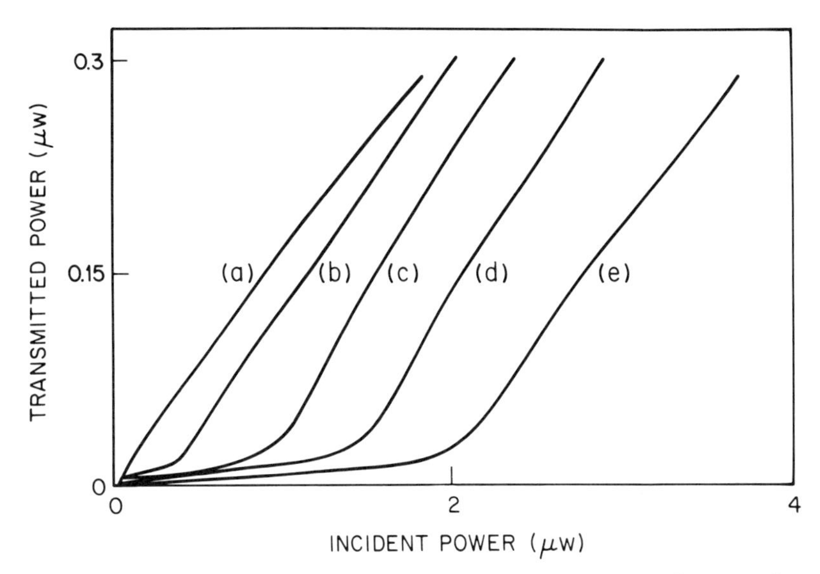

Figure 12. Transmitted optical power versus incident optical power for a SEED operated as an optical level shifter with drive currents of (a) 0, (b) 125, (c) 250, (d) 375, and (e) 500 nA (D. A. B. Miller et al., 1986b).

C. Optical Emitters

Devices such as light-emitting diodes or laser diodes that emit light because of the simultaneous presence of electrons and holes in the same material are often limited in how fast they can be turned off by the time taken for these carriers to be removed from the material. One proposed solution to this problem (Yamanishi and Suemune, 1983) is to use a confined structure such as a QW in which the carriers can be pulled to opposite sides of the well by applying a field perpendicular to the layers. With the carriers physically separated to some degree, the overlap between them and, hence, their luminescence strength will be reduced and the luminescence diminished. Note that this is accomplished not by removing the carriers completely from the material, but merely by moving them by a distance ~ 100 Å. Hence, there is a possibility that this technique might provide a fast way of modulating light-emitting devices.

As we have discussed, the physics of this effect is essentially the same as the QCSE, and has been modeled as such (Polland et al., 1985; Kan et al., 1986). It is now clear that the luminescence can indeed be modulated in this way. Applying the field does reduce the luminescence emission intensity and actually increases the overall luminescence emission time as would be expected (Polland et al., 1985; Kan et al., 1986). This reduction in luminescence intensity can be induced rapidly by sudden changes in the voltage across the sample, and it has been demonstrated that this can effectively turn off emission in times much less than the lifetime (Kan et al., 1986). The luminescence also tunes to longer wavelengths as the field is increased, as would be expected from the QCSE. A related effect has been seen in the emission from transistor structures (Levi et al., 1987).

VI. CONCLUSIONS

The basic mechanisms that govern QW electroabsorption are relatively clear, as are the relations and distinctions between QW and bulk effects. Of course, QWs only represent one of the first quantum-confined systems that we can control sufficiently to investigate such phenomena. Just as in other unusual physical effects in layered semiconductor structures, hopefully some of this physics will carry over into other confined systems as suitable technology develops.

It is also clear that real and novel devices can be made using the remarkable properties of QWs, and that these devices can operate under practical and attractive conditions. The ability to integrate optical and electronic functions to such an intimate degree that they are almost indistinguishable is a tantalizing goal for optoelectronics, offering the best of both electronic and optical worlds. The combination of integration by layered semiconductor growth and the new physical mechanisms, both electronic and optical, offered by quantum-confined structures may allow us to reach that goal.

REFERENCES

Ahn, D., and Chuang, S. L. (1986). *Phys. Rev. B* **34**, 9034–9037.

Alibert, C., Gaillard, S., Brum, J. A., Bastard, G., Frijlink, P., and Erman, M. (1985). *Solid State Commun.* **53**, 457–460.

Arakawa, Y., Larsson, A., Paslaski, J., and Yariv, A. (1986). *Appl. Phys. Lett.* **48**, 561–563.

Austin, E. J., and Jaros, M. (1985). *Appl. Phys. Lett.* **47**, 274–276.

Bar-Joseph, I., Klingshirn, C., Miller, D. A. B., Chemla, D. S., Koren, U., and Miller, B. I. (1987). Appl. Phys. Lett. **50**, 1011–1012.

Bastard, G. (1981). *Phys. Rev. B* **24**, 5693–5697.

Bastard, G. (1984). *Phys. Rev. B* **30**, 3547–3549.

Bastard, G., Mendez, E. E., Chang, L. L., and Esaki, L. (1983). *Phys. Rev. B* **28**, 3241–3245.
Borondo, F., and Sanchez-Dehesa, J. (1986). *Phys. Rev. B* **33**, 8758–8761.
Boyd, G. D., Miller, D. A. B., Chemla, D. S., McCall, S. L., Gossard, A. C., and English, J. H. (1987). *Appl. Phys. Lett.* **50**, 1119–1121.
Broido, D. A., and Sham, L. J. (1985). *Phys. Rev. B* **31**, 888–892.
Brum, J. A., and Bastard, G. (1985). *Phys. Rev. B* **31**, 3893–3898.
Chang, Y. C., and Schulman, J. N. (1983). *Appl. Phys. Lett.* **43**, 536–538.
Chemla, D. S., and Miller, D. A. B. (1985). *J. Opt. Soc. Am. B* **2**, 1155–1173.
Dow, J. D., and Redfield, D. (1970). *Phys. Rev. B* **1**, 3358–3371.
Ekenberg, U., and Altarelli, M. (1984). *Phys. Rev. B* **30**, 3569–3572.
Franz, W. (1958). *Z. Naturforsch. A* **13**, 484–489.
Glick, M., Reinhart, F. K., Weimann, G., and Schlapp, W. (1986). *Appl. Phys. Lett.* **48**, 989–991.
Goodhue, W. D., Burke, B. E., Nichols, K. B., Metze, G. M., and Johnson, G. D. (1986). *J. Vac. Sci. Technol. B* **4**, 769–772.
Hiroshima, T., and Lang, R. (1986). *Appl. Phys. Lett.* **49**, 639–641.
Hong, S., and Singh, J. (1986). *Appl. Phys. Lett.* **49**, 331–333.
Islam, M. N., Hillman, R. L., Miller, D. A. B., Chemla, D. S., Gossard, A. C., and English, J. H. (1987). *Appl. Phys. Lett.* **50**, 1098-1100.
Kan, Y., Yamanishi, M., Usami, Y., and Suemune, I. (1986). *IEEE J. Quantum Electron.* **QE-22**, 1837–1844.
Kash, J. A., Mendez, E. E., and Morkoc, H. (1985). *Appl. Phys. Lett.* **46**, 173–175.
Keldysh, L. V. (1958). *Zh. Eksp. Teor. Fiz. Pisma Red.* **34**, 1138–1141 (*Sov. Phys. JETP* **7**, 788–790).
Klipstein, P. C., Tapster, P. R., Apsley, N., Anderson, D. A., Skolnick, M. S., Kerr, T. M., and Woodbridge, K. (1986). *J. Phys. C* **19**, 857–871.
Knox, W. H., Miller, D. A. B., Damen, T. C., Chemla, D. S., Shank, C. V., and Gossard, A. C. (1986). *Appl. Phys. Lett.* **48**, 864–866.
Larsson, A., Yariv, A., Tell, R., Maserjian, J., and Eng, S. T. (1985). *Appl. Phys. Lett.* **47**, 866–868.
Larsson, A., Andrekson, P. A., Andersson, P., Eng, S. T., Salzman, J., and Yariv, A. (1986). *Appl. Phys. Lett.* **49**, 233–235.
Lederman, F., and Dow, J. D. (1976). *Phys. Rev. B* **13**, 1633–1642.
Levi, A. F. J., Hayes, J. R., Gossard, A. C., and English, J. H. (1987). Preprint.
Livescu, G., Miller, D. A. B., Henry, J. E., Gossard, A. C., and English, J. H. (1987), Paper ThU11, Conference on Lasers and Electro Optics, Baltimore, (Optical Society of America).
Luttinger, J. M. (1956). *Phys. Rev.* **102**, 1030–1041.
Mailhiot, C., and Smith, D. L. (1986). *J. Vac. Sci. Technol. B* **4**, 996–999.
Mendez, E. E., Bastard, G., Chang, L. L., Esaki, L., Morkoç, H., and Fischer, R. (1982). *Phys. Rev. B* **26**, 7101–7104.
Merkulov, I. A., and Perel, V. I. (1973). *Phys. Lett.* **45A**, 83–84.
Miller, D. A. B. (1984). *J. Opt. Soc. Am. B* **1**, 857–864.
Miller, D. A. B. (1985). United States Patent 4,546,244.
Miller, D. A. B., Chemla, D. S., Damen, T. C., Gossard, A. C., Wiegmann, W., Wood, T. H., and Burrus, C. A. (1984a). *Appl. Phys. Lett.* **45**, 13–15.

Miller, D. A. B., Chemla, D. S., Damen, T. C., Gossard, A. C., Wiegmann, W., Wood, T. H., and Burrus, C. A. (1984b). *Phys. Rev. Lett.* **53**, 2173–2176.

Miller, D. A. B., Chemla, D. S., Damen, T. C., Wood, T. H., Burrus, C. A., Gossard, A. C., and Wiegmann, W. (1984c). *Opt. Lett.* **9**, 567–569.

Miller, D. A. B., Chemla, D. S., Damen, T. C., Gossard, A. C., Wiegmann, W., Wood, T. H., and Burrus, C. A. (1985a). *Phys. Rev. B* **32**, 1043–1060.

Miller, D. A. B., Chemla, D. S., Damen, T. C., Wood, T. H., Burrus, C. A., Gossard, A. C., and Wiegmann, W. (1985b). *IEEE J. Quantum Electron.* **QE-21**, 1462–1476.

Miller, D. A. B., Chemla, D. S., and Schmitt-Rink, S. (1986a). *Phys. Rev. B* **33**, 6976–6982.

Miller, D. A. B., Weiner, J. S., and Chemla, D. S. (1986b). *IEEE J. Quantum Electron.* **QE-22**, 1816–1830.

Miller, D. A. B., Henry, J. E., Gossard, A. C., and English, J. H. (1986c). *Appl. Phys. Lett.* **49**, 821–823.

Miller, R. C., and Gossard, A. C. (1983). *Appl. Phys. Lett.* **43**, 954–956.

Miller, R. C., Kleinman, D. A., and Gossard, A. C. (1984). *Phys. Rev. B* **29**, 7085–7087.

Miyazawa, T., Tarucha, S., Ohmori, Y., Suzuki, Y., and Okamoto, H. (1986). *J. Appl. Phys. Japan* **25**, L200–L202.

Nagai, H., Yamanishi, M., Kan, Y., Suemune, I., Ide, Y., and Lang, R. (1986). Conf. Solid State Devices and Materials, Aug. 22–26, 1986, Tokyo, Japan.

Polland, H.-J., Schultheis, L., Kuhl, J., Goebel, E. O., and Tu, C. W. (1985). *Phys. Rev. Lett.* **55**, 2610–2613.

Schiff, L. I. (1968). *Quantum Mechanics*. McGraw-Hill, New York.

Singh, J., and Hong, S. (1986). *IEEE J. Quantum Electron.* **QE-22**, 2017–2021.

Tarucha, S., and Okamoto, H. (1986). *Appl. Phys. Lett.* **48**, 1–3.

Tarucha, S., Iwamura, H., Saku, T., and Okamoto, H. (1985). Japan. *J. Appl. Phys.* **24**, L442–444.

Tharmalingam, K. (1963). *Phys. Rev.* **130**, 2204–2206.

Van Eck., T. E., Chu, P., Chang, W. S. C., and Wieder, H. H. (1986). *Appl. Phys. Lett.* **49**, 135–136.

Wakita, K., Kawamura, Y., Yoshikuni, Y., and Asahi, H. (1985a). *Electron. Lett.* **21**, 338–340.

Wakita, K., Kawamura, Y., Yoshikuni, Y., and Asahi, H. (1985b). *Electron. Lett.* **21**, 574–576.

Wakita, K., Kawamura, Y., Yoshikuni, Y., Asahi, H., and Uehara, S. (1986a). *IEEE J. Quantum Electron.* **QE-22**, 1831–1836.

Wakita, K., Kawamura, Y., Yoshikuni, Y., and Asahi, H. (1986b). *Electron. Lett.* **22**, 907–908.

Weiner, J. S., Miller, D. A. B., Chemla, D. S., Damen, T. C., Burrus, C. A., Wood, T. H., Gossard, A. C., and Wiegmann, W. (1985). *Appl. Phys. Lett.* **47**, 1148–1150.

Weiner, J. S., Miller, D. A. B., Chemla, D. S., Gossard, A. C., and English, J. H. (1987a). *Electron. Lett.* **23**, 75–77.

Weiner, J. S., Miller, D. A. B., and Chemla, D. S. (1987b). *Appl. Phys. Lett.* **50**, 842–844.

Wood, T. H., Burrus, C. A., Miller, D. A. B., Chemla, D. S., Damen, T. C., Gossard, A. C., and Wiegmann, W. (1984). *Appl. Phys. Lett.* **44**, 16–18.

Wood, T. H., Burrus, C. A., Miller, D. A. B., Chemla, D. S., Damen, T. C., Gossard, A. C., and Wiegmann, W. (1985a). *IEEE J. Quantum Electron.* **QE-21**, 117–118.

Wood, T. H., Burrus, C. A., Tucker, R. S., Wiener, J. S., Miller, D. A. B., Chemla, D. S., Damen, T. C., Gossard, A. C., and Wiegmann, W. (1985b). *Electron. Lett.* **21**, 693–694.

Wood, T. H., Burrus, C. A., Gnauck, A. H., Wiesenfeld, J. M., Miller, D. A. B., Chemla, D. S., and Damen, T. C. (1985c). *Appl. Phys. Lett.* **47**, 190–192.

Yamamoto, H., Asada, M., and Suematsu, Y. (1985). *Electron. Lett.* **21**, 579–580.

Yamanaka, K., Fukunaga, T., Tsukada, N., Kobayashi, K. L. I., and Ishii, M. (1986). *Appl. Phys. Lett.* **48**, 840–842.

Yamanishi, M., and Suemune, I. (1983). *Japan. J. Appl. Phys.* **22**, L22–L24.

Zucker, J. E., Pinczuk, A., Chemla, D. S., Gossard, A. C., and Wiegmann, W. (1984). *Phys. Rev. B* **29**, 7065–7068.

14 OPTICAL AND OPTOELECTRONIC NONLINEARITY IN BISTABLE Si AND InP DEVICES

D. Jäger and F. Forsmann

INSTITUT FÜR ANGEWANDTE PHYSIK
UNIVERSITÄT MÜNSTER
D-4400 MÜNSTER, FEDERAL REPUBLIC OF GERMANY

I. INTRODUCTION

During the past years, the discovery of large optical nonlinearities of semiconductors has spurred both fundamental and applied research in optically bistable device technology. This section describes recent results on Si and semi-insulating InP at a wavelength of 1.06 μm and at room temperature. Special attention is given to artificial optical nonlinearities in

OPTICAL NONLINEARITIES
AND INSTABILITIES IN
SEMICONDUCTORS

ISBN 0-12-332915-9

hybrid systems where optoelectronic and electro-optic properties are integrated. As a result, a self–electro-optic effect takes place which leads to huge, externally controllable nonlinearities.

II. NONLINEAR OPTICAL PROPERTIES

At the wavelength $\lambda_0 = 1.064\ \mu m$ of the Nd–YAG laser, the quantum energy of the photons is 1.164 eV, which is slightly greater than the optical energy gap of 1.112 eV of Si at a temperature $T = 295$ K. As a result, a free-carrier plasma can be generated by the laser beam which can in turn alter the optical properties and establish an optical nonlinearity of the material. Obviously, this is a resonant effect which primarily depends on the physics of the second step, i.e. the influence of the free carriers on the optical properties.

Two mechanisms have been found in the past to be relevant for Si. Firstly, the free-carrier plasma gives directly rise to a change in the optical absorption and refraction which can be described by the classical Drude model. Experimentally, this kind of optical nonlinearity has been observed when using short optical pulses from a Q-switched laser system to study four-wave mixing (Jain et al., 1979), free-carrier gratings (Eichler and Massmann, 1982), and self-modulation by a Fabry–Perot cavity (Eichler et al., 1982). Secondly, the thermo-optical properties also seem to play an important role, especially when the measurements are performed by a cw laser. In that case, the lattice of the semiconductor is heated up by the recombining charge carriers. The resulting change of the refractive index then leads to optical bistability in Si (Eichler, 1983). Obviously, these two mechanisms are fundamental and can be applied to other semiconductors and appropriate laser systems.

Since the band gap of InP is 1.29 eV, the direct generation of photocarriers at 1.064 μm is comparatively small. The situation, however, can be changed by diffusion of Fe atoms which is usually done to get semi-insulating material (Glass et al., 1984). On the other hand, at high intensities of a Q-switched laser, Faradzhev et al. (1985) have recently observed bistability in n–InP by virtue of an avalanche breeding of additional photocarriers by the heated plasma.

For more details of these all-optical nonlinearities, the reader is referred to the literature and other parts of this book. In the following, we shall particularly concentrate on a general possibility to enhance optical nonlinearities by electrical means.

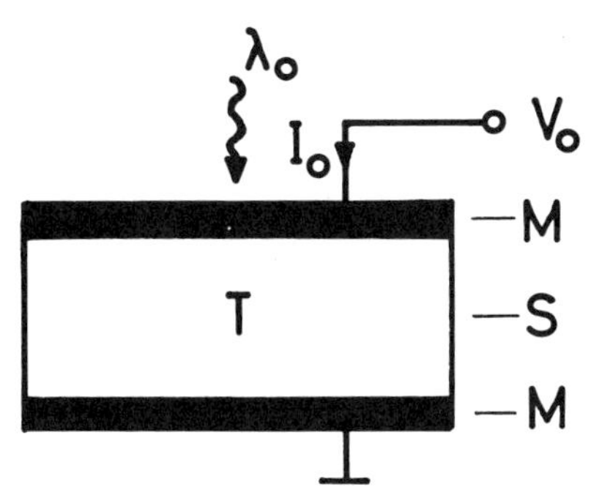

Figure 2. Experimental SEED devices. (a) Schottky diode on *n*-type Si. (b) A proposed microwave structure for ultrafast phenomena (see text). (1)–(4) denote different possibilities for the optical input with respect to the direction of current flow.

III. SELF–ELECTRO-OPTIC EFFECT AND OPTOELECTRONIC GAIN

The following concept of a self–electro-optic effect that leads to large nonlinearities is based upon the thermo-optical properties as an example, i.e., we assume that the absorption coefficient α and the refractive index n depend on the temperature T of the sample. In Fig. 1, a basic hybrid configuration is sketched to elucidate the principle: As a result of the optically absorbed power P_a, free charge carriers of density N are generated which give rise to a photocurrent I_{ph} and, in turn, to an electrical power P_e —Joule's heat—so that the temperature is changed. This alters the optical parameters and consequently an optical nonlinearity is established. Clearly, $P_e = I_{ph}(P_a)V_0$ is the optically induced power and is a function of P_a.

For the purpose of theoretical simplicity, we set in a first step

$$\alpha = \alpha_0 + \alpha' \Delta T \tag{1}$$

and

$$n = n_0 + n' \Delta T \tag{2}$$

where $\Delta T = T - T_0$ is the mean temperature rise of the sample. In a second step, we assume as a first-order approximation that ΔT is directly proportional to the total absorbed power to yield

$$\Delta T = R_h P_h \tag{3}$$

where $P_h = P_a + P_e$ and R_h is the usual thermal resistance (Jäger et al., 1985). As can be seen, with respect to the all-optical case with $P_e = 0$, we can introduce an optoelectronic gain $G = 1 + P_e/P_a$, i.e. the temperature rise is enhanced by the additional electrical power. Finally, the optical nonlinearity can now be estimated by using Eq. 3 in Eqs. 1 and 2, revealing the following properties. The nonlinearity is also enlarged by the gain factor G which can be controlled electrically by V_0 and where the $I_0(V_0)$ characteristic plays an important role. Moreover, a high internal photosensitivity $I_{ph}(P_a)$ will be advantageous.

This nonlinearity which can be described by the following simple scheme

$$P_{in} \rightarrow \overbrace{P_a \rightarrow N \rightarrow I_{ph}}^{P} \rightarrow \overbrace{P_e(V_0) \rightarrow \Delta T \rightarrow \alpha, n}^{M}$$

results from a two-step process: optoelectronic (P) and electro-optic properties (M) are combined via the thermal effect. P_{in} is the optical input power. Here, the dashed lines denote the just-mentioned all-optical nonlinearities. Obviously, other mechanisms that lead to an electro-optic effect are also imaginable. The applied electric or magnetic field, the carrier density, etc. then replace the parameter "temperature." Due to an integration of the optoelectronic and electro-optic properties, i.e. due to the simultaneous operation of the sample as a photodetector and a modulator, the devices are called self–electro-optic effect devices or SEEDs (Miller et al., 1984, and this volume). In the present case, we use this term both for absorptive and dispersive self-modulation. In the following, the nonlinearity is used to study bistability in special SEED devices where feedback is provided as shown by the dotted line in the scheme just described.

IV. BISTABILITY AND MULTISTABILITY

A. SEED Devices

Due to the additional choice of the metallic pattern and the physics of the metal–semiconductor contact, a huge variety of SEED devices can be imagined even if the semiconductor material is given. When designing a structure, it may be advantageous to start with special configurations that are well known from common photodetectors or electro-optic modulators. In Fig. 2, two examples are shown in order to give an impression. In particular, the properties of the Schottky SEED in Fig. 2(a) are directly connected with the depletion layer at the ground contact. Electrically, the dark current of this device is small. Optically, this layer serves as a back mirror, thus forming a Fabry–Perot interferometer (Airy formalism) when

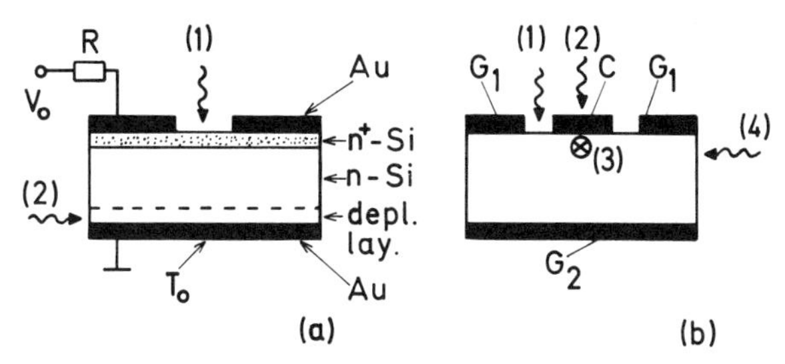

Figure 2. Experimental SEED devices. (a) Schottky diode on *n*-type Si. (b) A proposed microwave structure for ultrafast phenomena (see text). (1)–(4) denote different possibilities for the optical input with respect to the direction of current flow.

illumination at (1) is chosen and reflection at the semiconductor surface is considered. An optical input directly into the depletion layer at (2) will produce extremely efficient devices due to the high electric fields; wave-guiding properties may also be advantageous, leading to interesting concepts for integrated optics. From the electrical point of view, the structure in Fig. 2(b) will be suitable for high-speed operation. The sketched cross section is that of common microwave transmission lines—the center conductor C together with G_1 and G_2 form coplanar and microstrip lines, respectively. Those structures are currently under investigation for microwave integrated circuits (Jäger, 1985) and picosecond optoelectronics (Ketchen et al., 1986) and are well known from integrated optics technology. Again, different metal semiconductor contacts and illumination conditions can be chosen.

B. Bistability in Etalons

Experimental results of the reflected versus the input optical power for a Si Schottky Fabry–Perot interferometer are plotted in Fig. 3. A clear optical bistability and tristability are observed which can be traced back to a dispersive nonlinearity via $n(T)$ and a feedback by the cavity. As can be seen, the first switching occurs at $P_{\text{in}} \approx 300\ \mu\text{W}$. Hence, from the corresponding all-optical value of ≈ 0.5 W, a gain $G \approx 1000$ is estimated so that nearly the total amount of necessary power is provided electrically. Using $n' = 2.4 \times 10^{-4}\ \text{K}^{-1}$, the optical path length is increased by half a wavelength when the temperature is enhanced by 7.4 K and switching occurs. It should be noted that the optoelectronic, $I_0(P_{\text{in}})$, the electro-optic, $P_{\text{refl}}(V_0)$, as well as the electrical, $I_0(V_0)$, characteristics exhibit a similar multistable

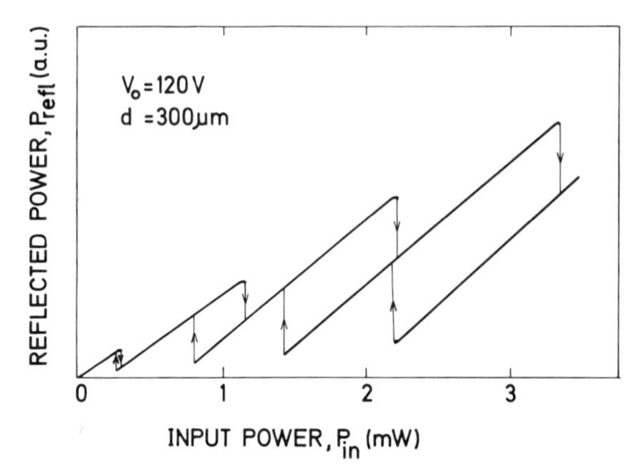

Figure 3. Optical bistability in a Schottky SEED of Fig. 2(a) with $R = 0$, illumination at (1). The resistivity of the substrate is 10–20 Ω cm. For more details, see Jäger and Forsmann (1987).

behavior as in Fig. 3. Fig. 4 shows an example of I_0 versus V_0 for different values of the optical input power. From this plot, one also judges the influence of the electrical circuit on the nonlinearity. One concludes that the switching points can be connected by the dashed lines of constant, opto-electronically induced electrical power, which of course are purely hyperbolic if the influence of the dark current can be neglected. The three

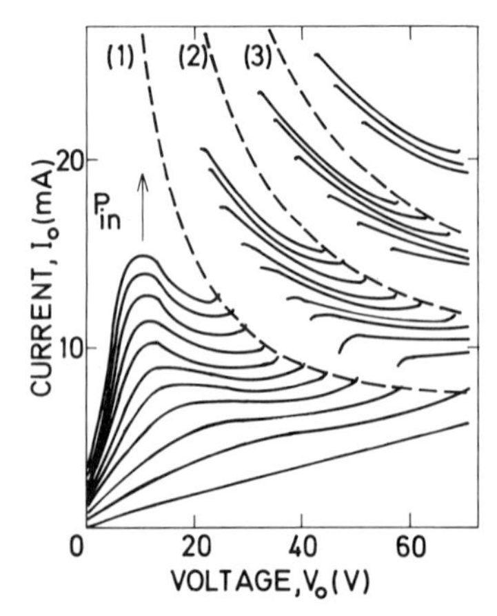

Figure 4. Measured current voltage characteristics of a SEED device of Fig. 2(a). Parameter of the curves is the optical input power, increasing from $P_{in} = 0$ to 12.5 mW in steps of 1.25 mW, as indicated by the arrow. Dotted lines represent constant electric power $P_e = 280$ mW (1), 580 mW (2), and 880 mW (3).

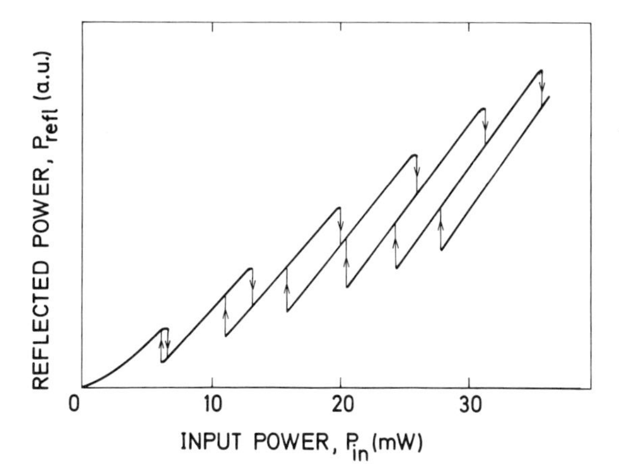

Figure 5. Optical bistability of a SEED device on semi-insulating InP, with experimental conditions as in Fig. 2(a).

different lines correspond to values that are multiples of 0.3 W, the required power between two steps.

To verify the underlying concept of optical nonlinearity, similar experiments as in Fig. 3 have been performed on InP:Fe substrat, (See Fig. 5.) Probably due to the impurity-induced band tailing, the photosensitivity is sufficiently high even at 1.064 μm, so that again bistability and a pronounced multistability can be detected. In this case $n' = 2.8 \times 10^{-4}$ K^{-1} and hence $\Delta T = 5.3$ K between each order.

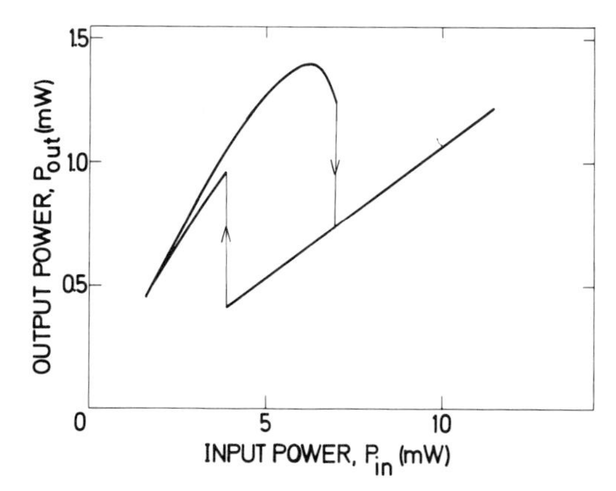

Figure 6. Bistability in a Schottky SEED of Fig. 2(a), case (2), i.e. optical input into the depletion layer. Input and output surfaces are wedge-shaped to prevent interferometer effects.

C. Self-Induced Absorption

If the presented SEED devices are just optically nonlinear elements as described above, bistability due to increasing absorption—$\alpha(T)$—should also be observable since the absorptive nonlinearity is also increased by the optoelectronic gain. Experimentally, we used the optical configuration (2) in Fig. 2(a). As can be seen from the transfer characteristics in Fig. 6, a clear bistability is obtained, again at milliwatt input powers in accordance with the underlying concept.

D. Dynamic Behavior

The dynamics of the SEED devices are basically determined by the three interlocking mechanisms: (i) Electronically, the generation of the photocarriers is determined by the temporal behavior of the optical signal whereas the plasma density is reduced by recombination and transport as diffusion or drift. (ii) Electrically, the temporal current flow depends on the external circuit as well as on the device properties such as the cut-off frequency of the Schottky diode. At high frequencies, propagation effects on the strip lines may also play a role. (iii) The thermal properties are mainly defined by the heat-generation process and the relaxation by diffusion. To describe this behavior, one can also use equivalent circuits even in the inhomogeneous case (Wedding and Jäger, 1985).

Experimentally, the switching times have been obtained from the temporal behavior of the reflected signal of a SEED device when the input signal is periodically modulated. From the results in Fig. 7, a switch-on time of about 100 μs and a switch-off time of about 250 μs are obtained. By

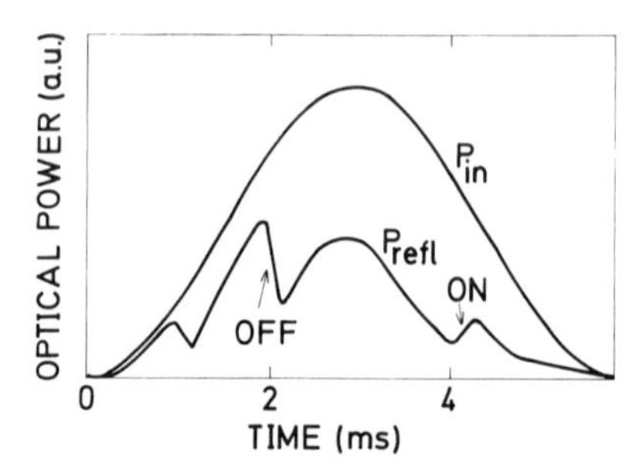

Figure 7. Dynamic behavior of a SEED device on Si material. A plot P_{refl} versus P_{in} reveals that two hysteresis cycles are recorded. The second order switching processes are marked.

changing the thermal boundary conditions, it is found that these time constants are mainly determined by the heat conduction. Finally, it should be mentioned that the SEED devices exhibit also instabilities such as self-pulsing (Wedding and Jäger, 1985).

V. SPECIAL APPLICATIONS

The special applications that can be foreseen result first from the selected materials. Since Si and InP are at present time basic materials for micro-electronics and optical communication techniques, respectively, the SEED devices are advantageous whenever a connection between electronic and optical circuits is desired. This is particularly true in case of digital techniques and binary or even multivalued logic. A special hybrid computer is imaginable. Additionally, the Nd–YAG laser is a well-established technology.

Further on, the nonlinear optoelectronic properties may be useful for specific optoelectronic devices such as a photodetector. Fig. 8 presents the characteristics of such a novel SEED detector. In Fig. 8(a), the short circuit current is plotted versus P_{in}. Fig. 8(b) shows a situation where at P_{in} = 7 mW, a differential gain is achieved yielding a photosensitivity of more

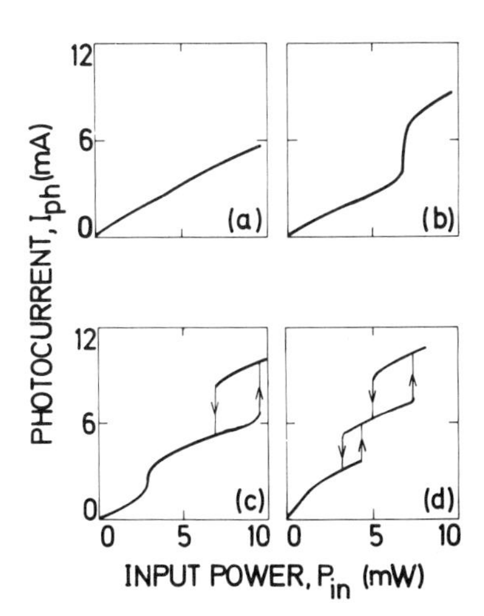

Figure 8. Photodetector characteristics of a Schottky SEED device of Fig. 2(a). (a)–(d) show the results for increasing voltage (V_0 = 0, 60, 120, and 180 V) and increasing dark current (I_d = 0, 0.85, 1.3, and 2.25 mA), respectively.

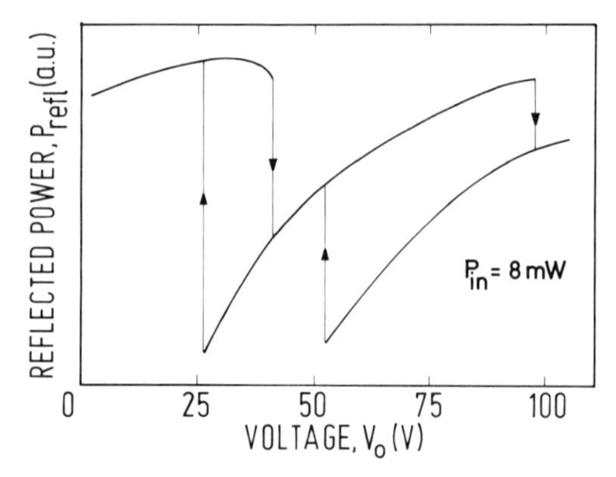

Figure 9. Modulator characteristics of a Si Schottky SEED of Fig. 2(a).

than 10 A/W. At larger values of the applied voltage, the Schottky SEED photodetector behaves as a threshold device where switching occurs between different states—see Figs. 8(c) and (d). These interesting and novel properties are optically induced in contrast to the usual photodetectors where, for example, an electrical gain by avalanche processes is utilized. An additional application results in case of instabilities, when a cw optical signal and a dc voltage produce optical as well as electrical pulses where, however, picosecond time scales would be most interesting.

The electro-optic properties of a SEED modulator are quite similar to the optoelectronic characteristics: high sensitivity and threshold possibilities (See Fig. 9.) We finally mention the interesting electrical properties of the devices where N- and S-shaped behaviors occur. Diodes with widely adjustable parameters can be realized and replace devices with purely electrical mechanisms.

REFERENCES

Eichler, H. J. (1983). *Opt. Commun.* **45**, 62–66.

Eichler, H. J., and Massmann, F. (1982). *J. Appl. Phys.* **53**, 3237–3242.

Eichler, H. J., Massmann, F., and Zaki, Ch. (1982). *Opt. Commun.* **40**, 302–306.

Faradzhev, B. G., Areshev, I. P., Stepanova, M. I., and Subashiev, V. K. (1985). *Sov. Tech. Phys. Lett.* **11**, 313–314.

Glass, A. M., Johnson, A. M., Olson, D. H., Simpson, W., and Ballmann, A. A. (1984). *Appl. Phys. Lett.* **44**, 948–950.

Jäger, D. (1985). *Int. J. Electron.* **58**, 649–669.
Jäger, D., and Forsmann, F. (1987). *Solid-State Electron.*, **30**, 67–71.
Jäger, D., Forsmann, F., and Wedding, B. (1985). *IEEE J. Quantum Electron.* **QE-21**, 1453–1457.
Jain, R. K., Klein, M. B., and Lind, R. C. (1979). *Opt. Lett.* **4**, 454–456.
Ketchen, M. B., Grischkowsky, D., Chen, T. C., Chi, C-C., Duling, I. N., III, Halas, N. J., Halbout, J-M., Kash, J. A., and Li, G. P. (1986). *Appl. Phys. Lett.* **48**, 751–753.
Miller, D. A. B., Chemla, D. S., Damen, T. C., Gossard, A. C., Wiegmann, W., Wood, T. H., and Burrus, C. A. (1984). *Appl. Phys. Lett.* **45**, 13–15.
Wedding, B., and Jäger, D. (1985). *Proc. SPIE, ECOOSA '84* **492**, 391–396.

15 OPTICAL BISTABILITY IN SEMICONDUCTOR LASER AMPLIFIERS

M. J. Adams, H. J. Westlake and M. J. O'Mahony

BRITISH TELECOM RESEARCH LABORATORIES
MARTLESHAM HEATH
IPSWICH IP5 7RE
ENGLAND

I. INTRODUCTION

It has recently become clear that optical bistability (OB) in semiconductor laser amplifiers has many potential advantages for applications in optical logic and signal processing. Foremost among these advantages may be listed (i) ready availability compared with some other bistable devices (see, e.g., Gibbs, 1985), (ii) wavelength compatibility with optical communica-

ISBN 0-12-332915-9

tions systems, (iii) inherent optical gain (for serial processing), and (iv) a combination of microwatt switching powers and nanosecond switching times leading to femtojoule optical switching energies (Sharfin and Dagenais, 1986a; Adams, 1986). In support of item (ii) on this list, it is worth noting that amplifier OB has been reported at 0.8 μm (Nakai et al., 1983), 1.3 μm (Sharfin and Dagenais, 1985), and 1.5 μm (Westlake et al., 1985), all of which are wavelengths employed for optical communications systems. The present chapter is concerned with items (iii) and (iv) of this list and seeks to give the physical reasons for these desirable attributes being found in laser amplifiers. Thus, in the next section, we discuss the physical mechanism of OB in amplifiers and then go on in section 3 to explore the important roles played by feedback and current injection in a resonant laser structure. Next, experimental results on 1.5-μm amplifiers are presented and compared with theoretical predictions. A final section summarises our findings and notes the scope for systems applications of OB in laser amplifiers.

II. PHYSICAL MECHANISM OF OB IN AMPLIFIERS

The mechanism of OB in laser amplifiers stems from the relatively strong dependence of refractive index (N) on electron concentration (n) for wavelengths close to the forbidden band-gap. This dependence (dN/dn) is responsible for many other aspects of semiconductor laser behaviour, e.g., wavelength chirping effects, phase and frequency modulation, and spectral linewidth of a single longitudinal mode. Perhaps the first intimations as to the significance of dN/dn came from the early observation of longitudinal mode spectra below threshold (Matthews et al., 1972). It was found that the characteristic spectral "comb" shifted in wavelength as the current was varied below threshold. This behaviour was interpreted as being due to refractive index changes (since the resonant modes are determined by the optical path length which depends on refractive index), and a detailed theoretical model was developed by Thompson (1972).

A second source of optical nonlinearity that also occurs in semiconductor lasers is the saturation of net optical gain with increasing input signal. In effect, this saturation comes about as a result of the dependence of material gain (g_m) on carrier density n. Once again, this dependence dg_m/dn is an important factor in determining such properties as threshold current, frequency and damping time of relaxation oscillations, amplitude modulation characteristics, and mode spectra. The theoretical analysis of material gain in semiconductors has been studied by many authors, notably Stern

(1976); a reliable experimental technique for deriving dg_m/dn from measurements of laser spectra below threshold was first established by Hakki and Paoli (1973).

The ratio of dN/dn and dg_m/dn is conventionally expressed in terms of a parameter b, defined as:

$$b = \frac{\dfrac{-4\pi}{\lambda}\left(\dfrac{dN}{dn}\right)}{\left(\dfrac{dg_m}{dn}\right)} \tag{1}$$

where λ is the wavelength. The quantity b is of relevance to the strength of optical confinement in gain-guided lasers (Kirkby et al., 1977), injection-locking properties of lasers (Lang, 1982), and the linewidth of a longitudinal mode (Henry, 1982), and it is variously known as the antiguiding parameter or the linewidth broadening factor. It can be calculated directly from first principles (Vahala et al., 1983) or measured experimentally by a range of methods. (For a review with numerous references to the literature, see Osinski and Buus, 1987.) Fig. 1 shows a recent experimental result for b as a function of photon energy near the band-gap, measured on two 1.5-μm InGaAsP laser diodes (Westbrook, 1986). The strong dependence on photon energy close to the band-gap is consistent with theoretical calculations (Vahala et al., 1983). The lasing photon energy for both devices was 0.81 eV (1.53 μm), and at this energy the value of b is close to 5. For 1.3-μm to 1.6-μm InGaAsP and 0.85-μm GaAs lasers, values of b are found to lie typically in the ranges 4 to 7 (Osinski and Buus, 1987) and 2.5 to 4, respectively.

The quantities dN/dn and dg_m/dn are closely related. Indeed, by simply differentiating the Kramers–Kronig relation (see, for example, Stern, 1964) with respect to carrier concentration n, we find:

$$\frac{dN(E)}{dn} = -\frac{\hbar c}{\pi}\int_0^\infty \frac{\dfrac{dg_m(E')}{dn}}{(E'^2 - E^2)}\,dE' \tag{2}$$

where E is the photon energy and the other symbols have their usual meaning. Readers who are familiar with calculations of nonlinear refraction in passive semiconductor samples will recognise Eq. (2) from the corresponding equation with gain replaced by absorption (Miller et al., 1981). The index change in the latter case comes about from the blocking of absorbing transitions by thermalised populations of electrons and holes which are created by absorption of an incident light beam. Thus, an

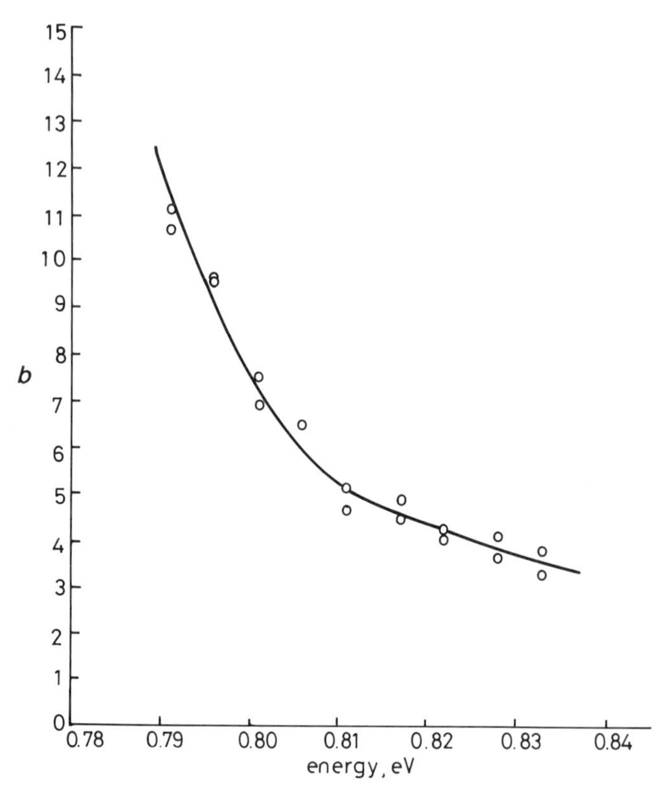

Figure 1. Linewidth enhancement factor b as a function of photon energy (two devices). (From Westbrook, 1986.)

increase of incident light intensity results in an increase of carrier concentrations and a consequent decrease of refractive index. In the case of the amplifier, by contrast, the carrier concentrations are injected electrically into the active region of the device and the incident light beam has the effect of depleting the carrier densities as a consequence of optical gain. Thus, for the amplifier increasing optical input leads to an *increase* in the refractive index.

III. EFFECTS OF A RESONANT CAVITY

The sources of optical nonlinearity discussed in the previous section are employed in a laser amplifier by means of a resonant cavity. The cavity may be of the Fabry–Perot (FP), distributed Bragg reflector (DBR), or distributed feedback (DFB) types. For pedagogical purposes, we shall consider here the FP cavity. In any case, the single-pass phase change ϕ in a cavity of length L is given by:

$$\phi = \phi_0 + \frac{2\pi L}{\lambda} \frac{dN}{dn} (n - n_1) \tag{3}$$

where n_1 is the electron concentration in the absence of an input signal, and ϕ_0 is the initial phase detuning as determined by the wavelength of the input optical signal. Strictly speaking, in a resonant cavity, the carrier concentration n is a function of position along the length, since the cavity contains forward- and backward-travelling waves each of which depends on position and is related to n via position-dependent rate equations. However, it turns out to be a very good approximation to assume a single average value I_{av} of the optical intensity in the cavity (Adams et al., 1985). Then the position-*independent* rate equation connecting n and I_{av} is given by:

$$\frac{dn}{dt} = \frac{j}{ed} - \frac{n}{\tau} - \frac{\Gamma g_m I_{av}}{E} \tag{4}$$

where t represents time, j the injected current density, e the electron charge, d the active layer thickness, τ the recombination lifetime, and Γ the fraction of the optical intensity confined to the active layer (the so-called confinement factor). For reference, a typical laser amplifier is illustrated in Fig. 2 with the relevant structural parameters indicated.

In order to find the mean optical intensity I_{av}, a straightforward longitudinal averaging procedure is carried out on the forward- and backward-tra-

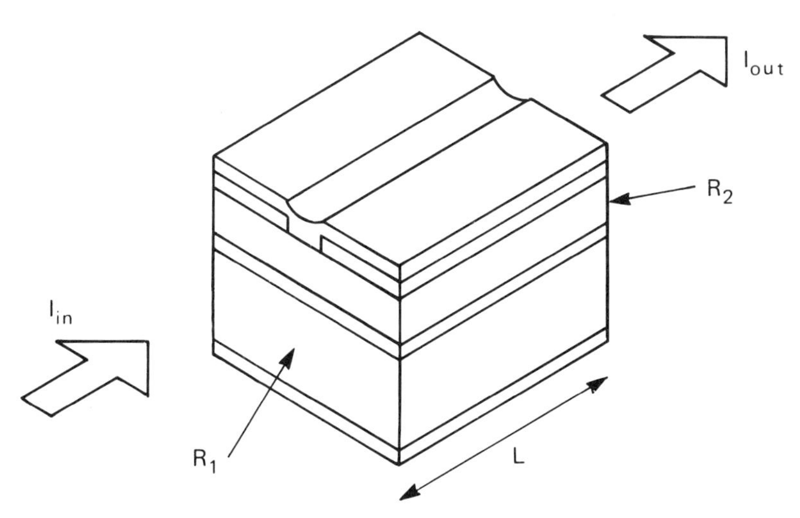

Figure 2. Schematic illustration of a semiconductor laser amplifier.

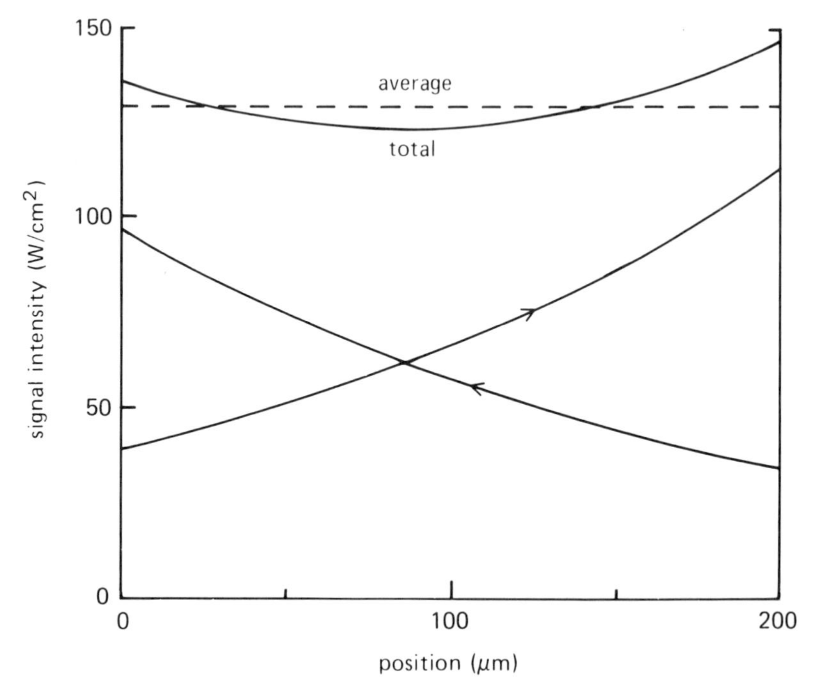

Figure 3. Signal intensity at resonance as a function of position along the cavity for forward and backward traveling waves. (From Adams et al., 1985.)

velling waves. This is illustrated in Fig. 3 (Adams et al., 1985) where the solid lines correspond to the travelling wave intensities and the total intensity as functions of position along the cavity, and the broken line denotes I_{av}. For this case of uncoated facets with reflectivity 30%, the total intensity varies only slowly along the cavity length. A self-consistent position-dependent calculation (Middlemast and Sarma, 1984) shows that the corresponding variation of electron concentration n along the length for this case is even smaller, in fact less than 0.2%. Hence, the approximation of a position-independent rate equation (Eq. 4) is justified.

If spontaneous emission is ignored (on the basis that the input intensity for bistability is usually significantly larger than the spontaneous emission at the wavelength of interest), then the average internal intensity I_{av} can be related to the input optical intensity I_{in} as follows (Adams et al., 1985):

$$I_{av} = I_{in} \frac{(1 - R_1)(1 + R_2 e^{gL})(e^{gL} - 1)}{\left[1 - (R_1 R_2)^{1/2} e^{gL}\right]^2 (1 + F \sin^2\phi) gL} \tag{5}$$

where

$$F = \frac{4(R_1R_2)^{1/2}e^{gL}}{\left[1 - (R_1R_2)^{1/2}e^{gL}\right]^2} \tag{6}$$

The corresponding relation between output and input intensities, I_{out}, I_{in}, respectively, is:

$$\frac{I_{\text{out}}}{I_{\text{in}}} = \frac{(1 - R_1)(1 - R_2)e^{gL}}{\left[1 - (R_1R_2)^{1/2}e^{gL}\right]^2(1 + F\sin^2\phi)} \tag{7}$$

In Eqs. 5 through 7, R_1 and R_2 are the input and output facet reflectivities, respectively (see Fig. 2), and g is the net gain per unit length experienced by the optical fields. The latter quantity is related to the material gain g_{m} by:

$$g = \Gamma g_{\text{m}} - \alpha \tag{8}$$

where α is an effective loss coefficient which includes scattering and absorption losses both inside and outside the active region of the amplifier. Finally, the material gain g_{m} is assumed to vary linearly with the carrier concentration n (Stern, 1976):

$$g_{\text{m}} = \frac{dg_{\text{m}}}{dn}(n - n_0) \tag{9}$$

where n_0 is the electron concentration at transparency.

With the aid of the definitions of b in Eq. 1, ϕ in Eq. 3, and g_{m} in Eq. 9, the rate equation (Eq. 4) can be rewritten in a more convenient form in terms of ϕ (Adams, 1985):

$$\frac{d\phi}{dt} = (\phi_0 - \phi)\left(1 + \frac{I_{\text{av}}}{I_{\text{s}}}\right) + \frac{g_0Lb}{2}\frac{I_{\text{av}}}{I_{\text{s}}} \tag{10}$$

In this equation, g_0 is the unsaturated material gain per unit length, and it will be evaluated as a given proportion of the material gain at threshold. Threshold is defined by the condition for the denominator of Eq. 5 to be zero on resonance, i.e., $2(gL)_{\text{th}} = -\ln(R_1R_2)$. The other new quantity appearing in Eq. 10 is the scaling intensity I_{s}, defined as:

$$I_{\text{s}} = \frac{E}{\Gamma\tau\dfrac{dg_{\text{m}}}{dn}} \tag{11}$$

The corresponding relation for net gain g in terms of ϕ is:

$$gL = \Gamma g_0L + \Gamma(\phi_0 - \phi)\frac{2}{b} - \alpha L \tag{12}$$

In the steady state ($d\phi/dt = 0$), Eqs. 10 and 12 reduce to (Adams et al., 1985a):

$$\phi = \phi_0 + \frac{g_0 L b}{2}\left(\frac{\dfrac{I_{av}}{I_s}}{1 + \dfrac{I_{av}}{I_s}}\right) \tag{13}$$

$$gL = \frac{\Gamma g_0 L}{1 + \dfrac{I_{av}}{I_s}} - \alpha L \tag{14}$$

The advantage of writing these relations in terms of the single-pass phase change ϕ rather than in terms of n is now revealed: for a given current drive condition relative to threshold (which fixes $g_0 L$), the strength of the optical nonlinearity is determined by the linewidth broadening factor b and the scaling intensity I_s which contains a number of material parameters. For example, it is clear from Eq. 13 that as the value of b increases, smaller values of I_{av}/I_s, e.g., smaller input intensities, are needed to achieve the same amount of nonlinear phase change.

The equations describing steady-state OB in an FP amplifier are thus Eqs. 5, 13, and 14. For a given normalised input intensity I_{in}/I_s, and unsaturated gain $g_0 L$, these equations must be solved self-consistently to find the steady-state values of ϕ, gL, I_{av}/I_s, and, hence, (from Eq. 7) the normalised output intensity I_{out}/I_s. The input parameters to be specified are ϕ_0, b, Γ, αL, R_1, and R_2. In relating the normalised intensities to real values, we need in addition the values of E, τ, and dg_m/dn; to relate to optical powers, we obviously need also the cross-sectional area of the amplifier. (See Fig. 2.) For a given laser structure, all of these input parameters are fixed, with the exception of the initial phase detuning ϕ_0 and the photon energy E. In fact, as we shall see, the value of ϕ_0 is especially important in determining OB characteristics in amplifiers, just as it is for the corresponding case of passive bistable etalons. (See, for example, Gibbs, 1985.) To illustrate this, Fig. 4 gives a plot of amplifier gain (I_{out}/I_{in}) versus ϕ_0 for an amplifier at 95% of lasing threshold and for five values of the normalised input intensity. With increasing optical input, the curves become further distorted from the conventional FP (low-input) case and possess multiple values of gain for a single value of ϕ_0 when the ratio I_{in}/I_s exceeds about 10^{-4}. This enables a crude estimate of the minimum optical power level for OB to be made: estimating I_s from Eq. 11 as about 10^6 W/cm^2 (using $E = 0.8$ eV, $\Gamma = 0.3$, $\tau = 1.7$ns, and dg_m/dN

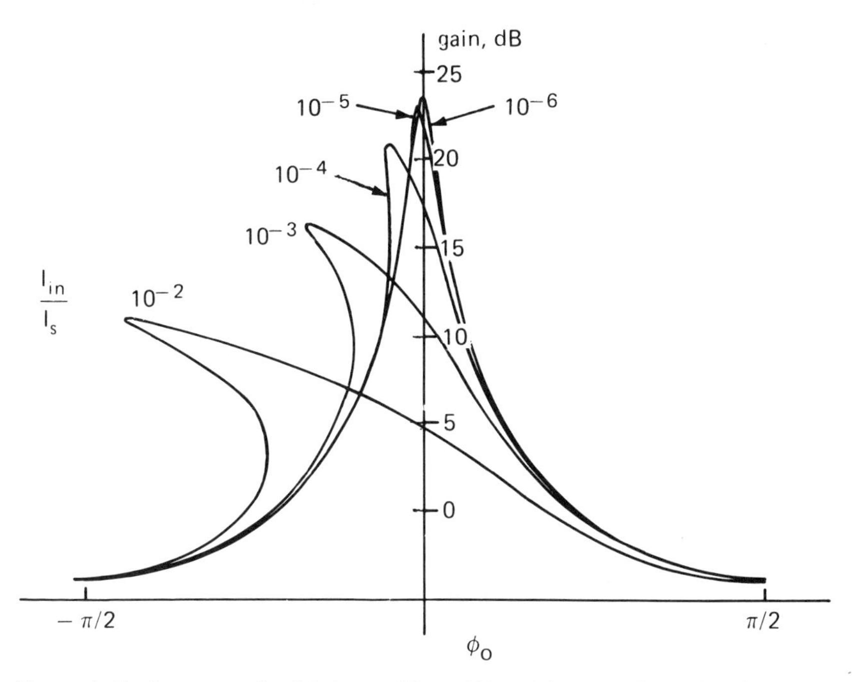

Figure 4. Tuning curves for OB in amplifiers. (After Adams et al., 1985.) The labeling parameter gives the value of normalised input intensity. (Copyright © 1985, IEEE.)

$= 2.7 \times 10^{-16}$ cm^2) (Westbrook, 1986) and taking a device-active area of 1 μm^2, it appears that OB should occur at input powers of order 1 μW. For comparison, tuning curves similar to those in Fig. 4 have been observed experimentally by several workers (Nakai et al., 1982; Otsuka and Kobayashi, 1983; Kuwahara et al., 1983; Adams et al., 1985a) at input power levels down to a few microwatts.

IV. OB MEASUREMENTS IN 1.5 μm AMPLIFIERS

The measurement of the transmission characteristics of a semiconductor laser amplifier provides valuable experimental results for comparison with theoretical prediction. In the measurements described here, an InGaAsP laser was configured as an amplifier with light launched into the active region of the device. This was done with a coupling loss of 5 dB using a tapered lens formed by fusing the end of a single-mode fibre. The amplifier, a double-channel planar buried heterostructure (DCPBH) device, had un-

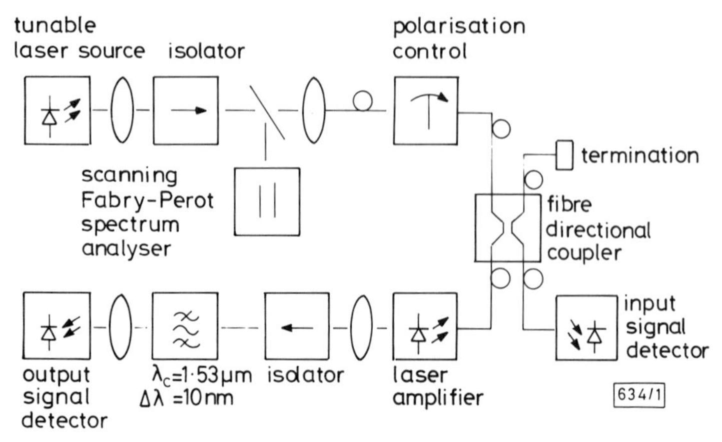

Figure 5. Amplifier characterisation equipment. (From Westlake et al., 1986.)

coated cleaved facets forming a 200-μm–long Fabry–Perot cavity. It was operated at about 95% of its lasing threshold and its temperature was held at 20°C with a Peltier heat pump operating in a control loop.

The setup used to characterise the amplifier is shown in Fig. 5, of which the major feature was the tunable laser source. This took the form of a grating tuned external cavity laser (Wyatt and Devlin, 1983) which produced a single optical line the width of which was in the order of kHz. The output wavelength could be varied, allowing the input wavelength to the amplifier to be matched to the peak resonance in the amplifier gain spectrum. Changes in input wavelength were observed using a scanning Fabry–Perot spectrum analyser which had a free spectral range of 7.5 GHz. The polarisation controller (Lefevre, 1980) set the state of polarisation to TE aı the input to the amplifier, the isolators controlled reflections in the system, and the directional coupler allowed signals representing both input and output intensity variations to be monitored.

Three sets of measured characteristics are presented here and are best dealt with under separate headings.

A. CW Characteristics

With the source laser operating with no applied modulation, the input intensity to the amplifier was varied slowly whilst the transfer function of the amplifier was plotted on a chart recorder. Several curves were plotted

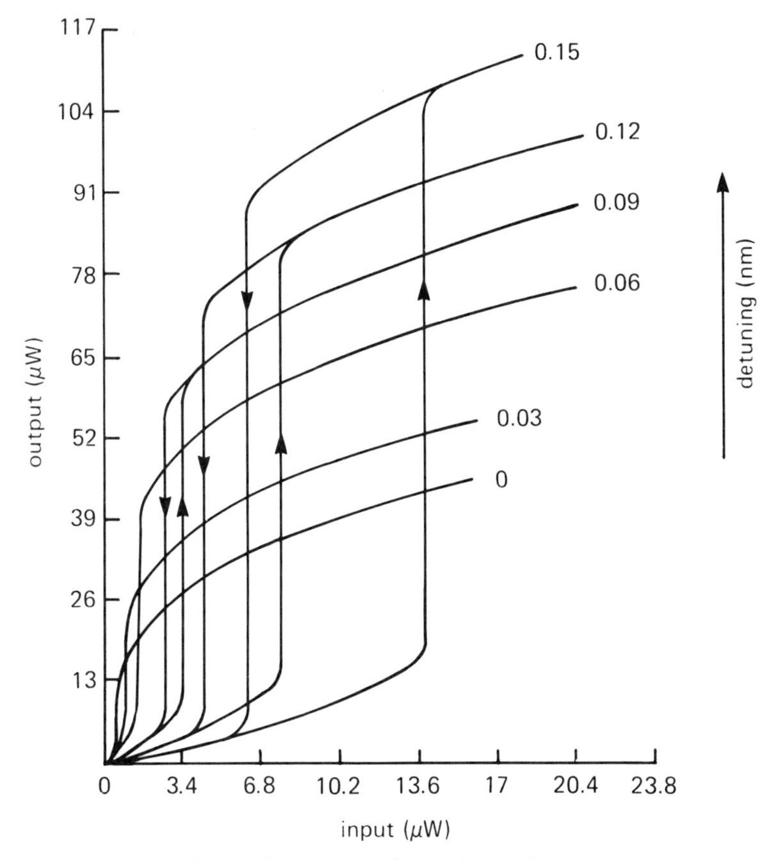

Figure 6. Measured steady-state hysteresis curves.

corresponding to source wavelengths both on the amplifier resonance and slightly to the long-wavelength side of it and these are shown in Fig. 6.

B. Dynamic Characteristics

Here, the source was directly modulated with a 10 MHz triangular waveform and tuned to operate at various wavelengths on the long-wavelength side of the amplifier resonance. The amplifier input and output signals shown in Fig. 7a were obtained when the source wavelength was initially detuned by 0.09 nm from the resonant mode peak.

The hysteresis characteristics of Fig. 7(b)–(d) were obtained by displaying the output of the amplifier against its input signal. Each plot corre-

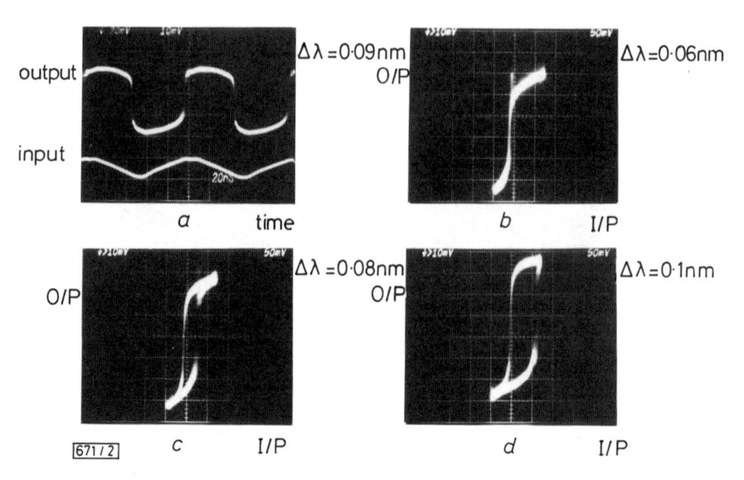

Figure 7. Variation of hysteresis with initial detuning: (a) real-time signals and (b), (c), (d) transfer functions. (From Westlake et al., 1985.)

sponds to a different detuning from resonance and shows the variation from nonlinearity to optical bistability. The mean input power coupled into the amplifier was -27 dBm, i.e. an input power range of zero to 8 μW (assuming complete extinction).

C. Operating Speed

For these measurements, the input intensity to the amplifier was modulated sinusoidally. The period of the modulation was varied and the hysteresis in the bistable switching process observed on an oscilloscope. As modulation frequencies approached 250 MHz, the hysteresis in the switching process was seen to close down rapidly, suggesting an operating-speed limitation. It is worth noting that at this frequency, the intensity of the input signal varies between maximum and minimum levels in 2 ns, a time similar to that taken for carrier recombination.

The loops shown in Fig. 8 were produced with an input power of 20 μW peak and an input wavelength detuned from resonance by 0.15 nm. A series of similar measurements at an input power of 3 μW peak and with smaller detuning produced closure at a similar input frequency. The intensity spike seen in this series of photographs is a well-known property of the switching process (Adams, 1985) and occurs when the phase in the laser cavity passes through a resonance.

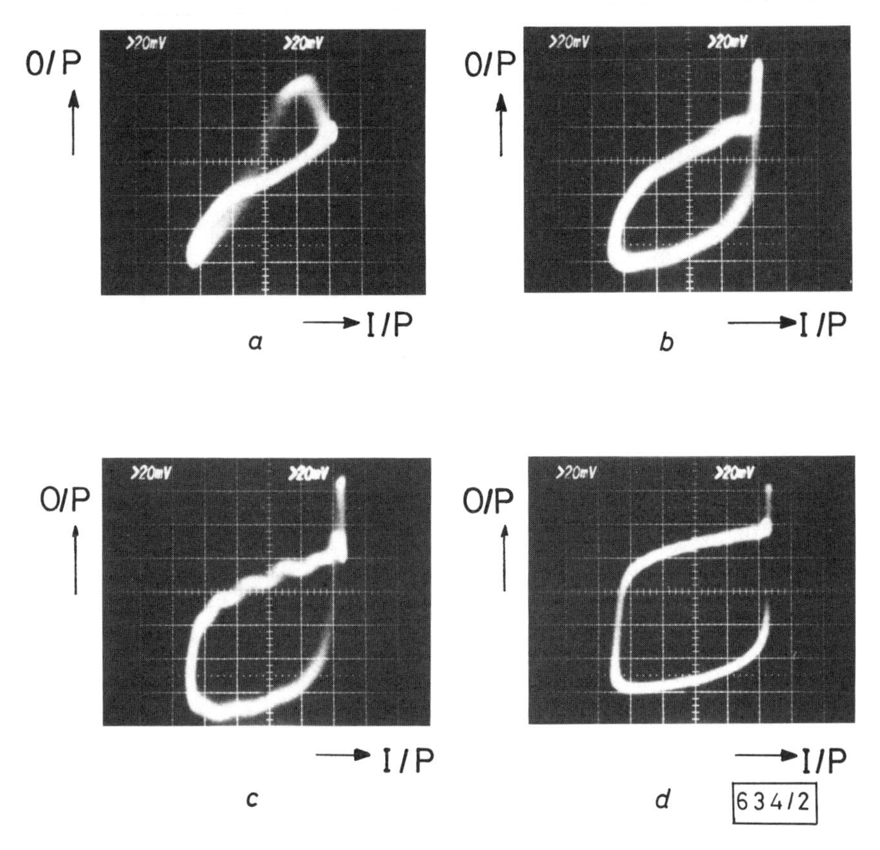

Figure 8. Measured hysteresis curves for input intensity varying sinusoidally with period (a) 4 ns, (b) 6 ns, (c) 8 ns, and (d) 20 ns. (From Westlake et al., 1986.)

V. COMPARISON WITH THEORY

In comparing results of the theory given in section 3 with the OB measurements discussed in section 4, we need to specify the material and structural parameters of the laser. For the device used in the experiments, the values of $\Gamma = 0.3$, $E = 0.8$ eV, $R_1 = R_2 = 0.38$, $\alpha L = 0.5$, $\tau = 1.7$ ns, and an active area cross section of 0.5 μm^2 were estimated. The value of dg_m/dn was taken from the measurements of Westbrook (1986) as 2.7×10^{-16} cm^{-2}. Thus, Eq. 11 yields a value of 9.3×10^5 W/cm^2 for the scaling intensity I_s. The value of g_0L at 95% of threshold is estimated as 4.65. Our approach to comparison with experiment is to leave the remaining free parameter, namely b, to be determined from the best fit of the theoretical plots to the experimental curves in Figs. 6 and 8. Thus, we seek a good agreement in terms of the steady-state characteristics for various detunings and for details of the transient response as given in Fig. 8(a)–(d).

It is pertinent here to comment on the application of the theory given in section 3 to time-dependent behaviour in amplifiers. The equations presented only contain an explicit time derivative for the electron concentration n or, equivalently, the single-pass phase change ϕ. In using Eqs. 5, 10, and 12 to calculate transient effects, therefore, we are using an adiabatic approximation where the optical fields respond instantaneously to changes in the electron density. In fact, this approximation is usually very good, since the cavity round-trip time is normally on the order of a few ps, whilst the carrier recombination time is of order ns.

Our best fit of the theoretical curves to the experimental results of Figs. 6 and 8 are shown in Figs. 9 and 10, respectively. In order to obtain these curves, the value of b was taken as 3.2 (Westlake et al., 1986), which is somewhat lower than the experimental value of about 5 at the lasing wavelength, as given by the measured values in Fig. 1. Whilst the reason for this discrepancy is not clear at present, it is felt that the convincing level of

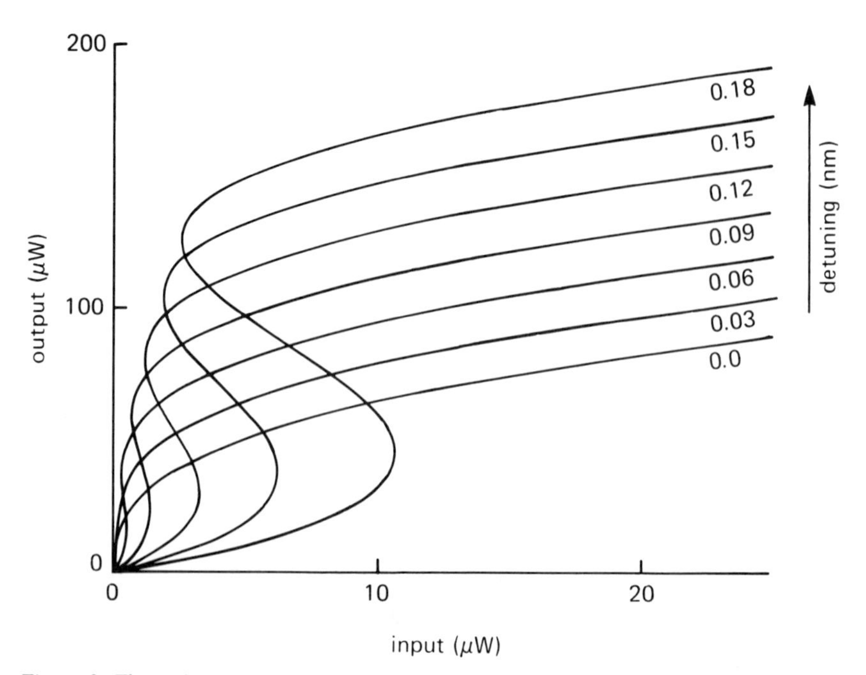

Figure 9. Theoretical steady-state hysteresis curves. Labeling parameter gives the value of initial wavelength detuning in nm.

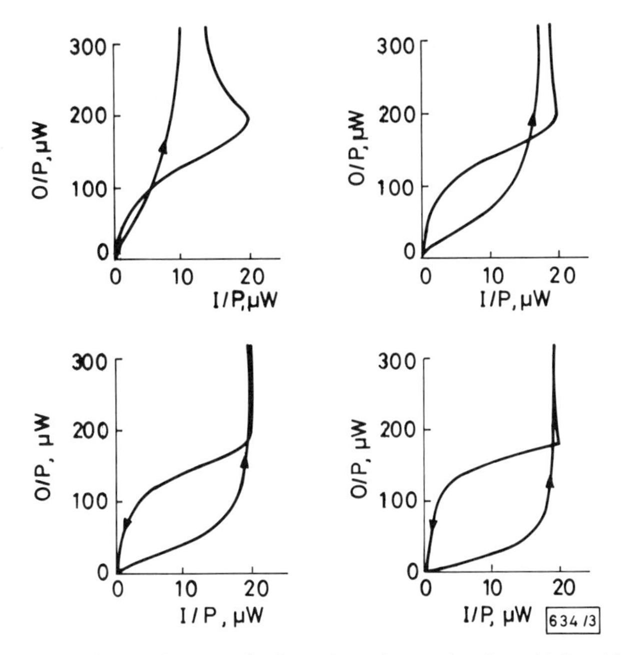

Figure 10. Theoretical hysteresis curves for input intensity varying sinusoidally with period (a) 4 ns, (b) 6 ns, (c) 8 ns, and (d) 20 ns. (From Westlake et al., 1986.)

agreement between experimental and theoretical characteristics strongly supports our model for amplifier bistability. Another point to note in comparing Fig. 6 with Fig. 9, and also Fig. 8 with Fig. 10, is that the wavelength detuning is a little greater for the theoretical plots than for the experimental curves. This shift of 0.03 nm required to give agreement is not considered significant, since this may be due to an uncertainty in experimentally measuring the zero-input resonant wavelength.

VI. CONCLUSION

In this chapter, we have discussed the basic physical mechanism of OB in semiconductor laser amplifiers and shown how a resonant cavity with gain can be used to yield switching energies of order femtojoules. Detailed measurements of OB characteristics, both in the steady state and under time-dependent input conditions, have been compared with theory. A satisfactory level of agreement has been obtained, although some discrepancy remains between the value of the material nonlinearity coefficient inferred from these comparisons and that found by direct measurement. Although the discussion here has centred on Fabry–Perot devices, it is worth noting that OB has also been reported in DFB laser amplifiers (Wyatt and Adams, 1986) at comparable levels of optical switching energy. The many attractive features of amplifier bistability offer wide scope for applications, while demonstrations of logic gates (Sharfin and Dagenais, 1986b), signal regeneration (Webb, 1986), and time-division switching (Suzuki et al., 1985) have already aroused considerable interest. The potential for integration of arrays of DFB amplifiers on a single chip for more sophisticated logic functions and signal processing offers further scope for OB applications.

ACKNOWLEDGMENTS

This work was partially supported by the Joint Optoelectronics Research Scheme. We are grateful to our colleagues in the devices division of British Telecom Research Laboratories for supplying the lasers, and we thank the directors of the Research and Technology Applications Departments, British Telecom, for permission to publish this chapter.

REFERENCES

Adams, M. J. (1985). *IEE Proc. J. Optoelectron.* **132**, 343–348.

Adams, M. J. (1986). *Int. J. Electron.* **60**, 123–142.

Adams, M. J., Collins, J. V., and Henning, I. D. (1985). *IEE Proc. J. Optoelectron.* **132**, 58–63.

Adams, M. J., Westlake, H. J., O'Mahony, M. J., and Henning, I. D. (1985a). *IEEE J. Quantum Electron.* **QE-21**, 1498–1504.

Gibbs, H. M. (1985). *Optical Bistability: Controlling Light with Light.* Academic Press, San Diego.

Hakki, B. W., and Paoli, T. L. (1973). *J. Appl. Phys.* **44**, 4113–4119.

Henry, C. H. (1982). *IEEE J. Quantum Electron.* **QE-18**, 259–264.

Kirkby, P. A., Goodwin, A. R., Thompson, G. H. B., and Selway, P. R. (1977). *IEEE J. Quantum Electron.* **QE-13**, 705–719.

Kuwahara, H., Chikama, T., and Nakagami, T. (1983). *Electron. Lett.* **19**, 295–297.

Lang, R. (1982). *IEEE J. Quantum Electron.* **QE-18**, 976–983.

Lefevre, H. C. (1980). *Electron. Lett.* **16**, 778–780.

Matthews, M. R., Dyott, R. B., and Carling, W. P. (1972). *Electron. Lett.* **8**, 570–572.

Middlemast, I., and Sarma, J. (1984). Presented at SILA '84, Cardiff, Wales, 27–28 March 1984.

Miller, D. A. B., Seaton, C. T., Prise, M. E., and Smith, S. D. (1981). *Phys. Rev. Lett.* **47**, 197–200.
Nakai, T., Ito, R., and Ogasawara, N. (1982). *Japan. J. Appl. Phys.* **21**, L680–L682.
Nakai, T., Ogasawara, N., and Ito, R. (1983). *Japan. J. Appl. Phys.* **22**, L310–L312.
Osinski, M., and Buus, J. (1987). *IEEE J. Quantum Electron.* **QE-23**, 9–29.
Otsuka, K., and Kobayashi, S. (1983). *Electron. Lett.* **19**, 262–263.
Sharfin, W. F., and Dagenais, M. (1985). *Appl. Phys. Lett.* **46**, 819–821.
Sharfin, W. F., and Dagenais, M. (1986a). *Appl. Phys. Lett.* **48**, 321–322.
Sharfin, W. F., and Dagenais, M. (1986b). *Appl. Phys. Lett.* **48**, 1510–1512.
Stern, F. (1964). *Phys. Rev. A* **133**, 1653–1664.
Stern, F. (1976). *J. Appl. Phys.* **47**, 5382–5386.
Suzuki, S., Terakado, T., Komatsu, K., Nagashima, K., Suzuki, A., and Kondo, M. (1985). Presented at IOOC-ECOC'85, Venice, Italy, 1–4 Oct. 1985.
Thompson, G. H. B. (1972). *Opto-electronics* **4**, 257–310.
Vahala, K., Chiu, L. C., Margalit, S., and Yariv, A. (1983). *Appl. Phys. Lett.* **42**, 631–633.
Webb, R. P. (1986). Presented at CLEO'86, San Francisco, California, 9–13 June 1986.
Westbrook, L. D. (1986). *IEE Proc. J. Optoelectron.*, **133**, 135–142.
Westlake, H. J., Adams, M. J., and O'Mahony, M. J. (1985). *Electron. Lett.* **21**, 992–993.
Westlake, H. J., Adams, M. J., and O'Mahony, M. J. (1986). *Electron. Lett.* **22**, 541–543.
Wyatt, R., and Adams, M. J. (1986). Presented at CLEO'86, San Francisco, 9–13 June 1986.
Wyatt, R., and Devlin, W. J. (1983). *Electron. Lett.* **19**, 110–112.

16 BISTABILITY IN SEMICONDUCTOR LASER DIODES

Ch. Harder

IBM RESEARCH DIVISION
ZURICH RESEARCH LABORATORY
8803 RÜSCHLIKON, SWITZERLAND

A. Yariv

T. J. WATSON LABORATORY
CALTECH
PASADENA, CALIFORNIA 91125

I. INTRODUCTION

Optical bistability in semiconductor laser diodes has received increased attention in the past few years. Numerous investigations have been directed

OPTICAL NONLINEARITIES
AND INSTABILITIES IN
SEMICONDUCTORS

ISBN 0-12-332915-9

at understanding the underlying physics and developing practical devices for opto-electronic signal processing, and encouraging progress has been made. The much simpler and also better understood passive optical bistable elements, such as the dispersive (Smith and Tomlinson, 1981; Miller, 1982; Gibbs *et al.*, 1976) or absorptive (Szöke *et al.*, 1969) Fabry–Perot etalons have only two macroscopic variables: light input and light output. In contrast, bistable laser diodes are active devices with light input, light output, pump current, and applied voltage as macroscopic variables. Because of these differences, bistable laser diodes not only display a much more complex behavior, but are also more versatile. A necessary optical bias, for instance, can be traded off against an electrical bias current. In addition, bistable laser diodes can have a large optical signal gain, can readily be coupled to electronic circuits, and are small and very efficient.

Bistability in semiconductor laser diodes due to numerous mechanisms has been reported since the first demonstration of the injection laser diode in 1962 (Adams, 1986). In this chapter, we review the main types of bistability in semiconductor lasers. We heuristically divide the reported cases of bistability into two classes. Bistable laser diodes of the first class ideally are single-mode lasers and the two stable states correspond to the laser being turned off (not lasing, only spontaneous emission) and switched on (lasing, emitting coherent light). Diodes of the second class ideally are two-mode lasers that can lase in either of the two modes but never in both at the same time. The two stable states are, for example, two longitudinal modes or two transverse modes of a laser diode and the total amount of stimulated emission is roughly the same in both states.

Bistable laser diodes of the first class have been successfully fabricated in both the 0.8-μm AlGaAs (Harder *et al.*, 1982a) and the 1.3-μm InGaAsP material system (Odagiri *et al.*, 1984). These bistable laser diodes have been obtained with a segmented contact as proposed by Lasher (1964) more than two decades ago to introduce a section with saturable absorption in the laser cavity. These lasers usually show a giant hysteresis in their characteristics and have been successfully operated as an optical memory cell in an optical time-division switching system (Suzuki *et al.*, 1986) with a switching speed of 2 ns and an optical switching power of 20 μW yielding the low power-delay product of 40 fJ.

Bistable laser diodes of the second class have been obtained with a wide variety of schemes. For example, a cleaved coupled cavity (C^3) laser diode (Tsang *et al.*, 1983) has been used to get wavelength bistability, and a twin stripe contact (White and Carroll, 1984) has been used to produce lateral-mode bistability. With the C^3 laser diode, logic operations on the basis of wavelength bistability have been shown with delays below one nanosecond by Tsang *et al.* (1983). Because the carrier distribution does not change appreciably between the two lasing states, the time response of these devices is not limited by carrier diffusion effects as it is the case for class-one bistability, and switching between the two states occurs with a shorter delay.

In the following chapter, we shall concentrate on class-one bistability, especially on the detailed characteristics of bistability in laser diodes with saturable absorption, and only briefly address different types of the class-two bistability in semiconductor lasers.

II. CLASS-ONE BISTABILITY IN SEMICONDUCTOR LASERS

A. Bistability in Semiconductor Lasers with Saturable Absorption

Conventional lasers have, independent of structural details, the same generic light output versus pump-power characteristics. Coherent light is only emitted if the pump power exceeds a threshold level; above the threshold, the light output is directly proportional to the additional pump power. If a saturable absorber is placed within the optical cavity, then the behavior of the laser will change dramatically. The laser becomes unstable, multistable, or, for a very specific configuration, bistable. A saturable absorber is a nonlinear passive element whose transmission coefficient depends on the input intensity. It is opaque at low intensities and becomes progressively more transparent with increasing intensities. The introduction of a saturable absorber into the laser cavity causes instabilities depending on the detailed characteristics of the gain medium, absorber medium and the optical cavity, such as saturation intensities, time constants and geometry (Abraham, 1983). As a matter of fact, saturable absorbers have been successfully used to induce repetitive Q-switching (Sorokin *et al.*, 1964), passive mode-locking (Mocker and Collins, 1965), chaos (Abraham, 1983) and bistability (Ruschin and Bauer, 1979) in essentially all laser systems, including semiconductor laser diodes (Van der Ziel *et al.*, 1981a, 1981b; Harder *et al.*, 1982a).

1. Historic Overview

In 1964, only two years after the first demonstration of the semiconductor laser diode, Lasher (1964) proposed a bistable laser on the basis of separate gain and absorbing sections. About the same time, Nathan *et al.*

(1965) measured bistability in a broad-area homojunction laser with a segmented contact creating separate gain and absorber sections. The lasers operated below 77° K and initial experiments were carried out to switch the bistable laser with a current pulse or an external light pulse. However, only a few lasers showed bistability whereas most devices were beset by unexplained pulsations. These instabilities were also observed by Basov (1968) and later by Lee and Roldan (1970), who linked the pulsations in broad-area homojunction lasers to a repetitive Q-switching mechanism. Double heterojunction laser diodes with a segmented contact and a gain-guided lateral mode were investigated ten years later by Goldobin *et al.* (1980), Ito *et al.* (1981) and Carney and Fonstad (1981), and, for the first time, these devices could be operated continuously at room temperature. In a series of papers, Kawaguchi and Iwane (1981) and Kawaguchi (1981, 1982) presented experimental results on InGaAsP double heterostructure lasers with a periodic stripe contact. Continuous-wave (CW) hysteresis and CW switching with an electrical and an optical pulse were observed when operating the devices below room temperature. Up to this point in time, all experiments had been done in gain-guided laser structures without a passive lateral waveguide, which complicated the interpretation of the results considerably.

Investigations of AlGaAs buried heterostructure laser diodes, which have a built-in single transverse mode waveguide, were published in several papers by Harder *et al.* (1981, 1982a, 1982b, 1983), and by Lau *et al.* (1982a, 1982b). The pulsations and bistability in these lasers were firmly explained on the basis of a negative differential resistance in the absorbing section which has an opto-electronic origin. Switching speeds between bistable states could be demonstrated with a time constant of 5 ns and an electrical switching energy as low as 6 pJ (Harder *et al.*, 1982a). We shall describe the characteristics of these lasers in more detail in Sections II.A.2. through II.A.4.

To increase the isolation between the gain and absorber sections, two antireflection-coated laser diodes were coupled by Stallard and Bradley (1983) to an external Fabry–Perot cavity and by McInerney *et al.* (1985) in an external ring cavity. Recently, Odagiri *et al.* (1984), Goto *et al.* (1984) and Suzuki *et al.* (1986) demonstrated a 1.3-μm InGaAsP–InP DC–PBH bistable laser diode with a proton-bombarded absorption section between the two contact pads. They successfully used the devices as optical memory cells in a time-division switching system. The "set" operation of the optical memory was achieved with an optical pulse of 10-μW power and 4-ns width, corresponding to the optical trigger energy of 40 fJ (Suzuki *et al.*, 1986).

2. Absorption in Laser Diodes

In this section, we shall describe experimental results of AlGaAs buried heterostructure diode lasers with a saturable absorber. The laser diodes with two separate sections, one with gain and one with absorption, can be readily obtained with a segmented contact as shown in Fig. 1. Gain and absorption can easily be adjusted by controlling the pump currents. To obtain bistable operation, the gain section is usually heavily forward-biased, whereas the absorber section is only slightly biased. The dependence of the gain on pump current in heavily forward-biased junctions is well known (Henry *et al.*, 1980), and we shall now have a closer look at the gain (or equivalently at the absorption) in a slightly biased junction.

Let us consider the following simplified case of absorption for light propagating along the intrinsic GaAs layer in a p-$Al_xGa_{1-x}As$/i-GaAs/n-$Al_xGa_{1-x}As$ diode (or short P-i-N diode) as shown in Fig. 2. The band-to-band absorption of the light with a photon energy close to the bandgap (corresponding to the lasing energy of this material) depends on the density of electrons and holes which are injected under bias from the contacts into the intrinsic GaAs region. We assume quasi neutrality (electron density equals hole density) and an absorption constant α which depends linearly

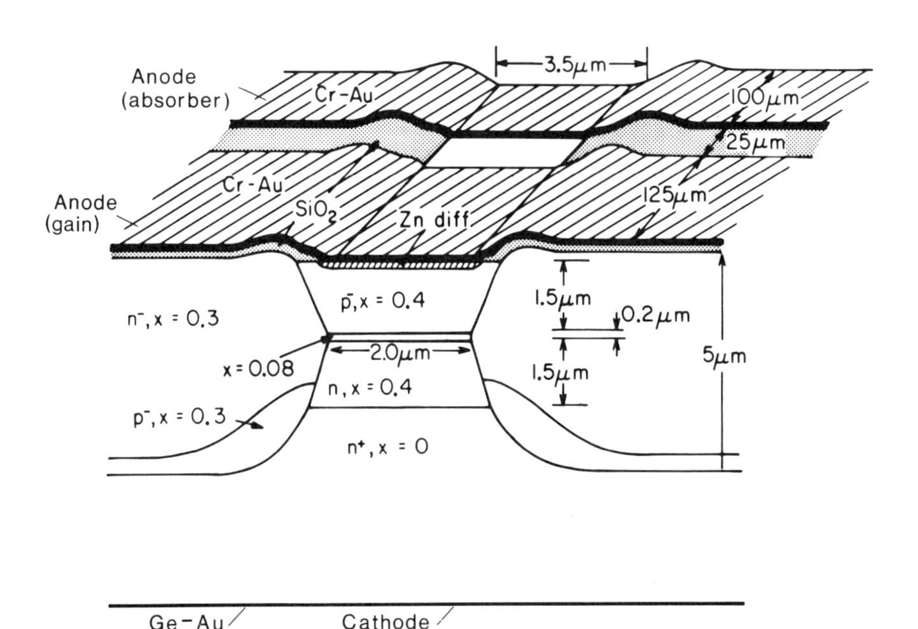

Figure 1. $Al_xGa_{1-x}As$ buried heterostructure laser with a common cathode and a segmented anode contact.

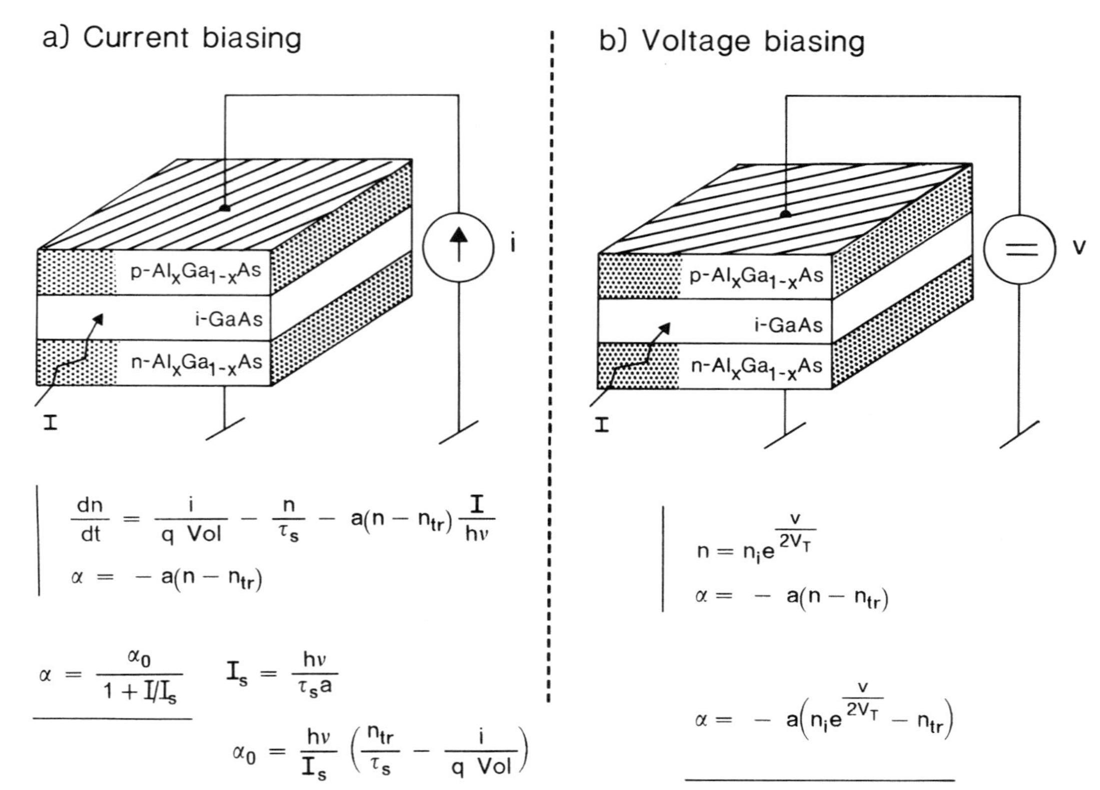

Figure 2. Simplified model of absorption in a biased p-type $Al_xGa_{1-x}As$/*i*-GaAs/*n*-type $Al_xGa_{1-x}As$ laser diode. The absorption is saturable for current biasing and unsaturable for voltage biasing.

on carrier density (Henry *et al.*, 1980).

$$\alpha = -a(n - n_{tr}). \tag{1}$$

The proportionality factor a is the differential gain constant and n_{tr} the carrier density corresponding to transparency. This expression for the absorption as a function of carrier density is, of course, identical with the usual gain expression, with the exception of the minus sign. Now, we separately consider the two cases of biasing the *P-i-N* junction with a current source and a voltage source.

For current biasing, the carrier density is given by the static solution of the following rate equation:

$$\frac{dn}{dt} = \frac{i}{q(Vol)} - \frac{n}{\tau_s} - a(n - n_{tr})\frac{I}{h\nu}. \tag{2}$$

The right side of the equation consists of the following three terms: (1) a pump term $i/[q(Vol)]$ with the injected current i, the magnitude of the electronic charge q, the volume of the absorbing section (Vol), (2) a spontaneous recombination term n/τ_s, which is inversely proportional to the spontaneous carrier lifetime τ_s, and (3) a pump term, $-a(n - n_{tr})I/h\nu$ (which is equal to α) due to the absorbed light with an intensity I and a photon energy $h\nu$. Solving these two equations for the absorption α as a function of the light intensity I, we obtain the following expression for $dn/dt = 0$:

$$\alpha = \frac{\alpha_0}{1 + \dfrac{I}{I_s}}. \tag{3}$$

The absorption is α_0 at low intensities and becomes bleached above the saturation intensity I_s,

$$I_s = \frac{h\nu}{\tau_s a} \tag{4}$$

$$\alpha_0 = \frac{h\nu}{I_s}\left[\frac{n_{tr}}{\tau_s} - \frac{i}{q(Vol)}\right]. \tag{5}$$

The *P-i-N* junction acts like a saturable absorber if biased with a current source. The amount of absorption α_0 can be adjusted with the externally applied pump current i, and the saturation intensity is inversely proportional to the gain constant and the spontaneous carrier lifetime.

When the *P-i-N* junction is biased with a voltage source, then the carrier density n in the undoped region is given as a function of the externally applied voltage v by

$$n = n_i e^{(v/2V_T)}. \tag{6}$$

We again assume quasi neutrality and Boltzmann statistics; V_T is the thermal voltage and n_i the intrinsic equilibrium carrier concentration. Under biasing with a voltage source, the absorption α does not depend on the light intensity I and is equal to

$$\alpha = -a\left[n_i e^{(v/2V_T)} - n_{tr}\right]. \tag{7}$$

The absorption cannot be saturated and the amount of absorption depends on the external voltage v. This is a direct consequence of the fact that the Fermi levels of electrons and holes can be electrically contacted via the cathode and anode electrodes.

When the *P-i-N* junction is biased with a current source with a shunt resistor (or a voltage source with a series resistor), an absorption characteristic between the saturable absorption of pure current biasing and pure voltage biasing can be obtained. In this discussion, we consider the *P-i-N* junction to be current-biased if the source impedance is much larger than the differential resistance of the junction, and voltage-biased if the source impedance is much smaller than the differential resistance of the junction. If the junction is only slightly biased (as is the case for the absorbing section), then the differential resistance will be relatively large and voltage and current biasing can be exercised. However, if the junction is heavily forward-biased, as is necessary to achieve gain, then the differential resistance will be small and voltage biasing will become impossible.

3. Static Characteristic

Typical dimensions and doping levels of an AlGaAs buried heterostructure laser (BH) with a split contact are shown in Fig. 1. The upper cladding layer is only lightly p-doped to increase the lateral parasitic resistance R_p between the two contact pads which is around 60 kΩ. In the following, it will become evident that good electrical isolation between the two segments is essential for the operation of this laser as a bistable element. Near-field and far-field measurements show that the laser is operating in the fundamental transverse mode under all operating conditions, demonstrating the effectiveness of the dielectric waveguide. Thus, the BH structure makes it possible to isolate and study the effects of inhomogeneous excitation on a device whose electronic and optical properties are stable and simple.

When both sections are connected, that is, under homogeneous excitation, the light-current characteristics display the conventional linear behavior above threshold as shown in Fig. 3, curve (a). For bistable operation, the absorber section (the section that is 100 μm long) is pumped with a

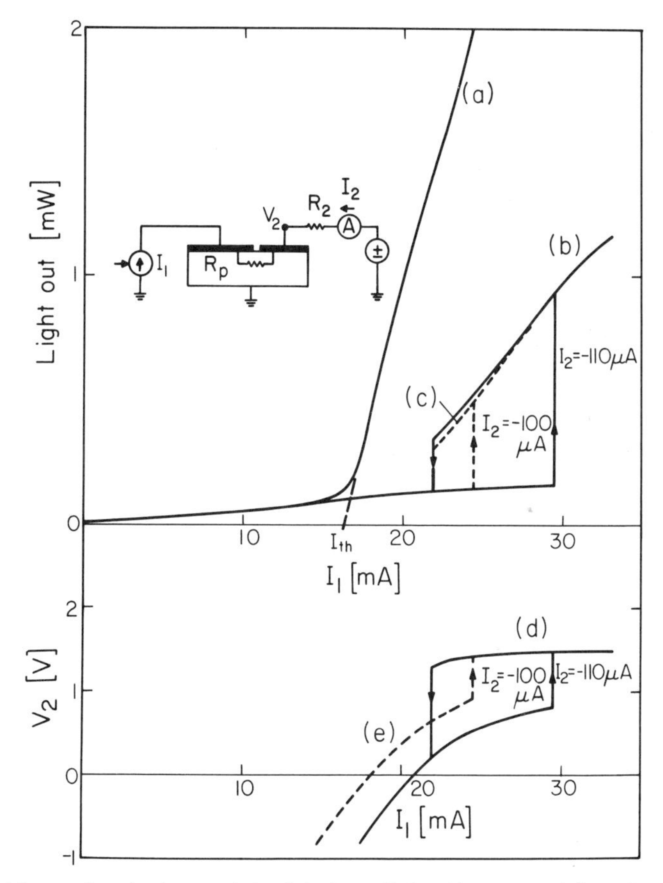

Figure 3. Measured static characteristic of the laser diode with a segmented anode contact. In the upper part, the light-current characteristic is shown as curve (a) for homogeneous pumping (both anode contacts are connected). I_{th} is the conventional threshold current. The characteristic is also shown for inhomogeneous pumping, curve (b) for $I_2 = -110$ μA and curve (c) for $I_2 = -100$ μA. The parasitic cross-talk resistance between the two anodes is $R_p = 60$ kΩ, and the current-source impedance is $R_2 = 200$ kΩ. In the lower half, the voltage V_2 across the absorber diode is shown as a function of I_1, the current through the gain diode for $I_2 = -110$ μA (curve d) and for $I_2 = -100$ μA (curve e). From Harder *et al.* (1982a); © 1982 IEEE.

constant current $I_2 = -110$ μA. The light output is shown in Fig. 3, curve (b) as a function of I_1, the current through the gain section. This characteristic displays a large hysteresis extending from $I_1 = 30$ mA (the switch-on current) to $I_1 = 22$ mA (the switch-off current). The characteristic is also shown for $I_2 = -100$ μA, curve (c), and the dependence of the bistable behavior on amount of saturable absorption (I_2) can be seen. Note that I_2 is negative, that is, carriers are extracted from the absorber section.

In the lower half of Fig. 3, the voltage V_2 across the absorbing section is shown [curves (d) and (e)] as a function of I_1. Note that the voltage V_2 is clamped at 1.45 V, the bandgap voltage of GaAs, when the device is lasing (when the absorption is bleached). From the treatment of the absorption in *P-i-N* diodes we have discussed, this is expected and it demonstrates the one-to-one correspondence of the voltage across the contacts and the quasi–Fermi-level separation of holes and electrons in the active region.

The current-voltage characteristic ($I_2 - V_2$) of the absorbing section is shown in Fig. 4 for different normalized currents I_1/I_{th} through the gain section. The negative differential slope in this characteristic is of particular interest and deserves special attention. This negative differential resistance is opto-electronic in origin, and the mechanism causing it can be understood with the following simple model as illustrated in Fig. 5. For simplicity, assume the bistable laser diode to consist of two parts, a gain section pumped with a constant current I_1 plus an absorber section which acts like a *P-i-N* photodiode within the optical cavity as has been discussed. The current I_2 through the photodiode consists of the following two terms: the normal diode current which depends exponentially on the applied voltage V_2 and a negative photo current I_{Ph} which is roughly proportional to the photon density in the active region. These photons are generated under

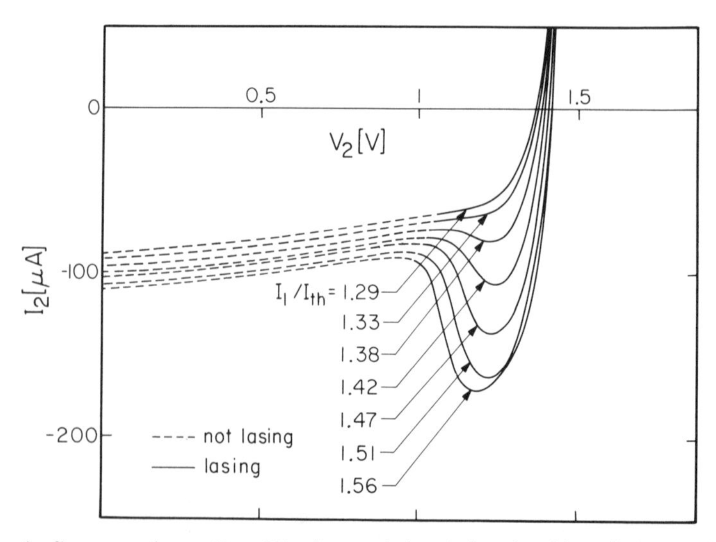

Figure 4. Current voltage ($I_2 - V_2$) characteristic of the absorbing diode for different normalized currents I_1/I_{th} through the gain section. I_{th} is defined in Fig. 3. From Harder *et al.* (1982a); © 1982 IEEE.

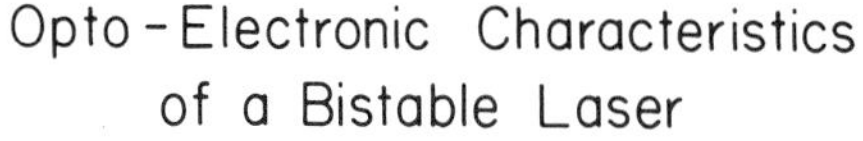

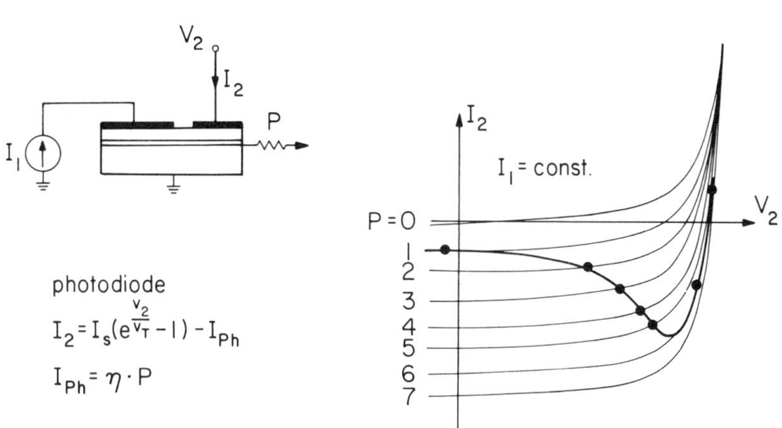

Figure 5. Simplified photodiode model to explain the current-voltage characteristic $I_2 - V_2$ of the absorber diode for a fixed current I_1 through the gain section. From Harder *et al.* (1982a); © 1982 IEEE.

the gain contact and guided via the buried heterostructure waveguide to the absorbing region under the second contact. This photodiode characteristic is drawn on the right side of Fig. 5 for eight different normalized photon densities, $P = 0$ to $P = 7$. We now explain how the measured $I_2 - V_2$ characteristic is produced. The gain section is biased with a fixed current I_1 and the absorbing section is biased with a voltage source V_2, thus introducing an unsaturable amount of absorption as was derived in Section 2. For zero voltage, $V_2 = 0$, the absorption of the photodiode is strong, thus suppressing stimulated emission, and only a small photocurrent due to spontaneous emission in the gain section is generated. Increasing the voltage V_2 reduces absorption and therefore the cavity losses as shown above in Section A. As soon as the total cavity losses are smaller than the gain, stimulated emission is initiated, and the photon density increases, which in turn generates a larger negative photo current in the absorbing section. Increasing V_2 further reduces the losses, which increases the stimulated emission and also the negative photocurrent I_{Ph}, thus producing the negative slope. Finally, at large voltages V_2, the absorbing section becomes transparent and the photo current decreases to zero. The positive exponential term representing the normal diode behavior now starts to dominate and I_2 increases very quickly. The specific curve as shown on the right side of Fig. 5 is obtained for a fixed gain current I_1. The whole set of curves as shown in Fig. 4 is obtained for different currents through the gain section. This qualitative argument is only useful to indicate the origin of the negative differential resistance, and a self-consistent analysis of the laser diode with a segmented contact is presented elsewhere (Harder *et al.*, 1982a).

In Section 2, we showed that the absorption characteristic of an isolated *P-i-N* diode depends on the biasing circuit. We shall now show that a laser diode that contains such an absorbing *P-i-N* diode can either be bistable or self-pulsating depending on the biasing circuit. The $I_2 - V_2$ characteristic is shown in Fig. 6 again along with the characterization of the source driving this section, the load line. This load line shows the voltage available to the absorber section as a function of the current through it. The state of the system satisfying all static circuit equations is given by the intersection of the load line with the characteristic of the device. The state P_1 in Fig. 6 is obtained by intersecting the characteristic of the absorber with a load line corresponding to a voltage source of $V = -0.85$ V and a series resistance of $R_L = 20$ kΩ. The gain section is driven with a normalized current of

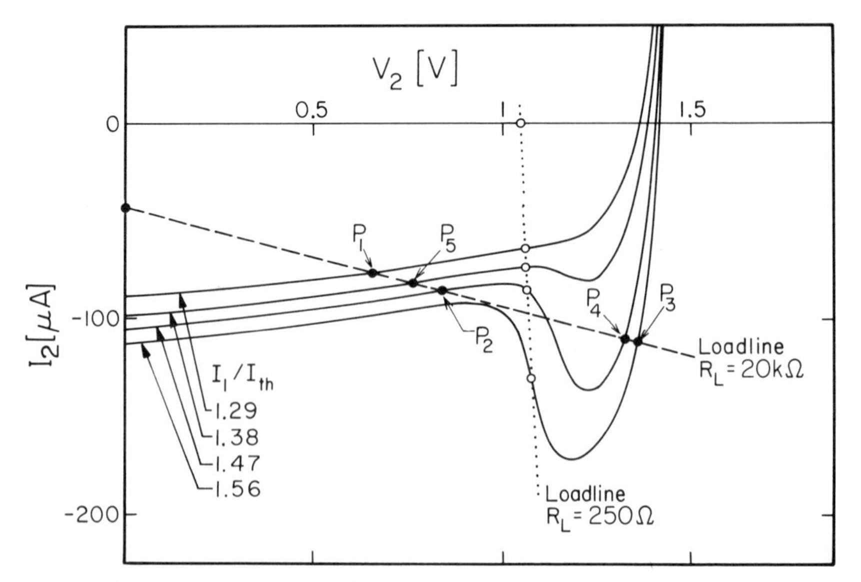

Figure 6. Current-voltage ($I_2 - V_2$) characteristic of the absorbing diode for different normalized currents I_1/I_{th} through the gain section. I_{th} is defined in Fig. 3. Also shown are two load lines corresponding to a biasing source of -0.85 V and $R_L = 20$ kΩ and 1.05 V and $R_L = 250$ Ω, respectively. From Harder *et al.* (1982a); © 1982 IEEE.

$I_1/I_{th} = 1.29$, with I_{th} being the threshold current of the homogeneously pumped laser. In the state P_1, the laser is switched off. Increasing the pump current I_1 causes the intersection point to move along the load line from P_1 to P_2, and at $I_1/I_{th} = 1.56$ to jump to P_3 since this is the only stable intersection of the load line with the characteristic of the absorbing section. In this state, the laser is switched on. A decrease of I_1 now causes the state to move back to P_4 and then to jump back to P_5, and the laser switches off again. A laser with a segmented contact whose absorbing section is biased with such a large load resistance will display a hysteresis in the light-versus-current characteristic and it is bistable. This is also obvious from the fact that the load line has three intersections, one unstable state and two stable ones, P_2 (switched off) and P_4 (switched on) with the $I_2 - V_2$ characteristic for $I_1/I_{th} = 1.47$.

The load line corresponding to a voltage source of $V = 1.05$ V and a series resistance of $R_L = 250\ \Omega$ is also shown in Fig. 6. This load line always has only one intersection with the characteristic and consequently the laser will not display bistability. With such a small load resistance on the absorbing section, the laser can be biased in the region of negative differential resistance of the absorber. This leads to electrical microwave oscillation and concomitant light-intensity pulsations.

The load resistance we have discussed is essentially R_2 in parallel with R_p, which is connected to the low-impedance forward-biased gain section, $R_L = R_p R_2/(R_p + R_2)$ and thus always smaller than R_2 or R_p. Since bistable operation is only obtained with a large R_L, both the source resistance R_2 and the parasitic resistance R_p have to be large ($R_p = 60$ kΩ). The experimental effect of the load resistance is shown in Fig. 7 for $R_L = 46$ kΩ, (Fig. 7, top) and $R_L = 1$ kΩ (Fig. 7, bottom). For a large load resistance, the characteristic indeed displays a large hysteresis and the light output is always stable. For a small load resistance, the characteristic has only a tiny hysteresis, essentially only a light jump, and the light output pulsates with 100% modulation depth at microwave frequencies. A tiny hysteresis and microwave oscillation are also observed with a large load resistance and a small separation resistance R_p obtained by attaching a small external resistor between the two top contacts. Since the negative differential resistance is not frequency-selective (at frequencies below a few GHz), the laser oscillates at the resonance of any frequency-selective mechanism such as the relaxation resonance (Harder *et al.*, 1981) or the round-trip time of an external cavity (Harder *et al.*, 1983). This mechanism could possibly explain the development of self-pulsations in aged lasers,

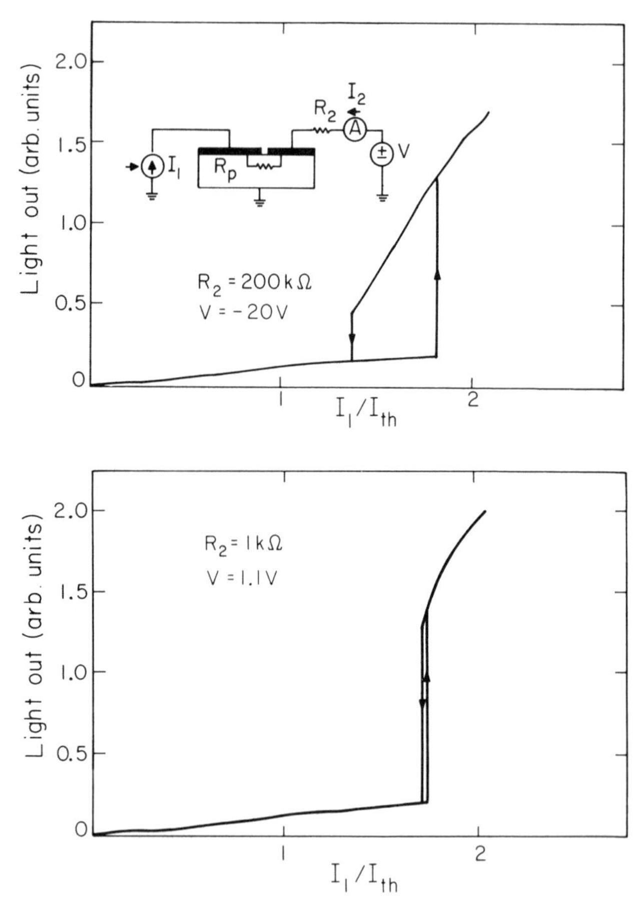

Figure 7. Light-current characteristic for two different biasing circuits of the absorber section. The characteristic in the upper half corresponds to biasing the absorber section with a large resistor of $R_2 = 200\ k\Omega$ (current biasing). The characteristic in the lower half corresponds to biasing the absorber section with a small resistor of $R_2 = 1\ k\Omega$ (voltage biasing). From Harder *et al.* (1982a); © 1982 IEEE.

with the inhomogeneous excitation caused by degraded contacts or centers with enhanced nonradiative recombination. These self-pulsations are not due to a repetitively Q-switched mechanism, but are the result of the opto-electronic negative differential resistance and do not result from a sublinear gain dependence which can be obtained for instance by damaging the crystal by proton bombardment (Van der Ziel *et al.*, 1981a; Henry, 1980).

4. Switching

The laser has two stable states when biased within the hysteresis loop. It can be switched on either by reducing the absorption or by increasing the gain for a short time by some means. This can be achieved by injecting a positive current pulse into the absorber section (absorber switching), by bleaching the absorber with an externally injected optical pulse (optical switching), or by increasing the current through the gain section (gain switching) for a short time. As an example, we shall now address the dynamics of the gain switching, especially the important issue of delay between trigger pulse and switching as a function of trigger amplitude. Such delays are observed in almost every bistable system including electronic Schmidt-triggers and pure optical systems, and it is well known that they can be reduced by increasing the trigger-pulse amplitude. From a thermodynamic point of view, the laser system undergoes a first-order phase transition during the switching (Scott *et al.*, 1975). As far as device

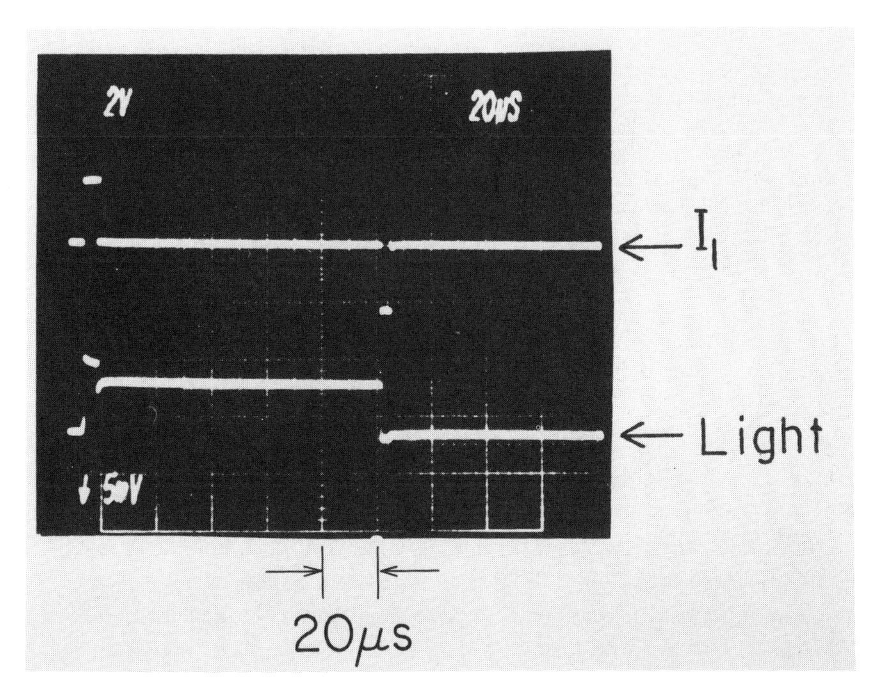

Figure 8. Gain switching of the bistable laser. The top trace is the current I_1 through the gain section and the lower trace is the measured light output. The horizontal scale is 20 μs per division. From Harder *et al.* (1982a); © 1982 IEEE.

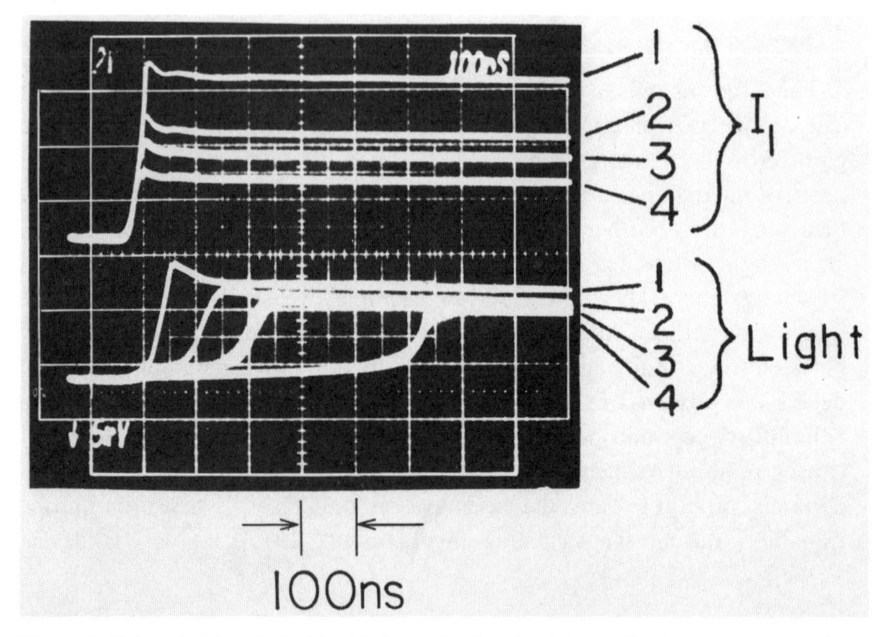

Figure 9. Gain switching of the bistable laser showing the delay of the light output after the trigger pulse. The upper four traces show four different trigger-current inputs (I_1) into the gain section. The lower four traces show the corresponding measured light outputs. The horizontal scale is 100 ns per division. For large trigger amplitudes (curve 1), the laser switches on within 30 ns, but for small trigger-current amplitudes (curve 4) the laser switches only after a 500-ns–long delay time. The switching time itself is essentially the same for these different trigger-pulse amplitudes, namely, 30 ns. From Harder *et al.* (1982a); © 1982 IEEE.

applications are concerned, one of the most important predictions gained from this point is that of the "critical slowing down" (Bonifacio and Meystre, 1979), i.e., an extremely slow response to a perturbation in the vicinity of the phase-transition point. The delay between trigger pulse and initiation of switching of the laser can be very long compared to any physical lifetime associated with the system, such as electron–hole recombination or RC-time constant, while the switching itself is relatively fast.

For bistable operation, the laser is biased at constant currents I_2 and I_1 corresponding to the middle of the hysteresis loop, i.e. for the device shown in Fig. 3, $I_2 = -110\ \mu A$ and $I_1 = 26$ mA. A small positive current pulse superimposed on I_1 switches the laser to the high (lasing) state, while a negative pulse switches the laser back to the off-state as shown in Fig. 8. A closer examination of the switch-on as shown in Fig. 9 reveals a time delay that is dependent on the amount of the trigger-pulse overdrive.

The carrier lifetime in the gain section (where carrier density is high) is in the order of a few nanoseconds, while the one in the low carrier-density absorber section is in the tens of nanoseconds. The dominating time constant is, unfortunately, the long one. Fig. 9 shows multiple traces of the light output (lower half) for four different trigger-current amplitudes (upper half) labeled 1 to 4. The smallest trigger pulse (I_1, trace 4) is very close to the threshold for switching, and the long delay after which the light output switches (light, trace 4) is evident: it is longer than 500 ns. After this long delay time, the light switches relatively fast, in 30 ns. As can be seen in Fig. 9, the delay can be reduced drastically by increasing the trigger-pulse current; for instance, for the largest trigger pulse (I_1, trace 1), the delay is less than 20 ns (light, trace 1). In fact, a switching delay of 5 ns can be achieved with a trigger pulse of 10 mA. This corresponds to a power delay product of 5 ns × 10 mA × 2 V = 100 pJ for gain switching. While the switch-on shows these interesting effects, the switch-off is usually fast and without delay. This arises from the fact that the switch-off process involves stimulated recombination of carriers at high carrier densities which is extremely fast. To observe the inherent switching behavior, the external parasitic capacitance at the absorbing section must be reduced to a minimum, since the switching is accompanied with a change of the voltage and, thus, charge across the absorber section, as can be seen in Fig. 3. In the delay experiment just described, the diode was isolated from the capacitance of the current source by placing a 250-kΩ resistor as close as possible. Experiments on absorber switching (injecting a current pulse into the absorber section) reveal a switch-on delay in the hundreds of nanoseconds for moderate overdrives and of 20 ns for large overdrives with a power delay product of 6 pJ. The interested reader will find a much more detailed discussion on these bistable lasers in the literature (Harder *et al.*, 1981, 1982a, 1982b, 1983; Lau *et al.*, 1982a, 1982b).

B. Semiconductor Lasers Coupled to an External Grating

Bistability in passive Fabry–Perot etalons filled with a material whose index of refraction depends on the intensity have been known for quite some time (Miller, 1982). The bistable characteristic is based on the nonlinear index of refraction and the macroscopic feedback from the optical resonator. Depending on the intensity-induced change of the index of refraction, the Fabry–Perot etalon is either in resonance with the frequency of the incoming light or out of resonance. A similar dispersive bistability has been observed in semiconductor lasers coupled to an external grating (Glas and Müller, 1982). In semiconductor lasers, the index of refraction depends on the carrier density (Henry *et al.*, 1981). It decreases with increasing carrier density, approximately with $-2 \times 10^{-20}\ \mathrm{cm}^3$ at the lasing wavelength, i.e., an injected carrier density of $10^{18}\ \mathrm{cm}^{-3}$ decreases the index of refraction by 0.02 in AlGaAs lasers (Henry *et al.*, 1981). Depending on the index of refraction, which changes with intensity through the change of the carrier density, one of the modes of the laser cavity coincides with the wavelength given by the external grating. By properly designing the external cavity, bistability has been observed (Glas and Müller, 1982) and a hysteresis in the light output versus drive current and versus tuning angle of the grating has been measured (Glas and Müller, 1982; Bazhenov *et al.*, 1982). These bistable laser systems belong to class one, and emit spontaneous light in the low state and coherent light at the tuning wavelength of the external grating in the on-state. By changing the length of the external cavity, the round-trip time can be changed and bistability or pulsations are observed. Bistability is obtained for a round-trip time equal to the carrier lifetime. Increasing the round-trip time leads to self-sustained pulsations and for very long round-trip times to an irregular time behavior (Glas *et al.*, 1983) reminiscent of chaos.

III. CLASS-TWO BISTABILITY IN SEMICONDUCTOR LASERS

A wide variety of bistable effects has been observed in laser diodes, and an excellent review has been published only recently by Adams (1986). Here, we shall only address the major characteristics of bistability observed in the longitudinal-mode spectrum, the transverse-mode pattern and the two polarization states. All observed effects belong to the previously defined second class, i.e., the diode is lasing in either of the two bistable states and the total amount of stimulated emission is approximately the same for both states. As a consequence, the gain and the carrier distribution are essentially the same in both states, and the switching time is not slowed down by carrier diffusion. Since the carrier relaxation time constant is also given by the shorter stimulated lifetime and not by the spontaneous lifetime, a much faster switching speed is expected. However, one should not neglect that the build-up time of a mode from spontaneous emission with little excess gain can also take in the order of one nanosecond. Unfortunately, the size of the hysteresis is usually small compared to the class-one devices, and it becomes more difficult to bias the class-two devices within the hysteresis loop.

A. Longitudinal-Mode Bistability

Bistability attributable to dispersive effects was observed relatively early in laser diodes with optical feedback from an external mirror (Lang and Kobayashi, 1980). The dependence of the index of refraction on carrier density (Henry *et al.*, 1981), which in turn depends on the optical power, causes the length of the laser-diode subcavity to change. This causes a phase difference between the subcavity modes and the fixed modes of the external cavity, which leads to a hysteresis in the light-current characteristics by jumping from one longitudinal mode of the external cavity to the next. The size of the hysteresis is rather small (Lang and Kobayashi, 1980) because the diode can lase in any of a large number of external cavity modes.

These effects become much more dramatic and the hysteresis becomes much wider in a shorter composite cavity, such as the cleaved coupled cavity (C^3) laser (Olsson *et al.*, 1984; Phelan *et al.*, 1986). For bistable operation, one section is biased above threshold whereas the other is biased as a low-loss nonlinear tunable Fabry–Perot cavity. The power is emitted in either of two longitudinal modes, but never in both simultaneously. Switching between the two modes can be achieved by applying a current pulse to the modulator. The bistability is due to the above-mentioned power-dependent dispersive effects in the resonant modulator and laser. Because both bistable states are well above threshold, the carrier lifetime is given by the faster stimulated lifetime and the switching has been observed with subnanosecond delays and power delay products of 4 pJ (Olsson *et al.*, 1984). Complete opto-electronic logic operations such as AND, OR, EXOR and INV based on wavelength bistability were demonstrated with subnanosecond delays (Tsang *et al.*, 1983).

Another type of longitudinal-mode bistability has been observed in laser diodes without external feedback (Nakamura *et al.*, 1978) due to strong coupling among longitudinal modes via the gain medium. The phenomenon has been attributed to gain suppression (Yamada and Suematsu, 1979; Kazarinov *et al.*, 1982), to spatial hole burning (Seki *et al.*, 1981) and to saturable absorption resulting from traps in the cladding layer (Copeland, 1980). Even though the hysteresis for this longitudinal-mode bistability is very narrow, this effect has been successfully used to reduce the so-called mode-hopping noise drastically (Biesterbos *et al.*, 1983).

B. Lateral-Mode Bistability

It is well known that the lateral intensity profile in stripe lasers depends on the index and gain profiles. Both index and gain depend through the real and imaginary parts of the index of refraction on the carrier density. The carrier density itself depends on the pump-current distribution and the intensity profile of the mode (Thompson, 1972). A self-consistent solution to this problem is usually the gain-guided mode for a narrow stripe contact and an index-guided mode for a wider contact stripe. The gain-guided mode is attributed to a high gain in the center which makes up for the radiation loss of the mode, and the index-guided mode is attributed to self-focusing which is caused by an increased index of refraction in the regions of high intensity owing to depletion of the injected carrier concentration (Thompson, 1972). With a twin-stripe contact laser (White and Carroll, 1984), a lateral pump-current distribution is generated which produces bistability between two lateral modes, the lowest-order index-guided mode and the lowest-order gain-guided mode. Switching between these two modes can be initiated by injecting a short current pulse into either of the two stripes, and transition times as short as 0.8 ns have been measured (White and Carroll, 1983a). Switching was also observed by injecting an external light pulse (White and Carroll, 1983b) and an all-optical flip-flop with a switch time of three nanoseconds was made. The complex wave-guiding mechanism of these twin-stripe lasers has been greatly simplified with the index-guided twin-ridge laser (MacLean *et al.*, 1986). The twin-ridge laser is designed to have the same intensity profile for the fundamental and the first-order modes and therefore no gain discrimination exists between them. Owing to the nonlinear index of refraction, bistability between these two modes is expected.

C. Polarization Bistability

The stimulated emission in conventional double-heterostructure lasers is polarized in the TE mode owing to the higher mirror reflectivity and confinement factor of the zero-order TE mode compared to that of the TM mode. However, the state of polarization of the radiation of lasers with internal stress is more complex. It has been observed that the polarization of stimulated emission can be changed from TE to TM mode in GaAs under an applied mechanical stress (Patel and Ripper 1973), and this effect has been linked to a higher optical gain for the TM mode under compressive stress. Recently, TE or TM emission of 1.3-μm InGaAsP lasers depending on the drive current has been observed (Craft *et al.*, 1984) and investigated as a function of temperature (Chen and Liu, 1984a). The polarization change which is accompanied with a wavelength change of 3 nm with temperature and drive current is attributed to a thermal stress effect in the InGaAsP active layer (Chen and Liu, 1984b). This effect can also lead to polarization bistability (with a concomitant wavelength bistability) as a function of the drive current (through heating) if the device is held exactly at the polarization transition temperature (Chen and Liu, 1985). Current-pulse–induced switching (Liu and Chen, 1985a, 1985b) has been used to demonstrate electro-optical AND, NAND, OR and NOR latching gates and different types of flip-flops. The input signal is a current pulse and the output signal is defined by the polarization direction, i.e., TE for a logic "1" and TM for the logic level "0." Since TE and TM modes are simultaneously accessible, the output and its complement are simultaneously available. Instrumentation-limited switching speeds of one nanosecond were measured with the laser held exactly at the polarization transition temperature.

IV. BISTABLE OPTO-ELECTRONIC CIRCUITS

Bistable opto-electronic circuits seem to be too slow and too bulky at first sight. This is not necessarily the case since laser diodes on GaAs or InP substrates can easily be integrated monolithically with high-speed electronic devices which themselves are much smaller and about as fast as the laser diode. Bistability in the light-current characteristic of a buried heterostructure AlGaAs laser diode monolithically integrated with a MESFET (translaser) has been observed (Lau and Yariv, 1984) owing to coupling of scattered laser light into the MESFET channel. Even though optical switching experiments revealed a switch-on time constant of 7 ns, much faster switching times are projected for an optimized design. The bistable opto-electronic circuit has been used to demonstrate basic signal-processing functions such as optical, electrical and microwave switching and also pulse-position and pulse-frequency demodulation of an optical signal.

Opto-electronic circuits consisting of a laser diode, photodiodes and resistors have been proposed for optical logic operations. Bistability is achieved by coupling the laser to a photodiode in a positive feedback path (Okumura *et al.*, 1985). Low-speed operation for a discrete version of an inverter, an AND gate and flip-flops has been verified. With etched mirrors, these circuits can be integrated monolithically and should lead to interesting and useful applications not only in optical communications and data processing such as repeaters, time division sampling and memory, but also in sophisticated optical systems with three-dimensional structures.

V. CONCLUSIONS

Since the demonstration of the first semiconductor laser diodes in 1962, a wide variety of experimental arrangements to produce bistability has been suggested and demonstrated. The research was not only spurred by theoretical interest (Lugiato *et al.*, 1978) but also by the quest for high-speed low-power opto-electronic memory cells and latching logic gates. In this review, we heuristically divide the observed bistability into two classes. Bistable laser diodes of the first class are ideally single-mode lasers with the two stable states corresponding to the laser being switched off (not lasing) and switched on (lasing). Diodes of the second class are ideally lasers with two modes with the two stable states corresponding to these two modes. These diodes lase in either of the two modes, but never in both at the same time. Bistable laser diodes of the second class have the interesting property that the switching is not slowed by carrier redistribution and carrier diffusion. Their ultimate switching delay is given by the buildup time of the mode from spontaneous emission which is in the subnanosecond range. The shortest switching delays can only be achieved with a large trigger-pulse amplitude. In bistable laser diodes, like in all bistable elements, the delay between trigger pulse and switching strongly depends on the amplitude of the trigger pulse. For very weak trigger pulses, the delay can be orders of magnitude longer than the characteristic time constants of the laser diode; this phenomenon is known as "critical slowing down." The speed of the switching transition itself is fast and relatively independent of the trigger pulse amplitude.

The usefulness of these bistable laser diodes as latching opto-electronic logic gates and as memory cells for optical signal processing has been demonstrated in various laboratories. Delay times below one nanosecond, switching powers of a few femtojoules and standby powers of some tens of milliwatts have been reported. The demonstration of a high-speed time-multiplexed optical highway using bistable laser diodes as optical memories

(Suzuki *et al.*, 1986) indicates the potential of these laser diodes for optical signal processing in future telecommunications networks.

We should like to thank Dr. Kam Lau and Dr. Bart Van Zeghbroeck for helpful discussions.

REFERENCES

Abraham, N. B. (1983). *Laser Focus*, May, 73–81.
Adams, M. J. (1986). *Int. J. Electron.* **60**, 123–142.
Basov, N. G. (1968). *IEEE J. Quantum Electron.* **4**, 855–864.
Bazhenov, V. Y., Bogatov, A. P., Eliseev, P. G., Okhotnikov, O. G., Pak, G. T., Rakhvalsky, M. P., Soskin, M. S., Taranenko, V. B., and Khairetdinov, K. A. (1982). *IEE Proc. Part I* **129**, 77–82.
Biesterbos, J. W. M., den Boef, A. J., Linders, W., and Acket, G. A. (1983). *IEEE J. Quantum Electron.* **19**, 986–990.
Bonifacio, R., and Meystre, P. (1979). *Opt. Commun.* **29**, 131–134.
Carney, J. K., and Fonstad, C. G. (1981). *Appl. Phys. Lett.* **38**, 303–305.
Chen, Y. C., and Liu, J. M. (1984a). *Appl. Phys. Lett.* **45**, 604–606.
Chen, Y. C., and Liu, J. M. (1984b). *Appl. Phys. Lett.* **45**, 731–733.
Chen, Y. C., and Liu, J. M. (1985). *Appl. Phys. Lett.* **46**, 16–18.
Copeland, J. A. (1980). *IEEE J. Quantum Electron.* **16**, 721–727.
Craft, D. C., Dutta, N. K., and Wagner, W. R. (1984). *Appl. Phys. Lett.* **44**, 823–825.
Gibbs, H. M., McCall, S. L., and Venkatesan, T. N. C. (1976). *Phys. Rev. Lett.* **36**, 1135–1138.
Glas, P., and Müller, R. (1982). *Opt. Quantum Electron.* **14**, 375–389.
Glas, P., Müller, R., and Klehr, A. (1983). *Opt. Commun.* **47**, 297–301.
Goldobin, I. S., Kurnosov, V. D., Luk'yanov, V. N., Semenov, A. T., Sapozhnikov, S. M., Shelkov, N. V., and Yakubovich, S. D. (1980). *Sov. J. Quantum Electron.* **10**, 1452–1453.
Goto, H., Nagashima, K., Suzuki, S., Kobayashi, K., Odagiri, Y., Ohta, Y., Kondo, M., and Komatsu, K. (1984). *Proc. Global Telecommun. Conference*, Atlanta, USA **2**, 880–884. IEEE.
Harder, Ch., Lau, K. Y., and Yariv, A. (1981). *Appl. Phys. Lett.* **39**, 382–384.
Harder, Ch., Lau, K. Y., and Yariv, A. (1982a). *IEEE J. Quantum Electron.* **18**, 1351–1361.
Harder, Ch., Lau, K. Y., and Yariv, A. (1982b). *Appl. Phys. Lett.* **40**, 124–126.
Harder, Ch., Smith, J. S., Lau, K. Y., and Yariv, A. (1983). *Appl. Phys. Lett.* **42**, 772–774.

Henry, C. H. (1980). *J. Appl. Phys.* **51**, 3051–3061.
Henry, C. H., Logan, R. A., and Merritt, F. R. (1980). *J. Appl. Phys.* **51**, 3042–3050.
Henry, C. H., Logan, R. A., and Bertness, K. A. (1981). *J. Appl. Phys.* **52**, 4457–4461.
Ito, H., Onodera, N., Gen-ei, K., and Inaba, H. (1981). *Electron. Lett.* **17**, 15–17.
Kawaguchi, H. (1981). *Electron. Lett.* **17**, 741–742.
Kawaguchi, H. (1982). *Appl. Phys. Lett.* **41**, 702–704.
Kawaguchi, H., and Iwane, G. (1981). *Electron. Lett.* **17**, 167–168.
Kazarinov, R. F., Henry, C. H., and Logan, R. A. (1982). *J. Appl. Phys.* **53**, 4631–4644.
Lang, R., and Kobayashi, K. (1980). *IEEE J. Quantum Electron.* **16**, 347–355.
Lasher, G. J. (1964). *Solid-State Electron.* **7**, 707–716.
Lau, K. Y., Harder, Ch., and Yariv, A. (1982a). *Appl. Phys. Lett.* **40**, 198–200.
Lau, K. Y., Harder, Ch., and Yariv, A. (1982b). *Appl. Phys. Lett.* **40**, 369–371.
Lau, K. Y., and Yariv, A. (1984). *Appl. Phys. Lett.* **45**, 719–721.
Lee, T. P., and Roldan, R. H. R. (1970). *IEEE J. Quantum Electron.* **6**, 339–352.
Liu, J. M., and Chen, Y. C. (1985a). *Electron. Lett.* **21**, 236–238.
Liu, J. M., and Chen, Y. C. (1985b). *IEEE J. Quantum Electron.* **21**, 298–306.
Lugiato, L. A., Mandel, P., Dembinski, S. T., and Kossakowski, A. (1978). *Phys. Rev. A* **18**, 238–254.
MacLean, D., White, I. H., Carroll, J. E., Armistead, C. J., and Plumb, R. G. (1986). *Proc. 12th European Conference on Optical Communication*, Barcelona, Spain, 229–232. EUREL.
McInerney, J., Reekie, L., and Bradley, D. J. (1985). *IEE Proc. Part. J*, **132**, 90–96.
Miller, D. A. (1982). *Laser Focus*, April, 79–84.
Mocker, H. W., and Collins, R. J. (1965). *Appl. Phys. Lett.* **7**, 270–273.
Nakamura, M., Aiki, K., Chinone, N., Ito, R., and Umeda, J. (1978). *J. Appl. Phys.* **49**, 4644–4648.
Nathan, M. I., Marinace, J. C., Rutz, R. F., Michel, A. E., and Lasher, G. J. (1965). *J. Appl. Phys.* **36**, 473–480.
Odagiri, Y., Komatsu, K., and Suzuki, S. (1984). *Proc. Conference on Lasers Electro Optics*, Anaheim, CA, Paper THJ3, 184–186. IEEE QEA and OSA.
Okumura, K., Ogawa, Y., Ito, H., and Inaba, H. (1985). *IEEE J. Quantum Electron.* **21**, 377–382.
Olsson, N. A., Tsang, W. T., Logan, R. A., Kaminow, I. P., and Ko, J-S. (1984). *Appl. Phys. Lett.* **44**, 375–377.
Patel, N. B., and Ripper, J. E. (1973). *IEEE J. Quantum Electron.* **9**, 338–341.
Phelan, P., Reekie, L., Bradley, D. J., and Stallard, W. A. (1986). *Opt. Quantum Electron.* **18**, 35–41.
Ruschin, S., and Bauer, S. H. (1979). *Chem. Phys. Lett.* **66**, 100–103.
Scott, J. F., Sargent, III, M., and Cantrell, C. D. (1975). *Opt. Commun.* **15**, 13–16.
Seki, K., Kamiya, T., and Yanai, H. (1981). *IEEE J. Quantum Electron.* **17**, 706–713.
Smith, P. W., and Tomlinson, W. J. (1981). *IEEE Spectrum*, June, 26–35.
Sorokin, P. P., Luzzi, J. J., Lankard, J. R., and Pettit, G. D. (1964). *IBM J. Res. Develop.*, April, 182–184.

Stallard, W. A., and Bradley, D. J. (1983). *Appl. Phys. Lett.* **42**, 858–859.
Suzuki, S., Terakado, T., Nagashima, K., Suzuki, A., and Kondo, M. (1986). *IEEE J. Lightw. Techn.* **4**, 894–899.
Szöke, A., Daneu, V., Goldhar, J., and Kurnit, N. A. (1969). *Appl. Phys. Lett.* **15**, 376–379.
Thompson, G. H. B. (1972). *Opto-electronics* **4**, 257–310.
Tsang, W. T., Olsson, N. A., and Logan, R. A. (1983). *IEEE J. Quantum Electron.* **19**, 1621–1625.
Van der Ziel, J. P., Tsang, W. T., Logan, R. A., and Augustyniak, M. (1981a). *Appl. Phys. Lett.* **39**, 376–378.
Van der Ziel, J. P., Tsang, W. T., Logan, R. A., Mikulyak, R. M., and Augustyniak, M. (1981b). *Appl. Phys. Lett.* **39**, 525–527.
White, I. H., and Carroll, J. E. (1983a). *Electron. Lett.* **19**, 337–339.
White, I. H., and Carroll, J. E. (1983b). *Electron. Lett.* **19**, 558–560.
White, I. H., and Carroll, J. E. (1984). *IEE Proc. Part H*, **131**, 309–321.
Yamada, M., and Suematsu, Y. (1979). *IEEE J. Quantum Electron.* **15**, 743–749.

17 INSTABILITIES IN SEMICONDUCTOR LASERS

K. A. Shore and T. Rozzi

SCHOOL OF ELECTRICAL ENGINEERING
UNIVERSITY OF BATH
UNITED KINGDOM

I. INTRODUCTION

A discussion of spatial and temporal instabilities in semiconductor lasers may be initiated by two contrasting objectives. On the one hand, optical communications systems have a requirement of stable optical sources. In the case of high bit-rate optical communications, this implies a need for

OPTICAL NONLINEARITIES
AND INSTABILITIES IN
SEMICONDUCTORS

ISBN 0-12-332915-9

mode-stabilised semiconductor lasers. Hence, every effort is made to eliminate the causes of spatial, temporal and spectral instability in lasers designed for such applications. However, in seeking to apply the semiconductor laser in the role of a high-speed optical switch or as an optical logic element, means are sought to utilise controllable instabilities in the device. To date, the emphasis in research has been on achieving stable lasers, but greater attention is now being given to the development of lasers for switching and logic functions. The design of such devices requires a description of the stability properties of the laser where both spatial and temporal aspects need to be considered.

Here we show how work initially directed at characterising spatial stability in conventional single-stripe semiconductor lasers has led to a framework for stability analysis in which the detailed behaviour of the device may be taken into account. Insights gained in the theoretical treatment of these problems have shown that multistripe devices are prominent candidates as devices exhibiting controllable instabilities. Attention has thus been directed at examining the utilisation of these multistripe lasers in which both electronic and optical means of stability control are available. The full potential of these devices, it is felt, is yet to be realized. In discussing these aspects, we include an outline of the analysis at each stage and summarise the main results obtained in respect to spatial and temporal stability. The exploitation of these results for switching and logic is noted where appropriate.

II. SPATIAL INSTABILITIES

A. Device Model

The cross-sectional geometry of a typical stripe-geometry laser is shown in Fig. 1, where the coordinate axes are also defined. (Light propagation is in the z-direction.) The mechanisms available for guiding the optical field in the transverse x-direction are discussed in the next subsection. It is assumed that owing to material changes in the y-direction, the optical field is confined by a real index guide. In the device, the active region has a thickness, d, and a width defined by the finite stripe-contact width, a, to which current injection into the active region is arranged.

The appropriate device-modelling equations for variations in the x-direction are (i) the carrier-diffusion equation:

$$D\frac{d^2N_0}{dx^2} + \frac{J}{ed} - BN_0^2 = gP_0(x) \tag{1}$$

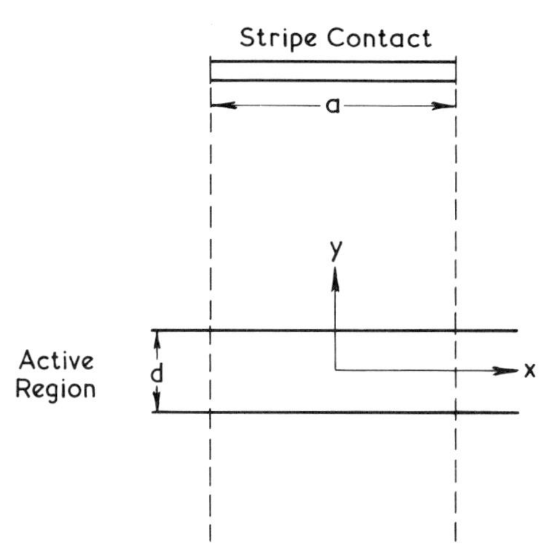

Figure 1. Schematic cross section for a stripe geometry laser.

and (ii) the appropriate form of the wave equation:

$$\frac{d^2\phi_0}{dx^2} + \left(\epsilon_0 k_0^2 - \beta_0^2\right)\phi_0 = 0 \tag{2}$$

The terms appearing in the equations are as follows. D is the diffusion coefficient. $N_0(x)$ is the carrier (electron) concentration. J is the injection current. B is the bimolecular recombination coefficient appropriate to low-doped devices. $P_0(x) = A_0^2|\phi_0(x)|^2$ is the optical field intensity. A_0 the field amplitude and ϕ_0 the modal field. ϵ_0 is the complex dielectric constant and β_0 is the longitudinal propagation constant.

$$g = \alpha N_0 - \beta \tag{3}$$

is the local gain.

The subscript zero appearing in these quantities denotes the fact that they represent the solutions to the device equations whose stability is to be analysed. The interaction between the equations is apparent since ϕ_0 from Eq. 2 appears explicitly in Eq. 1 and also N_0 from Eq. 1 appears in ϵ_0 as shown in Eq. 6.

The field normalisation is written

$$\int_{-\infty}^{\infty} \phi_0^2(x)\, dx = 1 \tag{4}$$

The photon density is defined as

$$S_0 = A_0^2 \int_{-\infty}^{\infty} |\phi_0|^2\, dx = \int_{-\infty}^{\infty} |\Phi_0|^2\, dx = A_0^2 \tag{5}$$

where Φ_0 is the total field and ϕ_0 the modal field.

B. Nonlinearities and Near-Field Shifts

Gain-guided stripe-geometry injection lasers commonly exhibit nonlinearities in their above-threshold light-current characteristics. It is found that the optical near field moves laterally across the output mirror facet as the nonlinear portion of the device light-current characteristic is encountered. In this process, the near field is displaced from an initially symmetric configuration with displacements of up to 1 μm being typical. Over a range of injection currents, the asymmetry of the near field increases with current, but for yet higher injection currents, a sudden snapback to a symmetric field occurs (Kerps, 1980).

Instabilities of the near field of this kind would result in significant changes in coupling efficiency in a laser-fibre optical communications system. It is thus essential, for such application, to stabilise the near field. This is now commonly achieved by ensuring that real index guiding is provided in the device structure and, hence, preventing near-field shifts. However, as will be discussed, it is also possible to consider a number of applications that exploit the observed nonlinearities and field instabilities. In this context, the requirement is for means of controlling and/or enhancing the nonlinearities.

It should be appreciated that two mechanisms are responsible for controlling the optical field in an active device such as the injection laser. The mechanisms may be represented by the real and imaginary parts of a complex dielectric constant and are commonly referred to as *gain guiding* and *refractive-index antiguiding*. To see the relationship between these mechanisms, we note that the change, $\Delta\epsilon$, of the dielectric constant due to variation in carrier concentration is given by

$$\Delta\epsilon = (-\rho + j)\frac{gn_0}{2k_0V} \tag{6}$$

Here, g is the local gain which is conventionally assumed to be linearly related to the carrier concentration. k_0 is the free space wave vector and $v = c/n_0$ is the velocity of light in the laser material. n_0 is the background refractive index.

The parameter of greatest interest in the present discussion is the antiguiding parameter, ρ, which measures the relative strengths of changes in the real and imaginary parts of the dielectric constant. (The parameter ρ is also referred to as the linewidth enhancement or α parameter.) The parameter, ρ, is a positive quantity whose value has been assumed variously to lie between about 0.5 and 6.0. It is clear from Eq. 3 that an increase in g

via an increase in the carrier concentration will imply a decrease in the real part of the dielectric constant. This will thus have a refractive-index antiguiding effect. Conversely, the imaginary part of the dielectric constant is increased with carrier concentration; this then is the gain-guiding mechanism. It is the competition between the two guiding mechanisms that is the cause of the near-field instability under consideration. The optical field in the device seeks a spatially stable position consistent with the two guiding mechanisms. Because of the nonlinearity of the interaction between the optical field and the carrier concentration, a spatially stable position need not correspond to a symmetric optical field. A stability analysis has shown that a stable asymmetric near field may exist even in a symmetric device (Shore and Rozzi 1981a).

In an ideal laser, such asymmetric near fields are degenerate: two asymmetric off-centre positions are possible, one being the mirror image of the other. In practice, slight nonuniformities in the device serve to break the degeneracy.

III. CONTROLLED INSTABILITIES AND THEIR APPLICATIONS

A. Electronically Controlled Instability

We have seen that in conventional stripe-geometry lasers, the optical near field may undergo a lateral movement off-centre to occupy a stable asymmetric position. In single-stripe devices, this behaviour is determined by the form of the local gain function via the complex dielectric constant. Gain guiding, which confines the optical field over a range of the light-current characteristics, may give way to real-index guiding for higher optical powers when the injection current is significantly increased. The fundamental difficulty, however, is that little control can be effected over the form of the gain function in single-stripe devices. This is unfortunate since the near-field shifts and nonlinearities that appear in stripe-geometry lasers are potentially useful. The position may be recovered by the rather simple expedient of constructing multistripe-geometry lasers and in particular twin-stripe lasers. In such devices, it is found that control can be exercised over the optical near field by means of independently controllable injection currents.

A schematic cross section of the twin-stripe laser is given in Fig. 2. The active region of the laser is now pumped by two injection currents. It is assumed that electrical isolation between the stripe contacts can be ensured

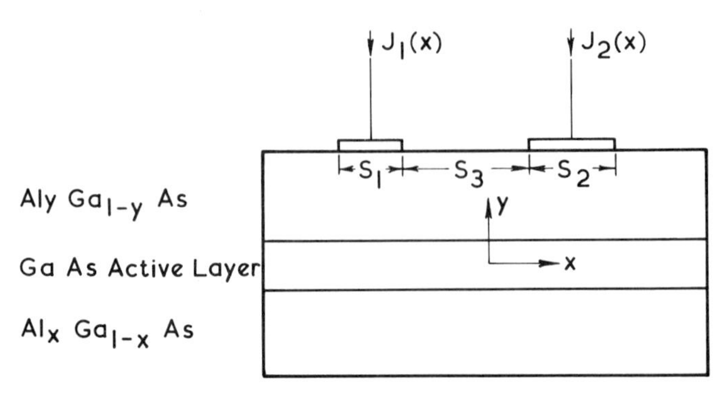

Figure 2. Twin-stripe laser geometry.

so that the currents J_1 and J_2 may be varied independently. An analysis of this device has been performed using a general model for semiconductor lasers (Shore and Rozzi 1981b). The essential requirement for carrying out the computer modelling of this configuration is the ability to include a current profile $J(x)$ which takes account of the currents J_1 and J_2. In all other respects, the solution of Eqs. 1 and 2 for twin-stripe devices is similar to the procedure followed for single-stripe devices (Shore et al. 1980). An alternative procedure for analysing these devices has been suggested by Katz (Katz 1983).

Computer analysis provided confirmation of the basic contention that controlled movement of the optical near field could be demonstrated through variation of the injection currents. The coupling between the stripes is affected by interstripe separation, carrier diffusion and current spreading effects (Shore and Hartnett 1982). The first of these is most easily adjusted at the device-design stage. The degree of electrical isolation between the stripes must be carefully considered to account for current leakage effects (Shore 1983, Shore 1984c). Finally, reference is made to calculations of the optical output power of the device as a function of the injection currents J_1 and J_2 (Shore et al. 1983). When the results of these calculations are displayed in the form of constant power contours in the J_1–J_2 plane, they provide confirmation of measured results developed to show the occurrence of bistability in twin-stripe lasers (White et al. 1982).

Practical application of the effects we have outlined have been discussed in two main areas. The earliest suggestions for using twin-stripe lasers were in view of a scanning function where controlled movement of the device far-field radiation pattern was of interest (Scifres et al. 1978). The rotation of the far-field pattern as injection currents are altered may be seen to be a consequence of the near-field shifts already discussed (Shore and Rozzi 1981b, Shore and Hartnett 1982). Further remarks on this application will be made in the discussion of optically controlled instability in the next subsection. The second area of interest, where exploitation of twin-stripe laser properties is under active investigation, has already been referred to and concerns the use of the device as an optical logic element. Bistability in these devices is well known by now, and switching times of the order of a fraction of a ns. have been demonstrated, but much work remains to be done in order to minimize overall operation times and other parameters, before useful ultrafast logic devices can be realised. It should be remarked for completeness that multistripe devices have received considerable attention recently due to their capability for providing large optical output powers. This aspect will not be pursued here since the phase-locking properties responsible for this feature of multistripe device characteristics do not belong to a discussion of optical instability.

B. Optically Controlled Instability

The previous subsection considered the scope offered by the twin-stripe laser in particular and multistripe lasers in general, for controlling near- and far-field patterns. The basis for electronic control of the optical field was the change in the gain profile in the device that could be effected by independent adjustment of the stripe currents. It is natural to enquire how the static properties of the device may be influenced by means of the injection of light into the active region of the laser. The superposition of the injected light on the lasing optical field causes a change in the lateral gain profile and, hence, the wave-guiding characteristics of the structure are altered. So that optical injection may have significant effects, it is necessary to be able to adjust the sensitivity of the device, i.e. to choose a gain profile that implies a potentially unstable optical lasing field. Since control of near-field position is offered by multistripe devices, it is in that context that optical injection effects are assessed.

In Fig. 3 we give a schematic diagram of the twin-stripe laser subject to optical injection into the active region below either or both of the stripe contacts. Significant changes in the lasing field intensity were demonstrated for relatively small optical injection powers (Shore 1982a, Shore 1984a). The latter observation was developed to show optically induced bistable action in the twin-stripe laser (Shore 1982b). It is believed that the latter effect will be of particular interest in the context of optical logic. As an

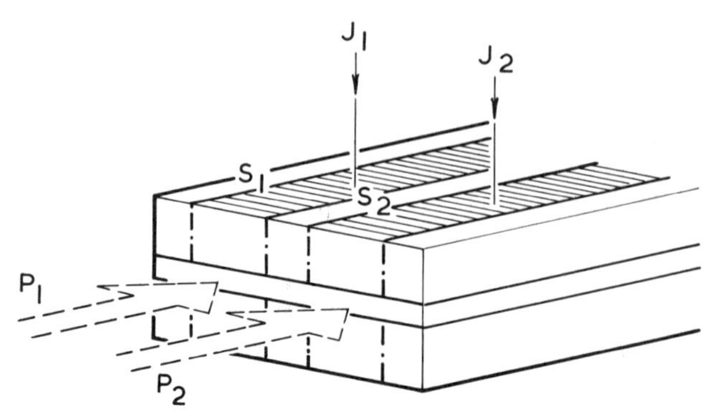

Figure 3. Optical injection into a twin-stripe laser.

all-optical approach to implementing bistability, it avoids recourse to electronic means of switching and thus is potentially fast in its response and also lends itself to integration in all-optical circuits.

Two other applications-oriented properties were identified. In the first, it was shown how the effects of near-field instability could be carefully tuned so as to produce optical limiter action in the device. In this case, the total light output from the device remained constant over a range of injected optical power (Shore 1982c). Secondly, the optically driven analog to the beam scanner already discussed was assessed. It was shown that rotation of the far field could indeed be achieved by optical means and, further, that the required optical injection power was not unreasonable. The calculations indicated that a far-field rotation of between 1° and 3° per mW of injected optical power may be achievable (Shore 1984b). Although such field rotations are less than what would be expected with electronically steered beam scanners, the ability to perform the scanning by optical means alone may still have attractions.

In this section, attention has been paid to the manner in which spatial instability may be exploited to develop the static properties of the semiconductor laser. In the next section, we consider temporal instabilities in laser diodes.

IV. TEMPORAL INSTABILITIES

In the preceding sections, spatial instabilities under steady-state conditions have been emphasized. We now briefly examine dynamic effects using a simplified spatial model. In the limit we may assume the existence of two reservoirs of carriers with carrier densities N_1 and N_2 and carrier lifetimes τ_{s_1} and τ_{s_2}, respectively. This model may represent two transverse or two longitudinal regions, two different types of recombination mechanisms, etc.

The photon density, S, of the laser mode is distributed over the reservoirs according to the fractions C_1 and C_2 such that $C_1 + C_2 = 1$. In the simplest model, C_1 and C_2 are constant in time.

We therefore consider the following system of rate equations.

$$\frac{ds}{dt} = \left[G - \frac{1}{\tau_p}\right]s \tag{7}$$

$$\frac{dN_1}{dt} = -\frac{N_1}{\tau_{s_1}} + j_1 - C_1 S g_1(N_1) + \frac{N_2 - N_1}{\tau} \tag{8}$$

$$\frac{dN_2}{dt} = -\frac{N_2}{\tau_{s_2}} + j_2 - C_2 S g_2(N_2) + \frac{N_1 - N_2}{\tau} \tag{9}$$

τ_p is the photon lifetime; τ is the diffusion time constant; $g_i(N_i)$ is the gain function in reservoir i. The gain functions in the two reservoirs are taken as different in general, allowing for the possibility of inhomogeneities. The optical gain G is given by

$$G = C_1 g_1(N_1) + C_2 g_2(N_2) \tag{10}$$

j_i are the nominal current densities in units of the electron charge. A stationary solution of these equations is obtained when the time derivatives are zero. Eliminating from these equations N_1 and N_2, we obtain a curve

$$S = S(j_1, j_2) \tag{11}$$

It is obvious from the rate equations that the solution $S = 0$ is stable below threshold. Instability can only arise for $S > 0$, however small S may be.

In order to investigate the stability of the system, the Hurwitz criterion is applied after linearising the rate equations around a stationary point. The region of potential instability in the J_1–J_2 plane results from such an analysis (Poh et al. 1979). A typical such region is shown in Fig. 4. A more precise determination of the instability region may be obtained by allowing the confinement factors C_i to vary in time and with injection current (Poh and Rozzi 1981).

It is emphasised again that instability in the context of this section refers to the dynamic behaviour of the device, e.g., self-sustained oscillation and temporal switching.

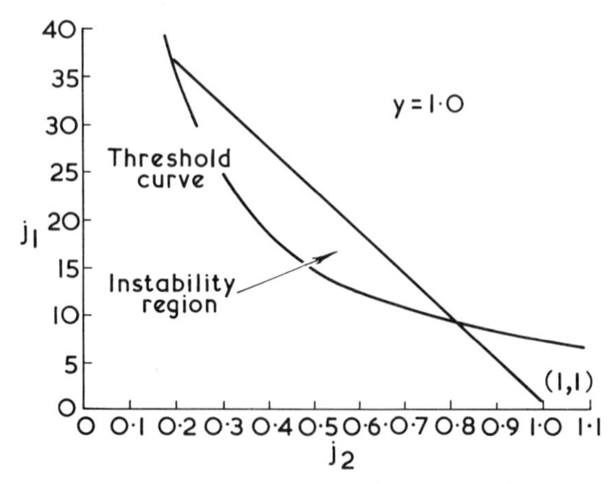

Figure 4. Instability region in the J_1–J_2 plane.

V. SPATIO-TEMPORAL INSTABILITIES

A. Transverse Mode Switching

We now turn to the descriptions of instabilities involving densities that are functions of *both* space and time. A simple illustration of a phenomenon that arises in this context may be obtained by considering the switching, at frequency ω, between two metastable states of the same photon density. In particular, the two states may be the degenerate off-centre optical fields in a gain-guided single-stripe laser discussed in Section 2.2.

It is assumed that solutions to the steady-state diffusion and wave Eqs. 1 and 2 are found such that an asymmetric modal field $\phi_a(x)$ is obtained with a corresponding carrier distribution $N_a(x)$. Under the postulate of degeneracy of the solution, a second steady solution is possible for which

$$\phi_b(x) = \phi_a(-x) \tag{12}$$

and

$$N_b(x) = N_a(-x) \tag{13}$$

A time-varying optical field is then taken to be

$$\phi(x, t) = a(t)\phi_a(x) + b(t)\phi_b(x) \tag{14}$$

In general, ϕ would satisfy the time-dependent wave equation

$$\nabla^2\phi - \frac{\epsilon}{c^2}\frac{\partial^2\phi}{\partial t^2} = \frac{2}{c^2}\frac{\partial\phi}{\partial t}\frac{\partial\epsilon}{\partial t} + \frac{1}{c^2}\phi\frac{\partial^2\epsilon}{\partial t^2} \tag{15}$$

We separate the time variation into two components, one of which is related to the optical carrier frequency ω_0, and the other being due to the "slow" variations of the field amplitudes:

$$\frac{\partial}{\partial t} = j\omega_0 + \frac{\partial}{\partial \tau} \tag{16}$$

Then the time-dependent wave equation becomes

$$\nabla^2 \phi + \frac{\omega^2}{c^2}\epsilon\phi = \frac{2j\omega_0}{c^2}\epsilon\left[\frac{\partial \phi}{\partial t} + \frac{1}{\epsilon}\frac{\partial \epsilon}{\partial \tau}\phi\right] \tag{17}$$

Also, by introducing a time-dependent diffusion equation, the oscillation frequency ω may be found (Shore and Rozzi 1983a).

The calculated oscillation frequency, ω, will be a complex quantity and, hence, real oscillations can only arise at isolated points at most, where the real part of ω can be taken as the oscillation frequency. Numerical calculation using typical parameters for the device indicate oscillation frequencies varying between 10 and 150 GHz (Shore and Rozzi 1982), thus implying optical switching times in the range of a few nanoseconds down to a few picoseconds (Shore and Rozzi 1983b). Hence, the generation of such oscillations would be of considerable interest in the context of high-speed optical switches.

B. Hopf Bifurcation Analysis

1. Motivation

The preceding example shows the need for a general method of spatio-temporal instability analysis. Inasmuch as the optical field can also be approximately described by the modes of a multilayer slab with stepped refractive index profile, the Hurwitz approach of section 4 also lends itself to an approximate description of the spatial as well as temporal evolution (Poh et al. 1979). One may require, however, the obtaining of a detailed spatial description by this method or even describing instabilities in multi-stripe or split-stripe lasers or more complicated types of instability where pulsations are associated with mode changes. Under these more general conditions, the approach described becomes rapidly intractable, due to the increasing number of coupled rate equations, one per region plus one for the photon density of each mode.

We show here how the techniques of Hopf bifurcation theory provide the required general technique for instability analysis. Moreover, whereas the

Hurwitz criterion determines just the regions of potential small signal instability, the Hopf algorithm provides a means for determining the critical points at which steady-state behaviour breaks into two oscillatory solutions. The existence of a Hopf bifurcation in the characteristics therefore implies the possibility of both bistable operation and self-sustained oscillations.

2. Perturbation Theory

Our approach is to develop from Eq. 17 a set of equations to describe the time evolution of the electromagnetic field. We proceed on the basis that a solution ϕ_0 of the stationary wave Eq. 1 is known.

The stationary solution will, in fact, have been obtained within a self-consistency scheme such that it satisfies both Eq. 1 and the stationary diffusion Eq. 2.

The self-consistency criteria will fix the amplitude A_0 of the lasing field ϕ_0 under stationary conditions. We introduce the notation $\phi_0 = A_0\phi_0$ and seek a representation of the time-dependent modal field ϕ. The time-dependent field is written as a perturbation of ϕ_0:

$$\Phi = \Phi_0 + \phi \tag{18}$$

The perturbation ϕ is expressed as a linear combination of orthogonal functions. It is natural to choose ϕ_0 as one basis function and to construct another basis function ϕ_1 orthogonal to ϕ_0 in a sense defined in Eq. 20. Using these two functions, we define

$$\phi(x, t) = a_0(t)\phi_0(x) + a_1(t)\phi_1(x) \tag{19}$$

where ϕ_1 is defined such that

$$\int_{-\infty}^{\infty} \phi_1^*(x)\phi_0(x)\,dx = 0 \tag{20}$$

We point out that in Eq. 19, the coefficients a_0 and a_1 include the time dependence of the perturbation. In general, a_1 is a complex quantity, $a_1' + ia_1''$, but a_0 may be taken as a real number. Two real rate equations involving a_0, a_1', a_1'' may be obtained together with rate equations for the carrier density using the time-dependent diffusion equation.

To describe the time evolution of the electron distribution, we introduce orthonormal functions $N_0(x)$ and $N_1(x)$. $N_0(x)$ is the normalized stationary electron distribution, i.e.

$$\int_{-\infty}^{\infty} N_0(x)N_0(x)\,dx = 1 \tag{21}$$

and $N_1(x)$ is defined such that

$$\int_{-\infty}^{\infty} N_0(x) N_1(x)\, dx = 0 \tag{22}$$

With these definitions we set

$$n(x, t) = v_0(t) N_0(x) + v_1(t) N_1(x) \tag{23}$$

This representation of the perturbation of the carrier concentration leads to two real rate equations for $v_0(t)$ and $v_1(t)$.

The specification of the optical field is completed by ensuring that the gain of the perturbed field matches the optical losses. This then gives rise to a first-order rate equation for the perturbation of the photon number, which is the fifth equation of the problem, as required for the five real variables $v_0, v_1, a_0, a_1', a_1''$.

3. Matrix Equations of Motion

By making the identifications

$$x_1 = v_0; \qquad x_2 = v_1; \qquad x_3 = a_0; \qquad x_4 = a_1'; \qquad x_5 = a_1''$$

the dynamic equations obtained by the method just outlined may be cast in the form

$$\frac{d\mathbf{x}}{dt} = \mathbf{A}\mathbf{x} \tag{24}$$

where

$$\mathbf{x}^T = [x_1, x_2, x_3, x_4, x_5] \tag{25}$$

and the elements of the matrix **A** are found from the perturbation analysis (Shore and Rozzi 1984).

4. Bifurcation Algorithm

To apply the techniques of Hopf bifurcation theory, it is required that the laser be described by an autonomous system of ordinary differential equations (Hassard et al. 1981).

$$\frac{d\mathbf{X}}{dt} = f(\mathbf{X}, S) \tag{26}$$

where **X** is an n-dimensional vector and s a real parameter. From this system of equations, the Jacobian matrix is formed:

$$A_{ij} = \frac{\partial f^i}{dx_j} \tag{27}$$

At a stationary point of the system (i.e. under steady-state conditions where $d\mathbf{X}/dt = 0$), the eigenvalues of the matrix $\mathbf{A}$ are found. The eigenvalues are ordered according to the magnitude of their real parts

$$\lambda_1' \geq \lambda_2' \geq \cdots \geq \lambda_n' \tag{28}$$

A critical point of the system is identified by finding a value S_c of the bifurcation parameter for which

$$\lambda_1(S) = \lambda_2^*(S) \tag{29}$$

for values of S in the neighbourhood of S_c.

At S_c it is required that λ_1 is purely imaginary and also that:

$$\begin{aligned} &\left.\frac{d\lambda_1'}{dS}\right|_{S=S_c} \neq 0 \\ &\lambda_1''(S_c) \neq 0 \\ &\lambda_j'(S_c) < 0, \qquad j = 2, n \end{aligned} \tag{30}$$

It is now required to apply this algorithm to the dynamics of the semiconductor laser.

It is appropriate also to note here that the formulation of the rate equations outlined above has utilized a particular set of basis functions defined in Eq. 20. A natural choice of basis functions has been made to relate to the physical problem that underlies the analysis, namely the occurrence of transverse mode switching in the device. The power of the Hopf bifurcation algorithm becomes apparent when consideration is given to the implications of the choice of alternative perturbation functions of the optical field. With suitable choices, it should be practicable to study the possibility of a variety of oscillations related to different perturbations, including those related to longitudinal effects.

5. Application to Laser Instabilities

We have already made the identification between the parameters $v_0, v_1, a_0, a_1', a_1''$, and a five-dimensional vector $\mathbf{X}$. Furthermore, equations of motion have been obtained in the form required by Eq. 26. In deriving the linearized form of the equations as given in Eq. 24, we have, in fact, defined the Jacobian matrix sought by the algorithm. Its eigenvalues must be found to follow through the Hopf bifurcation criteria.

The Hopf algorithm has established the existence of high-speed oscillations in the semiconductor laser (Shore and Rozzi 1984). In addition, the rather technical work of assessing the stability of the Hopf oscillations has been undertaken (Shore 1985a). The latter required the use of a canonical form for the dynamic system matrix (Shore 1985b). It is relevant to note that the oscillation frequency calculated for transverse mode switching (as described in section 5.1) is the same as that predicted by the Hopf bifurcation analysis.

In undertaking the Hopf analysis, it was perceived that the basic model was capable of quite general application and two particular lines of development were identified (Shore 1986a). Firstly, the theory was applied to more general laser structures such as twin-stripe lasers, demonstrating the existence of controllable bifurcations (Shore 1986b and c and 1987a), with the possibility of a nongeneric quasi-periodic transition to chaos (Shore 1987b). Secondly, alternative choices of perturbation functions can be made and the particular case of switching behaviour between fundamental and first-order transverse modes has been analysed (Shore and Rozzi 1986). In this latter case, certain simplifications of the dynamic model are obtained.

VI. CONCLUSIONS

Basic mechanisms for instabilities in semiconductor lasers have been discussed and a description has been given of the theoretical treatment of these instabilities. It has been shown that models can be restricted to spatial characterisation in the steady state or temporal characteristics with averaged spatial dependence or else they may include both temporal and spatial dependencies of the perturbation. In the latter case, the Hopf bifurcation method has been shown to be a powerful tool for instability analysis.

The strong nonlinearity in laser diodes lends itself to optically bistable operations which can be triggered by electrical or optical means. It appears possible also to realise clocking devices with frequencies above 100 GHz; however, the question remains open whether truly bistable operation can be achieved with such switching frequencies. Spatial instabilities have been presented as essential behaviour which lend themselves to possible exploitation in a number of optoelectronic devices such as limiters and spatial switches (Rozzi and Shore 1985). For the latter, it appears that a role exists in advanced communication systems and local area networks where relatively early practical realisation of these functions is anticipated.

ACKNOWLEDGMENT

The permission of the Optical Society of America to utilise material from Rozzi and Shore (1985) in the preparation of this contribution is gratefully acknowledged.

REFERENCES

Hassard, B. D., Kazarinoff, N. D., and Wan, Y. H. (1981). *Theory and Applications of Hopf Bifurcation*. Cambridge University Press, London, England.

Katz, J. (1983). "Electronic beam steering of semiconductor injection lasers: a theoretical analysis." *Appl. Opt.* **22**, 313.

Kerps, D. (1980). "Filament displacement and refraction losses in a stripe geometry AlGaAs d-h laser." *IEE Proc. Part I* **127**, 94.

Poh, B. S., and Rozzi, T. E. (1981). "Intrinsic instabilities in narrow stripe geometry lasers caused by lateral current spreading." *IEEE J. Quantum Electron.* **QE17**, 723.

Poh, B. S., Rozzi, T. E., and Velzel, C. H. (1979). "Single and dual filament self-sustained oscillations in dh injection lasers." *IEE Proc. Part I* **126**, 233.

Rozzi, T. E., and Shore, K. A. (1985). "Spatial and temporal instabilities in multi-stripe semiconductor lasers." *J. Opt. Soc. Am. B* **2**, 237

Scifres, D. R., Streifer, W., and Burnham, R. D. (1978). "Beam scanning with twin stripe injection lasers." *Appl. Phys. Lett.* **33**, 702.

Shore, K. A. (1982a). "Optically induced spatial instability in twin-stripe geometry lasers." *Opt. Quantum Electron.* **14**, 177.

Shore, K. A. (1982b). "Semiconductor laser bistable operation with an adjustable trigger." *Opt. Quantum Electron.* **14**, 321.

Shore, K. A. (1982c). "Optical limiter action in twin-stripe geometry lasers." *IEE Proc. Part I* **129**, 297.

Shore, K. A. (1983). "Above threshold current leakage effects in stripe-geometry injection lasers." *Opt. Quantum Electron.* **15**, 371.

Shore, K. A. (1984a). "Near field extinction in semiconductor lasers under optical injection." *Opt. Quantum Electron.* **16**, 157.

Shore, K. A. (1984b). "Radiation patterns for optically steered semiconductor laser beam-scanner." *Appl. Opt.* **23**, 1386.

Shore, K. A. (1984c). "Carrier diffusion and recombination influencing gain and current profiles in planar injection lasers." *J. Appl. Phys.* **56**, 1293.

Shore, K. A. (1985a). "Stability of self-pulsing due to Hopf bifurcation in semiconductor lasers." *IEEE J. Quant. Electron.* **QE-21**, 1249.

Shore, K. A. (1985b). "Optical switching properties of semiconductor lasers: analysis in canonical form." *IEE Proc. Part J* **132**, 85.

Shore, K. A. (1986a). "Hopf bifurcation analysis for semiconductor laser dynamics." In *Optical Instabilities*, R. W. Boyd, M. G. Rayner and L. M. Narducci, eds. Cambridge University Press, London, England, p. 271.

Shore, K. A. (1986b). "Instabilities in twin-stripe laser diodes." Society of Photo-Optical Instrument Engineers Quebec Symposium, topical meeting on optical chaos. Paper 667-16.
Shore, K. A. (1986c). "Multigigahertz switching at Hopf bifurcation in semiconductor lasers." Presented at Symposium on Nonlinear Guided Wave Optics and Fast Optical Switching. University of Glasgow, Scotland, 15 September 1986.
Shore, K. A. (1987a). "Optical and electronic control of Hopf bifurcation in twin-stripe semiconductor lasers." *IEE Proc. Part J* **134**, 51.
Shore, K. A. (1987b). "Nonlinear dynamics and chaos in semiconductor laser devices." *Solid-State Electron*, pp. 30, 59.
Shore, K. A., and Hartnett, P. J. (1982). "Diffusion and waveguiding effects in twin-stripe injection lasers." *Opt. Quantum Electron.* **14**, 169.
Shore, K. A., and Rozzi, T. E. (1981a). "Stability analysis of transverse modes in stripe-geometry injection lasers." *IEE Proc. Part I* **128**, 154.
Shore, K. A., and Rozzi, T. E. (1981b). "Near field control in multi-stripe geometry injection lasers." *IEEE J. Quantum Electron.* **QE-17**, 718.
Shore, K. A., and Rozzi, T. E. (1982). "Transverse mode oscillations at GHz frequencies in stripe-geometry lasers." *Opt. Quantum Electron.* **14**, 465.
Shore, K. A., and Rozzi, T. E. (1983a). "Switching frequency for transverse modes in stripe-geometry injection lasers." *Opt. Quantum Electron.* **15**, 497.
Shore, K. A., and Rozzi, T. E. (1983b). "Picosecond optical switching in semiconductor lasers." *Opt. Quantum Electron.* **15**, 549.
Shore, K. A., and Rozzi, T. E. (1984). "Transverse switching due to Hopf bifurcation in semiconductor lasers." *IEEE J. Quantum Electron.* **QE-20**, 246.
Shore, K. A., and Rozzi, T. E. (1986). "Bimodal switching behaviour in laser diodes." Society of Photo-Optical Instrument Engineers Quebec Symposium, topical meeting on optical chaos. Paper 667-21.
Shore, K. A., Rozzi, T. E., and In't Veld, G. H. (1980). "Semiconductor laser analysis: a general method for characterising devices of various cross-sectional geometries." *IEE Proc. Part I* **127**, 221.
Shore, K. A., Davies, N. G., and Hunt, K. (1983). "Constant power contours and bistability in twin-stripe injection lasers." *Opt. Quantum Electron.* **15**, 547.
White, I. H., Carroll, J. E., and Plumb, R. G. (1982). "Closely coupled twin-stripe lasers." *IEE Proc. Part I* **129**, 291.

INDEX

LIST OF MATERIALS